T0418800

CURRENT TRENDS AND FUTURE DEVELOPMENTS ON (BIO-) MEMBRANES

CURRENT TRENDS AND FUTURE DEVELOPMENTS ON (BIO-) MEMBRANES

Transport Phenomena in Membranes

Edited by

ANGELO BASILE
Hydrogenia, Genoa, Italy

KAMRAN GHASEMZADEH
Chemical Engineering Faculty, Urmia University of Technology, Urmia, Iran

ADOLFO IULIANELLI
Institute on Membrane Technology of the National Research Council of Italy (CNR-ITM), Rende (CS), Italy

ELSEVIER

Elsevier
Radarweg 29, PO Box 211, 1000 AE Amsterdam, Netherlands
The Boulevard, Langford Lane, Kidlington, Oxford OX5 1GB, United Kingdom
50 Hampshire Street, 5th Floor, Cambridge, MA 02139, United States

British Library Cataloguing-in-Publication Data
A catalogue record for this book is available from the British Library

Library of Congress Cataloging-in-Publication Data
A catalog record for this book is available from the Library of Congress

ISBN: 978-0-12-822257-7

For Information on all Elsevier publications
visit our website at https://www.elsevier.com/books-and-journals

Publisher: Susan Dennis
Acquisitions Editor: Kostas Marinakis
Editorial Project Manager: Bernadine A. Miralles
Production Project Manager: Joy Christel Neumarin Honest Thangiah
Cover Designer: Greg Harris

Typeset by MPS Limited, Chennai, India

Contents

4. Transport phenomena in electrodialysis/reverse electrodialysis processes 91

R. Zeynali, Kamran Ghasemzadeh and Angelo Basile

5. Transport phenomena in membrane distillation processes 111

Jianhua Zhang, Jun-De Li, Zongli Xie, Xiaodong Dai and Stephen Gray

6. Transport phenomena in dialysis processes 129

Marco Cocchi, Leone Mazzeo and Vincenzo Piemonte

7. Transport phenomena in pervaporation 165

Axel Schmidt and Jochen Strube

8. Transport phenomena in gas membrane separations 193

Foroogh Mohseni Ghaleh Ghazi, Mitra Abbaspour and Mohammad Reza Rahimpour

9. Transport phenomena in membrane contactor systems 209

Rahim Aghaebrahimian, Parisa Biniaz, Seyed Mohammad Esmaeil Zakeri and Mohammad Reza Rahimpour

10. Transport phenomena in drug delivery membrane systems 231

Sara A.M. El-Sayed and Mostafa Mabrouk

11. Transport phenomena in fixed and fluidized-bed inorganic membrane reactors 247

Alessio Caravella, Katia Cassano, Stefano Bellini, Virgilio Stellato and Giulia Azzato

12. Mass transport through capillary, biocatalytic membrane reactor 281

Endre Nagy and Imre Hegedüs

13. Transport phenomena in photocatalytic membrane reactors 309

Enrica Fontananova and Valentina Grosso

14. Transport phenomena in polymeric membrane reactors 325

Brent A. Bishop, Oishi Sanyal and Fernando V. Lima

List of contributors

Mitra Abbaspour Department of Chemical Engineering, Shiraz University, Shiraz, Iran

Rahim Aghaebrahimian Department of Chemical Engineering, Shiraz University, Shiraz, Iran

Giulia Azzato Department of Computer Engineering, Modelling, Electronics and Systems Engineering (DIMES), University of Calabria, Rende, Italy

Serena Bandini Department of Civil, Chemical, Environmental and Materials Engineering – DICAM, Alma Mater Studiorum, University of Bologna, Bologna, Italy

Angelo Basile Hydrogenia, Genoa, Italy

Stefano Bellini Department of Computer Engineering, Modelling, Electronics and Systems Engineering (DIMES), University of Calabria, Rende, Italy

Parisa Biniaz Department of Chemical Engineering, Shiraz University, Shiraz, Iran

Brent A. Bishop Department of Chemical and Biomedical Engineering, West Virginia University, Morgantown, WV, United States

Cristiana Boi Department of Civil, Chemical, Environmental and Materials Engineering – DICAM, Alma Mater Studiorum, University of Bologna, Bologna, Italy

Alessio Caravella Department of Computer Engineering, Modelling, Electronics and Systems Engineering (DIMES), University of Calabria, Rende, Italy; Institute on Membrane Technology – National Research Council (ITM-CNR), University of Calabria, Rende, Italy

Alessandra Carbone CNR ITAE, Messina, Italy

Katia Cassano Department of Computer Engineering, Modelling, Electronics and Systems Engineering (DIMES), University of Calabria, Rende, Italy

Marco Cocchi Faculty of Engineering, University Campus Biomedico of Rome, Rome, Italy

Xiaodong Dai Shengli College, China University of Petroleum, Dongying, P.R. China

Sara A.M. El-Sayed Refractories, Ceramics and Building Materials Department, National Research Centre, Dokki-Giza, Egypt

Enrica Fontananova Institute on Membrane Technology of the National Research Council (ITM-CNR), Rende (CS), Italy

Irene Gatto CNR ITAE, Messina, Italy

Kamran Ghasemzadeh Chemical Engineering Faculty, Urmia University of Technology, Urmia, Iran

Foroogh Mohseni Ghaleh Ghazi Department of Chemical Engineering, Shiraz University, Shiraz, Iran

Stephen Gray ISILC, Victoria University, Melbourne, VIC, Australia

Valentina Grosso Institute on Membrane Technology of the National Research Council (ITM-CNR), Rende (CS), Italy

Imre Hegedüs University of Pannonia, Research Institute of Biomolecular and Chemical Engineering, Laboratory of Chemical and Biochemical Processes, Veszprem, Hungary; Laboratory of Chemical and Biochemical Processes, Research Institute of Biomolecular and Chemical Engineering, University of Pannonia, Veszprem, Hungary

Jun-De Li College of Engineering and Science, Victoria University, Melbourne, VIC, Australia

Fernando V. Lima Department of Chemical and Biomedical Engineering, West Virginia University, Morgantown, WV, United States

Mostafa Mabrouk Refractories, Ceramics and Building Materials Department, National Research Centre, Dokki-Giza, Egypt

Leone Mazzeo Faculty of Engineering, University Campus Biomedico of Rome, Rome, Italy

Endre Nagy University of Pannonia, Research Institute of Biomolecular and Chemical Engineering, Laboratory of Chemical and Biochemical Processes, Veszprem, Hungary; Laboratory of Chemical and Biochemical Processes, Research Institute of Biomolecular and Chemical Engineering, University of Pannonia, Veszprem, Hungary

Enza Passalacqua CNR ITAE, Messina, Italy

Vincenzo Piemonte Faculty of Engineering, University Campus Biomedico of Rome, Rome, Italy

Elham Rahimpour University of Medical Sciences, Shiraz, Iran

Mohammad Reza Rahimpour Department of Chemical Engineering, Shiraz University, Shiraz, Iran

Oishi Sanyal Department of Chemical and Biomedical Engineering, West Virginia University, Morgantown, WV, United States

Axel Schmidt Institute for Separation and Process Technology, Clausthal University of Technology, Clausthal-Zellerfeld, Germany

Virgilio Stellato Department of Environmental Engineering (DIAM), University of Calabria, Rende, Italy

Jochen Strube Institute for Separation and Process Technology, Clausthal University of Technology, Clausthal-Zellerfeld, Germany

Zongli Xie CSIRO Manufacturing, Private Bag 10, Clayton South MDC, VIC, Australia

Seyed Mohammad Esmaeil Zakeri Department of Chemical Engineering, Shiraz University, Shiraz, Iran

R. Zeynali Chemical Engineering Faculty, Urmia University of Technology, Urmia, Iran

Jianhua Zhang ISILC, Victoria University, Melbourne, VIC, Australia

Preface

The main concept of "membrane technology" includes the intent of representing a separation infrastructure useful to replace the conventional separation processes in various industries. Membrane technology has indeed become a prominent separation technology over the past decennia. The most salient feature of membrane technology is that it works without the addition of chemicals, with a relatively low energy use and easy and well-arranged process conductions. Membrane technology is a generic term for a number of highly characteristic separation processes. These processes are of the same kind because in each of them a (different) membrane is used. Membranes are used more and more often for the creation of various processes such as water/wastewater treatment and gas separation processes. These potentialities are attracting special interests particularly in the countries in which extending this technology can help solve the existing issues related to the climate change and environmental pollution caused by the harmful emissions. Nevertheless, there is no a comprehensive reference to analysis of transport phenomena during various membrane processes. Hence the main aim of this book is, from one side, to describe the governing transport phenomena in various membrane processes regarding the scientific achievements in recent years; and, on the other side, to serve as a "one-stop" reference resource for important research accomplishments in the area of analysis of membrane processes based on mass transfer driving force. This book is anticipated to be an extremely valuable reference source for university and college faculties, professionals, postdoctoral research fellows, senior graduate students, and R&D laboratory researchers working in the area of modeling of membrane technologies. The various chapters were contributed by prominent researchers from industry, academia, and government/private research laboratories across the globe. Indeed, the book is an up-to-date record on the major findings and observations in the field of membrane processes.

In totality, there are 15 chapters. This book starts with Chapter 1 (Biniaz, Elham Rahimpour, Basile and Mohammad Reza Rahimpour), where the fundamentals of membrane technology are presented because, among all the conventional techniques under study, membrane processes have become highly popular in broad areas owing to the potential benefits of the technology and the novel developed membrane materials and systems. A membrane is a kind of a molecular sieve built in the structure of a film from several layers of material with small pores or fine mesh to allow the separation of small particles and molecules. Membranes work as a selective barrier that allows some particular substances to pass and holding others; they may also be used as a layer to immobilize phase interface, liquid, and solid fillers, or to imprint a template molecule for effective target analyte binding. This chapter tries to generally and briefly describe most of the basic concepts regarding membrane and membrane technology. Furthermore, recent advances in the process's development are examined too.

In Chapter 2 (Nagy and Hegedüs) the transport phenomena in ultrafiltration/microfiltration membranes are presented and discussed. In this work, the so-called "black box" model is extended by taking into account the mass transport process inside the membrane layer, as well. Accordingly, the developed model considers the simultaneous solute transport of macromolecules, microparticles, etc., across both the fluid polarization layer and the porous membrane layer. This model then makes it possible to express both the enhancement and the intrinsic enhancement factors, separately. Moreover, the outlet concentration, the interface concentration of the solute component, can be predicted by means of values of the boundary and the membrane layer mass transport properties. The model presented in this chapter takes into account the hindrance factor of the diffusion and convection, the partition coefficient between the two phases. In this context, the membrane performance is then illustrated by the help of some figures where it is shown, for example, that the membrane selectivity has a relative narrow water flux regime, in that the separation efficiency is effective, strongly depending on the ratio of the particle size and the pore size. The model can be also used for other hydrodynamic and osmotic pressure difference–driven membrane processes.

In Chapter 3 (Bandini and Boi), a general unique structural vision of transport phenomena in reverse osmosis (RO) and nanofiltration (NF) membranes is discussed. The most common models for RO (such as the solution–diffusion and/or the Spiegler–Kedem model) and for NF membranes (the extended Nernst–Planck equation) are introduced as particular cases of the general "statistical–mechanical theory" of membrane transport developed by Mason and Lonsdale in 1990. The use of that approach is recommended to develop a structural model when the physical meaning of the parameters is desired. The typical trends of solute rejection and of the total volume flux in RO processes are discussed, and the meaning of the model parameters is explained. NF modeling is presented according to the conditions of the porous vision of the Donnan-Steric-Pore-and-Dielectric-Exclusion (DSPM-DE) model. The complexity of the physical phenomena involved in the partitioning mechanisms is widely discussed: mechanisms of charge formation and of dielectric exclusion (image forces and Born partitioning) are described in detail. The general DSPM-DE model is adapted for the case of neutral solutes and for electrolyte mixtures: for each case, the basic equations are developed and the typical approximations are presented. The procedures for membrane parameter calculations are introduced, and a detailed discussion about the recommended correct method for data elaboration is presented. The final discussion is focused on the problems not yet completely solved as well as on the possible future trends.

Electrodialysis (ED) and reverse electrodialysis (RED) are the most commercially developed technologies for the desalination of brine water in last decades, and with the development of related technologies, both ED and RED will play an important role in an increasing number of fields. With respect to the importance of these two technologies, there are still a few studies that have looked for the issue of design and optimization of operation conditions in these technologies, and there are some barriers in these fields of technology. Nowadays, RED is one of the important technologies which leads to extracting of energy from the brine solutions on the basis of their gradient of salinity. The schemes of the ED and RED are feasible for developing flexible design and optimization of methods. In Chapter 4 (Zeynali, Ghasemzadeh and Basile) the transport phenomena in ED/RED application and advances in recent years are presented. Some of the discussed subjects for lightening the ED and RED application are desalination, energy conversion, desalination technology, water treatment, and some improvements to standard ED and RED. This chapter also summarizes the improvements of electrodes, feed solutions, membranes, and operation of membrane cell for ED and RED. Finally, a summarized comparison about the electrodes and solutions in the membrane cell of the ED and RED is shown.

In Chapter 5 (Zhang, Li, Xie, Dai and Gray), the transport phenomena in membrane distillation (MD) processes are studied. MD is a thermally driven membrane technology, and its mass transfer is coupled with the heat transfer. Membrane in MD acts as a barrier and only allows gas phase to pass through. This mass transfer is associated with evaporation that occurs at the membrane surface and requires latent heat. Hence the efficiency of the thermal energy transfer from bulky feed to the membrane surface will have significant influence on the mass transfer. In this chapter, the mechanisms of mass and heat transfer through the porous membrane in four major MD configuration are reviewed. Influence of membrane characteristics, membrane configuration, and module configuration on mass and heat transfers are discussed in detail as well.

The aspects of transport phenomena in artificial kidney are underlined in Chapter 6 (Cocchi, Mazzeo and Piemonte). Considering that hemodialysis is the most frequent renal replacement therapy for patients suffering from chronic renal failure and that dialyzer is an external artificial device able to receive the patient blood and filters toxins out of it by the means of membranes, this chapter illustrates the basic principles and governing equations related to the transport phenomena occurring in an artificial kidney device. In order to present the organ functionalities to be reproduced by the artificial device, a detailed description of natural kidney and its working principle is reported in the introduction section. The background section contains a classification of different technologies and systems for renal replacement therapy, while the mathematical models section reports both the design equations and a final real case study for regenerative dialysis therapy.

In Chapter 7 (Schmidt and Strube), the transport phenomena in a membrane system called pervaporation is considered and discussed. Pervaporation is a thermal separation process that uses the difference in permeability of at least two substances through a membrane for separation. As the only membrane process, a phase change from liquid feed to gaseous permeate occurs during pervaporation. Predictive prediction of pressure drop and mass and heat transfer are essential prerequisites for designing efficient pervaporation processes. After a brief introduction of the fundamentals, this chapter presents the most important models for describing the transport phenomena and finally demonstrates their application in a process simulation.

Chapter 8 (Ghazi, Abbaspour and Rahimpour) is dedicated to the transport phenomena in gas membrane separations.

In effect, membranes have the highest potential for gas separation, and membrane processes encompass a wide variety of gas separation applications and are now regarded as a cutting-edge separation technology for industrial applications. In this sector, a high degree of permeability combined with high selectivity of a particular gaseous species ensures exceptional performance of these systems. The cost-effectiveness of membrane separation technology is considerable; therefore for more widespread membrane applications, it is critical to understand transport processes in gas membrane separations and the causes of flux decline and the capacity to anticipate flux performance. Thus this chapter deals with mass transportation in the membrane process of gas separation. In addition, findings from many investigations carried out by numerous scholars were reviewed and discussed to study diversified facets of these complicated occurrences. Investigations of diffusion pathways in the membrane pore and assessment of transport resistances caused by membrane fouling will comprehensively explain the fouling phenomenon and applicable mass transfer mechanisms.

Chapter 9 (Aghaebrahimian, Biniaz, Zakeri and Rahimpour) is centered on the transport phenomena in membrane contactor (MC) systems. These are systems that can operate using a permeable membrane, particularly a hollow fiber membrane with high surface/volume ratio values, which separates two phases, such as gas and liquid or liquid and liquid, and is applied in various industrial areas such as gas absorption and wastewater treatment. In MCs, the role of the membrane is to offer an excellent contact and mass transfer between phases, and it has a barrier between the two phases without providing the selectivity to the components present in the process. Besides, the fluid components pass through the membrane pores mainly because of the mechanism of diffusion in almost all situations. This chapter provides a short review of applying membrane contactors with particular attention to the system's mass transfer phenomenon. The transfer phenomena are examined considering both the film theory and the resistance-in-series theory. Moreover, the wetting phenomenon as well as up-to-date approaches executed by researchers to increase mass transfer is also considered.

Owing to their importance in the delivery of molecules in drug delivery membranes, in Chapter 10 (El-Sayed and Mabrouk), both the transportation phenomena and the mechanism of transportation through membranes are discussed in detail. Therefore this work gives in the beginning background about the history, definition, and classification of membranes. Moreover, the preparation methods of these membranes and their applications are discussed according to the permeate size. As nanoporous membranes have massive application in the drug delivering field, a little focus on their advanced techniques and their relation to the drug transportation phenomena through membranes are demonstrated. Finally, the effect of surface modifications on the same phenomena of the nanoporous membranes is also discussed.

Chapter 11 (Caravella, Cassano, Bellini, Stellato and Azzato) aims at providing an overall (but nonexhaustive) overview on momentum, mass, and heat transport occurring in fixed- and fluidized-bed inorganic membrane reactors (MRs), discussing the mutual relationships among constitutive equations that describe the complex transport phenomena in these systems. As for the momentum transfer, several friction and drag models of literature are recalled along with the operating conditions in which they can be applied. As for the mass transport, a particular attention is paid to zeolite and metal membranes, which are the types of membrane mostly used. As for the heat transport, several correlations for the heat transfer coefficients of literature are also reported and briefly discussed.

In Chapter 12 (Nagy and Hegedüs), the mass transport through capillary biocatalytic MRs is discussed. The mass transport through a cylindrical membrane, independent of the transport direction, takes place from the lumen or from the shell side of catalytic MR and occurs in varying space. This affects the concentration change and accordingly the solute transport rate throughout the membrane. The aim of this chapter is to analyze how the varying space, the biochemical reaction rate, and its parameter values affect the concentration distribution, the inlet and the outlet mass transfer rate, the reactor performance under different operating conditions, focusing primarily on the effect of the local space coordinate. In particular, two important operating modes are discussed: the transport without sweeping phase and with that phase on the permeate side. Important difference between them can be noted as: the outlet concentration gradient is zero in the first case, while it can be larger than zero in the presence of sweeping phase. That means that there is no outlet diffusive flow during transport without sweeping phase. This operating mode is important operation for biocatalytic MRs. It is shown that the cylindrical effect can be essential, depending on the reaction rate; thus its role must not be taken out of consideration for the case of more exact evaluation of the experimental data and the real catalytic membrane performance.

Chapter 13 (Fontananova and Grosso) focuses on the mass transport phenomena of the main membrane operations involved on the transport phenomena in photocatalytic membrane reactors (PCMRs). Moreover, the key parameters that affect the process and the cooperative role of the membrane and the catalyst in the PCMR are also deeply discussed. Several case studies of PCMRs applied in wastewater treatment by advanced oxidation processes and fine chemical synthesis are also presented, highlighting advantages and existing imitations of these interesting integrated membrane operations.

Process intensification is an area of research aimed at reducing the size of chemical process units and maximizing their efficiency. MRs are an example of this promising technology and achieve improved efficiency through the combination of reactions and separations. However, the most effective membrane materials, such as palladium for hydrogen separation, are very expensive. This motivates the development of cost-effective membrane materials such as polymers. In Chapter 14, the authors (Bishop, Sanyal and Lima) present the first-principle equations for effectively modeling the energy and mass transport of polymeric membrane reactors, which are then used to model a nonisothermal, countercurrent water–gas shift MR system using a pure polybenzimidazole (PBI) membrane. Simulations are performed to study the operation of these polymeric membranes. The shortcomings of pure-PBI membranes are identified, and techniques to overcome the performance trade-off of these membranes, as reported in recent literature, are suggested. While being primarily focused on the cited chemical reaction membrane reactors, this work can be easily extended to other types of catalytic membrane reaction schemes.

In the field of polymer electrolyte fuel cells (PEFC), a key role is played by polymer electrolyte membranes. In the last chapter, Chapter 15 (Gatto, Carbone and Passalacqua), the main transport phenomena that occur in polymer electrolyte membranes are described. Moreover, the structural, chemical, physical, and electrochemical characteristics of the main proton and anionic exchange polymers, currently used in PEFC, are also reported.

The editors express their sincere and best gratitude to all the contributors who helped the realization of this volume, whose excellent support resulted in its successful completion. In particular, we thank all the reviewers a lot for taking their valuable time to make critical comments on each chapter. Last but not least, we would also like to thank the publisher, in particular special thanks to the responsibles at Elsevier, Liz Heijkoop, Narmatha Mohan, and Kostas Marinakis, for their great help.

Angelo Basile
Kamran Ghasemzadeh

CHAPTER

1

Fundamentals of membrane technology

Parisa Biniaz[1], *Elham Rahimpour*[2], *Angelo Basile*[3] *and Mohammad Reza Rahimpour*[1]

[1]Department of Chemical Engineering, Shiraz University, Shiraz, Iran [2]University of Medical Sciences, Shiraz, Iran [3]Hydrogenia, Genoa, Italy

Abbreviations

AEMs	anion-exchange membranes
BTX	benzene–toluene–xylene
CEMs	cation-exchange membranes
CP	concentration polarization
CVD	chemical vapor deposition
ED	electrodialysis
LLE	liquid–liquid extraction
LM	liquid membrane
MBR	membrane bioreactor
MCs	membrane contactors
MD	membrane distillation
ME	membrane extraction
MF	microfiltration
MMMs	mixed-matrix membranes
NF	nanofiltration
PVDF	poly(vinylidene fluoride)
RED	reverse electrodialysis
RO	reverse osmosis
SLME	supported liquid ME
UF	ultrafiltration
UV	ultraviolet
VOCs	volatile organic compounds

Current Trends and Future Developments on (Bio-) Membranes
DOI: https://doi.org/10.1016/B978-0-12-822257-7.00011-X

1.1 Introduction

Membrane-based separation processes are now extensively applied in our daily lives in various sectors, including energy, for example, biofuel production and upgrading, petrochemical, agriculture, food, medicine, pharmacy, and biotechnology industry, different environmental applications such as water, and wastewater treatment, desalination, soil, and air production. Compared to other traditional separation processes such as absorption, adsorption, and distillation, membrane technology is cost-efficient and straightforward, making it highly attractive and comprehensive in its applications.

Membrane processes offer a large number of benefits, including simplicity to scale up, low energy consumption, the capability to hybridize and combine with other processes, high intensity, continuous operation, and automatic operation. However, they have a few drawbacks as well, such as membrane fouling, low chemical stability, and limited lifetime, which can be addressed by recent advances in modifying membrane modules and novel membrane technologies (Kochkodan & Hilal, 2015). According to recent financial reports, global demand for membranes technology has reached \$15.6 billion, with an annual growth rate of 8% forecast in the future (Vandezande, 2015).

The membrane is the most critical element of membrane separation technology and has a direct effect on the efficiency and practical application of the process. A membrane is a thin sheet, layer, or film that serves as a selective barrier inside two phases: gas, liquid, or vapor. A membrane, in other words, is a permeable barrier that allows certain particles to pass through while holding others out. The characterizations of membrane module such as membrane pore size, shape, membrane surface characteristics (charge and porosity and /hydrophobicity), and the configuration of the membrane (geometry and dimensions) contribute to the proposed behavior (Biniaz, Makarem, & Rahimpour, 2020; Biniaz, Roostaie, & Rahimpour, 2021). The separation phenomenon via membrane is depicted in Fig. 1.1 as a general layout.

Membranes are utilized to separate particle–solvent, particle–solute, solute–solute, and solute–solvent, and selectivity refers to membranes' ability to differentiate among molecules or substances. Furthermore, the flux equation is the essential entity in any membrane process, since the ratio of fluxes involved determines the ideal selectivity (Ladewig & Al-Shaeli, 2017).

1.2 Membrane classification

There is a variety of standards utilized for classifying membranes. Membrane morphology (porosity and nonporosity), state, and shape are most commonly taken into account when

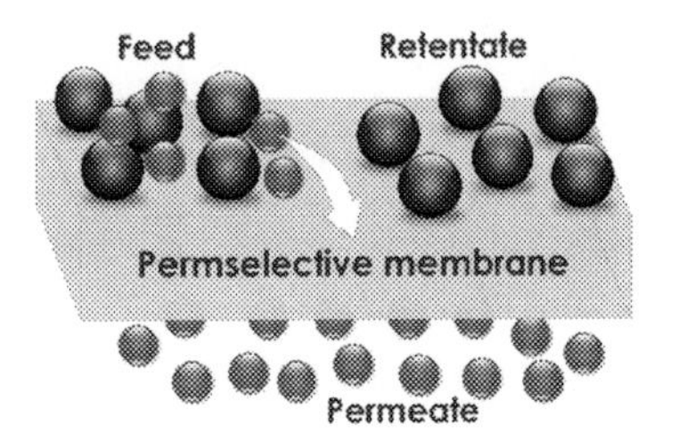

FIGURE 1.1 The separation phenomenon via membrane.

characterizing membranes. Table 1.1 represents a diagram of membrane classification according to these parameters (Jakubowska, Polkowska, Namieśnik, & Przyjazny, 2005).

Membranes may have a negative or positive charge, or may be neutral or bipolar. With regard to synthesis materials, membranes are classified into organic/polymer membranes, inorganic membranes, and mixed-matrix membranes (MMMs) (Kayvani Fard et al., 2018).

1.2.1 Organic membranes

The current membrane market is dominated by organic membranes with applications in both industrial and academic sectors. This is primarily because organic membranes offer tremendous benefits, such as low cost, high selectivity, and ease of processing. These kinds of membranes are generally categorized on the basis of their hydrophilicity or hydrophobicity. While hydrophilic membranes are less susceptible to fouling caused by various biorefinery feeds, they are more susceptible to deformation caused by temperature and pH fluctuations. The following examples are the most common types of organic polymers: poly(vinylidene fluoride) (PVDF), poly(ether sulfone), polysulfone, polyacrylonitrile, polytetrafluoroethylene, polycarbonate, polyimide, polyetherimide, polyamide, polydimethylsiloxane, polyphenylene oxide, polymethyl pentene, and cellulose acetate (Kang & Cao, 2014; Liu, Hashim, Liu, Abed, & Li, 2011). One of the most widely used polymer membranes is PVDF, which has received much attention from researchers and manufacturers in recent years (Baker, 2012; Mohshim, Mukhtar, Man, & Nasir, 2013; Sadeghi, Govinna, Cebe, & Asatekin, 2019). Polymeric membranes, on the other hand, have poor mechanical stability and fouling issues.

1.2.2 Inorganic membranes

Inorganic membranes have superior chemical, mechanical, and thermal stability and long lifetime, making them ideal for use in extreme environments, such as corrosive, chemical cleaning, and high-temperature environments (Benfer, Popp, Richter, Siewert, & Tomandl, 2001). The most common inorganic membranes are ceramic, silica, zeolite, palladium alloys, and molecular sieves/nanoporous carbon membranes. Various techniques such as the sol–gel process, slip casting, pyrolysis, and chemical vapor deposition have been recommended as commonly used methods to synthesize inorganic membranes. When it comes to morphology and structure, inorganic membranes are classified into

TABLE 1.1 Membrane classifications based on their morphology, state and shape. (Jakubowska et al., 2005).

State	Membrane							
	Morphology				Shape			
Solid	Porous			Nonporous	Planar		Tubes	
Liquid	Symmetric	Asymmetric		–	Tapes	Sheets	Capillary	Hollow fibers
Two phases	–	Composite	Dynamic	–	–			
Gel	–				–			

porous and nonporous/dense membranes. The former can include ceramic supports or porous metal and a second porous layer with a different structure and morphology on top. This group of membranes includes those with a variety of pore forms, such as straight pores having an equal diameter that extends from one side of the membrane to the other, conical-shaped pores with smaller pore diameters at the top than those at the bottom, pores with spongy shapes and regular structure. Metal, glass, zeolite, zirconia, alumina, carbon, cordierite, silicon, silicon carbide, titania, nitride, mullite, tin oxide, and mica are the most common porous inorganic membranes. The latter kind of inorganic membranes are made from either solid metal layers such as alloys, Ag, Pd, or solid electrolytes. The electrolyte module enables hydrogen and oxygen to diffuse through the membrane pores and ions to migrate oxides. Dense membranes may also have an immobilized liquid support layer (e.g., immobilized molten salts in porous steel or ceramic supports) that fills the membrane's pores and creates a semipermeable layer. Palladium, palladium alloys, nickel, silver, and stabilized zirconia are examples of dense membranes. Principally, dense inorganic membranes are applied to separate oxygen and hydrogen by charged particles. The type of substrate, species nature that separated, and physical and chemical interactions among the membrane and species play a crucial role in the efficiency of dense membranes. The density membranes' pore structure is determined by the synthesis protocol (Abels, Carstensen, & Wessling, 2013; Kayvani Fard et al., 2018). Table 1.2 enumerates the benefits and drawbacks of inorganic membranes in comparison to organic membranes.

1.2.3 Mixed matrix membranes

The final group of membranes is named MMMs. MMMs are hybrid membranes that improve the properties of membranes by having inorganic particulate materials (liquid,

TABLE 1.2 Benefits and drawbacks of inorganic membranes compared with organic membranes (Ladewig & Al-Shaeli, 2017).

Benefits	Drawbacks
Superior chemical, mechanical, and thermal stability over time at high temperatures	Capital cost is high
Having high resistance for use in harsh environments such as corrosive, chemical cleaning, and chemical degradation, and pH	Special dense membranes experience the phenomenon of embrittlement
Having high-pressure drops resistance	Membrane surface area per module volume is low
Microbiological degradation inertness	Challenging to achieve high selectivities in microporous membranes in large scale
Easy fouling cleaning	At low temperatures, the high hydrogen-selective (dense) membranes have a low permeability
Simple catalytic activation	Complex membrane for high-temperature module sealing
Long lifetime	

solid, or both solid and liquid) as the "filler" integrated into a polymeric material as the "matrix" as solid–polymer, liquid–polymer, and solid–liquid–polymer. Polymeric membranes, in most cases, are unable to surmount the polymer upper-bound limitation of permeability and selectivity. On the other hand, some inorganic membranes have much higher selectivity and permeability than organic membranes; however, they are often too expensive or difficult to manufacture on a wide scale. As a result, the primary goal of MMM fabrication is to address some of the disadvantages of polymeric and inorganic membranes by the combination of inorganic filler's superior mechanical properties with polymeric materials' high processability and low cost (Kickelbick, 2003; Li et al., 2013; Chung, Jiang, Li, & Kulprathipanja, 2007; Sanchez, Belleville, Popall, & Nicole, 2011). MMMs are made using various techniques, including surface coating and phase inversion (Zhang, Surampalli, & Vigneswaran, 2012; Zhang, Surampalli, Vigneswaran, Tyagi et al., 2012). Membranes made up of a mixture of inorganic and organic materials have better robustness, antifouling, and surface hydrophilicity properties than those without. Like any other early-stage technology, MMM science requires some commercial implementations and cost analysis to mature as a separation technology (Zhang & Surampalli, 2012).

A liquid membrane (LM) is a type of MMM produced from liquids that are immiscible with the feed phases and products and categorized into three basic LM configurations as bulk LM, emulsion LM, and supported/immobilized LM. In other words, the separating transport agent in LM is a carrier agent or liquid complexing immobilized or reinforced in a rigid solid with a porous structure. On the feed side, the liquid carrier fills the support medium's pores and interacts with the permeate. The resulting complex diffuses across the support/membrane, releasing the permeating agent on the product part while recovering the carrier agent, which diffuses back into the feed. Since the diffusion coefficients in liquids are usually higher in orders of magnitude than in polymers, LMs have a number of benefits, including (1) a higher flux and (2) high selectivity, which allows recognizing molecules precisely using carriers as the transport mechanism. However, LMs have several problems, and the biggest problem is stability. LMs require stability to function correctly, and if they are forced out of the pores or rupture in any way as a result of pressure differentials or turbulence, they do not work efficiently (Kang & Cao, 2014; Liu et al., 2011). Separation in LMs is accomplished by diffusion and complexing reactions. Gas separation and coupled transport, where ion transport separates metal compounds, are two of the most common uses for such membranes (Ersahin et al., 2012; Lu et al., 2016). Despite their high selectivity, LM lack physical stability, making them commercially unviable (Kocherginsky, Yang, & Seelam, 2007).

In membrane processes with MMMs, like some particular membrane-based separation technologies (e.g., solvent-extraction and solvent-stripping), membranes do not function as separation barriers. Instead, they are used as liquid and solid fillers, layers to immobilize phase interfaces or imprint a template molecule for effective target analyte binding. Extending the role of the membrane in membrane technology from a barrier to a substrate is critical and crucial because it leads to the development of novel membrane processes such as membrane contactors (MCs), perstraction [sort of membrane extraction (ME) process], advents separations, and quantitative biotechnology tools, develops membrane-based sensors, passive samplers domain (e.g., magnetic, flow, and electrical), and finally enables biochemical or chemical reactions to be combined with separation processes technologies (Zhang & Surampalli, 2012).

1.3 Membrane technology

Membrane technology refers to the science and engineering techniques used to move or reject substances, components, and species through membranes. It is used in a variety of industries because of its multidisciplinary nature. This process has a broader environmental and industrial application owing to its advantages as a clean separation technology that saves energy and can replace conventional separation processes such as filtration, ion exchange, distillation, and chemical treatment techniques. Other benefits include the ability to deliver high-quality products and device design versatility. The separation can be conducted continuously without additives, in mild conditions, and with almost low energy consumption by applying membrane technology (Zhang, Surampalli, & Vigneswaran, 2012; Zhang et al., 2012).

Furthermore, the technology can be used in conjunction with other separation processes to create hybrid processes that are more effective than traditional separation methods. However, this technology has some obstacles, such as concentration polarization (CP), low membrane lifetime, membrane fouling, low flux, and selectivity, which are unavoidable parts of the separation process. As an effective method, membrane technology is one of the most commonly used methods for separating oil–water wastewater or emulsions in food processing, desalination, pharmaceutical, and fuel cell industries. Compared to other separation technology, the membrane separation method has higher efficiency, consistent effluent quality, and lower energy consumption (Kumar, 2012).

A driving force, including a pressure gradient, concentration gradient, temperature gradient, or electrical gradient, causes the permeating compounds to move through the membrane (Table 1.3). This phenomenon is classified as the isothermal and nonisothermal operations and arises from membrane characterization such as membrane pore size and membrane pore shape. The isothermal process includes concentration-driven processes (ME and pervaporation), pressure-driven processes [reverse osmosis (RO), nanofiltration (NF), microfiltration (MF), and ultrafiltration (UF)], and finally, electrically driven processes [electrodialysis (ED) and electrophoresis]. In contrast, in the nonisothermal process, the driving force is thermal energy, such as the membrane distillation (MD) process.

Generally, in membrane technology, the permeation rate and transport mechanism are controlled by a number of factors. The driving force magnitude and the permeating molecule size compared with the available permanent size are two of these factors.

TABLE 1.3 Membrane technology classification according to the driving force.

Pressure-driven membrane technology	Membrane Concentration-driven membrane technology	Technology Electrically driven membrane technology	Thermally driven membrane technology
Reverse osmosis	Membrane extraction	Electrodialysis	Membrane distillation
Nanofiltration	Pervaporation	Electrophoresis	–
Ultrafiltration	Dialysis	–	–
Microfiltration	–	–	–

Furthermore, the chemical composition of the permeant and the material used to produce the membrane (dispersive, polar, ionic, and so on) may affect the separation process. The process conditions need to be carefully engineered. However, the membrane properties determine the performance limits.

1.3.1 Pressure-driven membrane technology

In the pressure-driven membrane separation process, the solution is divided into permeate and retentate phases by applying pressure as a driving force to the solution. In other words, the pressure differential between the feed and permeate sides of the membrane transports the solvent from the feed side to the permeate side.

On the basis of the membrane pore size and the essential transmembrane pressure, pressure-driven membranes are divided into three categories. As Fig. 1.2 clearly illustrates, the pore size range of 0.0001–0.001 μm, <0.5 nm, 35–100 bar regarded as RO; NF process has a pore size range of 0.001–0.01 μm, 0.5–10 nm,10–30 bar, UF method has pore size range of 0.01–0.1 μm, 1–10 bar, and finally MF technology has pore size range of 0.1–5 μm, 1–10 bar (Cui, Jiang, & Field, 2010).

Currently, RO is the most extensively used desalination and water treatment technology, and it is rapidly expanding (Lee, Arnot, & Mattia, 2011). RO is a pressure-driven membrane technology carried out with a semipermeable membrane that rejects dissolved components in the feed water. This rejection is due to the applied pressure to overcome osmotic pressure based on size exclusion, physical–chemical interactions among solvent, solute, and membrane, and charge exclusion.

The spirally wound module configuration is the most widely used membrane in RO. This configuration has a large specific membrane surface area, is easy to scale up, is interchangeable, has low replacement costs, and, most importantly, is the least expensive configuration to make from flat sheet thin film composite membrane (Li & Wang, 2010). Integrated membrane bioreactor (MBR) and RO systems are becoming more commonly used to remove emerging pollutants from urban wastewater. Concentrate from RO

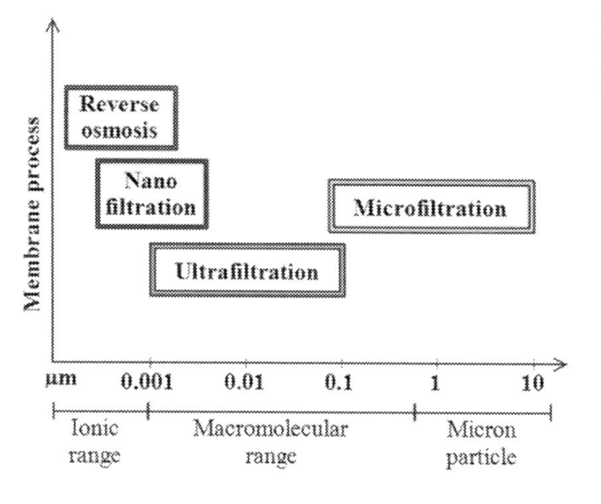

FIGURE 1.2 Schematic diagram of classifying pressure-driven membrane technology based on pore size (Biniaz et al., 2021) (with permission).

systems is mainly discharged to surface water, as opposed to other pressure-driven membrane processes (NF, MF, and UF), and the concentrate is mainly considered challenging to handle and dispose of owing to high concentrations (Joo & Tansel, 2015; Malaeb & Ayoub, 2011).

MF rejects OR separate particles with diameters ranging from 0.05–0.1 to 5 μm, and are commonly used to remove bacteria and protozoa. The MF membrane can also be used to extract suspended solids and minimize turbidity in water. UF is also a membrane filtration process that uses hydrostatic pressure to push water through a semipermeable membrane, similar to RO. The pore size range of UF membranes is 0.01–0.1 μm, and the molecular weight cut-off is commonly used to identify them. Viruses, metal hydroxides, emulsified oils, proteins, colloids, and other large molecular weight components are commonly removed from the water and other solutions using the UF process (Morao, Alves, & Cardoso, 2001; Shi, Tal, Hankins, & Gitis, 2014).

NF membranes have a pore size of 1 nm, which corresponds to a molecular weight cut-off of 300–500 Da, and have properties that put them somewhere between UF and RO. Because of the adsorption of charged solutes or dissociation of surface functional groups, NF membranes are also slightly charged in contact with the aqueous solution. Compared to RO membranes, NF membranes have a higher rejection of divalent ions, a lower rejection of monovalent ions, and a higher flux (Mohammad et al., 2015). Because of these characteristics, NF has been deployed in various applications, including water and wastewater treatment, biotechnology, pharmaceutical and, and food engineering (Van der Bruggen, Mänttäri, & Nyström, 2008). NF membrane–producing methods have also effectively fabricated membranes with improved selectivity, flux, and antifouling properties. In order to make nanocomposite membranes or thin-film composite, the primary process still relies on interfacial polymerization (Pendergast & Hoek, 2011). However, some issues with fouling control and mitigation continue to exist (Deen, 1987). Future research should concentrate on improving the ability to monitor, eliminate, and minimize fouling while creating better membranes and new innovative applications (Hilal, Al-Zoubi, Darwish, Mohamma, & Abu Arabi, 2004).

1.3.2 Concentration-driven membrane technology

Concentration-driven membrane processes, including ME, pervaporation, and dialysis, are briefly discussed in this section. The concentration gradient between feed and permeate in these processes causes a driving force for mass transport across the membrane.

1.3.2.1 *Membrane extraction*

ME techniques apply to various sample preparation methods and address conventional liquid–liquid extraction (LLE) issues using only a small amount of organic solvent and concentration gradient as a driving force. Furthermore, there are no emulsion issues in this process, and disposable products help to hold prices down (Gonçalves, Valente, & Rodrigues, 2017; Miró & Frenzel, 2004). ME is used in the separation of lactic acid, aromatic acid, and boric acid, the recovery of valeric acid from wastewaters, the separation of acid from salts (e.g., cadmium from phosphoric acid), the elimination of benzene–toluene–xylene (BTX) and volatile organic compounds (VOCs), and the identification of trace components like vitamins in vegetable oils, among other items (Gonçalves et al., 2017). A fully solid membrane is utilized to separate the donor and acceptor solutions in polymeric ME. Since silicone rubber is hydrophobic and extremely permeable to small hydrophobic molecules, it is mainly used in this process. The basis of selectivity is the difference in solubility and diffusion of different analytes into the polymer. Silicone rubber's solid nature makes it a flexible technique, allowing it to treat aqueous, organic, and gaseous samples. Modifying conditions in the acceptor phase, like analyte ionization, can improve selectivity. The predominant existence of silicone rubber means phase breakthrough is reduced, which is a notable benefit. The solid membrane's main drawback is that it prevents other functional groups (carriers), which can improve mass transfer and selectivity of the compounds of interest. Since there is no volatility associated with LM, it is an excellent method for extracting analytes in complex samples containing many organic materials like lipids (Jakubowska et al., 2005). Supported LM extraction (SLME) is a three-phase extraction technique and a flow-through ME process that can be used to prepare analytical samples and determine trace levels of inorganic and organic materials in water samples from which analytes are separated and extracted from a continuously flowing aqueous sample into another aqueous sample (Jönsson, 2012).

1.3.2.2 Pervaporation

The pervaporation technique is a membrane separation process for liquid mixture (or gas) stream in which the feed phase is partially vaporized throughout a membrane, leading to the accumulation of permeating substances in vapor form (Shao & Huang, 2007; Vandezande, 2015). In other words, a stream that contains two or more miscible substances is placed to one side of a nonporous polymeric membrane or a molecularly porous inorganic membrane, like zeolite membrane, while the other side is subjected to a vacuum or gas purge. The stream's components sorb into/onto the membrane and then pass through it and evaporate into the vapor flow. Subsequent to that, the resulting vapor is condensed. On the permeate side, an inert medium or low pressure is applied to remove the vaporized components. The difference in chemical potential, or the concentration gradient between fluid phases on opposite sides of a membrane, is the driving force of pervaporation (Figoli, Santoro, Galiano, & Basile, 2015; Wang, Wang, Wu, & Wei, 2020). The ratio of the diffusing components is determined by physicochemical interactions among the membrane material and diffusing species. A sorption–diffusion model describes the mass transport based on the affinity difference defined on the basis of the volatility difference in the conventional distillation process. Pervaporation has found widespread use in removing VOCs from wastewater, liquid hydrocarbon separation, removing water from glycerin, organophilic separations, bioethanol upgrading, and dehydration to speed up the esterification reaction. However, a phase change occurs in the pervaporation process, which uses the same amount of energy as distillation or evaporation (Crespo & Brazinha, 2015; Liu, Wei, & Jin, 2014). A mathematical model of the mass transfer in the pervaporation unit is needed to design hybrid processes that involve pervaporation as part of the separation strategy. Pervaporation models attempt to explain mass transfer via the membrane's selective layer. Because of the system's inherent complexity, the modeling of the pervaporation process at this stage necessitates certain assumptions, since it is still unclear where the transition from liquid to vapor occurs (Feng & Huang, 1997; Zhang & Drioli,

1995). PV membrane reactors have been extensively studied for assisting esterification reactions and fermentation. Membranes selectively extract the byproduct or the product itself from reaction mixtures to improve reversible reaction conversions and speeds (Amelio et al., 2016).

1.3.2.3 *Dialysis*

Dialysis is a separation process based on the membrane in which a concentration gradient causes smaller molecules to permeate from one side of the membrane (feed) to the other (receiving solution or dialysate). In the dialysis technique, a semipermeable membrane separates two different solutions (receiving solutions and feed), and because of the concentration gradient, small molecules below the membrane cut-off size diffuse into the dialysate solution, whereas larger molecules cannot. As a result, dialysis can be considered a highly efficient method for separating small molecules. The molecule diffusion rate through a dialysis membrane is a function of porosity and tortuosity, and it is also inversely proportional to the membrane's thickness; thus membranes used in the dialysis process should be thin enough to achieve a high rate of separation. Since the rate of molecule diffusion through a dialysis membrane is a function of tortuosity, porosity, and the thickness of the membrane, membranes used in the dialysis process should be thin enough to achieve a high separation rate (Kidambi et al., 2017; Nath, 2017). Dialysis is a spontaneous process, since the driving force is diffusion rate differences through the semipermeable membrane caused by differences in molecular size or concentration of solutes (species) (Camera-Rodaa, Loddob, Palmisanob, & Parrinob, 2019; Kidambi et al., 2017; Zhou, 2012). If the dialysate is not replaced with a fresh one on a regular basis, the concentration of diffusible solutes continues to equalize, lowering the driving force for separation. Dialysis is typically performed in mild operating conditions, such as ambient temperature, low-pressure drop through the membrane, and low shearing flow; thus it has been successfully used in some specialized research fields, such as life sciences, which require sensitive material separation (Lee & Koros, 2003).

While dialysis has its benefits, it also has two significant drawbacks: (1) it is a slow operation and (2) it has little selectivity. Since the concentration gradient is the driving force in dialysis, it takes a long time to complete, primarily when the concentration of solutes to be separated is low, as it is in most environmental situations. Owing to these significant disadvantages, industrial applications of traditional dialysis have decreased significantly compared to other faster and better separation methods such as UF or ED. However, in the medical sector, two major applications of dialysis, hemodialysis (artificial kidney) and blood oxygenators (artificial lungs) have progressed to the point that more than 100 million hemodialyzers are now in use worldwide (Kidambi et al., 2017; Zhou, 2012).

1.3.3 Electrically driven membrane technology

An external electrical gradient (electrical potential difference) is applied to drive charged species or ions across a membrane in an electrically driven membrane. ED and electrophoresis are standard technological procedures in this regard.

1.3.3.1 Electrodialysis and reverse electrodialysis

Ion-exchange membrane separation processes like ED and reverse electrodialysis (RED) are similar to regular dialysis in which neutral membranes are used. Low molecular weight solutes are removed from a solution (feed) using a semipermeable membrane in all of them. ED is an electrochemical membrane separation process using anion-exchange membranes (AEMs), cation-exchange membranes (CEMs), and an electrical potential difference as the driving force for separating salt ions and charged species from a feed solution (Campione, Gurreri et al., 2018; Strathmann, 2004). RED is a technology by which electricity is produced in accordance with the difference in the salinity level of two solutions like river water and seawater (Liu et al., 2020; Mei & Tang, 2018; Vermaas, Guler, Saakes, & Nijmeijer, 2012). Both ED and RED are considered electrochemical membrane processes using parallel AEMs and CEMs located near each other between two electrodes (Luo, Abdu, & Wessling, 2018). Fig. 1.3 illustrates a schematic diagram of an ED cell. As per fig. 6.9, the transfer of anions and cations occurs through AEMs and CEMs, which are put parallel to each other between an anode and a cathode, in the opposite direction utilizing a potential difference between two electrodes (anode and cathode) arising from a direct current. At the surface of the anode, oxidation (loss of electrons) occurs, while at the cathode edge, reduction (gain of electrons) occurs. It is worth mentioning that AEMs permit solely anions (a negatively charged ion) to pass. By contrast, CEMs allow only cations (a positively charged ion) to go across. Consequently, charged species are selectively separated, and two different solutions (low-salinity and high-salinity solutions) are produced

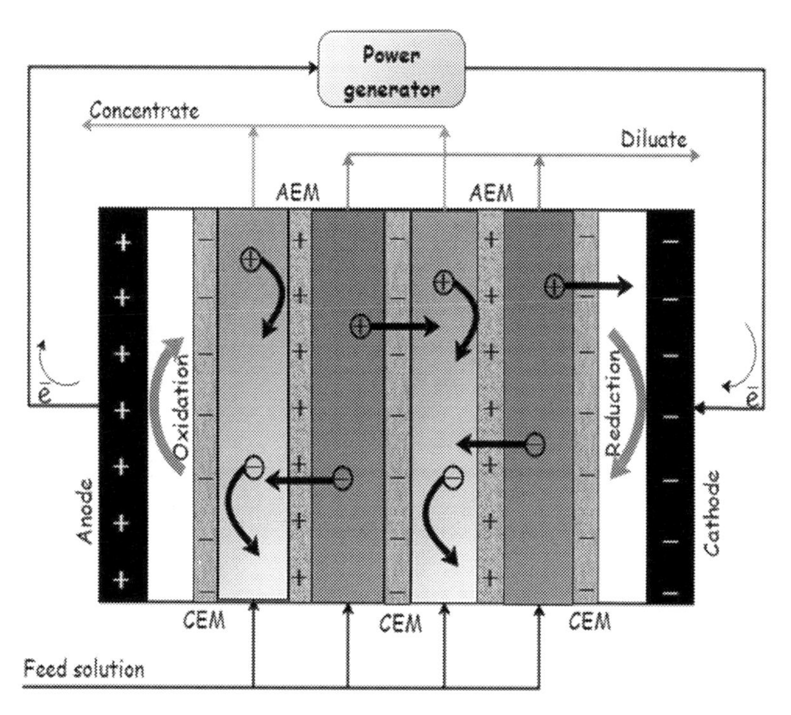

FIGURE 1.3 Schematic diagram of the electrodialysis process (Campione et al., 2018; Sun, Li et al., 2020).

in two compartments (concentrated and dilute compartments) (Roy, Moulik, Kamesh, & Mullick, 2020; Wang et al., 2017).

1.3.3.2 *Electrophoresis*

The term "electrophoresis" comes from Greek, with "electro" referring to the electrical current that adds energy to the electrons of a molecule's atoms and "phoresis" referring to particle movement. The electrophoresis process is a separation method that is often used to analyze biological or other polymeric samples. In other words, the movement and separation of charged particles (ions) under the influence of an electric field is referred to as electrophoresis. An electrophoretic device comprises two electrodes with opposite charges (anode and cathode) bound by an electrolyte. It is commonly used to analyze proteins and DNA fragment mixtures, and it is increasingly being used to analyze nonbiological and nonaqueous samples. Since electrophoresis does not alter the structure of a molecule and is highly sensitive to slight variations in molecular charge and size, it is an efficient analytical method (Rudge & Monnig, 2000).

1.3.4 Thermally driven membrane technology

Thermally driven membrane techniques, such as MD, are nonisothermal membrane processes with promising applications in the pharmaceutical, food, and textile industries, water and wastewater treatment plants, desalination, radioactive wastes, and toxic materials removal from aqueous solutions. Nonetheless, low productivity, low energy efficiency, and poor long-term stability are adversely affecting a large-scale industrial application process (Zare & Kargari, 2018).

1.3.4.1 *Membrane distillation*

MD is a separation technique based on a thermal gradient created over a microporous hydrophobic membrane. Simultaneously, low-grade heat or waste, such as solar, geothermal, tidal, wind, and nuclear energy, or low-temperature industrial flows, could carry out the process (Walton, Lu, Turner, Solis, & Hein, 2000). It is worth noting that the vapor pressure difference between the permeable hydrophobic membrane pores drives the process. In other words, nonvolatile compounds are kept on the retentate stream, while volatile vapor molecules are allowed to pass through the MD. The MD membrane's hydrophobic nature and surface tension are the fundamental causes of this phenomenon (Rahimpour, Kazerooni, & Parhoudeh, 2019).

A schematic graph of the MD method is shown in Fig. 1.4. As the diagram shows, volatile vapor molecules in the hot feed vaporized at the liquid/vapor interface will move through the membrane pores. On the other hand, the liquid feed is unable to pass through the pores of the membrane. The MD membrane's hydrophobic nature and surface tension are the fundamental causes of this phenomenon. Consequently, the liquid feed must avoid wetting the dry pores because it directly contacts the hydrophobic membrane. Then, different methods are used to collect or condense the permeated volatile vapors. Finally, entirely pure products are made theoretically free of solid, toxic chemicals, and nonvolatile contaminants (Tijing et al., 2015; Warsinger, Swaminathan, Guillen-Burrieza, Arafat, &

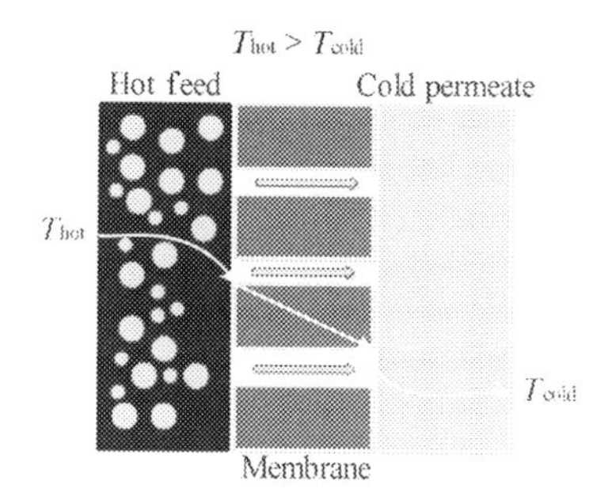

FIGURE 1.4 Schematic graph of the membrane distillation technology (Biniaz et al., 2019).

Lienhard, 2015). Few studies on a broad scale and long-term MD applications have been conducted because of many difficulties such as high energy consumption, scaling, fouling, and pore wetting. As a result, it is critical to develop novel membranes with specific properties such as low thermal conductivity, low mass transfer resistance, high thermal stability, and high chemical resistance, or a membrane with surface modifications to enhance MD efficiency and characteristics in order to reduce wetting and fouling phenomena, as well as increase permeate flow (Eykens, De Sitter, Dotremont, Pinoy, & Van der Bruggen, 2018; Swaminathan, Chung, Warsinger, AlMarzooqi, & Arafat, 2016).

Furthermore, the implementation of MD applications in wastewater treatment must deal with more organic and biological fouling and inorganic scaling. Therefore future research into the mechanisms of mixed fouling should be given careful and special attention (Biniaz, Torabi Ardekani, Makarem, & Rahimpour, 2019).

1.4 Concentration polarization term

CP is described as a phenomenon in which the concentration of a solute or particle near the membrane surface is higher than in bulk. More precisely, as Fig. 1.5 illustrates, any permeating organisms must first penetrate the stagnant boundary layer before passing through the membrane. As a result, the driving force available in the membrane is decreased, causing a decrease in flux.

All membrane filtration processes are subject to the CP phenomenon. It increases the concentration of solutes and substances at the membrane surface and enhances their breakthrough in the permeate phase. This not only raises the risk of fouling and degrades the permeate quality but also reduces the permeation rate in RO and NF due to increased osmotic pressure. The difference in permeability of the solvent and the solute causes CP. High solute and particle concentrations at the membrane's surface cause further back diffusion into the bulk until the rate of back diffusion equals the accumulation rate at the membrane surface, at which point a steady-state condition is reached. Increased crossflow velocity, a higher solute/ particle diffusion coefficient, and a higher temperature are examples of factors that increase back diffusion and minimize CP during the filtration process. Increased permeate flux or pressure filtration, on the other hand, raises CP. Various mathematical models were developed to describe the CP phenomenon during the membrane separation (Matsuura, 1993; Song & Elimelech, 1995). These models are: film theory,

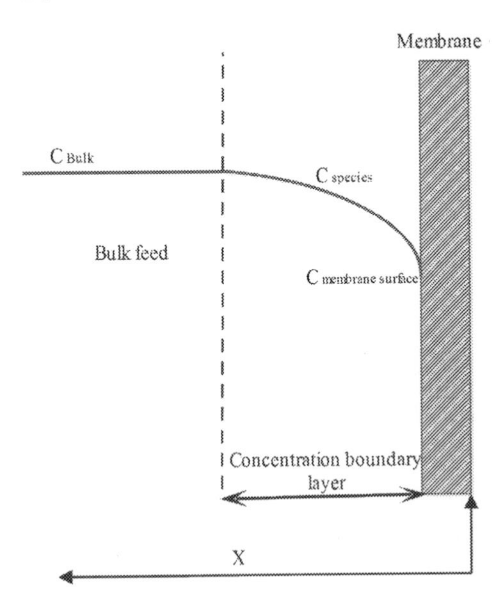

FIGURE 1.5 Schematic diagram of concentration polarization.

Spiegler–Kedem model, gel layer model, osmotic pressure model, theory of noninteracting particles, cake-enhanced CP (Shirazi, Lin, & Chen, 2010)

1.5 Membrane fouling

Membrane fouling is characterized as the deposition of suspended or dissolved substances on a membrane's external surfaces, pore openings, or within its pores, resulting in a loss of efficiency. Cake layer formation on the membrane surface, sludge flocs, or particles deposition on the membrane surface, adsorption of colloids or solutes or within membranes, and sludge flocs or particles deposition on the membrane surface are typical examples in this regard (Rudolph et al., 2019). Fouling is monitored in pressure-driven membrane processes on the basis of either a rise in pressure or reduction in flux over time. The usual and traditional fouling methods do not include details about the location, composition, or quantity of fouling. Because fouling is always cumulative, a performance loss is observed in the process. When fouling becomes permanent, the membrane is always not cleaned without chemicals, and the process of filtration or separation has to be stopped. The combination of membrane processes and various pretreatment alternatives may provide an effective fouling control and high-quality effluent. The most popular forms of membrane pretreatment used in water and wastewater treatment design are selected on the basis of the feed water supply, the nature of the existence contaminants, membrane materials, configuration of the module, operating conditions, and quality requirements for treated water. Adsorption, granular media biofiltration, peroxidation, and ion exchange are the most common types (Guo, Ngo, & Vigneswaran, 2012; Ibra et al., 2019; Nguyen, Roddick, & Fan, 2012). Fig. 1.6 shows external surface fouling and pore-blocking in MD technology.

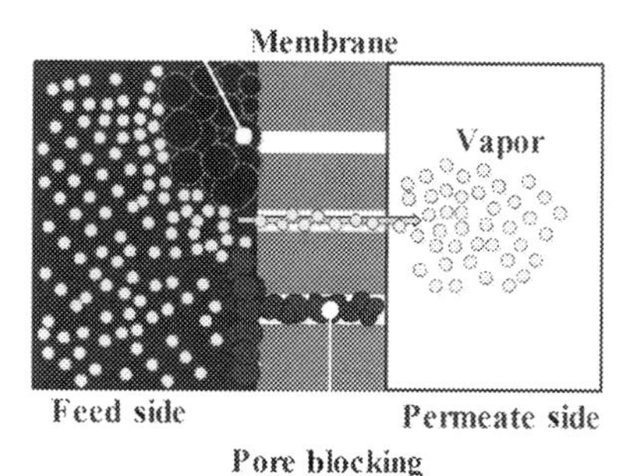

FIGURE 1.6 A schematic graph of surface fouling and pore-blocking in the membrane distillation technology (Biniaz et al., 2019).

1.6 New advances in membrane technologies

Membranes would undoubtedly become another solution for industrial challenges, according to growing research into the technology. On the other hand, new membrane technology would need a significant amount of research and development to be commercialized globally. The increasing worldwide competition in production has compelled the industry to upgrade the existing process designs and develop novel process designs because of the importance of the natural environment. As a result, the industry is emphasizing the production of new process designs based on alternative technologies. The future of membrane technology depends on continued research into membrane properties and basic aspects of transport phenomena in different membrane operations. Furthermore, fundamental and applied research such as membrane modification is required to develop novel membranes with improved characteristics and new membrane technologies.

1.6.1 Membrane modification

Conventional methods typically fabricate membranes, most of which contribute to relatively low porosity, membrane fouling, reduction in the permeate flux, membrane damage, early membrane replacement, or even shut down the process. Meanwhile, novel membrane production, which results in high porosity above 80% and interconnected open pore structures with high surface roughness, is developed to enhance membrane performance and provide high flux. One of the efficient methods to increase membrane antifouling is surface modification. The primary goal of membrane surface modification is to increase membrane hydrophilicity, improving membrane efficiency and reducing the adsorption of organic foulant on the membrane surface.

Hydrophilic polymers, via the formation of covalent bonds, are chemically modified onto membrane surfaces. Furthermore, by applying inorganic nanoparticles such as Fe_3O_4, Al_3O_4, ZrO_2, SiO_2, and TiO_2 into the membrane, the hydrophobic membranes can be converted into highly hydrophilic membranes (Laohaprapanon, Vanderlipe, Doma, & You, 2017; Yalcinkaya, Boyraz, Maryska, & Kucerova, 2020; Yalcinkaya, Siekierka, & Bryjak, 2017). Heating coating and hydrophilizing membrane surface by water-soluble solvents such as alcohols, acids, and mixtures of alcohols, acids, and water are additional examples of a surface modification technique.

Plasma surface modification of polymers, nanofiber production by electrospinning, and carbon nanotubes are other innovative production methods in this regard. Plasma treatments have been employed across a range of macro-and microporous materials for UF or MF to improve the membranes' performance and antifouling properties. Plasma-modified membranes are fabricated using vacuum filtration technique (Wu et al., 2004). First, the membranes are located in a chamber. A vacuum empties the chamber of air and creates a very low pressure inside. A small quantity of reactant gas, for instance, hydrogen, nitrogen, oxygen, inert (He, Ar), oxidative (CO_2, H_2O), or reductive gases (CF_3Cl, CF_2, NH_3), or a combination of these gases is then added into the chamber. An electrical charge, such as radio frequency electricity, is used to ionize the gas within the chamber, producing plasma ions. To put it another way, plasma glows are a complex environment in which ionized species, elementary particles, and ultraviolet (UV) radiation are simultaneously created and coexist and can initiate promotion and reconfiguration of chemical bonds and induce bond scission (Ostrikov, 2005; Ostrikov, Neyts, & Meyyappan, 2013). These activated compounds may react with the molecular structure of the outer layer through several reaction mechanisms and modify the same during specific modification time and particular chamber pressure.

Yang et al. (2013) invented plasma-modified ultrahigh carbon nanotubes membranes for exhibiting ultrahigh specific adsorption capacity for organic, metal contaminants, and salt (exceeding 400% by weight), which was two orders of magnitude greater than that detected in the current novel activated carbon-based water purification techniques. More importantly, single-layer porous graphene can be applied as an effective separation membrane because of its chemical and mechanical stability, flexibility, and, most importantly, its one-atom thickness. Graphene-based materials with water-permeable pores can be fabricated by creating either nanopores in graphene monolayers or two-dimensional channels that form between nanosheets of graphene oxide (Gong et al., 2019; Seo et al., 2018). An oxygen plasma etching process is used to create nanometer-sized pores in a graphene monolayer, which results in almost 100% salt rejection and accelerates water transport (Bo et al., 2017). Recently, Yang et al. (2019) reported novel hybrid membranes consisting of large-area graphene-nanomesh/single-walled carbon-nanotube using O_2 plasma exposure to fabricate outstanding mechanical strength while ultimately capturing the advantage of an atomically thin membrane.

1.6.2 Novel membrane reactors

Over the last few decades, researchers and industry experts have been particularly interested in membrane-based, simultaneous catalytic, and separation processes. In the future, this promising unit activity will be used to replace most convectional reactor processes. By simply combining a membrane separation unit with a chemical/biochemical reactor, a membrane separation unit may be used to adjust the reactant and substance concentration. This integration is often accomplished within the same unit. This technological definition has drawn significant worldwide research and process improvement activities over the last decades. Membrane reactors are (multiphase) reactors that combine catalytic reactions with membrane separation in one unit (Biniaz et al., 2020; Nagy, 2012, 2019).

MBRs and photoreactors are examples of recent advances of membrane reactors in process engineering. Separation of undesirable components can also occur during the dematerialization process via chemical conversion. Combining a membrane with a chemical reaction has shown tremendous benefits (Liu & Chen, 2002) that can be carried out in a reactor combined with a membrane module or in a membrane only. The catalytic membrane provides an efficacious tri-phase interaction of the gas, liquid, and active surface. The size of the catalytic layer pores can be modified according to reaction needs, with a narrow size distribution of pores, in the mesoporous or macroporous range. Researchers are exploring novel membrane technologies for addressing problems in fermentation and bioconversion of technically complex feeds, taking advantage of the special properties of both conventional MBRs and cell encapsulation processes. A typical example is a reverse membrane bioreactor in which bioconversion on a massive scale of complex feed containing highly suspended solids, inhibitors, and multisubstrates can be effectively handled (Meng et al., 2009; Stephenson, Brindle, Judd, & Jefferson, 2000).

Instead of having microorganisms suspended freely in the medium as in conventional MBRs, reverse MBRs use submerged membrane modules to house them among membrane layers, thereby providing a high density of local cells. Furthermore, they work on the basis of concentration-driven diffusion rather than the pressure-driven convection used in conventional MBRs (Arslan et al., 2021).

The membrane catalyst enables the reaction rate to be adjusted according to variable operational conditions such as different feed concentrations and flow rate. Employing the PV technique to cause an equilibrated reaction is one of the possible applications. According to the research conducted on membrane reactors, a significant percentage of catalytic membrane reactor applications include reversible reactions, while in a conventional reactor, a limited level of conversion is reached. When these reactions are carried out in a catalytic membrane, where one product can selectively permeate through the membrane and out of the reaction region, a much higher overall conversion is achieved than in a traditional reactor. In biotechnology and a variety of research fields, bioreactors with enzymes or whole cells immobilized are used (Cechetto, Di Felice et al., 2021; Westermann & Melin, 2009).

In various cases, a completed organic pollutant degradation in harmless and very small species without using chemicals, preventing sludge generation and disposal, is enabled by new technologies, such as the photocatalytic reaction. In these processes, the electronic excitation of a molecule or solid is caused by light absorption (usually ultraviolet light), which dramatically changes its ability to gain or lose electrons, promoting the decomposition of contaminants into harmless byproducts. Although MBR is an excellent alternative to traditional technologies, membrane fouling remains a severe and complex issue in the pressure-driven membrane process caused by pore blocks and the interaction between the foulants and membranes. Fouling in MBRs is primarily caused by soluble microbial materials, bacteria, and extracellular polymeric substances (Sohn et al., 2021).

1.6.3 Membrane contactors

MCs are innovative technology and new versions of the phase-contacting system used to improve mass transfer in liquid–liquid extraction and gas transfer processes. The principal function of the membranes is to serve as the interface between two phases without regulating the permeant rates. The principal driving force for the separation process in the system is the difference in chemical potential on both sides of the membrane (Li, Fane, Ho, & Matsuura, 2011). MCs are usually manufactured in a shell-and-tube form, with microporous thin hollow fiber membranes with pores that are small enough to prevent direct mixing of the flows on either side of the membrane because of capillary forces. In terms of physicochemical and morphological factors, MC systems must have nonwetting membranes with pore sizes in the range of 0.02–0.2 μm and high porosity (>75%–80%) to meet critical requirements of MC systems such as high interfacial area per unit volume and large volumetric mass transfer rate. Regular management of the optimal operating conditions that maximize the mass transfer coefficient while reducing the necessary membrane area is impossible because operating parameters such as concentration and temperature contribute to a problematic situation (Gugliuzza & Basile, 2013).

1.7 Conclusions and future trends

Membrane separation has become increasingly beneficial over traditional separation methods in recent years, thanks to economic and technical advances. Primary membrane and membrane technology principles and innovative concepts in the membrane operation were discussed in this chapter. Membrane and manufacturing process innovations must satisfy the intrinsic properties that define selectivity and flux for particular applications. Tunable and stable efficiency and membrane product sustainability over their entire life cycle are becoming increasingly essential.

An ideal membrane should have several specific features, including minimal thickness to maximize permeance, good mechanical strength, and uniform and narrow pore-size distribution for effective separation. In order to achieve these aims, plasma-modified nanoporous membranes of single- or few-atom thickness with excellent mechanical strength have been regarded as an excellent approach to fabricate ultrathin membranes with minimum transport resistance and maximum permeance and improve the long-term stability and permeability of membrane materials, consequently decreasing operational costs and process energy consumption. Therefore the design of a novel modified membrane is one of the most promising pathways to produce the next generation of membranes applied in industry. Therefore novel and environmentally friendly techniques are sought to produce surface energy–tunable membrane materials that can work at lower pressures to reduce the adverse impacts of physicochemical interactions while offering enhanced performance. Plasma technologies offer high-level platforms for a rapid functionalization of materials, allowing for the simultaneous tuning of surface energy and morphology.

The permeating compounds pass through the membrane because of a driving force, such as a pressure gradient, concentration gradient, electrical gradient, and temperature gradient. Membrane characterization, such as membrane pore size and membrane pore shape, which can cause this phenomenon, is divided into isothermal and nonisothermal operations.

Concentration-driven processes (ME and pervaporation), pressure-driven processes (RO, NF, MF, and UF), and electrically driven processes (ED and electrophoresis) are all part of the isothermal process. However, the driving force in a nonisothermal process, such as the MD process, is thermal energy. Membrane fouling is still a concern, despite the success of membrane processes in the separation and purification of organic compounds derived from biomass, owing to the complexity and high fouling propensity of bio-based streams. Fouling is a dynamic phenomenon that is an unavoidable part of any membrane system and has a negative impact on membrane efficiency. To satisfy the diverse and complicated requirements of industrial applications, modern online monitoring fouling devices must be built with faster data analysis and operational convenience.

Environmental awareness among the world's population is a growing phenomenon and finds its reflection in product development and manufacturing processes. In membrane technology, one can see initial steps in this direction with the replacement of hazardous solvents, the utilization of renewable materials for membrane production, and the reuse of membrane modules. Increasing the stability of organic polymer membranes and lowering the cost of inorganic membranes are two other examples. In the long term, several more advances in materials science would be required to create modern, advanced membranes.

Finally, treatment performance can be improved by combining treatment systems or using hybrid methods, leading to near-zero discharge. Treatment systems that are more compact and effective should be built further. While the production of long-lasting and energy-saving membranes has reduced operational costs, overall treatment costs are likely to rise because of increased concentrate treatment.

References

Abels, C., Carstensen, F., & Wessling, M. (2013). Membrane processes in biorefinery applications. *Journal of Membrane Science, 444*, 285–317.

Amelio, A., Van der Bruggen, B., Lopresto, C., Verardi, A., Calabro, V., & Luis, P. (2016). 14 – Pervaporation membrane reactors: Biomass conversion into alcohols. In A. Figoli, A. Cassano, & A. Basile (Eds.), *Membrane technologies for biorefining* (pp. 331–381). Woodhead Publishing.

Arslan, S., Eyvaz, M., Güçlü, S., Yüksekdağ, A., Koyuncu, İ., & Yüksel, E. (2021). Pressure assisted application of tubular nanofiber forward osmosis membrane in membrane bioreactor coupled with reverse osmosis system. *Journal of Water Chemistry and Technology, 43*(1), 68–76.

Baker, R. W. (2012). *Membrane technology and applications*. John Wiley & Sons. Available from http://doi.org/10.1002/9781118359686.

Benfer, S., Popp, U., Richter, H., Siewert, C., & Tomandl, G. (2001). Development and characterization of ceramic nanofiltration membranes. *Separation and Purification Technology, 22*, 231–237.

Biniaz, P., Makarem, M. A., & Rahimpour, M. R. (2019). Membrane reactors. In M. Benaglia, & A. Puglisi (Eds.), *Catalyst immobilization: Methods and applications* (pp. 307–324). Hoboken, NJ: Wiley.

Biniaz, P., T. Roostaie, M.R. Rahimpour (2021). 3 – Biofuel purification and upgrading: Using novel integrated membrane technology. In: M.R. Rahimpour, R. Kamali, M. Amin Makarem, M.K.D. Manshadi, (Eds.), *Advances in* bioenergy and microfluidic applications, Elsevier: (pp. 69–86). <https://doi.org/10.1016/B978-0-12-821601-9.00003-0>.

Biniaz, P., Torabi Ardekani, N., Makarem, M. A., & Rahimpour, M. R. (2019). Water and wastewater treatment systems by novel integrated membrane distillation (MD). *ChemEngineering, 3*(1), 8. Available from https://doi.org/10.3390/chemengineering3010008.

Bo, Z., Tian, Y., Han, Z. J., Wu, S., Zhang, S., Yan, J., …… Ostrikov, K. K. (2017). Tuneable fluidics within graphene nanogaps for water purification and energy storage. *Nanoscale Horizons, 2*(2), 89–98.

Camera-Rodaa, G., Loddob, V., Palmisanob, L., & Parrinob, F. (2019). Green synthesis of vanillin: Pervaporation and dialysis for process intensification in a membrane reactor. *Chemical Engineering Transactions, 75*, 1–6.

Campione, A., Gurreri, L., Ciofalo, M., Micale, G., Tamburini, A., & Cipollina, A. (2018). Electrodialysis for water desalination: A critical assessment of recent developments on process fundamentals, models and applications. *Desalination, 434*, 121–160.

Cechetto, V., Di Felice, L., Medrano, J. A., Makhloufi, C., Zuniga, J., & Gallucci, F. (2021). H_2 production via ammonia decomposition in a catalytic membrane reactor. *Fuel Processing Technology, 216*, 106772.

Chung, T.-S., Jiang, L. Y., Li, Y., & Kulprathipanja, S. (2007). Mixed matrix membranes (MMMs) comprising organic polymers with dispersed inorganic fillers for gas separation. *Progress in Polymer Science, 32*(4), 483–507.

Crespo, J.G., C. Brazinha (2015). 1 – Fundamentals of pervaporation. In: A. Basile, A. Figoli, M. Khayet (Eds.), *Pervaporation*, vapour permeation and membrane dist*illation*, Oxford, Woodhead Publishing: 3–17. <https://doi.org/10.1016/B978-1-78242-246-4.00001-5>.

Cui, Z.F., Y. Jiang R.W. Field (2010). Chapter 1 – Fundamentals of pressure-driven membrane separation processes. In: Cui, Z.F., Muralidhara, H.S. (Eds.), *Membrane technology*, Butterworth-Heinemann, Oxford (pp. 1–18). <https://doi.org/10.1016/B978-1-85617-632-3.00001-X>.

Deen, W. (1987). Hindered transport of large molecules in liquid-filled pores. *AIChE Journal, 33*(9), 1409–1425.

Ersahin, M. E., Ozgun, H., Dereli, R. K., Ozturk, I., Roest, K., & van Lier, J. B. (2012). A review on dynamic membrane filtration: Materials, applications and future perspectives. *Bioresource Technology, 122*, 196–206.

Eykens, L., De Sitter, K., Dotremont, C., Pinoy, L., & Van der Bruggen, B. (2018). Coating techniques for membrane distillation: An experimental assessment. *Separation and Purification Technology, 193*, 38–48.

Feng, X., & Huang, R. Y. M. (1997). Liquid separation by membrane pervaporation: A review. *Industrial & Engineering Chemistry Research, 36*(4), 1048–1066.

Figoli, A., S. Santoro, F. Galiano A. Basile (2015). 2 – Pervaporation membranes: Preparation, characterization, and application. In: A.Basile, A. Figoli,M. Khayet (Eds.), *Pervaporation, vapour permeation and membrane distillation*. Oxford, Woodhead Publishing: (pp. 19–63).

Gonçalves, L. M., Valente, I. M., & Rodrigues, J. A. (2017). Recent advances in membrane-aided extraction and separation for analytical purposes. *Separation & Purification Reviews, 46*(3), 179–194.

Gong, B., Yang, H., Wu, S., Xiong, G., Yan, J., Cen, K., ... Ostrikov, K. (2019). Graphene array-based anti-fouling solar vapour gap membrane distillation with high energy efficiency. *Nano-Micro Letters, 11*(1), 51. Available from https://doi.org/10.1007/s40820-019-0281-1.

Gugliuzza, A., & Basile, A. (2013). Membrane contactors: Fundamentals, membrane materials and key operations. In Angelo Basile (Ed.), , Handbook of membrane reactors (pp. 54–106). Woodhead Publishing:. Available from https://doi.org/10.1533/9780857097347.1.54.

Guo, W., Ngo, H. H., & Vigneswaran, S. (2012). *Fouling control of membranes with pretreatment. Membrane Technology and Environmental Applications* (pp. 533–580). American Society of Civil Engineers. Available from https://doi.org/10.1061/9780784412275.ch18.

Hilal, N., Al-Zoubi, H., Darwish, N. A., Mohamma, A. W., & Abu Arabi, M. (2004). A comprehensive review of nanofiltration membranes:Treatment, pretreatment, modelling, and atomic force microscopy. *Desalination, 170*(3), 281–308.

Ibrar, I., Naji, O., Sharif, A., Malekizadeh, A., Alhawari, A., Alanezi, A. A., & Altaee, A. (2019). A review of fouling mechanisms, control strategies and real-time fouling monitoring techniques in forward osmosis. *Water, 11*(4), 695.

Jakubowska, N., Polkowska, Ż., Namieśnik, J., & Przyjazny, A. (2005). Analytical applications of membrane extraction for biomedical and environmental liquid sample preparation. *Critical Reviews in Analytical Chemistry, 35*(3), 217–235.

Jönsson, J. (2012). Membrane-based extraction for environmental analysis. In Janusz Pawliszyn (Ed.), *Comprehensive sampling and sample preparation: Analytical techniques for scientists* (vol. 3, pp. 591–602). Oxford: Academic Press. Available from https://doi.org/10.1016/B978-0-12-381373-2.00104-67.

Joo, S. H., & Tansel, B. (2015). Novel technologies for reverse osmosis concentrate treatment: A review. *Journal of Environmental Management, 150*, 322–335.

Kang, G.-d, & Cao, Y.-M. (2014). Application and modification of poly (vinylidene fluoride)(PVDF) membranes–A review. *Journal of Membrane Science, 463*, 145–165.

Kayvani Fard, A., McKay, G., Buekenhoudt, A., Al Sulaiti, H., Motmans, F., Khraisheh, M., & Atieh, M. (2018). Inorganic membranes: Preparation and application for water treatment and desalination. *Materials, 11*(1), 74.

Kickelbick, G. (2003). Concepts for the incorporation of inorganic building blocks into organic polymers on a nanoscale. *Progress in polymer science, 28*(1), 83–114.

Kidambi, P. R., Jang, D., Idrobo, J. C., Boutilier, M. S., Wang, L., Kong, J., & Karnik, R. (2017). Nanoporous atomically thin graphene membranes for desalting and dialysis applications. *Advanced Materials, 29*(33), 1700277.

Kocherginsky, N., Yang, Q., & Seelam, L. (2007). Recent advances in supported liquid membrane technology. *Separation and Purification Technology, 53*(2), 171–177.

Kochkodan, V., & Hilal, N. (2015). A comprehensive review on surface modified polymer membranes for biofouling mitigation. *Desalination, 356*, 187–207.

Kumar, A. (2012). Fundamentals of membrane processes. *Membrane Technology and Environmental Applications*, 75–95. Available from https://doi.org/10.1061/9780784412275.ch03.

Ladewig, B., M.N.Z. Al-Shaeli (2017). Fundamentals of membrane processes. In: *Fundamentals of membrane bioreactors. Springer transactions in civil and environmental engineering* (pp. 13–37). Springer, Singapore. <https://doi.org/10.1007/978-981-10-2014-8_2>.

Laohaprapanon, S., Vanderlipe, A. D., Doma, B. T., Jr, & You, S.-J. (2017). Self-cleaning and antifouling properties of plasma-grafted poly (vinylidene fluoride) membrane coated with ZnO for water treatment. *Journal of the Taiwan Institute of Chemical Engineers, 70*, 15–22.

Lee, E. K., & Koros, W. J. (2003). Membranes, synthetic, applications. In R. A. Meyers (Ed.), *Encyclopedia of physical science and technology* (3rd ed., pp. 279–345). New York: Academic.

Lee, K. P., Arnot, T. C., & Mattia, D. (2011). A review of reverse osmosis membrane materials for desalination—Development to date and future potential. *Journal of Membrane Science, 370*(1–2), 1–22.

Li, D., & Wang, H. (2010). Recent developments in reverse osmosis desalination membranes. *Journal of Materials Chemistry, 20*(22), 4551–4566.

Li, H., Song, Z., Zhang, X., Huang, Y., Li, S., Mao, Y., ... Yu, M. (2013). Ultrathin, molecular-sieving graphene oxide membranes for selective hydrogen separation. *Science (New York, N.Y.), 342*(6154), 95–98.

Li, N. N., Fane, A. G., Winston Ho, W. S., & Matsuura, T. (2011). (Eds)*Advanced membrane technology and applications*. John Wiley & Sons. Available from http://doi.org/10.1002/9780470276280.

Liu, F., Hashim, N. A., Liu, Y., Abed, M. M., & Li, K. (2011). Progress in the production and modification of PVDF membranes. *Journal of Membrane Science, 375*(1–2), 1–27.

Liu, G., Wei, W., & Jin, W. (2014). Pervaporation membranes for biobutanol production. *ACS Sustainable Chemistry & Engineering, 2*(4), 546–560.

Liu, Q. L., & Chen, H. F. (2002). Modeling of esterification of acetic acid with *n*-butanol in the presence of $Zr(SO_4)_2 \cdot 4H_2O$ coupled pervaporation. *Journal of Membrane Science, 196*(2), 171–178.

Liu, X., He, M., Calvani, D., Qi, H., Sai Sankar Gupta, K. B., de Groot, H. J. M., ... Schneider, G. F. (2020). Power generation by reverse electrodialysis in a single-layer nanoporous membrane made from core–rim polycyclic aromatic hydrocarbons. *Nature Nanotechnology, 15*(4), 307–312.

Lu, D., Cheng, W., Zhang, T., Lu, X., Liu, Q., Jiang, J., & Ma, J. (2016). Hydrophilic Fe_2O_3 dynamic membrane mitigating fouling of support ceramic membrane in ultrafiltration of oil/water emulsion. *Separation and Purification Technology, 165*, 1–9.

Luo, T., Abdu, S., & Wessling, M. (2018). Selectivity of ion exchange membranes: A review. *Journal of Membrane Science, 555*, 429–454.

Malaeb, L., & Ayoub, G. M. (2011). Reverse osmosis technology for water treatment: State of the art review. *Desalination, 267*(1), 1–8.

Matsuura, T. (1993). *Synthetic membranes and membrane separation processes*. CRC Press.

Mei, Y., & Tang, C. Y. (2018). Recent developments and future perspectives of reverse electrodialysis technology: A review. *Desalination, 425*, 156–174.

Meng, F., Chae, S.-R., Drews, A., Kraume, M., Shin, H.-S., & Yang, F. (2009). Recent advances in membrane bioreactors (MBRs): Membrane fouling and membrane material. *Water Research, 43*(6), 1489–1512.

Miró, M., & Frenzel, W. (2004). Automated membrane-based sampling and sample preparation exploiting flow-injection analysis. *TrAC Trends in Analytical Chemistry, 23*(9), 624–636.

Mohammad, A. W., Teow, Y. H., Ang, W. L., Chung, Y. T., Oatley-Radcliffe, D. L., & Hilal, N. (2015). Nanofiltration membranes review: Recent advances and future prospects. *Desalination, 356*, 226–254.

Mohshim, D. F., Mukhtar, H. b, Man, Z., & Nasir, R. (2013). Latest development on membrane fabrication for natural gas purification: A review. *Journal of Engineering, 2013*, Article ID 101746, 7 pages, 2013. https://doi.org/10.1155/2013/101746.

Morao, A., Alves, A. B., & Cardoso, J. (2001). Ultrafiltration of demethylchlortetracycline industrial fermentation broths. *Separation and Purification Technology, 22*, 459–466.

Nagy, E. (2012). 8 - Membrane reactor. In: E. Nagy (Ed.), *Basic equations of the mass transport through a membrane layer*. Oxford, Elsevier: (pp. 193–211). <https://doi.org/10.1016/B978-0-12-416025-5.00008-9>.

Nagy, E. (2019). Chapter 13 - Membrane reactor. In E. Nagy (Ed.), *Basic equations of mass transport through a membrane layer* (2nd ed., pp. 369–380). Oxford: Elsevier.

Nath, K. (2017). *Membrane separation processes*. PHI Learning Pvt. Ltd.

Nguyen, T., Roddick, F., & Fan, L. (2012). Biofouling of water treatment membranes: A review of the underlying causes, monitoring techniques and control measures. *Membranes, 2*(4), 804–840.

Ostrikov, K. (2005). Colloquium: Reactive plasmas as a versatile nanofabrication tool. *Reviews of Modern Physics, 77*(2), 489.

Ostrikov, K., Neyts, E., & Meyyappan, M. (2013). Plasma nanoscience: From nano-solids in plasmas to nano-plasmas in solids. *Advances in Physics, 62*(2), 113–224.

Pendergast, M. M., & Hoek, E. M. (2011). A review of water treatment membrane nanotechnologies. *Energy & Environmental Science, 4*(6), 1946–1971.

Rahimpour, M.R., N.M. Kazerooni, M. Parhoudeh (2019). Chapter 8 - Water treatment by renewable energy-driven membrane distillation. In: Basile, Angelo, Cassano, Alfredo, Figoli, Alberto (Eds.), *Current* trends and future developments on (bio-) membranes, Elsevier: (pp. 179–211). <https://doi.org/10.1016/B978-0-12-813545-7.00008-8>.

Roy, A., Moulik, S., Kamesh, R., & Mullick, A. (2020). *Modeling in membranes and membrane-based processes*. John Wiley & Sons. Available from http://doi.org/10.1002/9781119536260.

Rudge, S. R., & Monnig, C. A. (2000). Electrophoresis techniques. *Separation and Purification Methods, 29*(1), 129–148.

Rudolph, G., Virtanen, T., Ferrando, M., Güell, C., Lipnizki, F., & Kallioinen, M. (2019). A review of in situ real-time monitoring techniques for membrane fouling in the biotechnology, biorefinery and food sectors. *Journal of Membrane Science, 588*, 117221.

Sadeghi, I., Govinna, N., Cebe, P., & Asatekin, A. (2019). Superoleophilic, mechanically strong electrospun membranes for fast and efficient gravity-driven oil/water separation. *ACS Applied Polymer Materials, 1*(4), 765–776.

Sanchez, C., Belleville, P., Popall, M., & Nicole, L. (2011). Applications of advanced hybrid organic–inorganic nanomaterials: From laboratory to market. *Chemical Society Reviews, 40*(2), 696–753.

Seo, D. H., Pineda, S., Woo, Y. C., Xie, M., Murdock, A. T., Ang, E. Y., ... Ostrikov, K. K. (2018). Antifouling graphene-based membranes for effective water desalination. *Nature Communications, 9*(1), 683.

Shao, P., & Huang, R. Y. M. (2007). Polymeric membrane pervaporation. *Journal of Membrane Science, 287*(2), 162–179.

Shi, X., Tal, G., Hankins, N. P., & Gitis, V. (2014). Fouling and cleaning of ultrafiltration membranes: A review. *Journal of Water Process Engineering, 1*, 121–138.

Shirazi, S., Lin, C.-J., & Chen, D. (2010). Inorganic fouling of pressure-driven membrane processes—A critical review. *Desalination, 250*(1), 236–248.

Sohn, W., Guo, W., Ngo, H. H., Deng, L., Cheng, D., & Zhang, X. (2021). A review on membrane fouling control in anaerobic membrane bioreactors by adding performance enhancers. *Journal of Water Process Engineering, 40*, 101867.

Song, L., & Elimelech, M. (1995). Theory of concentration polarization in crossflow filtration. *Journal of the Chemical Society, Faraday Transactions, 91*(19), 3389–3398.

Stephenson, T., Brindle, K., Judd, S., & Jefferson, B. (2000). *Membrane bioreactors for wastewater treatment*. IWA publishing.

Strathmann, H. (2004). *Ion-exchange membrane separation processes*. Elsevier.

Sun, Y., Li, J., Li, M., Ma, Z., Wang, X., Wang, Q., ... Gao, X. (2020). Towards improved hydrodynamics of the electrodialysis (ED) cell via computational fluid dynamics and cost estimation model: Effects of spacer parameters. *Separation and Purification Technology, 254*, 117599.

Swaminathan, J., Chung, H. W., Warsinger, D. M., AlMarzooqi, F. A., & Arafat, H. A. (2016). Energy efficiency of permeate gap and novel conductive gap membrane distillation. *Journal of Membrane Science, 502*, 171–178.

Tijing, L. D., Woo, Y. C., Choi, J.-S., Lee, S., Kim, S.-H., & Shon, H. K. (2015). Fouling and its control in membrane distillation—A review. *Journal of Membrane Science, 475*, 215–244.

Van der Bruggen, B., Mänttäri, M., & Nyström, M. (2008). Drawbacks of applying nanofiltration and how to avoid them: A review. *Separation and Purification Technology, 63*(2), 251–263.

Vandezande, P. (2015). 5 - Next-generation pervaporation membranes: Recent trends, challenges and perspectives. In A. Basile, A. Figoli, & M. Khayet (Eds.), *Pervaporation, vapour permeation and membrane distillation* (pp. 107–141). Oxford: Woodhead Publishing.

Vermaas, D., Guler, E., Saakes, M., & Nijmeijer, K. (2012). Theoretical power density from salinity gradients using reverse electrodialysis. *Energy Procedia, 20*, 170–184.

Walton, J., H. Lu, C. Turner, S. Solis, H. Hein (2000). *Solar and waste heat desalination by membrane distillation.*

Wang, L., Wang, Y., Wu, L., & Wei, G. (2020). Fabrication, properties, performances, and separation application of polymeric pervaporation membranes: A review. *Polymers, 12*(7), 1466.

Wang, Q., Gao, X., Zhang, Y., He, Z., Ji, Z., Wang, X., & Gao, C. (2017). Hybrid RED/ED system: Simultaneous osmotic energy recovery and desalination of high-salinity wastewater. *Desalination, 405*, 59–67.

Warsinger, D. M., Swaminathan, J., Guillen-Burrieza, E., Arafat, H. A., & Lienhard, J. H., V (2015). Scaling and fouling in membrane distillation for desalination applications: A review. *Desalination, 356*, 294–313.

Westermann, T., & Melin, T. (2009). Flow-through catalytic membrane reactors—Principles and applications. *Chemical Engineering and Processing: Process Intensification, 48*(1), 17–28.

Wu, Z., Chen, Z., Du, X., Logan, J. M., Sippel, J., Nikolou, M., ... Hebard, A. F. (2004). Transparent, conductive carbon nanotube films. *Science (New York, N.Y.), 305*(5688), 1273–1276.

Yalcinkaya, F., Boyraz, E., Maryska, J., & Kucerova, K. (2020). A review on membrane technology and chemical surface modification for the oily wastewater treatment. *Materials, 13*(2), 493.

Yalcinkaya, F., Siekierka, A., & Bryjak, M. (2017). Preparation of fouling-resistant nanofibrous composite membranes for separation of oily wastewater. *Polymers, 9*(12), 679.

Yang, H. Y., Han, Z. J., Yu, S. F., Pey, K. L., Ostrikov, K., & Karnik, R. (2013). Carbon nanotube membranes with ultrahigh specific adsorption capacity for water desalination and purification. *Nature Communications, 4*, 2220.

Yang, Y., Yang, X., Liang, L., Gao, Y., Cheng, H., Li, X., ... Duan, X. (2019). Large-area graphene-nanomesh/carbon-nanotube hybrid membranes for ionic and molecular nanofiltration. *Science (New York, N.Y.), 364*(6445), 1057–1062.

Zare, S., & Kargari, A. (2018). 4 - Membrane properties in membrane distillation. In V. G. Gude (Ed.), *Emerging technologies for sustainable desalination handbook* (pp. 107–156). Butterworth-Heinemann.

Zhang, S., & Drioli, E. (1995). Pervaporation membranes. *Separation Science and Technology, 30*(1), 1–31.

Zhang, T.C., R.Y. Surampalli (2012). Novel membrane-separation techniques and their environmental applications. In: *Membrane technology and environmental applications*: (pp. 696–726). American Society of Civil Engineers (ASCE). Doi:10.1061/9780784412275.ch23

Zhang, T.C., R.Y. Surampalli, S. Vigneswaran (2012). The values of membrane science and technology: Introduction and overview. In: *Membrane technology and environmental applications* (pp. 1–40), American Society of Civil Engineers (ASCE). DOI:10.1061/9780784412275.ch01

Zhang, T. C., Surampalli, R. Y., Vigneswaran, S., Tyagi, R. D., Leong Ong, S., & Kao, C. M. (2012). (Eds.) *Membrane technology and environmental applications.* American Society of Civil Engineers.

Zhou, J. L. (2012). 1.18 - Sampling of humic and colloidal phases in liquid samples. In J. Pawliszyn (Ed.), *Comprehensive sampling and sample preparation* (pp. 335–348). Oxford: Academic Press.

CHAPTER

2

Transport phenomena in ultrafiltration/microfiltration membranes

Endre Nagy and Imre Hegedüs

Laboratory of Chemical and Biochemical Processes, Research Institute of Biomolecular and Chemical Engineering, University of Pannonia, Veszprem, Hungary

Abbreviations

G graphene
GO graphene oxide
rGO reduced graphene oxide
ILs ionic liquids
MF microfiltration
UF ultrafiltration

Nomenclature

C concentration in the membrane layer, kg/m^3, mol/m^3
C_p the outlet reactant concentration, kg/m^3, mol/m^3
D diffusion coefficient, m^2/s–
J mass transfer rate, kg/m^2s, mol/m^2s
k mass transfer coefficient, m/s
N number of sublayers, ($N = 500$ for both layers)
p pressure, Pa
r radius, size
R rejection coefficient,–
Y dimensionless local coordinate, –
v volume flow rate, m^3/m^2s

Current Trends and Future Developments on (Bio-) Membranes
DOI: https://doi.org/10.1016/B978-0-12-822257-7.00013-3

Greek

δ thickness of the transport layers, m
Φ partition coefficient,–
ξ hindrance factor

Superscript

o bulk
m membrane inlet or polarization layer's outlet surface

Subscript

b bulk surface of the fluid layer
c convective
d diffusive
i ith sublayer
j boundary or membrane layer
L fluid boundary layer
m membrane selective layer and its outlet surface
p pore
s solute

2.1 Introduction

Ultrafiltration (UF) and microfiltration (MF) membrane separation processes are frequently used industrial membrane separation technologies. Both of them are pressure-driven, size-excluding filtration processes, where nanoparticles in a fluid phase are removed from the continuous phase. Membranes are selective borders and thanks to their selectivity, depending on its structure, that is, on porosity, pore size, tortuosity, they can allow some types of particles (macromolecules) to pass through them, but they do not allow it for other types of materials, depending mainly on their size. Configurations of membranes used for MF processes should be hollow fiber, tubular, spiral-wound, or flat-sheet membranes. Two main types of configuration of UF/MF process are cross flow and dead-end processes.

UF and MF as separation technologies are distinguished from each other by the pore size of the membrane applied for separation processes. The pore size of MF membranes ranges between about 0.1 and 5–10 μm (Fig. 2.1). This pore size could retain particles with hydrodynamic diameter above about 700–1000 kDa, the size of which is about 0.1–0.2 μm (He & Niemeyer, 2003). The size range, from 0.1 to 10 μm, corresponds to the size range of cells (yeasts, bacteria, and some cells of multicellular organisms, e.g., red blood cells) (Fig. 2.1).

The pore size of UF membranes ranges about 0.01–0.1 μm (10–100 nm). UF membranes could retain solutes, colloid particles, nanoparticles, with particle size between 5 and 150 nm (Castro-Muñoz, Boczkaj, Gontarek, Cassano, & Fíla, 2020; Lutz, 2015). This particle size range corresponds to about 1–100 kDa molecular weight of particles. Typical examples for particles in the nature between these molecular weight are viruses or biopolymers (e.g., proteins, polysaccharides, DNA, etc.; molecular weight of albumin is about 66 kDa) (Huter & Strube, 2019; Lutz, 2015) (Fig. 2.1). Therefore the best examples for application of UF are separation of products of bioprocesses, for example, pharmaceutical bioprocesses, as biologically active products or mammalian cells, yeasts, bacteria, enzymes as "biocatalyzers" of bioreactor (Becker, 2015; Lutz, 2015).

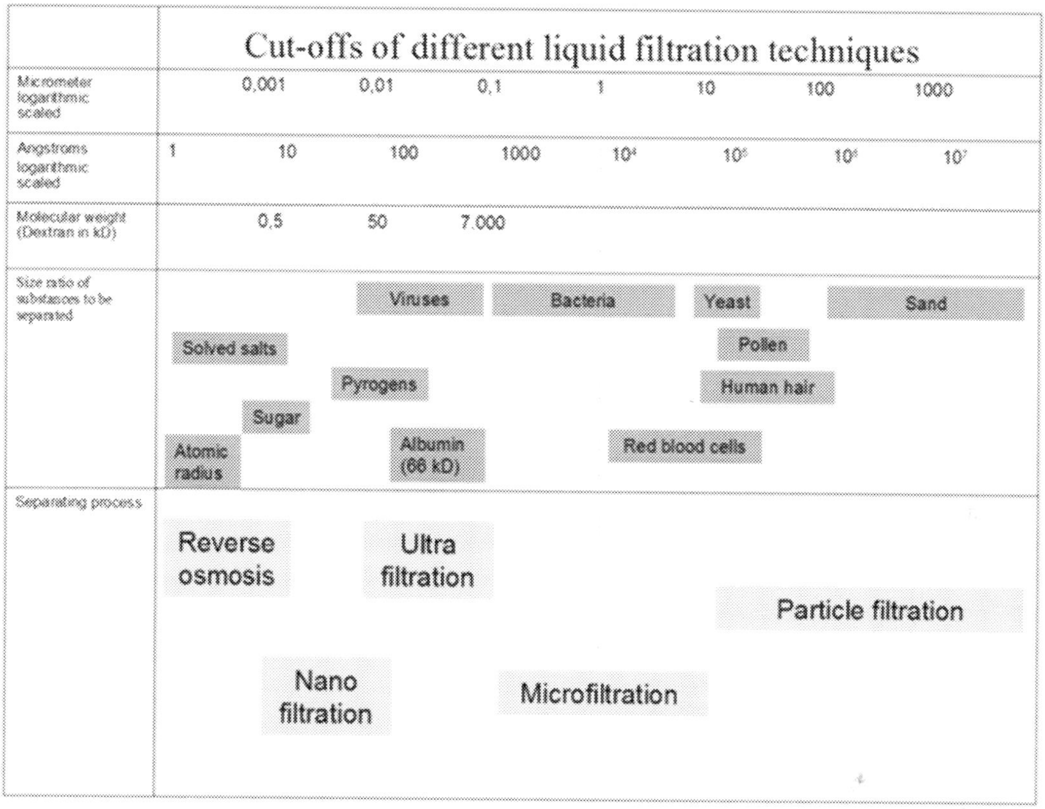

	Cut-offs of different liquid filtration techniques
Micrometer logarithmic scaled	0,001 0,01 0,1 1 10 100 1000
Angstroms logarithmic scaled	1 10 100 1000 10^4 10^5 10^6 10^7
Molecular weight (Dextran in kD)	0,5 50 7.000
Size ratio of substances to be separated	Viruses, Bacteria, Yeast, Sand, Solved salts, Pollen, Pyrogens, Human hair, Sugar, Atomic radius, Albumin (66 kD), Red blood cells
Separating process	Reverse osmosis, Ultra filtration, Particle filtration, Nano filtration, Microfiltration

FIGURE 2.1 Cutoffs of different liquid filtration processes.

MF processes are used in wide areas of industry, for example, in wastewater technology, desalination, pharmaceutical industry, biotechnology, and food industry (Anis, Hashaikeh, & Hila, 2019). Most important research activities recently focused on wastewater treatment, in elimination of fouling during the processes or in fabrication of more effective membranes (Anis et al., 2019). For example, the number of research papers in the area of wastewater treatment reached 795 between 2009 and 2018 (Anis et al., 2019). Contrary to it, UF process is often used for drinking water purification, protein concentration, for example, recovery of enzymes from bioreactor after the biocatalytic processes, for their reuse in a next production cycle. It is frequently used in water cleaning, for example, for disinfection and removal of pathogenic microorganism (Lee, n.d.), or in seawater reverse osmosis (Tabatabai, 2014).
Generally, MF and UF technology should be used in industry as follows:

- *Purification processes*, where contaminants are rejected from the pure products, for example, water purification, where impurities are separated from drinking water (Al Aani, Mustafa, & Hilal, 2020).
- *Concentration processes*, by which valuable parts or particles of a mixture are concentrated (retained) by membrane, while solvent or other not-valuable by-products

might pass through the membrane (e.g., concentration and purification enzymes for its reuse after the end of the biocatalytic membrane process) (Yang et al., 2019).

- Other *separation processes*, by which the valuable component(s) are removed from the reaction mixture, where they are selectively transported through the membrane (e.g., products of fermentation are separated by membrane permeation in the bioreactor, while the biocatalysts as, for example, enzymes or bacteria are retained) (Huter & Strube, 2019).

Membrane materials used for UF and MF processes could be usually:

- *hydrophilic* membranes (they are often used for wastewater treatment) (Nasrollahi, Aber, Vatanpour, & Mahmoodi, 2019),
- but in some cases they are *hydrophobic* membranes (e.g., polysulfone or graphene-based membranes) (Sabet, Soleimani, Mohammadian, & Hosseini, 2019).

Structure of UF/MF membranes is usually porous material, which can be symmetric (isotropic) or asymmetric (anisotropic) (Asad, Sameoto, & Sadrzadeh, 2020). Asymmetric (anisotropic) membranes involve a support layer with relatively great pores, which provide the membrane mechanical stability and a very thin active layer, which has not very small pores The pore size of this active layer is responsible for separation and it determines the cutoff value of the membrane. The separation mechanism of porous membranes is based on molecular sieving or size-exclusion mechanism, which does not allow the larger particles, macromolecules, larger than the pore size, to pass through this active membrane layer. The mass transport through the active layer takes place by diffusion, amid the chemical potential difference and additionally, by convection due to the hydraulic and or osmotic pressure difference across the membrane. In the case of MF, the diffusive flow can be neglected because of the high convective flow, since role of diffusion plus convection must be taken into account during UF processes. This transport is discussed in detail in the theoretical part of this chapter (see Section 2.2).

Membranes could be characterized by some characteristic properties as, for example, by pore-size, molecular weight cutoff, porosity, tortuosity, and its thickness. The transport rate during UF and MF processes could essentially be limited by some external phenomena as: for example, the fouling (transport through it is not discussed in this study) and/or concentration polarization (Baker, 2004; Chew, Kilduff, & Belfort, 2020).

1. Fouling as main mass transfer limiting phenomena could reduce both selectivity and permeability of the membrane separation and reduction of its effect is essential for the effective usage of the separation; unfortunately, the effect of this phenomenon is hard to predict. Bacchin, Si-Hassen, Starin, Clifton, and Aimar (2002) surveyed the mass transport model for colloid separation. Recently Tow and Lienhard (2016) recommended a description of the solute transport through fouling layer. This model was then applied for osmotic pressure-driven membrane process (Nagy, Hegedüs, Tow, & Lienhard, 2018). Fouling results with the interaction among solutes such as suspended colloids, dissolved organic matters, and proteins, with each other or with the membrane material, and therefore it can often create hindrance in the solute transport (Bacchin et al., 2002; Benavente et al., 2016).
2. Concentration polarization is often caused by the selective transport of some inert particles and/or dissolved species in the fluid laminar (without dispersion) transport layer. The retained species accumulate in the solution near to the membrane inlet

surface and lower the permeation driving force; the extent of the accumulated species concentration depends on the type of solute(s) (i.e., small, medium, and large effects by colloids, proteins and salt, respectively). Concentration polarization layer is a highly concentrated solute layer in which the retained components attempts to diffuse back toward the bulk fluid phase and causes a decline in filtration efficiency (see Subchapter 2.2).

2.2 On membrane material and its preparation

The important and updated methodologies of the membrane preparations and its materials are briefly surveyed in the following subsections. It contains the lately developed methods which can essentially change the membrane separation properties and thus the membrane performance.

2.2.1 Membrane preparation by conventional materials

One of the most frequently used chemical type of materials for UF and MF are organic polymers.

1. *Polymeric membranes* could be composed by natural polymers (e.g., cellulose, chitosan, etc.), modified natural polymers (e.g., regenerated cellulose, cellulose acetate, etc.), or synthetic polymeric materials (e.g., polyacrylonitrile, polysulfone, etc.) These conventional polymeric materials are usually linear polymers (see Fig. 2.2/II) or spatial polymers without branches (Fig. 2.2/IV).

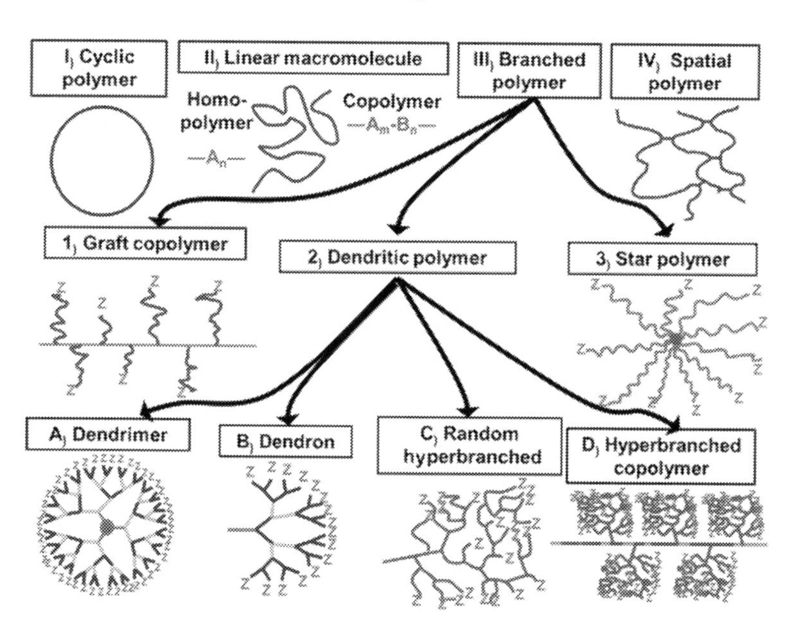

FIGURE 2.2 Distribution of polymeric materials by branching types.

Nowadays, some new types of polymers as highly branched polymers (Fig. 2.2/III) have been developed and some results show that highly branched polymeric structures could also enhance the selectivity and the effectiveness of UF and MF processes. These new polymeric materials contain numerous branches and high number of functional groups on the end of branches (Fig. 2.2/III). Main types of branched polymers are (1) graft copolymers where a long polymeric chain contains side-chains; (2) dendritic polymers with tree-branch-like branches; (3) star polymers, where numerous polymeric chains are originated from a central core (Fig. 2.2/III/$1_)$, $2_)$, $3_)$).

Dendritic polymers are highly branched and they could subdivided: (A) dendrimers, which are symmetric, monodispersed spheres and their branches, are started from a central core with numerous functional groups on the surface; (B) dendrons, which are highly branched spheroid particles (their branches are also originated from one central core) with functional groups on their branch-ends (Fig. 2.2/A and B), while (C) randomly hyperbranched polymers where branches are not ordered and they have a random polymeric structure and (D) hyperbranched copolymers, where these polymers are grafted to a linear polymer. Both hyperbranched polymers and copolymers could contain also numerous functional groups on their branch-ends (Fig. 2.2/C and D). On the other hand, these materials could be nontoxic, biocompatible and biodegradable materials, therefore their application in water cleaning is obvious (Sajid, Nazal, Ihsanullah, Baig, & Osman, 2018). For example, removing heavy metal ions by complexation (Borbély & Nagy, 2009) with primary amine groups of highly branched poly(amidoamine) (PAMAM) polymers, dendrimers, or dendrons in the membrane could be very effective process (Zhang, Liu, Yang, Zhou, & You, 2018).

2. Other widely used types of UF and MF membranes are *ceramic membranes*. These materials are usually applied in industrial/municipal wastewater treatment and drinking water treatment because ceramic membranes have better fouling resistance, higher permeability and longer lifetime than other, for example, polymeric membranes. Moreover, ceramic membranes could degrade contaminants and clean itself (Li, Sun, Lu, Ao, & Li, 2020). Ceramic materials are usually mixtures of alumina, silica, zirconia and other oxides. Main preparation steps usually are preparation of slurry, and then it is shaping and sintering (Ishak, Hashim, Othman, Monash, & Zuki, 2017). Ceramic membranes are usually combined with nanomaterials to improve selectivity, permeability and to avoid its fouling. For example, Çelik, Çelik, Flahaut, and Suvaci (2016) synthesized alumina oxide – graphene oxide hybrid membrane and its heat stability and conductivity values are increased.
3. *Ionic liquids* (ILs) as new materials could also be effectively enhanced the filtration processes (Foong, Wirzal, & Bustam, 2020). ILs are liquid salts, which are stable at room temperature and they usually contain organic compound(s) as partner ion. Contrary to traditional organic solvents, ILs have very low vapor pressure (they are not flammable), have a good electrochemical stability and high conductivity (Foong et al., 2020). Thanks to the ionic interactions between compounds of ILs, they have tunable physical and chemical features, therefore ILs could be used as solvent agents of hardly insoluble materials, for example, separation of lignin compounds from lignocellulose easily possible by alkylbenzenesulfonate ILs (Tan et al., 2009). Enzymes, DNA, and RNA are

not only solved in ILs but also their operational and heat stability are also increased (Elgharbawy, Moniruzzama, & Goto, 2020). While the first and second generations of ILs are toxic, the third generation of ILs are biocompatible and biodegradable (Egorova, Gordeev, & Ananikov, 2017). ILs could also be used as reaction media for organic catalysis (Fehér, Tomasek, Hancsók, & Skoda-Földes, 2018) or as separation media for metal ions, for example, separation of small amounts of iron, alumina, titanium, calcium, sodium, and silicon ions from red mud (Binnemans, Pontikes, Jones, Van, & Blanpain, 2013; Koók et al., 2017).

During last few years, ILs play more and more important role in membrane technology, for example, in *supporting IL membranes,* where the porous membrane support is fulfilled by IL. These membranes can separate different organic compounds from each other, for example, Kamaz, Vogler, Jebur, Sengupta, and Wickramasinghe (2020) separated *cis*- and *trans*-stilbene, obtained different mass transfer coefficients, applying imidazonium-based ILs, which were trapped in the pores of polypropylene membrane.

Membranes for IL application are polymeric membranes, which build of conventional monomers and repeated units of IL monomers. For example, Sengupta, Kumar, Kamaz, Jebur, and Wickramasinghe (2019) synthesized poly IL UF membrane using imidazonium rings as IL monomers which has antimicrobial features.

2.2.2 Membrane preparation by nanosized materials

New materials have been synthesized during the last few years that could improve the selectivity and the efficiency of the separation processes. Nanosized and nanofuncionalized materials as additives for nanocomposites of membrane materials could change dramatically some principal features of membranes and represent new, promising physicochemical properties of these membranes. During the last few years, wide range of materials have been developed based on nanomaterials and these materials increase advantageously the characteristics of the *carbon nanotube* membrane and they resulted in more economic and efficient separation technologies (Song et al., 2018).

- *Carbon-based nanomaterials* (fullerenes, carbon nanotubes, graphene, etc.) as nanocomposite materials could essentially change the characteristics of membranes (selectivity, permeability, etc.) and therefore nanocomposite membranes have increased effectiveness for numerous applications (Roy et al., 2020; Song et al., 2018).
- Fullerenes are "zero-dimensional" molecules, which is building up by pure carbon atoms (Fig. 2.3A) and their derivatives (fullerenol, carboxy-fullerenes, arginine-derivatives, etc.) are good additives of membranes and can increase dramatically their porosity and selectivity (Dmitrenko et al., 2019).
- Carbon nanotubes membranes are one of the most frequently investigated nanomaterial-based membranes, which get about half of research articles during the last decade (Roy et al., 2020). Carbon nanotubes contain one or more one-atom thick carbon tubes. These carbon tubes have cylindrical structure with about 1 nm diameter. Carbon atoms in the tube have sp^2 hybridization and the carbon nanotube is building up in a hexagonal structure with integrated pentagonal rings (Fig. 2.3).

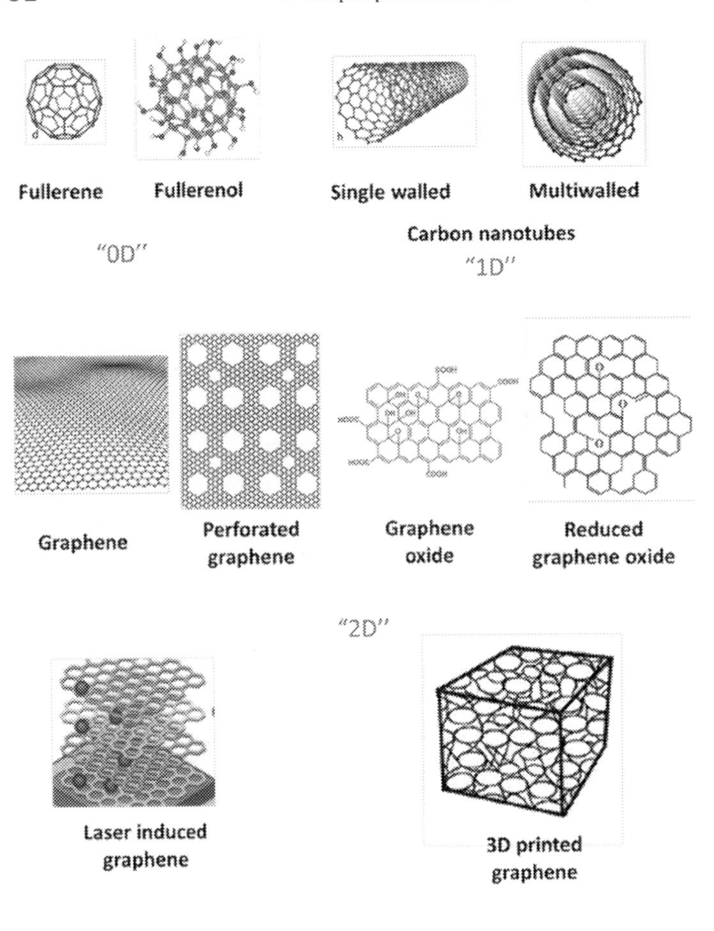

FIGURE 2.3 Carbon-based nanomaterials (A) "0D" fullerenes and its derivatives; (B) "1D" Carbon nanotubes; (C) "2D" Graphene and its derivatives; (D) "3D" laser-induced grapheme.

Two main types of carbon nanotubes are: single walled and multiwalled-carbon-nanotubes. Carbon nanotubes have promising physical properties: their tensile strength is two order of magnitudes higher than that of the steel, moreover they are very elastic, because their Young modulus is five time higher than that of steel (Mittal, Dhand, Rhee, Park, & Lee, 2015). Other promising feature of carbon nanotubes is their low density (it is one quarter of the density of steel). Their high surface area/volume ratio (10–20 m^2/g) makes them also effective for enhance or change physical and chemical properties of composite materials as nanosized component. Integration of carbon nanotubes into the membrane matrix could change dramatically the membrane properties: its hydrophobic property and permeability could increase but the intensity of fouling is decreasing (Roy et al., 2020). For example, nanocomposite UF membrane, containing polyphenylesulfone/multiwalled-carbon-nanotubes/polyvinylpyrrolidone/a-methyl-2-pyrrolydon, has four times higher permeability than that of membranes without carbon nanotubes and it could effectively remove heavy metal ions. It can remove 72%–98% of Pb^{2+}, Hg^{2+}, and Cd^{2+} ions from their water-based solutions (Chandrashekhar et al., 2019).

Carbon nanotubes could reduce fouling building up process, as it was observed in the case of, for example, sulfonated carbon nanotubes-blended polyvinylindine fluoride UF membranes, applying it as bioreactors (fouling recovery ratio of the membrane could reach more than 83%) (Ayyaru, Pandiyan, & Ahn, 2019). Sensor-functionalized carbon nanotubes integrated into membrane material could enhance the effectiveness of removal of polycyclic aromatic hydrocarbons from wastewater (Akinpelu et al., 2019). Carbon nanotubes could be conductive, for example, carbon nanotubes composed with alumina nanoparticles could conduct electrostatic negative repulses and remove humic acid foulants from the membrane (Mao et al., 2019).

Graphene-based materials (namely graphene, perforated graphene, graphene oxide, and reduced graphene oxide) play an important role as membrane components in separation technology (Song et al., 2018). Graphene (G) is one carbon atom thick, two-dimensional layer, which contains carbon atoms only in sp^2 hybridization and hexagonal atomic arrangement. Contrary to its subnanometric thickness, graphene is extremely strong material with outstanding elasticity. Thanks to its high specific surface area (2600 m^2/g), graphene could absorb easily various chemical components. This layer can be used in nanofiltration and reverse osmosis technologies (Sabet et al., 2019). Perforated graphene contains tunable uniform, small pores (Cho, Droudian, Wyss, Schlichting, & Park, 2018), therefore it is ideal membrane additive component for UF, as well (Song et al., 2018). While graphene is a hydrophobic material, graphene oxide (GO) is the oxidized form of graphene, which contains –OH, =O, and –COOH functional groups, bound chemically to its surface, and therefore its hydrophobic character is lowered. Reduced graphene oxide (rGO) containing membranes have promising antibacterial activity (Zhang et al., 2019).

Laser-induced graphene layer has porous, three dimensional structure, which could reduce the microbial fouling in membranes, due to its strong antimicrobial activity (Singh et al., 2017). Composition of laser-induced graphene and graphene oxide in membrane results in fine-tuned hydrophobic-hydrophilic membrane, low molecular weight cutoff value (even 90 kDa) with high flux and good antifouling property (Guirguis et al., 2019; Song et al., 2018; Thakur, Singh, Thamaraiselvan, Kleinberg, & Arnusch, 2019).

The next step toward creation of UF membranes with more specific selectivity is the functionalization of nanomaterials in membrane. For example, functionalization nanomaterials in membranes with biologically active materials (e.g., cyclodextrins, calyxarenes, collagene fibers, DNA, polypeptides or enzymes, etc., as biofunctionalized nanomaterials) are ideal potential absorbents for monocyclic and polycyclic hydrocarbons (Basak, Hazra, & Sen, 2020). Other nanosized materials, for example, nanosilica (Kaleekkal et al., 2018), manganese dioxide nanospheres (Sri Abirami et al., 2019), titanium dioxide dopes (Romadhoni, Hidayat, Andina, & Iqbal, 2019), etc. could highly enhance the porosity of polymeric membranes.

2.2.3 Other membrane preparation methods

Not only new materials as composites but also *new technological solutions* for UF are developed during the last few years, which can be more economic and effective than classical ones.

1. *Electronspun membranes*, based on new nanomaterial fabrication method (electronspinning), are nanosized fibers with high surface area to volume ratio (Foong et al., 2020). During the electrospinning, a very thin jet of polymer solution is created and charged by high voltage and this jet is solidified and nanosized (even less than three nanometer of diameter) fibers are obtained. Electronspun membranes could remove, for example, heavy metal ions by physical interactions (affinity) or electrostatic interactions (e.g., chelation and complexation). ILs could also be used as membrane components, for example, by combination with electronspun materials in membranes. For example, cellulose is usually insoluble in water or organic solvents but soluble in some ILs and this feature could be used for creation of cellulose acetate membrane by electrospinning (Lee, Jeong, Kang, Lee, & Park, 2009).
2. *Polymer-enhanced UF* combines the advantages of UF and macromolcular complexation for the separation process (Huang & Feng, 2019). For example, heavy metal ions can be separated from their aqueous solvent by technology of complexation reaction of metal ions with the chelate-building components of the membrane material. For example, poly(amidoamine) dendrimers and dendrons contain numerous, even more than one hundred, primary amino groups at the end of their branches (see Z signs in Fig. 2.2/A and B). These amino groups could interact with heavy metal ions (e.g., with Cu^{2+} ions) by complexation reaction and separate them from wastewater even at low concentration values (Ertürk, Gürbüz, Tülü, & Bozdoğan, 2018).
3. *Predeposited dynamic membrane filtration* is also a new technology, which is based on a dynamic membrane that contains a layer of the deposited particles on the membrane surface and this layer acting as a secondary membrane (Anantharaman, Chun, Hua, Chew, & Wang, 2020). In some cases, this process is more economical one, it is technologically more simple than the conventional UF methods, for example, Zhang, Wei, Yong, Liu, and Liu (2018) successfully cleaned oily seawater by dynamic membrane of silica–alumina support layer deposited by yttrium, which can work at lover pressure than conventional reverse osmosis membrane; the total cost of the separation process was reduced by 22%.

2.3 Theoretical part

UF/MF is an external pressure-driven filtration process, during which the separation is occurred by size exclusion and additionally diffusive plus convective transport of the microparticles, to be separated during the process, can be taking place. Accordingly, the convective flow can play important role in the transport in this process. It is well known that the convective flow does not induce separation effect on the transport of particles. On the other hand, the often applied membrane for separation is capillary membrane with rather low lumen radius, due to its advantageous surface/volume ratio. In this study, the solute transport through flat-sheet membranes is discussed, taking into account the effect of the fluid mass transport resistance in the laminar boundary layer (often called as polarization layer), as well. Let us look the mass transport through a flat-sheet asymmetric porous membrane layer and taking into account the component and solvent transport both the external, fluid boundary layer and also through the asymmetric membrane layer.

The main point of this study is to show the simultaneous effect of the membrane and the fluid transport properties. Nagy, Kulcsar, and Nagy (2011) discussed firstly the simultaneous transport of the transport layers during nanofiltration, which is then extended, and applied to UF process. The effect of the mass transports and its effect on the concentration distribution are analyzed in the next section, in details.

2.3.1 The mass transport through a flat-sheet membrane

The concentration distribution with notations of the important concentrations is illustrated by Fig. 2.4. Note, the interface concentrations are equal to each other, on the two sides of the boundary layer and the membrane layer. This means that it is supposed that there is no remarkable solubility of the transported solute components in the membrane matrix. The membrane matrix behaves as an inert phase. If there is nevertheless interconnection between the matrix and solute components, it can easily be taken into account with change of the interface concentrations. Values of them will be different, namely, $C_m^* = \Phi C_L^*$ with Φ is the partition coefficient. Shapes of the concentration distribution curves are concave for the boundary layer and convex one for the membrane layer. This phenomenon is the consequence of the fact that the convective and the diffusive flows have reverse direction in the boundary layer, while they are concurrent in the membrane layer. However, the tendencies of the solute concentrations are different. It increases in the boundary layer while it decreases in the membrane layer. Accordingly the differential mass balance equations have the same form. The general differential mass balance expression, for hydraulic pressure difference-driven membrane process, for the membrane boundary layer, taking into account that the direction of the component and solvent transport, for both layers, is as:

$$-D\frac{d^2C}{dy^2} + v\frac{dC}{dy} = 0 \tag{2.1}$$

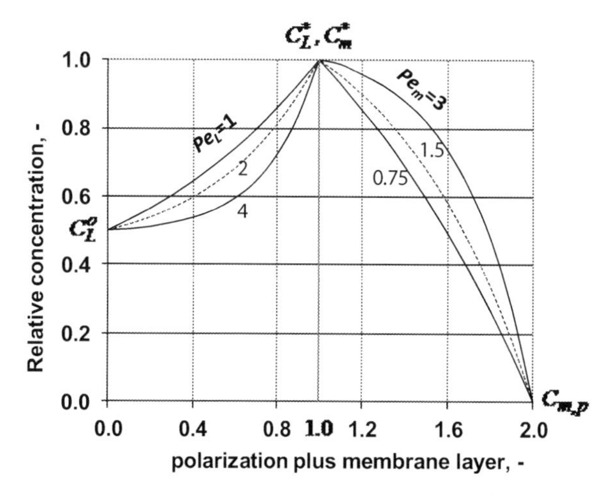

FIGURE 2.4 The concentration distribution of solute transport and notations through microporous membrane with transmembrane, accordingly at different Peclet-numbers.

Its dimensionless form, considering the local coordinate, is as ($Y = y/\delta$):

$$-\frac{d^2C}{dY^2} + Pe\frac{dC}{dY} = 0 \tag{2.2}$$

The general solution of expression defined by Eq. (2.2) is as follows:

$$C = Te^{PeY} + S \tag{2.3}$$

The boundary conditions are illustrated in Fig. 2.1. Thus, for example, for the boundary layer they are as:

$$C = C^o \quad \text{at} \quad Y = 0 \tag{2.4}$$

and

$$C = C_L^* \quad \text{at} \quad Y = 1 \tag{2.5}$$

Note that Eq. (2.5) does not involve any restriction for the outlet boundary layer. It allows the diffusive flow and the convective outlet flow into the membrane and out of the membrane surface. Thus the value of the outlet fluid concentration is determined mostly by the operating conditions, as flow rate of the sweeping phase, determining the hydrodynamic conditions on the permeate side. Accordingly, the value of the outlet concentration can essentially be deviated from that obtained by restriction induced by Eq. (2.12). This restrictive expression is reasoned by the fact, that there is no sweeping phase on the permeate side of the membrane, accordingly $dC/dy = 0$ at the outlet membrane surface, at $Y = \delta_m$. One can simply get the concentration distribution along the fluid boundary (polarization) layer with boundary conditions defined by Eqs. (2.4) and (2.5), as:

For the fluid boundary layer $\left(Pe_L = \frac{v_o \delta_L}{D_L} \equiv \frac{v_o}{k_L}\right)$:

$$C = \frac{C^o - C_L^*}{1 - e^{Pe_L}} e^{Pe_L Y} + \frac{C_L^* - C^o e^{Pe_L}}{1 - e^{Pe_L}} \tag{2.6}$$

Likewise, the concentration distribution be computed for the membrane layer, using suitable boundary conditions (see Fig. 2.4); $C_{m,p}$ is the membrane concentration at the outlet membrane surface; the concentration of the permeate side fluid boundary layer, on membrane outlet surface is as: $C_p = C_{m,p}/\Phi$):

$$C = \frac{C_m^* - C_{m,p}}{1 - e^{Pe_m}} e^{Pe_m Y} + \frac{C_{m,p} - C_m e^{Pe_m}}{1 - e^{Pe_m}} \tag{2.7}$$

For calculation of the membrane Peclet number, we have applied the model of Bowen and Welfort (2002). They have taken into account two characteristic features of the flowing solute and solvent: (1) the solute component is transporting through the pores of the selective layer, only and the solute component has different concentrations in the membrane and the fluid layers; (2) the diffusive and the convective transports are hindered by the pores wall, thus, for example, the real volumetric convective velocities are different in the boundary and membrane layer, $v_L \neq v_m$, namely $v_L = \xi_c v_m / \varepsilon$, where ξ_c is the hindrance factor for convection, ε denotes the phase holdup of the selective membrane layer (it is assumed that the surface porosity of the membrane is equal to the of the bulk selective layer).

Accordingly, for example, the value of Pe_m can be predicted as: $Pe_m = \frac{\xi_c v_L/\varepsilon}{D_L \xi_D/(\delta_m \tau/\varepsilon)} = \frac{\xi_c v_L \delta_m \tau}{D_L \xi_D}$ (where ξ_c and ξ_D are the dimensionless hindrance factors for the convective and the diffusive flows, respectively; D_L means the bulk fluid diffusion coefficient; τ is the tortuosity factor; ε is the membrane phase holdup; δ_m is the real thickness of the membrane selective layer, m). Accordingly, the physical mass transfer coefficients for the both phases are defined:

$$k_L = \frac{D_L}{\delta_L}; k_m = \frac{D_L \xi_d}{\delta_m \tau/\varepsilon} \tag{2.8}$$

The hindrance factor for the convective flow is predicted by expression published by Bowen and Mohammad (1998), as: $\xi_c = (2-\Phi)(1+0.054\lambda - 0.988\lambda^2 + 0.44\lambda^3)$, while that for the diffusive flow: $\xi_D = 1 - 2.3\lambda + 1.154\lambda^2 + 224\lambda^3$ where $\Phi = (1-\lambda)^2$ and $\lambda = r_s/r_p$ (r_s denotes the solute radius and r_p means the effective pore radius). The mass transfer rate, for example, for the boundary layer will be, as ($dC/dY = Pe_L T$ at $Y = 0$):

$$J = \frac{D_L}{\delta_L}\left[-\left.\frac{dC}{dY}\right|_{Y=0} + Pe_L(T+S)\right] \equiv k_L^o Pe_L S = v_L S \tag{2.9}$$

and thus:

$$J = v_L \frac{C_L^* - C^o e^{Pe_L}}{1 - e^{Pe_L}} \tag{2.10}$$

The value of S can be got easily taking into account Eqs. (2.3) and (2.6) (for polarization layer) or Eq. (2.7) (for the selective membrane layer). The solute transfer rate can be expressed for the membrane layer (Bowen & Welfort, 2002) as:

$$J = \xi_c v_L \frac{C_{m,p} - C_m^* e^{Pe_m}}{1 - e^{Pe_m}} \tag{2.11}$$

The outlet convective flow, in case of without sweeping phase on the permeate side, one should define the following expression for the mass transfer rate C_p means the fluid concentration on the outlet membrane surface as (note $C_p = C_{m,p}/\Phi$):

$$J = v_L C_p \tag{2.12}$$

Eqs. (2.10) to (2.13) should be equal to each other, accordingly the last expression should mean strong restriction; it is obvious that the outlet concentration is determined by the mass transport process through the transport layers. On the other hand, Eq. (2.12) means that there is not diffusive flow on the outlet interface, namely it value is equal to zero, as it is illustrated in Fig. 2.4, due to the absence of sweeping phase. Let us create link between the interface concentrations of the two transport layers. According to the concentration distribution in the polarization layer, it can rightly be assumed, that the solute concentration is lower in the membrane selective layer than that in the fluid boundary layer. We are using the expression introduced by Bowen and Welfort (2002), namely $C_m = C_L\Phi$, accordingly the connection between concentrations at the membrane inlet surface is $C_m^* = \Phi C_L^*$, while at the outlet membrane surface is $C_{m,p} = \Phi C_p$, with $\Phi = (1-\lambda)^2$. Let us express the interface concentration between the inlet fluid and membrane phases, C_L^* and C_m^* as:

$$C_L^* = C_p\left(1 - e^{\mathrm{Pe}_L}\right) + C^o e^{\mathrm{Pe}_L} \tag{2.13}$$

and ($C_m^* = \Phi C_L^*$ and $C_{m,p} = \Phi C_p$)

$$C_L^* = \frac{\Phi C_p - \xi_c C_p\left(1 - e^{\mathrm{Pe}_m}\right)}{\Phi e^{\mathrm{Pe}_m}} \tag{2.14}$$

Then the equality of the above two expressions, [Eqs. (2.13) and (2.14)], give the relative value of the outlet concentration, or the enhancement, taking into account the simultaneous effect of the both transport layers, as:

$$E \equiv \frac{C_p}{C^o} = \frac{\xi_c \Phi e^{\mathrm{Pe}_L + \mathrm{Pe}_m}}{(1 - \xi_c\Phi)(e^{\mathrm{Pe}_m} - 1) + \xi_c \Phi e^{\mathrm{Pe}_L + \mathrm{Pe}_m}} \tag{2.15}$$

The intrinsic enhancement, given for the membrane layer, can easily be obtained by Eq. (2.14), applied the transport data of the membrane layer, as:

$$E_o = \frac{C_p}{C_L^*} = \frac{\Phi e^{\mathrm{Pe}_m}}{\Phi - \xi_c(1 - e^{\mathrm{Pe}_m})} \tag{2.16}$$

According to Eq. (2.15), the outlet concentration can directly be expressed/predicted by measurable parameters, which is not possible using the "black box" model. The value of rejection coefficient, R, can then be calculated by means of the above expression, that is:

$$R = 1 - \frac{C_p}{C^o} = 1 - \frac{\xi_c \Phi e^{\mathrm{Pe}_L + \mathrm{Pe}_m}}{(1 - \xi_c\Phi)(e^{\mathrm{Pe}_m} - 1) + \xi_c \Phi e^{\mathrm{Pe}_L + \mathrm{Pe}_m}} \tag{2.17}$$

Accordingly, one can get, from Eq. (2.17), as:

$$R = 1 - \frac{C_p}{C^o} = \frac{\left(1 - \xi_c\Phi\right)\left(e^{\mathrm{Pe}_m} - 1\right)}{(1 - \xi_c\Phi)(e^{\mathrm{Pe}_m} - 1) + \xi_c \Phi e^{\mathrm{Pe}_L + \mathrm{Pe}_m}} \tag{2.18}$$

Or applying Eq. (2.13), one gets for the intrinsic enhancement, with the application of the concentration distribution in the polarization layer, only as:

$$E_o = \frac{1}{1 - e^{\mathrm{Pe}_L}}\left(1 - \frac{C^o e^{\mathrm{Pe}_L}}{C_L^*}\right) = \frac{1}{1 - e^{\mathrm{Pe}_L}}\left(1 - \frac{E_o e^{\mathrm{Pe}_L}}{E}\right) \tag{2.19}$$

After reconstruction Eq. (2.19), one gets back the known expression defined for the "black box" model (Nagy, 2019, p. 202) as:

$$E_o = \frac{E}{E(1 - e^{\mathrm{Pe}_L}) + e^{\mathrm{Pe}_L}} \tag{2.20}$$

Replacing back the value of C_L^* defined by Eq. (2.14) into Eq. (2.13), or from Eq. (2.19), namely the following expression is obtained:

$$E_o \equiv \frac{C_p}{C_L^*} = \frac{1}{1 - e^{\mathrm{Pe}_L}}\left(1 - \frac{C^o e^{\mathrm{Pe}_L} \Phi e^{\mathrm{Pe}_m}}{\Phi C_p - \xi_c(1 - e^{\mathrm{Pe}_m})}\right) \tag{2.21}$$

As it is expected we have got back Eq. (2.20) proving the correctness of Eqs. (2.14), (2.18). Note that Eq. (2.15) gives in limiting case (when ξ_c and Φ tends to be unity) the expression given by Eq. (2.20), as it given in literature (Baker & Welfoot, 2002; Nagy, 2019, p. 202) for the "black box" model. In limiting case of Eq. (2.21), when (when ξ_c and Φ and e^{Pe_m} tend to be unity), the enhancement can be got back as the known expression of the "black box" model, as:

$$E = \frac{E_o e^{\mathrm{Pe}_L}}{1 + (e^{\mathrm{Pe}_L} - 1)E_o} \tag{2.22}$$

2.4 Results and discussions

The solute and solvent transport in this hydraulic pressure-driven membrane process through porous membrane layer is analyzed. The basic expressions of the UF process are discussed in the literature almost by "black box" model that does not analyze the transport resistance of the membrane layer (Baker, 2004). Its effect is practically involved in the outlet product concentration additionally. Accordingly, important parameter values as, for example, the enhancement of the mass transfer rate or the intrinsic enhancement factor cannot be expressed by separate expression. Thus their values can only be known by experimental data. Presently Nagy & Hegedüs (2020) extended in a study the known parameters as enhancement, E, or intrinsic enhancement, E_o taking into account the transport resistance of solute component(s) across the membrane layer, as well. How then the membrane affects the E, ($E = C_p/C^o$) and E_o ($E_o = C_p/C_m$) values, the transport rate will be briefly be shown, illustrated and analyzed. The concentration distribution is shown in Fig. 2.4, obtained by the boundary conditions defined by Eqs. (2.4) and (2.5), for comparison of the results, plotted in this study in the following subsections, obtained by the model involving the outlet mass transfer rate defined by Eq. (2.12), which allows convective outlet flow, only. This figure illustrates the concentration distribution at three different values of Pe_L and Pe_m, predicted by Eqs. (2.6) and (2.7), respectively. The curves' shapes are as it is expected, namely concave in the polarization layer, and convex in the membrane layer. The curvature of the curves strongly depends on the values of the Peclet-number. Comparing these data to those plotted in Fig. 2.5, they are similar, though the curvature of curves can be different, depending mainly on the Pe-number of the phases. The essential difference of concentration distribution in the membrane layer is caused by the high Pe_m-number; it can reach value of 20, with the change of the λ value.

2.4.1 Simultaneous transport through the polarization and membrane layer, across a membrane

The two most important expressions applied in the case of "black box" model, which gives the function between E_o and E are written here by Eqs. (2.20) and (2.22), respectively (Baker, 2004; Nagy, 2019, p. 204). Taking into account the simultaneous transport resistances of both the transport layers, enhancement and the intrinsic enhancement can be

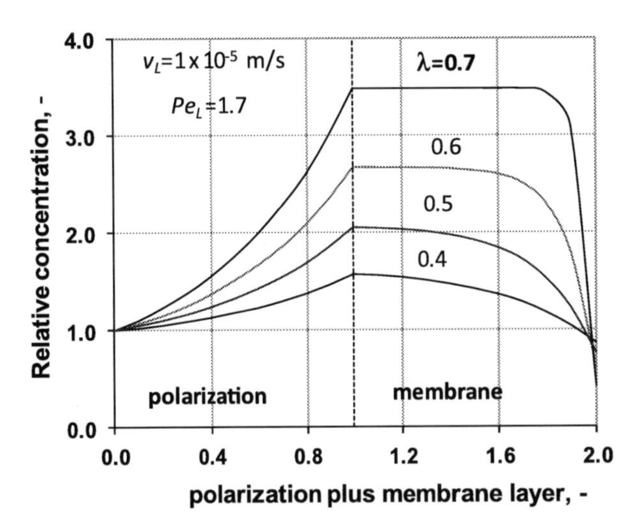

FIGURE 2.5 Concentration distribution across the fluid polarization and the porous membrane layer, at different values of the ratio of the macromolecule and pore size. [$D_L = 0.6 \times 10^{-10}\,\text{m}^2/\text{s}$; $M = 1\times10^5$ D; $\delta_L = 10\,\mu\text{m}$; $\delta_m = 1\,\mu\text{m}$; $v_o = 1\times10^{-5}\,\text{m/s}$ ($v_o = J_w$)]; Note, that the actual value of k_m is strongly affected by the hindrance factors and because of it, the value of Pe_m, as well.

expressed independently from each other, as they are expressed by Eqs. (2.15) and (2.21). This is an essential difference between our model and the "black box" models. Accordingly one can predict the outlet and the interface concentrations applied the values of the mass transport parameters, as for example, Pe_L, Pe_m, D_L/δ_L, D_m/δ_m, Φ. Accordingly, one can then easily predict the membrane performance, taking into account the hindrance effect of the polarization layer in the feed fluid phase.

2.4.1.1 Two-layer concentration distribution

The concentration distribution during the diffusive plus convective mass transport in case of pressure-driven, size-exclusion filtration process is a result of complex transport process. Especially, the macromolecule solute transport is affected by several characteristic, structural membrane properties, as for example, hindrance factors, ξ_c, ξ_D, partition coefficient, Φ. Typical concentration distribution curves are shown in Fig. 2.5 (it is important to note that the membrane concentration is plotted by the equilibrium concentration according to expression of $C_m = C_L\Phi$, thus these concentrations are equal to each other at the internal interface). Note that the value of the membrane Peclet number is strongly affected by the hindrance factor of the diffusion process, values of the hindrance factor change between about 0.04 and 0.3, with decreasing value of r_s/r_o, which values can essentially modify values of the membrane diffusion coefficient, and so the membrane mass transfer coefficient, and consequently the value of Pe_m. Note that the basic value of the diffusion coefficient, in the membrane pores, is considered to be equal to that in the bulk fluid phase (see Eq. 2.8). Consequence of the high Pe_m-numbers (between 3 and 21) is the peculiar shape of the curves in the membrane layer. It might be worth to note, that the diffusive and the convective flows can have opposite (boundary layer) and the same direction (membrane layer).

2.4.1.2 Change in values of couple transport parameters

Owing to their important role in the values of Pe_m, the next figure (Fig. 2.6) illustrates how the values of couple transport parameters change as a function of λ. Both the

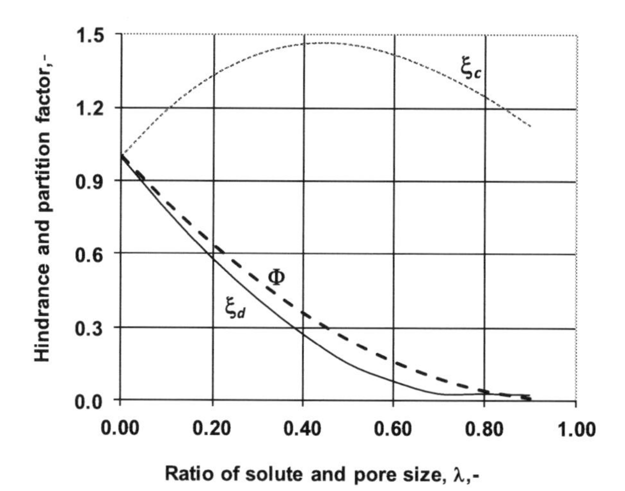

FIGURE 2.6 Parameter values, which alter the value of the membrane mass transfer coefficient, as a function of the ratio of solute and pore size.

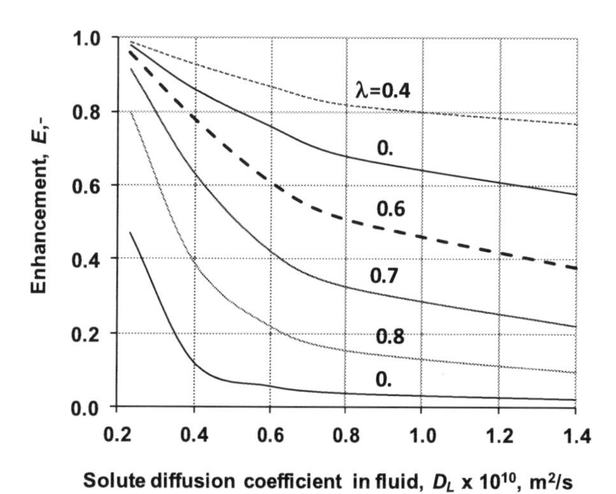

FIGURE 2.7 Effect of the solute diffusion coefficient on the outlet concentration, more exactly on the enhancement. ($\delta_L = 10\ \mu m$; $\delta_m = 1\ \mu m$; $v_o = 1 \times 10^{-5}$ m/s ($v_o = J_w$); $E = C_p/C^o$).

diffusive hindrance factor, ξ_D, and the partition coefficient, Φ, can decrease more than one order of magnitude as a function of the ratio of the solute and pore size. On the other hand, the convective hindrance factor rather moderately changes.

Size of the macromolecules, colloids, and microparticles is a determining factor in the transport process through a membrane layer with given pore size in the filtering processes. He and Niemeyer (2003) studied the diffusion coefficient of proteins between about 10 kDa and 1000 kDa molecule size. According the authors, the diffusion coefficient varies between 10×10^{-11} and 2×10^{-11} m^2/s, in this molecule size range. Fig. 2.7 illustrates the effect of the diffusion coefficient on the enhancement factor, at different values of the λ parameter. Both the diffusion coefficient and the values of λ, significantly alter the membrane performance at the investigated parameter value range. The enhancement gradually decreases with the increase in the D_L values, though the change's gradient lowers with the increase of the diffusion coefficient, at constant value of λ. The most important parameter is the λ parameter, which decisively affects the transport resistance across the membrane layer. The effect of λ on the outlet concentration is even stronger, at given value of D_L.

2.4.1.3 *The effect of the specific water flux*

The specific volume (or linear) flow rate of the fluid phase is also an important factor, which essential for separation performance of the particles. Its value depends firstly on the hydrodynamic pressure difference between the two sides of the membrane, and obviously its value is altered by the membrane properties, as well. The effect of the hydrodynamic pressure difference on the water flux can be estimated by means of the Hagen–Poiseuille equation ($v = r_p^2 \Delta p / 2\mu\delta_m$; Bird, Steward, & Lightfoot, 1960), since the osmotic pressure difference is determined by the ion concentration in the solutes (Nagy, 2019). The convective velocity also transports the solute molecules; accordingly it affects the concentration gradient across the porous membrane layer. Its role on the transport process crucially important, especially in case of pressure-driven membrane processes as the UF process. Two figures illustrate the effect of the water flux on the membrane separation. The effect of the water flux on the relative values of the outlet concentrations, at different values of λ parameter is illustrated in Fig. 2.8. It is remarkable that the value of enhancement has minimum values as a function of the water flux. This means that the effectiveness of the separation can depend on the water flux, as well. This can mean that the external effect caused by the hydrodynamic pressure difference should carefully be chosen in order to reach the effective selectivity of the filtration process. This figure well illustrates that the minimum value of C_p/C^o is shifted in direction of the lowering water flux values as a function of λ parameter. There is a rather wide range of the water flux, at its lower and higher values, as well, where the effectiveness of separation is low, in which water flux range this process is not effective, at all. Generally it can be stated that there is rather a narrow water flux range, where the filtration offers acceptable selectivity.

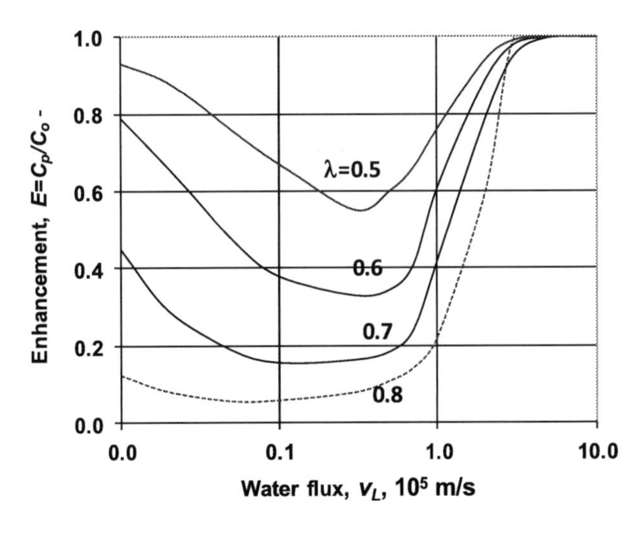

FIGURE 2.8 Enhancement as a function of the water flux (convective velocity), at different values of λ parameter. ($D_L = 0.6 \times 10^{-10}$ m^2/s; $M = 1 \times 10^5$ D; $\delta_L = 10$ μm; $\delta_m = 1$ μm).

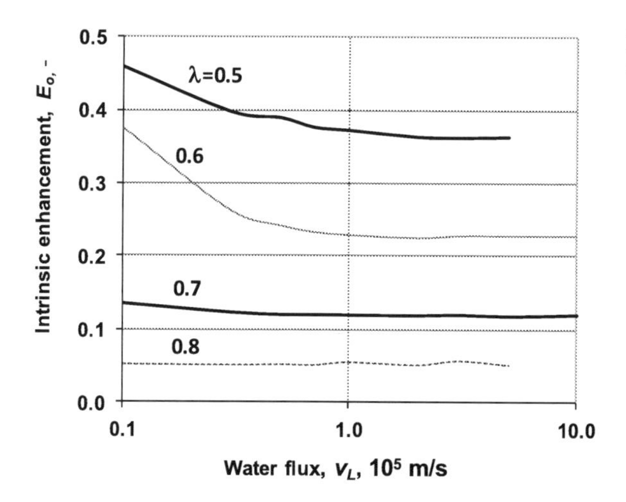

FIGURE 2.9 The change of the intrinsic enhancement at different λ values.

The effect of the intrinsic enhancement is illustrated in Fig. 2.9, at different values of λ parameter. The change in the value of C_p/C_L^* is rather moderate as a function of the water flux, but it depends relative significantly at values of λ parameter. At $v_o=(1-10)\times 10^{-5}\ m^3/m^2s$, both the Pe_L and Pe_m values are relatively high, their values can reach values of 50 at higher values of λ. Their high exponential values dominate in the values of C_p and C_L^*, thus the value of intrinsic enhancement remain practically constant in this water flux regime.

2.5 Conclusion and future trends

The so-called "black box" model was extended by taking into account the membrane mass transport, as well. Accordingly, the developed model takes into account the simultaneous solute transport of macromolecules, microparticles, etc., across both the fluid polarization layer and the porous membrane layer. This model then makes it possible to express both the enhancement and the intrinsic enhancement factors, individually. Likewise the outlet concentration, the interfacial concentration of the solute component can be predicted by means of values of the boundary and the membrane layer mass transport properties. This model takes into account the hindrance factor of the diffusion and convection, the partition coefficient between the two phases. The membrane performance was then illustrated by some figures. It is shown that the membrane selectivity has a relative narrow water flux regime, in that the separation efficiency is effective, strongly depending on the ratio of the particle size and the pore size. This model can be used for other hydraulic and osmotic pressure-driven membrane processes, in them there is not used sweeping phase on the membrane permeate side.

Both MF and UF apply porous membrane for separation of solvent and small molecules from larger molecules, for example, macromolecules, biopolymers, cells, bacteria, etc. UF/MF processes are used in wide areas of industry, for example, in wastewater technology, desalination, pharmaceutical industry, biotechnology, food industry, etc. Costs of these processes are rather high, which should be decreased with the improvement of the membrane separation properties. This means more precise pore size, water flux increase, development fouling resistant membranes, etc. Two new methodologies of membrane preparations seem to be promising ones, namely the incorporation of nanomaterials in the membrane active layer, producing thin film nanocomposite membrane, and preparing electrospun membrane by nanosized fibers. The nanoparticles can change the hydrophobic-hydrophilic character of membrane (e.g., graphene oxide), can essentially increase the water flux (e.g., zeolite) or can improve the heavy metal removal (e.g., electrospun membrane). On the other hand, they can increase the fouling resistance of the membrane, which is basically important to provide more efficient operating conditions during industrial processes. These new membrane preparation methods are likely to offer essentially cheaper separation processes in the next future.

Acknowledgment

The Hungarian National Development Agency, grants OTKA 116727 and NKFIH-1158–6/2019, is gratefully acknowledged for the financial support.

References

Akinpelu, A. A., Ali, M. E., Johan, M. R., Saidur, R., Qurban, M. A., & Saleh, T. A. (2019). Polycyclic aromatic hydrocarbons extraction and removal from wastewater by carbon nanotubes: A review of the current technologies, challenges and prospects. *Process Safety and Environmental Protection*, *122*, 68–82. Available from https://doi.org/10.1016/j.psep.2018.11.006.

Al Aani, S., Mustafa, T. N., & Hilal, N. (2020). Ultrafiltration membranes for wastewater and water process engineering: A comprehensive statistical review over the past decade. *Journal of Water Process Engineering*, *35*, 101241. Available from https://doi.org/10.1016/j.jwpe.2020.101241.

Anantharaman, A., Chun, Y., Hua, T., Chew, J. W., & Wang, R. (2020). Pre-deposited dynamic membrane filtration – A review. *Water Research*, *173*, 115558. Available from https://doi.org/10.1016/j.watres.2020.115558.

Anis, S. F., Hashaikeh, R., & Hila, N. (2019). Microfiltration membrane processes: A review of research trends over the past decade. *Journal of Water Process Engineering*, *32*, 100941. Available from https://doi.org/10.1016/j.jwpe.2019.100941.

Asad, A., Sameoto, D., & Sadrzadeh, M. (2020). Chapter 1 - Overview of membrane technology. In M. Sadrzadeh, & T. Mohammadi (Eds.), *Nanocomposite membranes for water and gas separation* (pp. 1–28). Elsevier Inc. Available from https://doi.org/10.1016/b978-0-12-816710-6.00001-8.

Ayyaru, S., Pandiyan, R., & Ahn, Y. H. (2019). Fabrication and characterization of anti-fouling and non-toxic polyvinylidene fluoride - sulphonated carbon nanotube ultrafiltration membranes for membrane bioreactors applications. *Chemical Engineering Research and Design*, *142*, 176–188. Available from https://doi.org/10.1016/j.cherd.2018.12.008.

Bacchin, P., Si-Hassen, D., Starov, V., Clifton, M. J., & Aimar, P. (2002). A unifying model for concentration polarization and gel-layer formation and particle deposition in cross-flow membrane filtration of colloidal suspensions. *Chemical Engineering Science*, *57*(1), 77–91. Available from https://doi.org/10.1016/S0009-2509(01)00316-5.

Baker, R. W. (2004). *Membrane technology and applications* (2nd ed.). Chichester: John Wiley & Sons.

Basak, G., Hazra, C., & Sen, R. (2020). Biofunctionalized nanomaterials for in situ clean-up of hydrocarbon contamination: A quantum jump in global bioremediation research. *Journal of Environmental Management*, *256*, 109913. Available from https://doi.org/10.1016/j.jenvman.2019.109913.

Becker, F. G. (2015). No 主観的健康感をh' Ut在宅高齢者における 健康関連指標に関する共分散構造分析.

Benavente, L., Coetsier, C., Venault, A., Chang, Y., Causserand, C., Bacchin, P., & Aimar, P. (2016). FTIR mapping as a simple and powerful approach to study membrane coating and fouling. *Journal of Membrane Science, 520*, 477–489. Available from https://doi.org/10.1016/j.memsci.2016.07.061.

Binnemans, K., Pontikes, Y., Jones, P. T., Van Gerven, T., Blanpain, B. (2013). Recovery of rare earths from industrial waste residues: A concise review. *Third international slag valorisation symposium* (pp. 191–205).

Bird, R. B., Steward, W. R., & Lightfoot, E. N. (1960). *Transport phenomena*. New York: John Wiley& Sons.

Borbély, G., & Nagy, E. (2009). Removal of zinc and nickel ions by complexation-membrane filtration process from industrial wastewater. *Desalination, 240*(1–3), 218–226. Available from https://doi.org/10.1016/j.desal.2007.11.073.

Bowen, W. R., & Mohammad, A. W. (1998). Diafiltration by nanofiltration: Prediction and optimisation. *AIChE Journal. American Institute of Chemical Engineers, 44*(8), 1799–1812. Available from https://doi.org/10.1002/aic.690440811.

Bowen, W. R., & Welfort, J. S. (2002). Modelling the performance of membrane nanofiltration – Critical assessment and model development. *Chemical Engineering Science, 57*(7), 1121–1137. Available from https://doi.org/10.1016/S0009-2509(01)00413-4.

Castro-Muñoz, R., Boczkaj, G., Gontarek, E., Cassano, A., & Fíla, V. (2020). Membrane technologies assisting plant-based and agro-food by-products processing: A comprehensive review. *Trends in Food Science & Technology, 95*, 219–232. Available from https://doi.org/10.1016/j.tifs.2019.12.003.

Çelik, Y., Çelik, A., Flahaut, E., & Suvaci, E. (2016). Anisotropic mechanical and functional properties of graphene-based alumina matrix nanocomposites. *Journal of the European Ceramic Society, 36*(8), 2075–2086. Available from https://doi.org/10.1016/j.jeurceramsoc.2016.02.032.

Chandrashekhar, M. N., Nayak, M., Isloor, A., Inamuddin, M., Lakshmi, B., Marwani, H. M., & Khan, I. (2019). Polyphenylsulfone/multiwalled carbon nanotubes mixed ultrafiltration membranes: Fabrication, characterization and removal of heavy metals Pb^{2+}, Hg^{2+}, and Cd^{2+} from aqueous solutions. *Arabian Journal of Chemistry, 13*(3), 4661–4672. Available from https://doi.org/10.1016/j.arabjc.2019.10.007.

Chew, J. W., Kilduff, J., & Belfort, G. (2020). The behavior of suspensions and macromolecular solutions in crossflow microfiltration: An update. *Journal of Membrane Science, 601*, 117865. Available from https://doi.org/10.1016/j.memsci.2020.117865.

Cho, K., Droudian, A., Wyss, R. M., Schlichting, K. P., & Park, H. G. (2018). Multifunctional wafer-scale graphene membranes for fast ultrafiltration and high permeation gas separation. *Science Advances, 4*(11). Available from https://doi.org/10.1126/sciadv.aau0476.

Dmitrenko, M. E., Penkova, A. V., Kuzminova, A. I., Atta, R. R., Zolotarev, A. A., Mazur, A. S., ... Ermakov, S. S. (2019). Development and investigation of novel polyphenylene isophthalamide pervaporation membranes modified with various fullerene derivatives. *Separation and Purification Technology, 226*, 241–251. Available from https://doi.org/10.1016/j.seppur.2019.05.092.

Egorova, K. S., Gordeev, E. G., & Ananikov, V. P. (2017). Biological activity of ionic liquids and their application in pharmaceutics and medicine. *Chemical Reviews, 117*, 7132–7189. Available from https://doi.org/10.1021/acs.chemrev.6b00562.

Elgharbawy, A. A. M., Moniruzzama, M., & Goto, M. (2020). Recent advances of enzymatic reactions in ionic liquids: Part II. *Biochemical Engineering Journal, 154*, 107426. Available from https://doi.org/10.1016/j.bej.2019.107426.

Ertürk, A. S., Gürbüz, M. U., Tülü, M., & Bozdoğan, A. E. (2018). Evaluation of Jeffamine® core PAMAM dendrimers for simultaneous removal of divalent heavy metal ions from aqueous solutions by polymer assisted ultrafiltration. *Acta Chimica Slovenica, 65*(1), 65–74. Available from https://doi.org/10.17344/acsi.2017.3485.

Fehér, C., Tomasek, S., Hancsók, J., & Skoda-Földes, R. (2018). Oligomerization of light olefins in the presence of a supported Brønsted acidic ionic liquid catalyst. *Applied Catalysis B: Environmental, 239*, 52–60.

Foong, C. Y., Wirzal, M. D. H., & Bustam, M. A. (2020). A review on nanofibers membrane with amino-based ionic liquid for heavy metal removal. *Journal of Molecular Liquids, 297*, 111793. Available from https://doi.org/10.1016/j.molliq.2019.111793.

Guirguis, A., Maina, J. W., Kong, L., Henderson, L. C., Rana, A., Li, L. H., ... Dumée, L. F. (2019). Perforation routes towards practical nano-porous graphene and analogous materials engineering. *Carbon, 155*, 660–673. Available from https://doi.org/10.1016/j.carbon.2019.09.028.

He, L., & Niemeyer, B. (2003). A novel correlation for protein diffusion coefficients based on molecular weight and radius of gyration. *Biotechnology Progress, 19*(2), 544–548. Available from https://doi.org/10.1021/bp0256059.

Huang, Y., & Feng, X. (2019). Polymer-enhanced ultrafiltration: Fundamentals, applications and recent developments. *Journal of Membrane Science*, *586*, 53–83. Available from https://doi.org/10.1016/j.memsci.2019.05.037.

Huter, M. J., & Strube, J. (2019). Model-based design and process optimization of continuous single pass tangential flow filtration focusing on continuous bioprocessing. *Processes*. Available from https://doi.org/10.3390/pr7060317.

Ishak, N. F., Hashim, N. A., Othman, M. H. D., Monash, P., & Zuki, F. M. (2017). Recent progress in the hydrophilic modification of alumina membranes for protein separation and purification. *Ceramics International*, *43*(1 Part B), 915–925. Available from https://doi.org/10.1016/j.ceramint.2016.10.044.

Kaleekkal, N. J., Radhakrishnan, R., Sunil, V., Kamalanathan, G., Sengupta, A., & Wickramasinghe, R. (2018). Performance evaluation of novel nanostructured modified mesoporous silica/polyetherimide composite membranes for the treatment of oil/water emulsion. *Separation and Purification Technology*, *205*, 32–47. Available from https://doi.org/10.1016/j.seppur.2018.05.007.

Kamaz, M., Vogler, R. J., Jebur, M., Sengupta, A., & Wickramasinghe, R. (2020). π Electron induced separation of organic compounds using supported ionic liquid membranes. *Separation and Purification Technology*, *236*, 116237. Available from https://doi.org/10.1016/j.seppur.2019.116237.

Koók, L., Nemestóthy, N., Bakonyi, P., Göllei, A., Rózsenberszki, T., Takács, P., ... Bélafi-Bakó, K. (2017). On the efficiency of dual-chamber biocatalytic electrochemical cells applying membrane separators prepared with imidazolium-type ionic liquids containing $[NTf_2]^-$ and $[PF_6]^-$ anions. *Chemical Engineering Journal*, *324*, 296–302. Available from https://doi.org/10.1016/j.cej.2017.05.022.

Lee, K. Y., Jeong, L., Kang, Y. O., Lee, S. J., & Park, W. H. (2009). Electrospinning of polysaccharides for regenerative medicine. *Advanced Drug Delivery Reviews*, *61*, 1020–1032. Available from https://doi.org/10.1016/j.addr.2009.07.006.

Li, C., Sun, W., Lu, Z., Ao, X., & Li, S. (2020). Ceramic nanocomposite membranes and membrane fouling: A review. *Water Research*, *175*, 115674. Available from https://doi.org/10.1016/j.watres.2020.115674.

Lutz, H. (Ed.), (2015). *Ultrafiltration for bioprocessing*. Woodhead Publishing. Available from https://doi.org/10.1016/c2013-0-18176-7.

Mao, H., Qiu, M., Zhang, T., Chen, X., Da, Y., Jing, W., & Fan, Y. (2019). Robust CNT-based conductive ultrafiltration membrane with tunable surface potential for in situ fouling mitigation. *Applied Surface Science*, *497*, 143786. Available from https://doi.org/10.1016/j.apsusc.2019.143786.

Mittal, G., Dhand, V., Rhee, K. Y., Park, S. J., & Lee, W. R. (2015). A review on carbon nanotubes and graphene as fillers in reinforced polymer nanocomposites. *Journal of Industrial and Engineering Chemistry*, *21*, 11–25. Available from https://doi.org/10.1016/j.jiec.2014.03.022.

Nagy, E. (2019). *Basic equations of mass transport through a membrane layer*. Amsterdam: Elsevier.

Nagy, E., & Hegedüs, I. (2020). Diffusive plus convective mass transport, accompanied by biochemical reaction, across capillary membrane. *Catalysts*, *10*, 1115.

Nagy, E., Hegedüs, I., Tow, E. W., & Lienhard, V. J. H. (2018). Effect of fouling on performance of pressure retarded osmosis (PRO) and forward osmosis (FO). *Journal of Membrane Science*, *565*, 450–462. Available from https://doi.org/10.1016/j.memsci.2018.08.039.

Nagy, E., Kulcsar, E., & Nagy, A. (2011). Mass transport for nanofiltration: Coupled effect of the polarization and membrane layers. *Journal of Membrane Science*, *368*, 215–222. Available from https://doi.org/10.1016/j.memsci.2010.11.046.

Nasrollahi, N., Aber, S., Vatanpour, V., & Mahmoodi, N. M. (2019). Development of hydrophilic microporous PES ultrafiltration membrane containing CuO nanoparticles with improved antifouling and separation performance. *Materials Chemistry and Physics*, *222*, 338–350. Available from https://doi.org/10.1016/j.matchemphys.2018.10.032.

Romadhoni, A., Hidayat, P., Andina, V. R., & Iqbal, R. M. (2019). Synthesis, characterization, and performance of tio 2 -n as filler in polyethersulfone membranes for laundry waste treatment. *Jurnal Sains dan Seni ITS*, *8*.

Roy, K., Mukherjee, A., Maddela, N. R., Chakraborty, S., Shen, B., Li, M., ... Garciá Cruzatty, L. C. (2020). Outlook on the bottleneck of carbon nanotube in desalination and membrane-based water treatment-A review. *Journal of Environmental Chemical Engineering*, *8*(1), 103572. Available from https://doi.org/10.1016/j.jece.2019.103572.

Sabet, M., Soleimani, H., Mohammadian, E., & Hosseini, S. (2019). Graphene utilization for water desalination process. *Defect and Diffusion Forum*, *391*, 195–200. Available from https://doi.org/10.4028/www.scientific.net/DDF.391.195.

Sajid, M., Nazal, M. K., Ihsanullah., Baig, N., & Osman, A. M. (2018). Removal of heavy metals and organic pollutants from water using dendritic polymers based adsorbents: A critical review. *Separation and Purification Technology*, *191*, 400–423. Available from https://doi.org/10.1016/j.seppur.2017.09.011.

Sengupta, A., Kumar, S. E., Kamaz, M., Jebur, M., & Wickramasinghe, R. (2019). Synthesis and characterization of antibacterial poly ionic liquid membranes with tunable performance. *Separation and Purification Technology*, *212*, 307–315. Available from https://doi.org/10.1016/j.seppur.2018.11.027.

Singh, S. P., Li, Y., Be'Er, A., Oren, Y., Tour, J. M., & Arnusch, C. J. (2017). Laser-induced graphene layers and electrodes prevents microbial fouling and exerts antimicrobial action. *ACS Applied Materials & Interfaces*, *9*, 18238–18247. Available from https://doi.org/10.1021/acsami.7b04863.

Song, N., Gao, X., Ma, Z., Wang, X., Wei, Y., & Gao, C. (2018). A review of graphene-based separation membrane: Materials, characteristics, preparation and applications. *Desalination*, *437*, 59–72. Available from https://doi.org/10.1016/j.desal.2018.02.024.

Sri Abirami, S. M., Divya, K., Selvapandian, P., Mohan, D., Rana, D., & Nagendran, A. (2019). Permeation and antifouling performance of poly (ether imide) composite ultrafiltration membranes customized with manganese dioxide nanosphers. *Materials Chemistry and Physics*, *231*, 159–167. Available from https://doi.org/10.1016/j.matchemphys.2019.04.023.

Tabatabai, S. A. A. (2014). Coagulation and ultrafiltration in seawater reverse osmosis pretreatment. *Dissertation Master of Science in Water Supply Engineering UNESCO-IHE*. Iran: Institute for Water Education.

Tan, S. S. Y., MacFarlane, D. R., Upfal, J., Edye, L. A., Doherty, W. O. S., Patti, A. F., ... Scott, J. L. (2009). Extraction of lignin from lignocellulose at atmospheric pressure using alkylbenzenesulfonate ionic liquid. *Green Chemistry: An International Journal and Green Chemistry Resource*, *11*, 339–345. Available from https://doi.org/10.1039/b815310h.

Thakur, A. K., Singh, S. P., Thamaraiselvan, C., Kleinberg, M. N., & Arnusch, C. J. (2019). Graphene oxide on laser-induced graphene filters for antifouling, electrically conductive ultrafiltration membranes. *Journal of Membrane Science*, *591*, 117322. Available from https://doi.org/10.1016/j.memsci.2019.117322.

Tow, E. W., & LienhardV, J. H. (2016). Quantifying osmotic membrane fouling to enable comparisons across diverse processes. *Journal of Membrane Science*, *511*, 92–107. Available from https://doi.org/10.1016/j.memsci.2016.03.040.

Yang, Q., Luo, J., Guo, S., Hang, X., Chen, X., & Wan, Y. (2019). Threshold flux in concentration mode: Fouling control during clarification of molasses by ultrafiltration. *Journal of Membrane Science*, *586*, 130–139. Available from https://doi.org/10.1016/j.memsci.2019.05.063.

Zhang, G., Zhou, M., Xu, Z., Jiang, C., Shen, C., & Meng, Q. (2019). Guanidyl-functionalized graphene/polysulfone mixed matrix ultrafiltration membrane with superior permselective, antifouling and antibacterial properties for water treatment. *Journal of Colloid and Interface Science*, *540*(2019), 295–305. Available from https://doi.org/10.1016/j.jcis.2019.01.050.

Zhang, Y., Wei, S., Yong, M., Liu, W., & Liu, S. (2018). $Y_xSi_{1-x}O_2$-SO_3H self-assembled membrane formed on phosphorylated $Y_xSi_{1-x}O_2/Al_2O_3$ for oily seawater partial desalination and deep cleaning. *Journal of Membrane Science*, *556*, 384–392. Available from https://doi.org/10.1016/j.memsci.2018.04.010.

Zhang, X., Liu, D., Yang, L., Zhou, L., & You, T. (2015). Self-assembled three-dimensional graphene-based materials for dye adsorption and catalysis. *Journal of Materials Chemistry A*, *3*, 10031–10037. Available from https://doi.org/10.1039/C5TA00355E.

CHAPTER

3

Transport phenomena in reverse osmosis/nanofiltration membranes

Serena Bandini and Cristiana Boi

Department of Civil, Chemical, Environmental and Materials Engineering – DICAM, Alma Mater Studiorum, University of Bologna, Bologna, Italy

Abbreviations

General notation

0^- feed/membrane interface, feed side
0^+ feed/membrane interface, membrane side
δ^- membrane/permeate interface, membrane side
δ^+ membrane/permeate interface, permeate side
$\underline{\nabla}$ gradient
$|_T$ isothermal conditions

Latin symbols

a_i activity (–)
c mole concentration mol/m^3
c_w^o pure water mole concentration mol/m^3
D_{iM} diffusion coefficient inside the membrane of *i*-species m^2/s
D_{ij} interdiffusion coefficient inside the pore m^2/s
D_{ip} hindered diffusivity inside the pore m^2/s
D_{iw}^0 unconfined diffusivity in water (m^2/s)
e elementary charge (= 1.602*10^{-19}) C
F Faraday constant (= 96485) C/mol
I $=\frac{1}{2}\sum_{i=1}^{n} z_i^2 c_i$ ionic strength mol/m^3
$J_i(J_s)$ mole flux of *i*-species (solute) mol/m^2/s
J_v total volume flux m/s
k_B Boltzmann constant (= 1.381 × 10^{-23}) J/K
k_D viscous flow parameter m^2
K_s solute permeability m/s
K_{ic} convective hindrance factor (–)
K_{id} diffusive hindrance factor (–)

Current Trends and Future Developments on (Bio-) Membranes
DOI: https://doi.org/10.1016/B978-0-12-822257-7.00003-0

l_p solvent permeability coefficient as defined in Eq. (3.11) $m^2/s/Pa$
L_{pw} hydraulic membrane permeability m/s/Pa
L_p membrane permeability m/s/Pa
N number of species
n_{ix} x-component of molar flux vector of *i*-species $mol/m^2/s$
$\underline{n}$ molar flux vector $mol/m^2/s$
N_A Avogadro number ($= 6.023 \times 10^{23}$)/mol
p pressure Pa
$\overline{p}$ pressure inside the pore Pa
Pe Pèclet number as defined in Eq. (3.13) (–)
Pe_i hindered Pèclet number as defined in Eqs. (4.11) and (5.12) (–)
P_s solute permeability as defined in Eq. (3.13) m/s
r_B parameter defined in Tables 3.3 and 3.6 m
r_i Stokes radius m
r_p average pore radius (average slit half-thickness) m
R_{real} real rejection (–)
S solubility (–)
T_0 reference temperature K
$\underline{v}$ velocity m/s
$\overline{V}$ partial molar volume m^3/mol
$\tilde{V}_w^o$ molar volume of pure water m^3/mol
x' axial coordinate across the membrane (accounting porosity and tortuosity) (Fig. 3.1) m
X volume membrane charge density mol/m^3
Y_i parameter defined in Eq. (4.3)–Table 3.4 (–)
z active layer thickness as defined in Fig. 3.1) m
z_i ionic valence (–)

Greek letters

α_i' bugger factor (–)
γ_i activity coefficient
γ_{DE} parameter defined in Table 3.3 (–)
Γ_i partitioning coefficient at the membrane/external solutions interfaces (–)
δ effective membrane thickness, accounting for tortuosity and porosity m
ΔP pressure difference across the membrane (external phases) Pa
ΔP_{eff} effective pressure difference Pa
ΔW_{DE} dielectric exclusion excess energy as defined in Eqs. (3.18)–(3.19) (–)
ΔW_{im} excess energy due to image forces, defined in Table 3.3 (–)
ΔW_{Born} excess energy due to Born partitioning defined in Table 3.3 (–)
$\Delta\pi$ osmotic pressure difference Pa
$\Delta\psi_D$ Donnan potential (–)
ε_0 vacuum permittivity ($= 8.854 \times 10^{-12}$) $C^2/J/m$
ε_r dielectric constant (–)
ε_{rm} membrane dielectric constant (–)
ε_{rp} dielectric constant of the pore solution (–)
ε_{rs} dielectric constant of the bulk solution (–)
ε/τ porosity to tortuosity ratio (–)
ϕ_i steric partitioning coefficient, defined in Table 3.2 (–)
η dynamic viscosity inside the pore Pa.s
η^0 unconfined dynamic viscosity Pa.s
κ^{-1} Debye length m
λ_i $= r_i/r_p$ (–)
μ_i^{el} electrochemical potential J/mol
σ_v, σ_{vi} Staverman reflection coefficient defined in Eqs. (3.12) and (3.22) (–)

σ_s, σ_{si} solute reflection coefficient defined in Eqs. (3.12) and (4.7) (–)
ω solute permeability coefficient as defined in Eq. (3.11) $m^2/s/Pa$
ψ dimensionless electrostatic potential (–)
Ψ electrostatic potential V

Subscripts and superscripts

i ion or uncharged solute
0 at the feed/membrane interface
δ at the membrane/permeate interface
asym asymptotic conditions
inside inside the pore
RS reference state
M membrane
s solute
tot total
w water

3.1 Introduction

Membrane processes are typically nonequilibrium kinetic processes that must keep a driving force through the membrane, which should have the right selectivity. Reverse osmosis (RO) and NF are mainly pressure-driven processes operating exclusively with liquid streams.

NF membranes were introduced into the market after RO membranes and initially sold as "loose membranes," with low NaCl rejection. They were developed in the second half of the 1980s, after thin film composite membrane manufacturing was established.

Membrane morphology is the main difference between RO and NF membranes: the latter ones are characterized by a thin film on top of the composite membrane (typically polyamide supported on polysulfone) endowed with micropores in the range from 0.3 to 1 nm. The amphoteric behavior of the membrane material allows the formation of a membrane charge that depends on the feed pH and the electrolyte solution: the membrane acts as a charged functional layer, which allows to separate electrolyte solutions on the basis of their ion valence and to obtain a relatively complete retention of neutral substances such as complex sugars (Afonso & de Pinho, 2000; Atra, Vatai, Bekassy-Molnar, & Balint, 2005; Bandini, Drei, & Vezzani, 2005; Bargeman, Vollenbroek, Straatsma, & Schroen, 2005; Bargeman, Westerink, Guerra Miguez, & Wessling, 2014; Bouchoux, Roux-de Balmann, & Lutin, 2005; Bowen & Mukhtar, 1996; Boy, Roux-de Balmann, & Galier, 2012; Chandrapala et al., 2016; Condom, Larbot, Youssi, & Persin, 2004; Cuartas-Uribe et al., 2009; Freger, Arnot, & Howell, 2000; Hall, Starov, & Lloyd, 1997; Lipp, Gimbel, & Frimmel, 1994; Mazzoni & Bandini, 2006; Mazzoni, Bruni, & Bandini, 2007; Mucchetti, Zardi, Orlandini, & Gostoli, 2000; Nilsson, Trägårdh, & Őstergren, 2008; Nyström, Kaipia, & Luque, 1995; Paugam, Taha, Dorange, Jaouen, & Quéméneur, 2004; Quin, Oo, Lee, & Coniglio, 2004; Szoke, Patzay, & Weiser, 2002; Tanninen & Nystrom, 2002; Timmer, van der Horst, & Robbertsen, 1993; Tsuru, Urairi, Nakao, & Kimura, 1991; Vellenga & Tragardh, 1998; Wang, Wang, & Wang, 2002; Xu & Lebrun, 1999).

In the case of neutral solutes, rejection is mainly related to size exclusion effects (typically NF membranes are listed with a molecular weight cut-off in the range from 150 to 1000 Dalton). However, the small pore sizes are rather effective in the fractionation of mixtures: xylose from glucose, glucose from maltose, mixtures of fructo-oligosaccharides as representative examples (Bandini & Nataloni, 2015; Catarino, Minhalma, Beal, Mateus, & de Pinho, 2008; Feng, Chang, Wang, & Ma, 2009; Goulas, Kapasakalidis, Sinclair, Rastall, & Gradison, 2002; Li, Li, Chen, & Chen, 2004; Pinelo, Jonsson, & Meyer, 2009; Qi, Luo, Chen, Hang, & Wan, 2011; Rizki, Janssen, Boom, & Van der Padt, 2019; Sjöman, Mänttäri, Nyström, Koivikko, & Heikkilä, 2007; Sjöman, Mänttäri, Nyström, Koivikko, & Heikkilä, 2008; Zhang, Yang, Zhang, Zhao, & Hua, 2011).

For electrolyte solutions, on the other end, rejection is remarkably affected by numerous electric and electrostatic phenomena arising as a consequence of the pore size and the membrane charge: Donnan partitioning and dielectric-exclusion (DE) phenomena are to be accounted for, in addition to steric exclusion (Bandini & Vezzani, 2003; Bowen & Mukhtar, 1996; Bowen & Welfoot, 2002a; Szymczyk & Fievet, 2005; Yaroshchuk, 2000, 2001). The type and valence of the ionic species (symmetric or nonsymmetric electrolytes), the membrane material, and the operative conditions such as pH and ionic strength values existing in the feed side are important to determine the membrane selectivity (Bandini & Vezzani, 2003; Bandini et al., 2005; Bowen & Mukhtar, 1996; Bowen & Welfoot, 2002a; Condom et al., 2004; Hagmeyer & Gimbel, 1998, 1999; Mazzoni & Bandini, 2006; Mazzoni et al., 2007; Nilsson et al., 2008; Paugam et al., 2004; Quin et al., 2004; Szoke et al., 2002).

RO and NF processes are widely used today by several industries; among those the food, pharmaceutical, chemical, and petrochemical industries can be considered, in addition to the use of RO by municipalities for water purification.

Mathematical models are fundamentals for process development and design; nowadays, RO modeling is well established and simple models as the solution–diffusion model are implemented in process simulators that are freely downloadable. This is not true for NF for which simple models are not available. Indeed, process engineers would rather have simple models, with few adjustable parameters that are independent of the operating conditions. However, the current development level of NF models, even if advanced, does not satisfy all the needs of the process engineer (Bandini & Bruni, 2010; Oatley et al., 2012; Van der Bruggen, Manttari, & Nystrom, 2008; Wang & Lin, 2021; Yaroshchuk, Bruening, & Zholkovskiy, 2019).

The comprehension of the transport phenomena involved in mass transfer through membranes started in the middle of the 18th century, with the research of Nollet in 1747 on osmotic phenomena. Despite a few milestones such as Fick's law of diffusion, Graham studies on gas permeation, and Nernst–Planck research on electrolyte transport (in the second half of the 18th century), the first models to describe the mass transport in RO membranes, originally known as hyperfiltration membranes, were developed in the 1960s with the Kedem–Katchalsky and the Spiegler–Kedem models (Soltanieh & Gill, 1981; Spiegler & Kedem, 1966) and with the well-known solution–diffusion model by Lonsdale, Merten, and Riley (1965).

In the following 30 years, several models were proposed to describe the RO process, for instance, the "frictional model," the "preferential adsorption-capillary model," the "finely porous model," and the "highly porous model" in which the equations are, in some respects, equivalent even if presented with different notation. These models were developed with the intention to address additional phenomena as to overcome the limitations of two the fundamental RO models, the Spiegler–Kedem and the solution–diffusion model, a complete and detailed review was presented by Soltanieh and Gill (1981).

The real breakthrough was the model of Mason and Lonsdale (1990), who introduced a statistical–mechanical theory of membrane transport. Their model is a trade-off between the plurality of the existing models and the need of having a theory and a method that can enable the general description of transport phenomena in largely different membranes, in terms of both material and function, independently on the state of aggregation of the streams. They demonstrated that models typically developed for single processes, such as ultrafiltration (UF), RO, electrodialysis, gas separations, etc., are only particular cases of a more general transport model, written according to the principles of nonequilibrium thermodynamics (Bird, Stewart, & Lightfoot, 2007; Seader & Henley, 2006). The molar flux of each species (both solvent and solute) is expressed by accounting for (1) the driving forces for all the diffusive mechanisms (ordinary diffusion, pressure diffusion, forced diffusion, and thermal diffusion), (2) the viscous flow contribution, as well as (3) the interdiffusion among the moving species. The model for a specific process can then be derived by the general equations by introducing the corresponding boundary conditions and by applying the proper simplifications.

With regard to RO, the statistical–mechanical approach by Mason and Lonsdale also demonstrated the physical meaning of the parameters of the Spiegler–Kedem model, which were derived according the principles of the irreversible processes thermodynamics and, at the same time, it confirmed the principles and the validity limits of the solution–diffusion model. A synthetic documentation of it will be reported in the next sections.

Nowadays, the Mason and Lonsdale model with three parameters, or alternatively, the Spiegler–Kedem model, is recognized by all the researchers as a satisfactory model to describe mass transfer in RO membranes. Conversely, modeling of transport and partitioning phenomena in NF membranes is still an open problem, which has not been completely solved yet, even if the first studies date back to the 1960s–70s (Dresner, 1972; Hoffer & Kedem, 1967; Jitsuhara & Kimura, 1983; Simons & Kedem, 1973). Original descriptions of ion transport in NF were based on phenomenological approaches derived from irreversible processes thermodynamics (Dresner, 1972; Hoffer & Kedem, 1967; Simons & Kedem, 1973). The membrane is treated as a black box in which slow processes occur near equilibrium and no hypothesis was made on the transport mechanisms; however, it is impossible to obtain information from these models about the flow mechanism across the membrane, or about the partitioning mechanisms at the interfaces. The extension to multicomponent systems is highly complex; however, some authors were able to successfully apply these kind of models (Ahmad, Chong, & Bhatia, 2005; Garba, Taha, Cabon, & Dorange, 2003; Garba, Taha, Gondrexon, & Dorange, 1999; Gupta, Hwang, Krantz, & Greenberg, 2007; Perry & Linder, 1989; Wang, Tsuru, Nakao, & Kimura, 1995; Yaroshchuk et al., 2019; Yaroshchuk, Martinez-Llado, Llenas, Rovira, & de Pablo, 2011).

Structural models for NF were introduced as a simplification of the Space Charge Model as proposed by Teorell, Meyer and Sievers (Tsuru et al., 1991; van der Host, Timmers, Robbertsen, & Leenders, 1995; Wang et al., 1995). The milestone in NF modeling is represented by the work by Bowen and Mukhtar (1996), who introduced the DSPM in which (1) steric partitioning and Donnan equilibrium were accounted for as the main partitioning phenomena, (2) ionic transport across the membrane was described by the extended Nernst–Planck equation. The ion concentration and electric potential profiles were assumed as radially homogeneous in the pore and a fixed membrane charge was considered uniform in the membrane volume, according to a Freundlich isotherm.

From 1996 to 2005, research focused on the identification of the transport phenomena occurring across the membrane, and above all on the determination of the partitioning phenomena between the membrane and the external phases. All the models developed later are extensions and improvements of the original DSPM model, which included the DE as remarkable partitioning phenomena (Bandini & Vezzani, 2003; Bowen & Welfoot, 2002a; Szymczyk & Fievet, 2005).

Moreover, from 2002 to date, studies have been devoted to the identification of the minimum number of necessary parameters for the model and, above all, to finding their values, by studying different methods of data processing and analysis (Bandini & Morelli, 2017; Bandini & Vezzani, 2003; Bowen & Welfoot, 2002a; Cavaco Morão, Szymczyk, Fievet, & Brites Alves, 2008; Deon, Dutournie, Limousy, & Bourseau, 2009; Déon, Escoda, Fievet, & Salut, 2013; Escoda, Deon, & Fievet, 2011; Escoda, Lanteri, Fievet, Deon, & Szymczyk, 2010; Oatley et al., 2012; Oatley-Radcliffe, Williams, Barrow, & Williams, 2014; Silva et al., 2016; Szymczyk & Fievet, 2005). These studies are parallel and cannot disregard the membrane charge determination and the comprehension of its forming mechanisms (Afonso, 2006; Afonso, Hagmeyer, & Gimbel, 2001; Ariza & Benavente, 2001; Bruni & Bandini, 2008; Childress & Elimelech, 1996; Déon et al., 2013; Ernst, Bismark, Springler, & Jekel, 2000; Escoda et al., 2010, 2011; Fievet, Aoubiza, Szymczyk, & Pagetti, 1999; Fievet, Szymczyk, Aoubiza, & Pagetti, 2000; Fridman-Bishop, Tankus, & Freger, 2018; Hagmeyer & Gimbel, 1998, 1999; Oatley et al., 2012; Oatley-Radcliffe et al., 2014; Peeters, Mulder, & Strathmann, 1999; Schaep & Vandecasteele, 2001; Szymczyk, Fatin-Rouge, Fievet, Ramseyer, & Vidonne, 2007; Szymczyk, Fievet, & Bandini, 2010; Szymczyk, Fievet, Mullet, Reggiani, & Pagetti, 1998; Takagi & Nakagaki, 1990; Takagi & Nakagaki, 1992; Tay, Liu, & Sun, 2002; Teixeira, Rosa, & Nystrom, 2005; Zhao, Xing, Xu, & Wong, 2005).

Today the dispute is still ongoing between the choice of phenomenological models (Yaroshchuk et al., 2019) and of structural models in which the transport and partitioning mechanisms are related to the various driving forces of the process, as well as to the physicochemical properties of the membrane and of the solutions (Bandini & Bruni, 2010; Oatley et al., 2012; Wang & Lin, 2021).

Most of the models that today have major success are structural models, in which the key points can be identified as membrane characterization, description of transport phenomena across the membrane, and the identification of partitioning phenomena at the membrane/external phase interfaces. All these aspects are closely linked and depend on the material and the membrane morphology.

The membrane has a selective microporous layer as demonstrated by means of scanning electron microscopy (SEM) image analysis and other techniques (Bowen & Doneva, 2000; Bowen, Mohammad, & Hilal, 1997; Déon et al., 2013; Otero et al., 2006, 2008), and the mean pore sizes are only one order of magnitude greater than atomic dimensions. Owing to the amphoteric properties of the polymer, the membrane surface has an electric charge that depends on the pH of the contact solution and the type and concentration of electrolytes. Electrokinetic measurements of streaming potential (Afonso et al., 2001; Ariza & Benavente, 2001; Bruni & Bandini, 2008; Childress & Elimelech, 1996; Déon et al., 2013; Ernst et al., 2000; Hagmeyer & Gimbel, 1998, 1999; Oatley et al., 2012; Oatley-Radcliffe et al., 2014; Peeters et al., 1999; Szymczyk et al., 1998, 2007, 2010; Tay et al., 2002; Teixeira et al., 2005; Zhao et al., 2005), membrane potential measurements (Déon et al., 2013; Escoda et al., 2010, 2011; Fievet et al., 1999, 2000; Takagi & Nakagaki, 1990, 1992), and titration experiments (Afonso, 2006; Schaep &

Vandecasteele, 2001) are the experimental evidence of this characteristic, which is observed both for polymeric and ceramic materials. Unlike ion exchange membranes, specific and non-specific adsorption of ions onto the membrane can reverse the zeta-potentials from negative to positive values and the points of zero charge can be greatly affected by salt concentration (Ariza & Benavente, 2001; Bruni & Bandini, 2008; De Lint, Biesheuvel, & Verweij, 2002; Fridman-Bishop et al., 2018; Mazzoni & Bandini, 2006; Mazzoni et al., 2007; Takagi & Nakagaki, 1990, 1992, 2003; Teixeira et al., 2005; Tsuru et al., 1991); the extent of the volume charge is, however, remarkably lower than the typical values of ion exchange membranes.

Mass transfer of ionic species is described by the extended modified Nernst-Planck equation in which diffusion, convection, and electromigration across the membrane are evaluated assuming a mono-dimensional problem along the axial coordinate of the pores and accounting of the hindered transport through narrow pores, as a consequence of solute–membrane interactions (Bandini & Bruni, 2010; Oatley et al., 2012; Wang & Lin, 2021). The porous vision has also been applied in a systematic manner to the description of the total volume flux (Bandini & Morelli, 2017).

The role of membrane charge and of DE phenomena have been recognized as fundamental in ionic partitioning. Owing to the small pore dimensions (nanoscale) and to the low dielectric properties of polymeric materials with respect to the dielectric constant of water solutions, numerous phenomena occur simultaneously, making the problem extremely complex to be described. The continuous vision is often criticized: (1) the nanoscale dimensions affect the water dipole and the ion solvation, with effects on the viscosity inside the pores (Bowen & Welfoot, 2002a; Oatley et al., 2012; Yaroshchuk, 2000), (2) the effect of the pore walls reduce, with drag factors, both the convective and the diffusive motion of the solutes across the membrane (Deen, 1987; Ferry, 1936), (3) the Stokes radius could not be fully representative of the species dimensions inside the pores (Bandini & Morelli, 2017; Kiso et al., 2010; Santos et al., 2006; Van der Bruggen, Schaep, Maes, & Vandecasteele, 1999). Moreover, the charge inside the membrane is not homogeneous since it could depend on the concentration distribution along the pore length and this could cause an electrokinetic potential inside the pore (Yaroshchuk et al., 2019).

Therefore the problem complexity is not only mathematical (which can be easily overcome today by a good numerical approach) but rather physical. Although the phenomena are well understood, the problem schematization and the calculation of the relevant adjustable parameters is not well defined in a general way. However, in the case of neutral solutes [such as oligosaccharides (Bandini & Morelli, 2017; Rizki et al., 2019)] and simple electrolytes mixtures (Déon et al., 2013; Escoda et al., 2011), the solution seems to be very close.

The objectives of this chapter are twofold:

1. summarize the fundamentals of the Mason and Lonsdale model and report the most important results for RO models;
2. state the general equations of the DSPM-DE model and describe all the transport and partitioning phenomena studied; it will be demonstrated that the NF models currently in use are a particular case of the general model as well.

To better understand the NF theory, the discussion will be focused on two representative case studies: (1) neutral solutes and (2) multicomponent mixtures of electrolytes. Finally, some considerations on future perspective will conclude this chapter.

3.2 Statistical–mechanical model by Mason and Lonsdale

The general model introduced by Mason and Lonsdale (1990) is reported in this section, with the aim to show that most of the phenomenological and structural models developed for RO and NF processes are only particular cases of it. In addition, for some cases, the specific approximations and simplifications will be highlighted.

In accordance with the principles of nonequilibrium thermodynamics (Bird et al., 2007; Seader & Henley, 2006), the velocity of each species i (both solvent and solute or ionic species) in a multicomponent mixture can be expressed as depending on all the diffusive mechanisms (ordinary diffusion, pressure diffusion, forced diffusion and thermal diffusion) as well as on the viscous flow contribution.

Eq. (3.1) represents the statistical–mechanical model in which thermal diffusion and pressure diffusion (not included in the electrochemical potential) are neglected for simplicity. The notation has been slightly modified with respect to the original version in Mason and Lonsdale (1990), a list of symbols is reported in the *abbreviations*.

$$\sum_{j=1}^{N} \frac{c_j}{cD_{ij}}(\underline{v}_i - \underline{v}_j) = -\frac{1}{RT}\underline{\nabla}\mu_i^{el}\Big|_T - \frac{\alpha_i' k_D}{\eta D_{iM}}\underline{\nabla}p \tag{3.1}$$

Eq. (3.1) is a vector equation that should be written for each permeating compound i. In the case of membranes, the membrane also needs to be accounted among the species j in the sum, thereby obtaining the coefficients D_{iM} in addition to the coefficients D_{ij}, which represent the diffusion coefficients between i and j in the multicomponent mixture across the membrane itself ($D_{ij} = D_{ji}$).

The molar flux vector ($\underline{n}_i$) of each permeating species can then be calculated with the corresponding velocity, by accounting $\underline{n}_i = c_i\underline{v}_i$; $\underline{\nabla}\mu_i^{el}|_T$ is the isothermal chemical or electrochemical potential gradient as defined in Eq. (3.2).

$$\underline{\nabla}\mu_i^{el}\Big|_T = RT\underline{\nabla}\ln a_i^{RS}\Big|_T + \overline{V}_i^{RS}\underline{\nabla}p + z_iF\underline{\nabla}\Psi \tag{3.2}$$

in which "RS" represents the reference state (pressure, composition) at which the activity is also calculated, and $\overline{V}_i^{RS}$ is the corresponding value of the partial molar volume.

The left hand side of Eq. (3.1) accounts for the interdiffusion among the permeating species and the diffusion inside the membrane. The second term of the right hand side of Eq. (3.1) accounts for the contribution of the total pressure gradient inside the membrane represented by a viscous motion across the pores according to a Darcy-like flow: k_D is a viscous flow parameter characteristic of the membrane structure, η is the viscosity of the solution inside the membrane, and α_i' is a dimensionless parameter [tagged as "bugger factor" by Mason and Lonsdale (1990)] which will be recognized later as a convective hindrance factor (for NF).

Even if RO membranes are typically regarded as dense membranes, they show microporosities and in the IUPAC material classification they are considered to be "microporous materials." This justifies the porous vision of the model, which treats RO membrane as microporous.

The statement of a general model should contain equations to describe (1) solute transport across the membrane, (2) water or total flux across the membrane, (3) solute (and solvent) partitioning at the interfaces between the membrane and the external phases.

3.3 Water partitioning: the osmotic equilibrium

RO and NF membranes are typically asymmetric with a selective skin supported on a highly porous support. Although SEM imaging provides valuable evidence regarding the membrane structure, it is difficult to determine with good precision the thickness of the selective layer as well as the shape and the tortuosity of the microporosities. The thickness of the selective layer is generally referred to as an effective thickness (δ accounting the tortuosity and the porosity), which should be determined as an adjustable parameter.

Fig. 3.1A shows a schematization of the selective layer, simplified as a unidimensional layer with microporosities: it is generally accepted that the effective driving force for the process is located in this layer, owing to the highly porosity of the support (Rautenbach & Albrecht, 1989).

In RO and NF modeling, water is the solvent and is considered a continuous medium for which the membrane is not a physical barrier. As a consequence, at the interfaces between the membrane and the external phases the equilibrium conditions that impose the equality of the chemical potentials are in order.

Accounting for the osmotic pressure definition represented by Eq. (3.3), in which the water activity is expressed with reference to the state of pure liquid water at 1 bar,

$$\underline{\nabla}\pi = -\frac{RT\underline{\nabla}\ \ln a_w|_T}{\tilde{V}_w^o} = -c_w^o\underline{\nabla}\mu_w|_T + \underline{\nabla}p \tag{3.3}$$

The following equations should be considered at the interfaces, with reference to the notation of Fig. 3.1A:

$$\begin{aligned}
\mu_w(0^-) = \mu_w(0^+) &\Rightarrow \frac{\mathrm{RTln}a_w(0^-)}{\tilde{V}_w^o} + p(0^-) = \frac{\mathrm{RTln}a_w(0^+)}{\tilde{V}_w^o} + \overline{p}(0^+) \\
\mu_w(\delta^-) = \mu_w(\delta^+) &\Rightarrow \frac{\mathrm{RTln}a_w(\delta^-)}{\tilde{V}_w^o} + \overline{p}(\delta^-) = \frac{\mathrm{RTln}a_w(\delta^+)}{\tilde{V}_w^o} + p(\delta^+)
\end{aligned} \tag{3.4}$$

Since the osmotic pressure can be interpreted as a fictitious force, Eq. (3.4) is equivalent to the force balances at the feed/membrane and at the membrane/permeate interfaces, as represented by Eq. (3.5) in which the corresponding osmotic pressure differences, $\Delta\pi_0$ and $\Delta\pi_\delta$, are introduced.

$$\begin{aligned}
&\Delta P = p(0^-) - p(\delta^+) \\
&p(0^-) - \overline{p}(0^+) = \pi(0^-) - \pi(0^+) = \Delta\pi_0 \\
&p(\delta^+) - \overline{p}(\delta^-) = \pi(\delta^+) - \pi(\delta^-) = \Delta\pi_\delta
\end{aligned} \tag{3.5}$$

3.4 Reverse osmosis models

The basic equations for the most used RO models are summarized in this section, with the aim to show their derivation from the general statistical–mechanical model by Mason

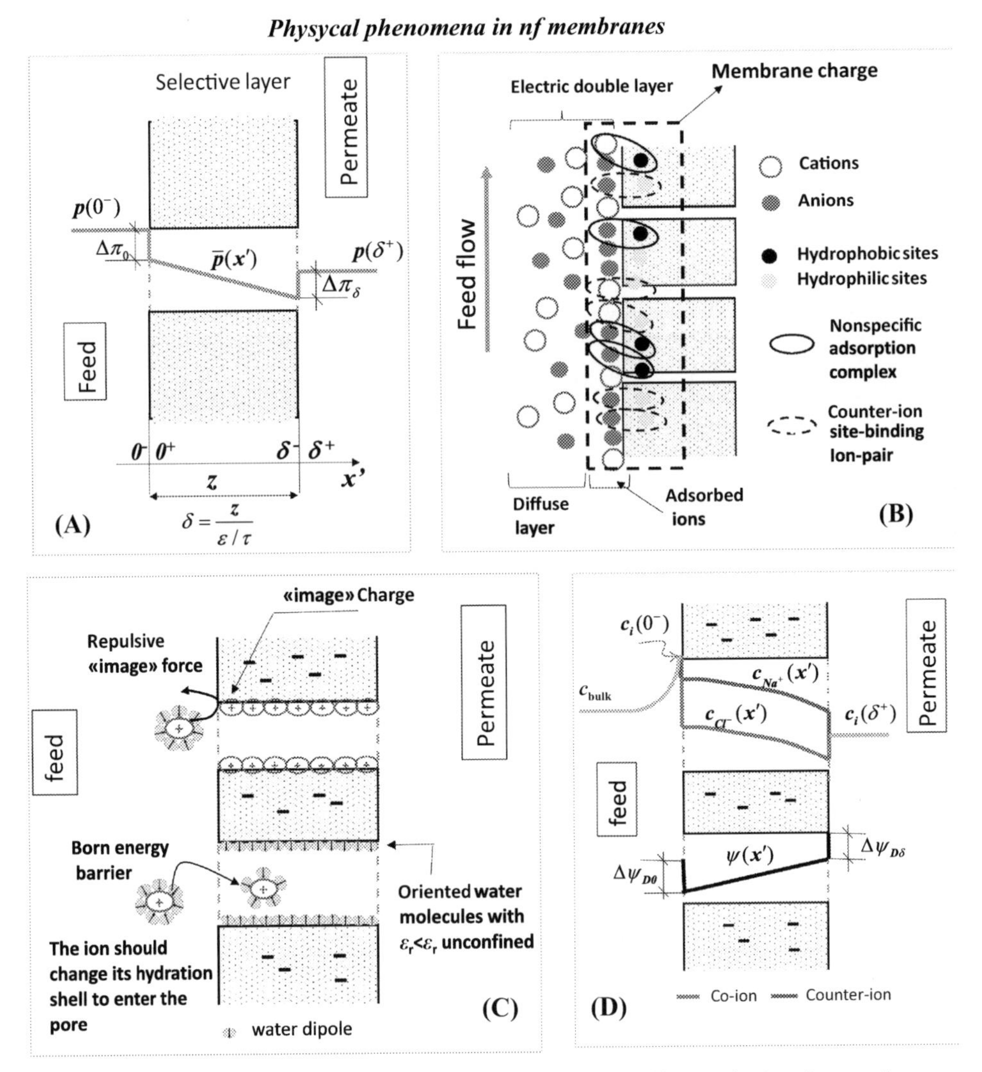

FIGURE 3.1 Physical phenomena in NF membranes. (A) osmotic equilibria at the interfaces and pressure profile along a pore; (B) mechanism of membrane charge formation according to the adsorption amphoteric model (Bruni & Bandini, 2008; Szymczyk et al., 2010), showing the nonspecific adsorption on hydrophobic sites and counter-ion site-binding onto the dissociated hydrophilic sites; (C) mechanisms of dielectric exclusions: "image forces" and Born partitioning; (D) concentration and dimensionless electric potential profiles for NaCl–water solutions in a negatively charged membrane.

and Lonsdale. The solution–diffusion model and the Spiegler–Kedem model will be considered. The main aim of this section is to point out the importance of knowing the basic elements of a model to be able to use it properly and critically.

The main features of the models and recommendations for their use will be also introduced.

3.4.1 The solution–diffusion model

The solution–diffusion model was originally introduced by Lonsdale et al. (1965). It can be derived as a particular case of Eq. (3.1), under the following hypotheses:

1. the transport mechanism inside the membrane is purely diffusive (the convective contributions are neglected)
2. solute–solvent interdiffusion is neglected
3. the membrane is electrically neutral
4. the membrane is a high rejection membrane and the permeating stream is highly diluted
5. isothermal conditions.

The model equations are summarized in Table 3.1, while the main steps to obtain the model itself are reported in the following.

In the case of binary systems (water = "w" and solute = "s"), in which the solute is considered a neutral compound, Eq. (3.1) can be translated into Eq. (3.6). The detailed derivation of the model equations is reported in Appendix.

$$\underline{n}_w = -c_w \frac{D_{\text{wM}}}{\text{RT}} \nabla \underline{\mu}_w \Big|_T \simeq -\frac{D_{\text{wM}}}{\text{RT}} \left[\nabla \underline{\overline{p}} - \nabla \underline{\pi}\right]$$

$$\underline{n}_s = -c_s \frac{D_{\text{sM}}}{\text{RT}} \nabla \underline{\mu}_s \big|_T \simeq -D_{\text{sM}} \nabla \underline{c}_s \tag{3.6}$$

in which the effect of osmotic pressure on the water flux and the diffusive character of the solute flux are self-explanatory.

The equations can be easily adapted to a unidimensional problem, as the one depicted in Fig. 3.1A; we can obtain the following set of equations:

$$\begin{aligned} n_{\text{wx}} &= -\frac{D_{\text{wM}}}{\text{RT}} \left[\frac{d\overline{p}}{dx'} - \frac{d\pi}{dx'}\right] \\ n_{\text{sx}} &= -D_{\text{sM}} \frac{dc_s}{dx'} \end{aligned} \tag{3.7}$$

in which x' represents a sort of effective axial coordinate accounting for porosity and tortuosity.

Eq. (3.7) should be coupled with the solute/solvent partitioning at the membrane interfaces, in order to obtain relationships involving the feed and permeate concentration values.

With regard to the solute partitioning, the solubility coefficient is assumed to be a constant value at both interfaces, as represented by Eq. (3.8), to account for the membrane barrier effect:

$$S_s = \frac{c_s(0^+)}{c_s(0^-)} = \frac{c_s(\delta^-)}{c_s(\delta^+)} \tag{3.8}$$

With regard to water partitioning, Eq. (3.5) can be used.

Eq. (3.7) can be easily integrated along the membrane thickness, by accounting that under steady-state conditions, the x components of the molar fluxes are constant with time

TABLE 3.1 Basic equations of the models for reverse osmosis processes: binary systems (water + solute).

Three-parameter model (Spiegler–Kedem type)

Parameters

$L_p, P_s, \sigma_v = \sigma_s$

Total volume flux

$$J_v = L_p(\Delta P - \sigma_v \Delta\pi) \tag{1.1}$$

Solute flux

$$J_s = c_s(0^-)(1-\sigma_s)J_v + \frac{(1-\sigma_s)J_v(c_s(0^-) - c_s(\delta^+))}{e^{\mathrm{Pe}} - 1} \tag{1.2}$$

$$\mathrm{Pe} = \frac{(1-\sigma_s)J_v}{P_s} \tag{1.3}$$

Solute real rejection

$$R_{\mathrm{real},s} = 1 - \frac{c_s(\delta^+)}{c_s(0^-)} = 1 - \frac{J_s/J_v}{c_s(0^-)} = \frac{(e^{\mathrm{Pe}} - 1)\sigma_s}{e^{\mathrm{Pe}} - \sigma_s} \tag{1.4}$$

Asymptotic rejection

$R^{\mathrm{asym}}_{\mathrm{real},s} = \lim_{J_v \to \infty} R_{\mathrm{real},s} = \sigma_s$

At very low fluxes

$$\left.\frac{dR_{\mathrm{real},s}}{dJ_v}\right|_{Jv \to 0} = \frac{\sigma_s}{P_s} \tag{1.5}$$

Two-parameter model (solution–diffusion model)

Parameters

L_{pw}, K_s

Total volume flux

$$J_v = L_{\mathrm{pw}}(\Delta P - \Delta\pi) \tag{1.6}$$

Solute flux

$$J_s = K_s(c_s(0^-) - c_s(\delta^+)) \tag{1.7}$$

Solute real rejection

$$R_{\mathrm{real},s} = 1 - \frac{c_s(\delta^+)}{c_s(0^-)} = 1 - \frac{J_s/J_v}{c_s(0^-)} = \frac{J_v}{J_v + K_s} \tag{1.8}$$

Asymptotic rejection

$R^{\mathrm{asym}}_{\mathrm{real},s} = \lim_{J_v \to \infty} R_{\mathrm{real},s} = 1$

At very low fluxes

$$\left.\frac{dR_{\mathrm{real},s}}{dJ_v}\right|_{\mathrm{Jv} \to 0} = \frac{1}{K_s} \tag{1.9}$$

in which

$\Delta\pi = \pi(0^-) - \pi(\delta^+)$

and along the membrane. Assuming that the total volume flux across the membrane is coincident with the water volume flux, the final Eq. (3.9) can be obtained:

$$J_v \simeq \frac{n_{wx}}{c_w} = \frac{D_{wM}}{RTc_w\delta}(\Delta P - \Delta\pi) = L_{pw}(\Delta P - \Delta\pi)$$
$$J_s = n_{sx} = \frac{D_{sM}S_s}{RT\delta}(c_s(0^-) - c_s(\delta^+)) = K_s(c_s(0^-) - c_s(\delta^+)) \quad (3.9)$$

in which

$$\Delta\pi = \pi(0^-) - \pi(\delta^+);\ \delta = \frac{z}{(\varepsilon/\tau)}$$

The hydraulic permeability L_{pw} and the solute permeability K_s are defined straightforwardly, referring to the total membrane area; they should be considered to be adjustable parameters, although their meaning is self-explanatory from Eq. (3.9).

The calculation of the local rejection can be performed by considering that the asymmetric morphology of RO membranes (like NF membranes) allows the application of the "cross flow hypothesis," which considers the mass transfer resistances in the support and in the permeate side as negligible with respect to the mass transfer in the skin. The local composition of the permeate, which determines the driving forces in Eq. (3.9), corresponds to the concentration of the permeating stream and is remarkably independent of the concentration existing downstream the membrane, whatever the module "flow pattern." The evidence of this assumption is both experimental and theoretical, as it has been documented by Rautenbach and Albrecht (1989) for several case studies.

The expression of the rejection and the corresponding value of the asymptotic rejection are reported in Table 3.1.

3.4.2 The three-parameter model

The most famous three-parameter model for RO is the one developed by Spiegler and Kedem (1966) as an improvement of the previous Kedem–Katchalsky model (Soltanieh & Gill, 1981). They are both phenomenological models developed on the basis of irreversible process thermodynamics; however, such models can be also derived from the statistical–mechanical model by Mason and Lonsdale (1990).

In this section, we discuss only the main features of the model, the complete demonstration is reported in Mason and Lonsdale (1990).

In the case of binary systems, in which the solute is considered as a neutral compound, Eq. (3.1) can be translated into Eq. (3.10), under the following hypotheses:

1. the membrane is electrically neutral
2. isothermal conditions

$$\frac{c_s}{cD_{ws}}(\underline{v}_w - \underline{v}_s) + \frac{\underline{v}_w}{D_{wM}} = -\frac{1}{RT}\underline{\nabla}\mu_w\Big|_T - \frac{\alpha'_w k_D}{\eta D_{wM}}\underline{\nabla p}$$
$$\frac{c_w}{cD_{sw}}(\underline{v}_s - \underline{v}_w) + \frac{\underline{v}_s}{D_{sM}} = -\frac{1}{RT}\underline{\nabla}\mu_s\Big|_T - \frac{\alpha'_s k_D}{\eta D_{sM}}\underline{\nabla p} \quad (3.10)$$

in which five coefficients can be observed ($D_{wM}, D_{sM}, D_{sw}, \alpha'_w k_D, \alpha'_s$), representing parameters related to diffusive transport, to steric hindrance and to the viscous flow inside the "pores," respectively.

After a few algebraic passages, Mason and Lonsdale obtained the following set of equations [Eq. (11)], representing the volumetric mean velocity (velocity of the center of volume) ($\underline{v}_{tot}$) and the solute molar flux vector ($\underline{n}_s$) containing four parameters:

$$\begin{aligned} \underline{v}_{tot} &= \overline{V}_w \underline{n}_w + \overline{V}_s \underline{n}_s = -l_p(\underline{\nabla}\overline{p} - \sigma_v \underline{\nabla}\pi) \\ \underline{n}_s &= -(c_w \overline{V}_w)\omega \underline{\nabla}\pi + c_s\left[1 - (c_w \overline{V}_w)\sigma_s\right]\underline{v}_{tot} \end{aligned} \quad (3.11)$$

Such parameters depend on operative conditions (temperature and concentration) and are algebraic combinations of the transport parameters previously introduced in Eq. (3.10).

The equations can be again easily adapted to a unidimensional problem, as represented in Fig. 3.1A; under the hypothesis that the membrane is a high rejection membrane and the permeating stream is highly diluted, according to Van't Hoff equation ($\underline{\nabla}\pi \simeq RT\underline{\nabla}c_s$), Eq. (3.12) can be finally obtained:

$$\begin{aligned} J_v &= -l_p\left[\frac{d\overline{p}}{dx'} - \sigma_v \frac{d\pi}{dx'}\right] \\ J_s &= -\omega RT \frac{dc_s}{dx'} + (1 - \sigma_s)J_v \end{aligned} \quad (3.12)$$

Parameters l_p and ω are the total permeability and the solute permeability coefficients, respectively, while σ_v and σ_s are the Staverman and the solute reflection coefficients, respectively.

Eq. (3.12) exactly match with the equations originally introduced by Spiegler–Kedem (1966), when $\sigma_v = \sigma_s$ (Mason & Lonsdale, 1990); they can be used with three or four parameters. The diffusive and convective contributions in determining the solute flux are self-explanatory.

Since in RO membranes the reflection coefficients typically range from 0.8 to 0.99, being mostly greater than 0.9, the model can be used as a three-parameter model and integrated by assuming all the parameters as constant values along the membrane and with the operative conditions.

From this point on, Mason and Lonsdale (in agreement with Spiegler–Kedem) integrated the differential Eq. (3.12) across the membrane, by also invoking the finite difference approximation to relate the total volume flux to the pressure difference. Finally, the three-parameter model can be represented using Eq. (3.13), also reported in Table 3.1, together with the corresponding relationships for the real rejection.

$$\begin{aligned} J_v &= L_p(\Delta P - \sigma_v \Delta\pi) \\ J_s &= c_s(0^-)(1 - \sigma_s)J_v + \frac{(1 - \sigma_s)J_v(c_s(0^-) - c_s(\delta^+))}{e^{Pe} - 1} \\ Pe &= \frac{(1 - \sigma_s)J_v}{P_s}; P_s = \frac{\omega RT}{\delta}; \sigma_v = \sigma_s \end{aligned} \quad (3.13)$$

in which L_p and P_s are the membrane permeability and the solute permeability, respectively.

It is interesting to observe that for high-rejection membranes ($\sigma_S \rightarrow 1$), for which relatively low-volume fluxes are measured, the Peclèt number is much lower than unity; Eq. (3.13) degenerate into Eq. (3.14), which reminds of the original Kedem–Katchalsky model (Soltanieh & Gill, 1981):

$$\mathrm{Pe} << 1 \Rightarrow \lim_{\mathrm{Pe}\to 0} \frac{(1-\sigma_S)\mathrm{Jv}}{e^{\mathrm{Pe}}-1} = P_s \Rightarrow J_s = c_s(0^-)(1-\sigma_S)J_v + P_s(c_s(0^-) - c_s(\delta^+)) \tag{3.14}$$

In passing, we can also observe that when $\sigma_S = 1$, the solution–diffusion model is obtained.

As a conclusive remark, it is important to highlight the twofold relevance of the statistical–mechanical model by Mason and Lonsdale. On the one hand, it allows one to understand the physical meaning of the phenomenological parameters of the traditional models and to critically account for the corresponding approximations; on the other hand, it gives a good effective tool to develop new models for RO (suitable for multicomponent mixtures for instance) and for new processes, such as NF. The use of this model will certainly lead to more complicated equations to be solved, since sets of differential equations will require numerical approach; however, nowadays it should be no longer regarded as a limiting step.

3.4.3 Conclusive remarks and recommendations for reverse osmosis models

Although simple models are always preferred and valuable, nowadays the available computing power allows to appreciate more accurate complex models. Nevertheless, the solution–diffusion model is still the most used model owing to its simplicity and broad validity.

This model is widely implemented in process simulators (https://www.suezwater technologies.com/resources/winflows; https://imsdesign.software.informer.com), for example, to characterize the membranes in industrial spiral modules for RO applications, where high rejection membranes are typically used.

All parameters present in RO models are considered adjustable parameters to be calculated with a fitting procedure on a sufficient number of experimental data. Although the Mason–Lonsdale model supports that the parameters depend on operative conditions, the experimental evidence of the last 50 years clearly documented that:

1. membrane permeability greatly depends on temperature and is rather constant in a wide range of pressure (up to 80–100 bar). An Arrhenius-type dependence with temperature was originally assumed, which is consistent with a viscous flow across the pores as it will be demonstrated for NF membranes (see Section 3.5.1.2);
2. solute permeability and reflection coefficient can be considered to be constant values in a wide composition and pH range.

For a specific membrane (i.e., for a specific L_p) and for a specific solute (i.e., for a specific P_s and σ_s) Eq. (1.4) of Table 3.1 predicts a unique curve relating the real rejection with the Péclet number, which can be finally interpreted as a unique curve as a function of the total volume flux. In addition, when the solution–diffusion model is considered, Eq. (1.8) itself shows a unique curve rejection versus total flux.

The corresponding behaviors are represented in Fig. 3.2 for the three-parameter model. Apparently, the solute reflection coefficient corresponds to the asymptotic rejection, whereas the slope of the rejection curve versus volume flux at very low fluxes is related to the solute permeability. A fitting procedure on experimental data allows the parameters calculation; it is recommended to use data at very low fluxes to estimate the solute permeability.

The typical trends of total volume flux along the transmembrane pressure and of the real rejection versus the volume flux are represented in Fig. 3.3, for the simple case of the solution–diffusion model, for different values of the hydraulic and of the solute permeability. Obviously, the model always predicts complete asymptotic rejection, whereas the slope of the rejection curve at very low fluxes is also in this case related to the solute permeability. It is evident that the solution–diffusion model should be recommended only to simulate performances of very high rejection membranes.

In principle, nothing forbids the use of the three-parameter model also for NF. Nevertheless, in NF the curve of rejection versus J_v is not unique for the electrolytes cases, but its behavior strongly depends on pH and on concentration (Quin et al., 2004; Bandini et al., 2005; Nilsson et al., 2008; Condom et al., 2004; Mazzoni & Bandini, 2006; Mazzoni et al., 2007), with different trends for monovalent electrolytes, symmetrical and nonsymmetrical. In the case of multiionic mixtures, the ions present in lower amount can also show negative rejection values; when this happens, there are no more advantages in using the three-parameter model and often it is impossible to fit rejection data with Eq. (1.4)–Table 3.1, hence the need to develop specific models for NF.

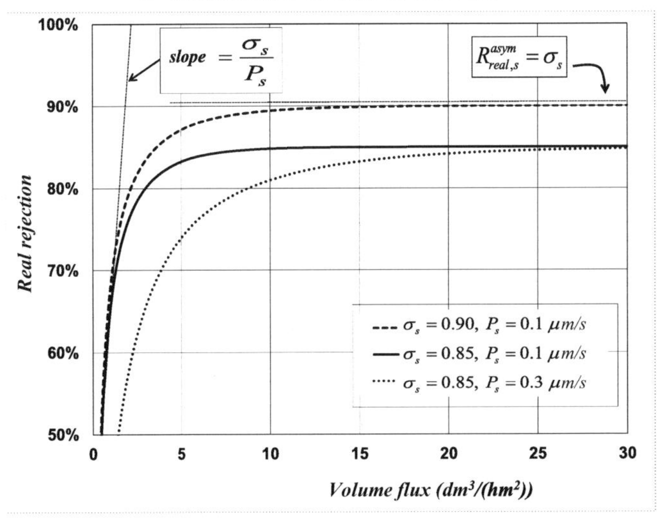

FIGURE 3.2 Three-parameter model. Behavior of the real rejection versus the total volume flux according to Eq. (1.4)—Table 3.1 and representation of the corresponding parameters [Eq. (1.5)—Table 3.1].

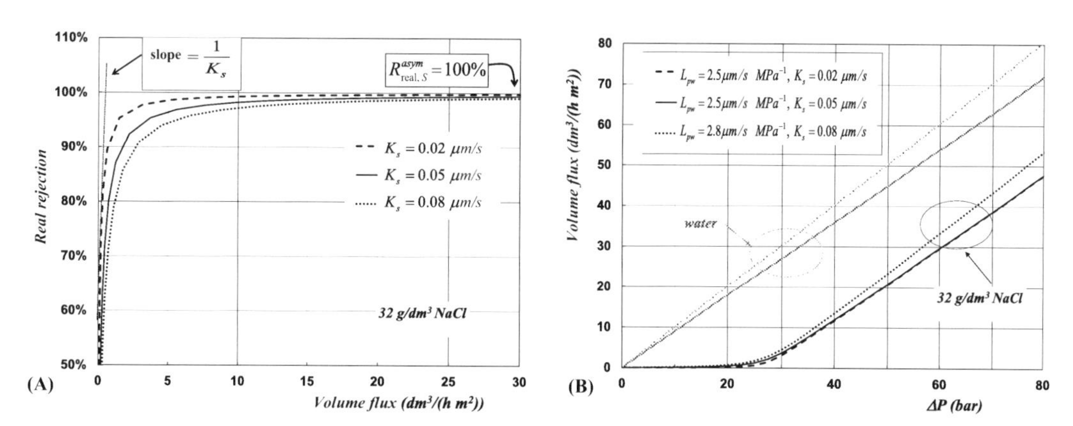

FIGURE 3.3 Solution–diffusion model. Solution–diffusion RO model for 32 g/dm^3 NaCl–water solutions. (A) real rejection versus total volume flux at various solute permeabilities; (B) total volume flux versus transmembrane pressure: comparison between pure water and salt solution cases.

3.5 Nanofiltration modeling: the porous vision of the Donnan-steric-pore-and-dielectric-exclusion model for aqueous solutions

The general physical problem of NF modeling consists of three key steps:

1. description of the transport phenomena across the membrane active layer;
2. description of partitioning phenomena at the interfaces; and
3. schematization of the membrane and of the membrane charge.

The description of the three key steps of NF modeling suffers from some critical points, among those the pore-size determination and the membrane charge evaluation.

The first critical point is the pore size: small nanoscale dimensions of the average pore size (in the range from 0.3 to 1.5 nm) give rise to some phenomena, such as DE (which is not relevant neither in RO nor in UF (Yaroshchuk, 2000) and drag forces hindering the species motion along the narrow pores (Deen, 1987; Ferry, 1936). There are also a few recognized effects on the modification of the water dipole, as the well documented increase in the dynamic viscosity with a corresponding decrease of the dielectric constant inside the pore with respect to the unconfined values (Bowen & Welfoot, 2002a; Yaroshchuk, 2000). Finally, the pore size gives a nonnegligible sieving effect, although it is difficult to define a pore shape (cylindrical, slit-like, fractal network structure (Wang & Lin, 2021), or other) and to obtain pore size distributions (Bowen & Welfoot, 2002b; Montesdeoca, Janssen, Boom, & Van der Padt, 2019), which might be easily included in the model equations.

The second critical point is the membrane charge: owing to the amphoteric behavior of the polymeric material, the membranes, when in contact with electrolytic solutions, acquire a superficial charge that is the sum of a fixed charge and of a charge due to specific and nonspecific adsorption (Bruni & Bandini, 2008). A schematic of the phenomena is depicted in Fig. 3.1B, in which are shown the ion-hydrophobic sites complexes and the counter-ion site binding ion pairs with the dissociated hydrophilic sites. These phenomena occur in the electric double layer at the feed membrane interface and have been highlighted by a multitude of zeta-potential measurements (Ariza & Benavente, 2001; Childress & Elimelech, 1996; Hagmeyer & Gimbel, 1999; Szymczyk et al., 2007, 2010; Teixeira et al., 2005). However, the charge measurements obtained by electrokinetic methods cannot be considered representative of the effective membrane charge, since they are taken at atmospheric pressure in absence of transmembrane flux. Moreover, since the membrane is porous, the membrane charge is likely present on the whole membrane volume, that is, on the whole pore surface. To the best of our knowledge, there is still no direct method to measure the membrane charge of an NF membrane being operated in NF conditions.

The membrane charge is fundamental on the partitioning of charged/ionic species, both for the Donnan equilibrium effect and for the DE effect caused by the image forces (Yaroshchuk, 2000).

Today, the DSPM-DE model is the most widely accepted model to describe NF. It is an extension of the original DSPM introduced by Bowen and Mukhtar (1996), with the inclusion of dielectric phenomena as further partitioning mechanisms (Bowen & Welfoot, 2002a; Bandini & Vezzani, 2003; Szymczyk & Fievet, 2005).

The model relies on the following general assumptions:

1. The active layer is modeled as a porous medium with a sieve effect for solutes: partitioning phenomena occur at the pore mouth, whereas transport phenomena occur inside and across the pores in a unidimensional direction. The pore morphology is unknown.
2. The solvent (water) is considered as a continuous medium endowed by a density, a viscosity, and a dielectric constant.
3. Viscous flow is accepted to describe convective motion across the pore; however, hindered convection and diffusion are accounted to describe the transport of solute species in a confined space of the same order of magnitude of the molecular size.
4. Interdiffusion across the membrane is neglected in view of the rather diluted solutions inside the membrane.
5. Partitioning at the interfaces is attributed to the superposition of steric partitioning, DE phenomena and to Donnan equilibrium, accounting for nonideality of the solutions and for the osmotic equilibrium.
6. The volumetric membrane charge density is generally uniform in the membrane.

3.5.1 Mass transfer across the membrane pores

The general transport equation of a solute species i in a single pore of the DSPM-DE model is typically represented by the extended Nernst–Planck equation, Eq. (3.15), in which the molar flux vector ($\underline{n}_i$) can be expressed as a function of the electrochemical potential gradient accounting for the convective contribution.

$$\underline{n}_i = - c_i \frac{D_{ip}}{RT} \underline{\nabla} \mu_i^{el}\big|_T + K_{ic} c_i \underline{v}_{tot} \tag{3.15}$$

The convective contribution can be in turn represented by Eq. (3.16) in which the pressure gradient is referred to inside the membrane:

$$\underline{v}_{tot} = - \frac{k_D}{\eta} \underline{\nabla p} \tag{3.16}$$

It is evident that Eqs. (3.15) and (3.16) match exactly with the statistical–mechanical theory of Mason and Lonsdale (as represented by Eq. (3.1), with the permutation of the diffusion coefficient across the membrane (D_{iM}) with the hindered diffusivity of the specie inside the pore ($D_{ip} = K_{id} D_{iw}^0$) and of the coefficient (α_i') with the hindrance factors for convection (K_{ic}). (D_{ip}) should be calculated at the temperature and composition existing inside the membrane.

Eq. (3.15) can be adapted to a unidimensional problem, referring to the scheme of Fig. 3.1A, to relate the molar flux across the membrane of each species (J_i) to each driving force contained in the electrochemical potential, as documented by Eq. (3.17):

$$J_i = - c_i D_{ip} \frac{d\ln\gamma_i}{dx'} - D_{ip} \frac{dc_i}{dx'} - z_i c_i D_{ip} \frac{F}{RT} \frac{d\Psi}{dx'} - \frac{c_i D_{ip} \overline{V}_i}{RT} \frac{d\overline{p}}{dx'} + K_{ic} c_i J_v \tag{3.17}$$

in which J_v is the total volume flux across the membrane (volumetric flow per unit of total membrane area) and the axial coordinate x' accounts also the membrane tortuosity and porosity.

Apparently, Eq. (3.17) accounts for the activity contribution, molecular diffusion, electromigration, pressure diffusion and for the convective contribution across the membrane pores, respectively. The role of each contribution in determining the solute flux is obviously different. The first term is generally neglected, since $d\ln\gamma_i = d\gamma_i/\gamma_i \simeq 0$, while the fourth term has been evaluated in (Bowen & Welfoot, 2002a) as absolutely negligible on the rejection of mineral ions, and not very sensitive (lower than 2%) on the rejection of uncharged solutes as sugars. That conclusion will be accounted for in the reformulation of the problem for the cases studies discussed in Section 3.6.1.

The meaning and relevance of hindrance factors K_{ic} and K_{id} have been widely documented (Anderson & Quinn, 1974; Bowen & Sharif, 1994; Bungay & Brenner, 1973; Dechadilok & Deen, 2006; Deen, Satvat, & Jamieson, 1980; Deen, 1987; Faxen, 1922; Nakao & Kimura, 1982; Paine & Scherr, 1975), K_{ic} is a sort of drag factor, whereas K_{id} quantifies the pore wall effect onto the decrease of the solute–solvent diffusion coefficient below its value in the free bulk solution (D_{iw}^0). Hindrance factors are related to the hydrodynamic coefficients (λ_i) according to relationships obtained by solving the motion problem of a spherical species inside cylindrical pores and/or slit-like pores of infinite length, in which λ_t is assumed as the Stokes radius to pore size ratio. A collection of the various relationships proposed to calculate hindrance factors is reported in Table 3.2.

TABLE 3.2 Hindrance factors.

Cylindrical pores	References
$\phi_i = (1-\lambda_i)^2$	
$K_{ic} = (2-\phi_i)^{\#}\left(1-\frac{2}{3}\lambda_i^2 - 0.163\lambda_i^3\right)$ $K_{id} = \left(1 - 2.1044\lambda_i + 2.089\lambda_i^3 - 0.948\lambda_i^5\right)$ Centerline approximation $0 \le \lambda_i < 0.4$	Anderson and Quinn (1974)
$K_{ic} = (2-\phi_i)^{\#}\frac{K_{i,s}}{(\pi)}; K_{id} = \frac{6\pi}{K_{i,t}}$ $K_{i,t} = \frac{9}{4}\pi^2\sqrt{2}(1-\lambda_i)^{-5/2}\left[1+\sum_{n=1}^{2} a_n(1-\lambda_i)^n\right] + \sum_{n=0}^{4} a_{n+3}\lambda_i^n$ $K_{i,s} = \frac{9}{4}\pi^2\sqrt{2}(1-\lambda_i)^{-5/2}\left[1+\sum_{n=1}^{2} b_n(1-\lambda_i)^n\right] + \sum_{n=0}^{4} b_{n+3}\lambda_i^n$ $a_1 = \frac{-73}{60}, a_2 = \frac{77.293}{50.400}, a_3 = -22.5083, a_4 = -5.6117,$ $a_5 = -0.3363, a_6 = -1.216, a_7 = 1.647$ $b_1 = \frac{7}{60}, b_2 = \frac{-2.227}{50.400}, b_3 = 4.0180, b_4 = -3.9788,$ $b_5 = -1.9215, b_6 = 4.392, b_7 = 5.006$ Centerline approximation $0 \le \lambda < 1$	Bungay and Brenner (1973)
$K_{ic} = (2-\phi_i)^{\#}\left(1.0 + 0.054\lambda_i - 0.988\lambda_i^2 + 0.441\lambda_i^3\right)$ $K_{id} = 1.0 - 2.30\lambda_i + 1.154\lambda_i^2 + 0.224\lambda_i^3$ Centerline approximation $0 \le \lambda_i < 0.8$	Bowen and Sharif (1994)
$K_c = \frac{\phi_i\left(1 + 2\lambda_i - \lambda_i^2\right)\left(1 - \frac{2}{3}\lambda_i^2 - 0.20217\lambda_i^5\right)}{\left(1 - 0.75857\lambda_i^5\right)}$ $K_d = \frac{\phi_i\left(1 - 2.105\lambda_i + 2.0865\lambda_i^3 - 1.7068\lambda_i^5 + 0.72603\lambda_i^6\right)}{\left(1 - 0.75857\lambda_i^5\right)}$ Centerline approximation $0 \le \lambda_i < 0.9$	Nakao and Kimura (1982)
$K_{ic} = \left(\frac{1 + 3.867\lambda_i - 1.907\lambda_i^2 - 0.834\lambda_i^3}{1 + 1.867\lambda_i - 0.741\lambda_i^2}\right)$ $K_{id} = \frac{\left(1 + a_1\lambda_i \ln\lambda_i - a_2\lambda_i + a_3\lambda_i^2 + a_4\lambda_i^3 - a_5\lambda_i^4 + a_6\lambda_i^5 + a_7\lambda_i^6 - a_8\lambda_i^7\right)}{\phi_i}$ $a_1 = \frac{9}{8}$; $a_2 = 1.56034$; $a_3 = 0.528155$; $a_4 = 1.91521$; $a_5 = 2.81903$; $a_6 = 0.270788$; $a_7 = 1.10115$; $a_8 = 0.435933$ Cross-Sectional average $0 \le \lambda_i < 1$	Dechadilok and Deen (2006)

Slit-like pores	References
$\phi_i = (1-\lambda_i)$	
$K_{ic} = \frac{1}{2}\left(3 - \phi_i^2\right)\left(1 - \frac{\lambda_i^2}{3} + O\left(\lambda_i^3\right)\right)$ $K_{id} = 1 - 1.004\lambda_i + 0.418\lambda_i^3 + 0.21\lambda_i^4 - 0.169\lambda_i^5 + O\left(\lambda_i^6\right)$ Centerline approximation $0 \le \lambda_i < 1$	Deen (1987), Faxen (1922)

Slit-like pores	References
$K_{ic}=\dfrac{(1-b_1\lambda_i^2+b_2\lambda_i^3-b_3\lambda_i^4+b_4\lambda_i^5-b_5\lambda_i^6+b_6\lambda_i^7)}{\phi_i}$	Dechadilok and Deen (2006)
$K_{id}=\dfrac{(1+a_1\lambda_i\ln\lambda_i-a_2\lambda_i+a_3\lambda_i^3-a_4\lambda_i^4+a_5\lambda_i^5)}{\phi_i}$	
$a_1=\dfrac{9}{16}$; $a_2=1.19358$; $a_3=0.4285$; $a_4=0.3192$; $a_5=0.08428$.	
$b_1=3.02$; $b_2=5.776$; $b_3=12.3675$; $b_4=18.9775$; $b_5=15.2185$; $b_6=4.8525$. Cross-Sectional avarage $0\le\lambda_i<0.95$	

$(2-\sigma_i)$ accounts for a parabolic profile for solvent velocity.

3.5.1.1 *Solute partitioning*

Partitioning of a species at an interface is stated by the thermodynamic equilibrium that imposes the equality of chemical (electrochemical) potentials. The case of a membrane interface should consider an isothermal interface but not isobaric (owing to the osmotic pressure effects) and that the solute entrance is hindered by a sieve effect. By assuming the validity of the superposition of electric and DE effects, Eqs. (3.18) and (3.19) can be considered to define the partitioning coefficients of an ionic species at the feed–membrane and at the membrane–permeate interfaces.

$$\Gamma_{i0}=\frac{c_i(0^+)}{c_i(0^-)}=\phi_i\frac{\gamma_i(0^-)}{\gamma_i(0^+)}\exp(-z_i\Delta\psi_{D0})\exp(-z_i^2\Delta W_{DE0})\exp\left(\frac{\overline{V}_i\Delta\pi_0}{RT}\right) \tag{3.18}$$

$$\Gamma_{i\delta}=\frac{c_i(\delta^-)}{c_i(\delta^+)}=\phi_i\frac{\gamma_i(\delta^+)}{\gamma_i(\delta^-)}\exp(-z_i\Delta\psi_{D\delta})\exp(-z_i^2\Delta W_{DE\delta})\exp\left(\frac{\overline{V}_i\Delta\pi_\delta}{RT}\right) \tag{3.19}$$

Apparently, five contributions are responsible for the ion partitioning: steric exclusion, represented by the coefficient ϕ_i (see Table 3.2), nonideality of the solution, represented by the activity coefficients γ_i, Donnan equilibrium, through $\Delta\psi_D$, DE, through ΔW_{DE}, and the pressure difference at the interface [Eq. (3.5)].

3.5.1.1.1 Donnan equilibrium and dielectric exclusion

Donnan equilibrium plays a relevant role in the case of charged membranes (Fig. 3.1B): it always offers favorable partitioning for counter-ions and unfavorable partitioning for coions.

The mechanism of DE of ions from NF membranes is due to a series of concomitant electric effects; it is typically recognized that two main phenomena are involved: the "image forces" and the "Born exclusion." The DE partition is always unfavorable for any ion, regardless of its sign; it might be remarkably relevant also in the case of uncharged membranes, whereas electric exclusion related to Donnan equilibrium becomes negligible.

The "image forces" phenomenon arises as a consequence of a polarization charge located at the interface between two different dielectric media: the polymeric membrane and the liquid solution. Since the dielectric constant of the liquid is much higher than the corresponding value of the membrane, the polarization charge assumes the same sign of the ion, which is represented as a force, image of the ion itself, acting as a repulsion for ion entrance. A scheme is reported in Fig. 3.1C, in which the phenomenon is represented as a "fictitious" charge on the pore wall of the same sign of the ion (the "image"), which rejects the ion itself. DE has been widely studied by Yaroshchuk (2000), who provided a comprehensive description of the mechanism and developed simple relationships to express the ion polarization energy in order to describe the repulsion in cylindrical and in slit-like pore geometry. For both geometries, the effect of the image forces is located at the pore mouth, assuming the separation effect of the polarization charges along the membrane pores to be negligible.

The summary of the relationships used by several authors (Bowen & Welfoot, 2002a; Bandini & Vezzani, 2003; Szymczyk & Fievet, 2005) to express the excess solvation energies is reported in Table 3.3.

The Born exclusion is due to the variation of the solvent dielectric properties inside the membrane pores (ε_{rp}) with respect to the external bulk values (ε_{rs}), which are caused by an alteration of the solvent structure when the solvent is confined in small narrow pores. Several authors (Bowen & Welfoot, 2002a; Brovchenko & Geiger, 2002; Hall et al., 1997; Hartnig, Witschel, & Spohr, 1998; Senapati & Chandra, 2001; Spohr, Hartnig, Gallo, & Rovere, 1999; Szymczyk & Fievet, 2005; Yaroshchuk, 2000; Zhang, Davis, Kroll, & White, 1995). Bowen and Welfoot (2002a) ascribed this variation to a rigid orientation of the water dipole occurring very close to the pore walls; other authors (Deon et al., 2009; Déon et al., 2013; Escoda et al., 2011; Oatley et al., 2012; Oatley-Radcliffe et al., 2014; Szymczyk & Fievet, 2005) demonstrated that it was dependent on the electrolyte type and on the composition.

The difference ($\varepsilon_{rp} < \varepsilon_{rs}$) implies a different ion solvation inside the membrane with respect to the condition existing in the unconfined space. Since the solvation energy of an ion is higher at low values of the dielectric constant, the ion should overcome an energy barrier to enter the pore, as it is represented in Fig. 3.1C, which is typically described with a solvation energy barrier as suggested by the Born model [Eqs. (3.1)–(3.2) of Table 3.3].

The original relationship developed by Born accounted for the influence of the solvent on the solute radius; as a simplification, some authors assume a constant value for the ionic radius, generally taken as coincident with the Stokes radius, although other authors suggested instead the choice of different dimensions, such as the Pauling radius (as Born did) or the radius of the cavity formed by the ion in the solvent (Szymczyk & Fievet, 2005).

Apart from the greater mathematical complexity of the equations describing the "image force" repulsion with respect to the simpler Born partitioning, it is important to observe that ΔW_{im} depends on the Debye length inside the pores that means on the ionic strength inside the pores. Its evaluation requires to select/define the pore shape and to calculate a mean pore radius. Conversely, the Born energy depends on the solvent properties (ε_{rp}, ε_{rs}) and on the ionic radius only, but it does not depend on the pore geometry nor on the ionic concentration inside the membrane.

TABLE 3.3 Dimensionless excess solvation energies for dielectric exclusion phenomena.

Dielectric exclusion

$$\Delta W_{DE} = \Delta W_{im} + \Delta W_{Born}$$

Born exclusion

$$\Delta W_{Born,i}(0) = \frac{e^2}{8\pi\varepsilon_0 k_B T r_i} \cdot \left(\frac{1}{\varepsilon_{rp}(0^+)} - \frac{1}{\varepsilon_{rs}}\right) \quad (3.1)$$

$$\Delta W_{Born,i}(\delta) = \frac{e^2}{8\pi\varepsilon_0 k_B T r_i} \cdot \left(\frac{1}{\varepsilon_{rp}(\delta^-)} - \frac{1}{\varepsilon_{rs}}\right) \quad (3.2)$$

Image force exclusion

Cylindrical pores

$$\Delta W_{im}(0) = \frac{2r_B(0^+)}{\pi r_p}\int_0^{\infty} \frac{K_0(k)K_1(v(0)) - \tilde{\beta}_0(k)K_0(v(0))K_1(k)}{I_1(v(0))K_0(k) + \tilde{\beta}_0(k)I_0(v(0))K_1(k)}dk \quad (3.3)$$

$$\Delta W_{im}(\delta) = \frac{2r_B(\delta^-)}{\pi r_p}\int_0^{\infty} \frac{K_0(k)K_1(v(\delta)) - \tilde{\beta}_\delta(k)K_0(v(\delta))K_1(k)}{I_1(v(\delta))K_0(k) + \tilde{\beta}_\delta(k)I_0(v(\delta))K_1(k)}dk \quad (3.4)$$

in which

$$\tilde{\beta}_0(k) = \frac{k}{v(0)}\frac{\varepsilon_{rm}}{\varepsilon_{rp}(0^+)}; \ \tilde{\beta}_\delta(k) = \frac{k}{v(\delta)}\frac{\varepsilon_{rm}}{\varepsilon_{rp}(\delta^-)}$$

$v^2(0) = k^2 + (r_p\kappa(0^+))^2$; $v^2(\delta) = k^2 + (r_p\kappa(\delta^-))^2$;
I_0, I_1, K_0, K_1: modified Bessel functions

Image force – slit-like pores

$$\Delta W_{im}(0) \cong -r_B(0^+)\frac{1}{r_p}\ln\left[1 - \gamma_{DE}(0^+)\exp(-2r_p\kappa(0^+))\right] \quad (3.5)$$

$$\Delta W_{im}(\delta) \cong -r_B(\delta^-)\frac{1}{r_p}\ln\left[1 - \gamma_{DE}(\delta^-)\exp(-2r_p\kappa(\delta^-))\right] \quad (3.6)$$

in which

$$r_B(0^+) = \frac{F^2}{8\pi\varepsilon_0\varepsilon_{rp}(0^+)RT\,N_A}; \ r_B(\delta^-) = \frac{F^2}{8\pi\varepsilon_0\varepsilon_{rp}(\delta^-)RT\,N_A}$$

$$\gamma_{DE}(0^+) = \frac{1 - \varepsilon_{rm}/\varepsilon_{rp}(0^+)}{1 + \varepsilon_{rm}/\varepsilon_{rp}(0^+)}; \ \gamma_{DE}(\delta^-) = \frac{1 - \varepsilon_{rm}/\varepsilon_{rp}(\delta^-)}{1 + \varepsilon_{rm}/\varepsilon_{rp}(\delta^-)}$$

$$\kappa^{-1}(0^+) = \frac{1}{F}\sqrt{\frac{\varepsilon_0\varepsilon_{rp}(0^+)RT}{2I(0^+)}}; \ \kappa^{-1}(\delta^-) = \frac{1}{F}\sqrt{\frac{\varepsilon_0\varepsilon_{rp}(\delta^-)RT}{2I(\delta^-)}}$$

3.5.1.2 *Total flux and membrane permeability*

The expressions that are typically used to describe the volume flux originate from combinations of phenomenological models characteristic of RO, where the assumptions of viscous motion inside the pores have been incorporated. A systematic analysis of the problem has been recently presented in (Bandini & Morelli, 2017), and the main results will be described in the following.

By considering a unidimensional problem and with reference to the scheme of Fig. 3.1A, Eq. (3.16) can be integrated on the two internal membrane interfaces accounting that under steady-state conditions the total volume flux is constant and assuming that the viscous flow parameter (k_D) is constant along the pore length.

A linear pressure profile is obtained straightforwardly. By introducing the osmotic equilibrium reported as Eq. (3.5), the pressure difference inside the membrane ($\Delta\overline{p}$) can be related to the total pressure difference kept across the membrane (ΔP) at the external membrane, thereby obtaining the relationships (3.20):

$$\left.\begin{array}{l} J_v = -\dfrac{k_D}{\eta}\dfrac{d\overline{p}}{dx'} \Rightarrow J_v = \dfrac{k_D}{\eta\delta}\Delta\overline{p} \\ \Delta P = p(0^-) - p(\delta^+) \\ \quad = \Delta\overline{p} + \Delta\pi_0 - \Delta\pi_\delta \end{array}\right] \Rightarrow \begin{array}{l} J_v = L_p(\Delta P - \sigma_v \Delta\pi) \\ L_p(T, c_{\text{inside}}) = \dfrac{k_D}{\eta(T, c_{\text{inside}})\delta};\ \sigma_v = \dfrac{\Delta\pi_0 - \Delta\pi_\delta}{\Delta\pi} \end{array} \tag{3.20}$$

in which L_p represents the membrane permeability and η represents the dynamic viscosity of the solution inside the membrane. The correspondence of Eq. (3.20) with the phenomenological approach represented by Eq. (3.13) is self-explanatory.

First, an explanation of the meaning of the Staverman reflection coefficient can be obtained. When Van't Hoff equation is used to calculate the osmotic pressure (generally valid in NF of many substances), the following relationships can be easily derived:

$$\begin{array}{l} \dfrac{\Delta\pi_0 - \Delta\pi_\delta}{RT} \simeq \displaystyle\sum_{i=1}^{N} c_i(0^-)(1-\Gamma_{i0}) - \sum_{i=1}^{N} c_i(\delta^+)(1-\Gamma_{i\delta}) \\ \Delta\pi_0 - \Delta\pi_\delta \simeq \displaystyle\sum_{i=1}^{N} \pi_i(0^-)(1-\Gamma_{i0}) - \sum_{i=1}^{N} \pi_i(\delta^+)(1-\Gamma_{i\delta}) \end{array} \tag{3.21}$$

and the final relationship for the volume flux is then obtained as:

$$\begin{array}{l} J_v = L_p(\Delta P - \sigma_v \Delta\pi);\ \sigma_v \Delta\pi \simeq \displaystyle\sum_{i=1}^{N} \sigma_{vi} \Delta\pi_i \\ \underline{\sigma_{vi} \simeq 1 - \Gamma_{i0}\pi_i(0^-) - \Gamma_{i\delta}\pi_i(\delta^+)} \\ \qquad\qquad \Delta\pi_i \end{array} ;\ \Delta\pi_i = \pi_i(0^-) - \pi_i(\delta^+) \tag{3.22}$$

Second, the porous vision details the membrane permeability as a geometrical-transport parameter. For instance, in case of Hagen–Poiseuille motion across a cylindrical pore, $k_D = r_p^2/8$, whereas in the case of a slit-like geometry, $k_D = r_p^2/3$. However, none of these relationships can be used as a predictive tool of the membrane permeability, since neither the cylindrical nor the slit-like geometry match with the actual pore size of a NF membrane. Moreover, they do not account for the pore-size distribution, and the value of the effective thickness cannot be known with precision. However, the results of Eq. (3.20) can be used as an indication of which are the parameters determining the membrane permeability, which is a clear combination of geometrical characteristics and of operative conditions contained in the viscosity of the solution flowing inside the pores (η).

With regard to the viscosity inside the pore, the assumption of bulk solvent properties may not be valid within narrow pores. Bowen and Welfoot (2002a) proposed to increase the water viscosity inside the pore of a factor of 10 with respect to the corresponding unconfined value ($\eta_w/\eta_w^0 = 10$) as a consequence of a rigid solvent structure close to the pore wall; the same ratio can be estimated for aqueous solutions as well.

Eq. (3.20) can be also used to express the pure water flux, and the corresponding hydraulic permeability (L_{pw}) can be defined straightforwardly. When the membrane geometrical parameters are not affected by composition nor by the solution type (no swelling or other effects), the following relationships hold true (Bandini & Morelli, 2017):

$$L_p(T, c_{\text{inside}}) \cdot \eta(T, c_{\text{inside}}) = \frac{k_D}{\delta} \simeq \text{const} \Rightarrow$$
$$L_p(T, c_{\text{inside}}) \cdot \eta^0(T, c_{\text{inside}}) = L_{\text{pw}}(T_0) \cdot \eta_w^0(T_0) \quad (3.23)$$

Eq. (3.23) is a reconfirmation of an Arrhenius-type trend typically observed in RO membranes, contained in the viscosity dependence on temperature. Eq. (3.23) can be used to estimate the permeability and the hydraulic permeability as well, once an experimental value of L_{pw} is available at a reference temperature T_0. Conversely, it can be also used to quantify possible swelling effects, as a deviation from the behavior represented by the equation itself.

3.6 Application of Donnan-Steric-pore-and-dielectric-exclusion modeling in nanofiltration: case studies

The following discussion will be focused on the cases of neutral solutes and of electrolytes mixtures. The general DSPM-DE model is adapted for each case: the basic equations are developed, and the typical approximations are presented. The procedures for membrane parameters' calculations are introduced and a detailed discussion on the correct method for data elaboration is finally presented.

3.6.1 Neutral solutes

In the case of neutral solutes, Eq. (17) can be simplified by neglecting the variation of the activity coefficient along the axial coordinate; it can be re-elaborated in order to obtain a relationship for the concentration gradient, as reported in Eq. (4.2) of Table 3.4. That equation can be integrated along the membrane effective thickness, coupled with the partitioning Eqs. (4.18) and (4.19), where only steric exclusion (ϕ_i) and nonideality of the solution (γ_i) are considered. The relatively simple expression for the real rejection can be finally obtained, as represented by Eq. (4.6) of Table 3.4.

Accounting also for Eqs. (3.20)–(3.22) to describe the total volume flux, the complete set of equations reported in Table 3.4 can be obtained, which characterize the DSPM-DE model of neutral solutes, sometimes abbreviated as Steric Pore Model (Bandini & Morelli, 2017).

Apparently, for a specific membrane (that means for a known value of the hydraulic permeability), the solute rejection depends on the pore geometry, on the hydrodynamic coefficients λ_i, included in the hindrance factors, and on the solute concentration through γ_i. Pe_i has the meaning of a hindered Péclet number that is independent of the effective

TABLE 3.4 Basic equations and approximations of the Donnan-steric-pore-and-dielectric-exclusion model for neutral solutes.

Parameters

Pore geometry, L_p, δ, λ_i

Solute flux

$$J_i = J_v c_i(\delta^+) \tag{4.1}$$

Concentration gradient through the membrane:

$$\frac{dc_i}{dx'} = \frac{J_v}{D_{\text{ip}}}\left[(K_{\text{ic}} + Y_i)c_i - c_i(\delta^+)\right] = \frac{\text{Pe}_i}{\delta}\left[c_i - \frac{c_i(\delta^+)}{K_{\text{ic}} + Y_i}\right] \tag{4.2}$$

$$Y_i = \frac{D_{\text{ip}}\overline{V}_i}{\text{RT}}\frac{\eta}{k_D} = \frac{D_{\text{ip}}\overline{V}_i}{\text{RT}}\frac{1}{L_p \cdot \delta} \tag{4.3}$$

Partitioning at the membrane/external solutions interfaces

$$\Gamma_{i0} = \frac{c_i(0^+)}{c_i(0^-)} = \phi_i \frac{\gamma_i(0^-)}{\gamma_i(0^+)} \simeq \phi_i \text{ feed/membrane} \tag{4.4}$$

$$\Gamma_{i\delta} = \frac{c_i(\delta^-)}{c_i(\delta^+)} = \phi_i \frac{\gamma_i(\delta^+)}{\gamma_i(\delta^-)} \simeq \phi_i \text{ membrane/permeate} \tag{4.5}$$

Real rejection

$$R_{\text{real},i} = 1 - \frac{c_i(\delta^+)}{c_i(0^-)} = 1 - \frac{(K_{\text{ic}} + Y_i)\Gamma_{i0}}{1 - [1 - (K_{\text{ic}} + Y_i)\Gamma_{i\delta}]e^{-\text{Pe}_i}} \simeq 1 - \frac{K_{\text{ic}}\phi_i}{1 - (1 - K_{\text{ic}}\phi_i)e^{-\text{Pe}_i}} \tag{4.6}$$

Asymptotic rejection

$$R_{\text{real},i}^{\text{asym}} = 1 - (K_{\text{ic}} + Y_i)\Gamma_{i0} = \sigma_{\text{si}} \simeq 1 - K_{\text{ic}}\phi_i \tag{4.7}$$

Total flux

$$J_v = L_p \Delta P_{\text{eff}} = L_p(\Delta P - \sigma_v \Delta\pi)$$
$$\sigma_v \Delta\pi \simeq \sum_{i=1}^{N} \sigma_{\text{vi}} \Delta\pi_i; \sigma_{vi} = 1 - \frac{\Gamma_{i0}\pi_i(0^-) - \Gamma_{i\delta}\pi_i(\delta^+)}{\Delta\pi_i} \simeq 1 - \phi_i \tag{4.8}$$

Membrane permeability (no pore swelling with temperature nor concentration)

$$L_p(T, c_{\text{inside}}) \cdot \eta^0(T, c_{\text{inside}}) = L_{\text{pw}}(T_0) \cdot \eta_w^0(T_0) \tag{4.9}$$

Hydraulic permeability (no pore swelling with temperature)

$$L_{\text{pw}}(T)\eta_w^0(T) = L_{\text{pw}}(T_0)\eta_w^0(T_0) \tag{4.10}$$

in which

$$\text{Pe}_i = \frac{(K_{\text{ic}} + Y_i)J_v\delta}{D_{\text{ip}}} \simeq \frac{K_{\text{ic}}J_v\delta}{D_{\text{ip}}} \tag{4.11}$$

$$D_{\text{ip}} = K_{\text{id}} D_{\text{iw}}^0(T, c_{\text{inside}}) \tag{4.12}$$

$$c_{\text{inside}} = \sum_{i=1}^{N} \frac{c_i(0^+) + c_i(\delta^-)}{2} \simeq \sum_{i=1}^{N} \phi_i \frac{c_i(0^-) + c_i(\delta^+)}{2} \tag{4.13}$$

$$\Delta\pi = \pi(0^-) - \pi(\delta^+)$$

$\phi_i, K_{\text{ic}}, K_{\text{id}}$ are functions of λ_i as defined in Table 3.2

membrane thickness δ, owing to the porous vision intrinsic in Eq. (3.23). Moreover, it is important to observe that the asymptotic rejection is also independent of the membrane thickness.

Remarkably, the model predicts a unique curve, which relates rejection as a function of Pe_i (Eq. (4.6)–Table 3.4), according to an equation quite similar to the corresponding relationship of the three-parameter model for RO (Eq. (1.4)–Table 3.1). It is easy to demonstrate that, when $\Gamma_{i0} \simeq \Gamma_{i\delta}$, Eq. (4.6) degenerates into Eq. (1.4) accounting for $\sigma_{si} = [1 - (K_{ic} + Y_i)\Gamma_{i0}]$. Thus the general validity of Spiegler–Kedem model is interestingly evident.

The DSPM-DE model for neutral solutes is typically used in approximated forms, depending on the chemical properties of the molecules. Many studies on the NF of oligosaccharides solutions are reported in the literature (Bandini & Morelli, 2017, 2018; Bandini & Nataloni, 2015; Bargeman et al., 2005, 2014; Bouchoux et al., 2005; Catarino et al., 2008; Escoda et al., 2011; Feng et al., 2009; Goulas et al., 2002; Li et al., 2004; Pinelo et al., 2009; Qi et al., 2011; Rizki et al., 2019; Sjöman et al., 2007, 2008; Zhang et al., 2011). In these cases, owing to the relatively low molar fraction of the solutes, the partitioning coefficients are solely related to steric hindrance effects, and the contribution of the pressure gradient inside the membrane, represented by the term Yi of Table 3.4, is often neglected.

Although the general equation [Eq. (4.6)–Table 3.4] is rather elegant, the role of the parameter Yi seems negligible. An accurate evaluation of Yi is not easy to perform, since the partial molar volume of the solute can be affected by the confinement. However, for a typical polyamide membrane with glucose–water solutions at 30°C (Bandini & Morelli, 2017), assuming for $\overline{V} = 110\ \text{cm}^3/\text{mol}$ as suggested by Bowen and Welfoot (2002a), Yi can be estimated close to 0.11, with a corresponding value of $K_{ic} = 0.70$ and an asymptotic rejection of 98.9%. Whereas for negligible Yi ($Yi = 0$) the asymptotic rejection is predicted as 98.7%.

Finally, it is important to observe that the model allows the evaluation of the reflection coefficients. In Bandini and Morelli (2017), the extent of the differences between the Staverman and the solute reflection coefficients as a function of the hydrodynamic coefficients were presented in a detailed discussion. At $\lambda_i \geq 0.8$, which is a typical value of oligosaccharides, the parameters approach the same values, and, as a consequence, the Staverman coefficient can be estimated as the asymptotic rejection. Conversely, at $\lambda_i = 0.4$, Eqs. (4.8)–Table 3.4 predicts $\sigma_s = 0.3$ and $\sigma_v = 0.4$ for a slit-like pore geometry, while it predicts $\sigma_s = 0.52$ and $\sigma_v = 0.67$ for a cylindrical pore geometry. It is recommended to account for the precise values, especially in the evaluation of the effective driving force (ΔP_{eff}) at very low fluxes.

3.6.1.1 Statement of the problem

The basic equations of the DSPM-DE model for neutral solutes are summarized in Table 3.4. A set of N + 1 equations [N equations as (4.6) for each solute] and the total flux [Eq. (4.8)] should be solved, assuming as known variables the feed conditions [concentration $c_i(0^-)$, pH, temperature] and the pressure difference across the membrane ΔP.

The model structure allows an easy extension to dilute multicomponent solutions, since the multicomponent interdiffusion inside the membrane is neglected.

The model contains morphological parameters (membrane pore geometry and the effective membrane thickness, δ), a transport parameter (the hydraulic membrane permeability, L_{pw}), and the hydrodynamic coefficient of the solute (λ_i), which includes an overall information about the dimension of the molecule moving inside the narrow pores of the membrane.

The hydrodynamic coefficients are traditionally calculated as the ratio between the Stokes radius of the molecule and the mean pore size (r_p) of the membrane. That description introduces the additional geometrical parameter r_p, which should account also for the pore size distribution and requires the knowledge of the Stokes radius.

The use of Stokes radius can be criticized since the molecular shape is not generally spherical and the hydration shell can be affected by temperature, composition and by the presence of electrolytes (salting-out). In Bandini and Morelli (2017), the problem has been bypassed avoiding the introduction of the pore radius, and assuming the hydrodynamic coefficient of the solute as a binary interaction parameter including solute/membrane interactions during the motion along the membrane pores.

Finally, the study presented in Bandini and Morelli (2017) demonstrated that when no membrane swelling occurs with temperature nor composition, the membrane permeability (L_p) can be calculated by Eqs. (4.9)—Table 3.4 after the experimental determination of the hydraulic permeability at a reference temperature [$L_{pw}(T_0)$].

The model equations of Table 3.4 can be solved to simulate membrane performances, when the model parameters are known. The model parameters, however, cannot be predicted basing on chemical-physical and geometrical properties; they must be considered as adjustable parameters to be determined by a fitting procedure on experimental data. The type of experimental data and the minimum number of trials necessary for membrane characterization is still an open challenge, which is addressed by various authors in different ways. A critical review has been presented in Bandini and Morelli (2017).

3.6.1.1.1 Parameters calculations procedures

The calculations of parameters lay on the availability on experimental data of real rejection versus volume flux. As to cover a broad range of rejection values, the data must be obtained at different concentrations and temperatures in a pressure field interval as large as possible. For several neutral single solutes, such as oligosaccharides, it has been found that the data can be plotted on a single curve, generally independent of the solute concentration and pH, whereas a slight dependence on the temperature has been observed. A schematic drawing of this case is reported in Fig. 3.2A.

In the case of mixtures with electrolytes, however, an effect of the salt composition has been observed by some authors (salting-out effect).

Following the theoretical basis represented by Eq. (4.6)—Table 3.4, it is simple to observe that the asymptotic rejection depends exclusively on (λ_i), whereas the slope of the curves at low flux values depends on the combined effect of the effective thickness and of the hindered diffusivity, which itself can depend on temperature and/or composition.

To avoid misunderstandings in the conclusions, it is fundamental to adopt a method for data processing that allows to understand and/or distinguish and/or separate the effects, as introduced in Bandini and Morelli (2017). The criterion requires the determination of the effective pressure difference, which can be calculated accounting for the concentration polarization effects. From a plot of the experimental volume flux versus ΔP_{eff}, a linear trend is typically obtained and the corresponding slope can be calculated, as represented in Fig. 3.4B.

NEUTRAL SOLUTES

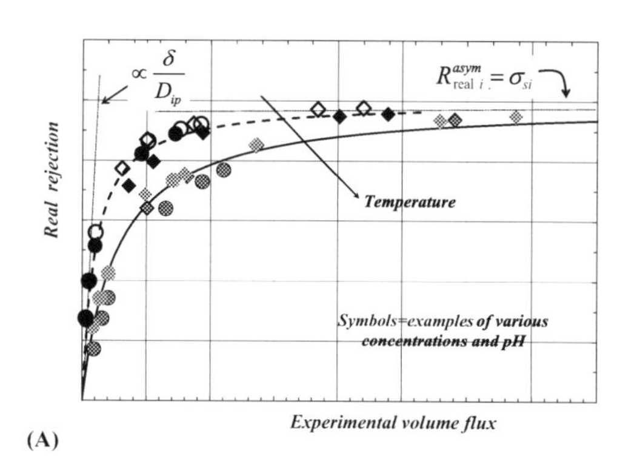

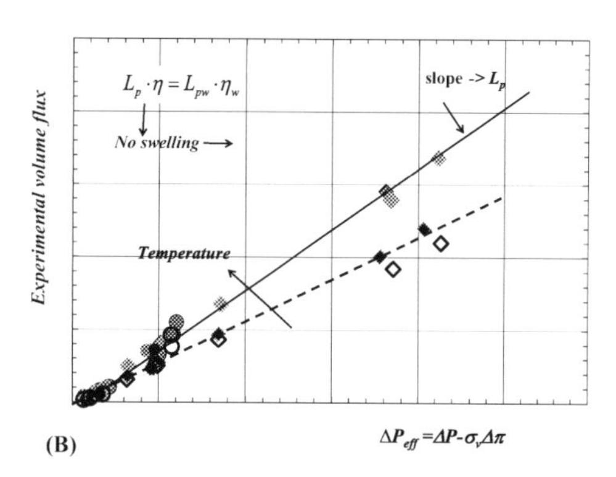

FIGURE 3.4 Neutral solutes. Temperature effect on NF membranes performances with neutral solutes: qualitative example of data elaboration and meaning of the parameters reported in Table 3.4. (A) real rejection data versus experimental volume flux; (B) experimental volume flux versus the effective driving force.

The comparison of data reported in Figs. 3.4A and B allows to draw the correct conclusion on the effect of temperature and of the solute type (for example) on the membrane permeability. When the L_p value, calculated as the slope of the data plotted as in Fig. 3.4B, matches with the value calculated according to the Eqs. (4.9)–(4.10)–Table 3.4, it can be easily concluded that no swelling effects occur. As a consequence, it can be concluded that the geometrical parameters of the membrane are not affected by the operative conditions [as described by relationships (23)]; therefore the shape of rejection curves of Fig. 3.4A cannot be ascribed to different values of the membrane thickness, or, more commonly, to a variation of the parameter (k_D/δ) but on the temperature effect on the hydrodynamic coefficients. That conclusion is also in agreement with the behavior of the sugar hydration number that typically decreases with temperature.

When the comparison is negative, information on the swelling presence can be derived. A detailed discussion has been reported in Bandini and Morelli (2017), and reference should be made to that paper for a complete documentation. The criterion can be obviously applied for membrane characterization with oligosaccharides and electrolytes mixtures as well.

The key steps required for the membrane characterization and for the parameter calculation in the case of neutral solutes, with the corresponding recommended experiments, are therefore the following:

1. perform experiments with pure water at two different temperatures and calculate the hydraulic permeability (L_{pw}) by Eq. (4.8) and Table 3.4: check if swelling occurs by using the relationships of Eq. (4.10) and Table 3.4;
2. perform experiments with single solutes to obtain a rather complete rejection versus flux curve;
3. compare data plotted in the format (real rejection versus experimental flux) with the same data plotted in the format (experimental flux versus effective pressure difference): calculate the membrane permeability and check if the solution affects the membrane parameters and quantify the swelling, if it is the case;

4. select the pore geometry (cylindrical or slit-like);
5. calculate the hydrodynamic coefficients and membrane thickness, by fitting the couple (λ_i,δ) on the rejection versus flux curve, according to Eq. (4.6)—Table 3.4.

Finally, the λ_i value obtained at room temperature is typically used also to estimate a mean pore radius, which is necessary to process rejection data with electrolyte solutions (as it will be discussed in the next section).

3.6.2 Electrolyte solutions

In the case of electrolytes solutions, the problem is typically reformulated by accounting for Eq. (3.17) with some simplifications. Eq. (3.24) is used to describe the solute transport across the membrane, in which the effect of pressure gradient along the membrane is generally negligible, owing to the relatively low value of the partial molar volume of the ionic species [see Bowen and Welfoot (2002a) for details].

$$J_i \simeq -D_{\mathrm{ip}}\frac{dc_i}{dx'} - z_i c_i D_{\mathrm{ip}}\frac{F}{\mathrm{RT}}\frac{d\Psi}{dx'} + K_{\mathrm{ic}}c_i J_v \tag{3.24}$$

In addition, since low molecular weight solutes greatly permeate the membrane, the contributions of the osmotic pressure at the interfaces can also be neglected in the solute partitioning Eqs. (3.18—3.19).

3.6.2.1 Statement of the problem

The basic equations of the DSPM-DE model for electrolytes solutions are summarized in Table 3.5. A set of N + 2 equations [N equations as Eq. (5.2) for each ionic species], one equation to calculate the electric potential gradient Eq. (5.3) and the total flux [Eq. (5.8)—Table 3.5] should be solved, assuming as known variables the feed conditions [concentration $c_i(0^-)$, pH, temperature] and the pressure difference across the membrane ΔP.

Eq. (5.3) can be obtained by performing the derivative of Eq. (5.4) with respect to the axial coordinate of the membrane, under the hypothesis of constant volumetric charge (X) along the membrane.

Depending of the solution concentration, activity coefficients can be evaluated according to the Debye—Hückel theory or its improvements by Davies, as it is summarized in Table 3.6. The ideal solution hypothesis is to be excluded for electrolytes solution, whereas experience shows that for NaCl solutions the limiting Debye—Hückel theory might be a good approximation; in the case of multivalent nonsymmetric electrolytes, more advanced models should be used, such as the Bromley equations (Zemaitis, Clark, Rafal, & Scrivner, 1986).

The numerical solution of the general equations of the DSPM-DE model leads to calculate the total volume flux and the concentration profile of each species along the membrane pore; the Donnan potentials at the interfaces, the electric potential profile along the axial coordinate, as well as real rejection of each ionic species can be obtained straightforwardly.

The solution of the general problem requires the knowledge of morphological and transport parameters, which describe the ion/membrane and the ion/water interactions in

TABLE 3.5 Basic equations of Donnan-steric-pore-and-dielectric-exclusion model for electrolytes solutions.

Parameters

pore geometry, $r_p, \delta, L_{pw}, r_i, \varepsilon_{rp}, X$

Solute flux

$$J_i = J_v c_i(\delta^+) \quad (5.1)$$

Concentration gradient through the membrane:

$$\frac{dc_i}{dx'} = \frac{J_v}{D_{ip}}\left[K_{ic}c_i - c_i(\delta^+)\right] - \frac{z_i c_i F}{RT}\frac{d\Psi}{dx'} = \frac{Pe_i}{\delta}\left[c_i - \frac{c_i(\delta^+)}{K_{ic}}\right] - \frac{z_i c_i F}{RT}\frac{d\Psi}{dx'} \quad (5.2)$$

Potential gradient through the membrane:

$$\frac{d\Psi}{dx'} = \frac{\sum_{i=1}^{N} z_i \frac{J_v}{D_{ip}}\left[K_{ic}c_i - c_i(\delta^+)\right]}{\frac{F}{RT}\sum_{i=1}^{n} z_i^2 c_i} = \frac{\sum_{i=1}^{N} z_i \frac{Pe_i}{\delta}\left[c_i - \frac{c_i(\delta^+)}{K_{ic}}\right]}{\frac{F}{RT}\sum_{i=1}^{n} z_i^2 c_i} \quad (5.3)$$

Electroneutrality conditions:

$$\sum_{i=1}^{N} z_i c_i + X = 0 \quad (5.4)$$

$$\sum_{i=1}^{N} z_i c_i(\delta^+) = 0 \quad (5.5)$$

Partitioning at membrane/external solutions interfaces:

$$\Gamma_{i0} = \frac{c_i(0^+)}{c_i(0^-)} = \phi_i \frac{\gamma_i(0^-)}{\gamma_i(0^+)} \exp(-z_i \Delta\psi_{D0}) \exp(-z_i^2 \Delta W_{DE0}) \quad (5.6)$$

$$\Gamma_{i\delta} = \frac{c_i(\delta^-)}{c_i(\delta^+)} = \phi_i \frac{\gamma_i(\delta^+)}{\gamma_i(\delta^-)} \exp(-z_i \Delta\psi_{D\delta}) \exp(-z_i^2 \Delta W_{DE\delta}) \quad (5.7)$$

Total flux

$$J_v = L_p \Delta P_{\text{eff}} = L_p(\Delta P - \sigma_v \Delta\pi)$$
$$\sigma_v \Delta\pi \simeq \sum_{i=1}^{N} \sigma_{vi} \Delta\pi_i; \ \sigma_{vi} = 1 - \frac{\Gamma_{i0}\pi_i(0^-) - \Gamma_{i\delta}\pi_i(\delta^+)}{\Delta\pi_i} \quad (5.8)$$

Membrane permeability (no pore swelling with temperature nor concentration)

$$L_p(T, c_{\text{inside}}) \cdot \eta^0(T, c_{\text{inside}}) = L_{pw}(T_0) \cdot \eta_w^0(T_0) \quad (5.9)$$

Hydraulic permeability (no pore swelling with temperature)

$$L_{pw}(T)\eta_w^0(T) = L_{pw}(T_0)\eta_w^0(T_0) \quad (5.10)$$

Real rejection

$$R_{\text{real},i} = 1 - \frac{c_i(\delta^+)}{c_i(0^-)} \quad (5.11)$$

(Continued)

TABLE 3.5 (Continued)

in which

$$Pe_i = \frac{J_v K_{ic} \delta}{D_{ip}} \quad (5.12)$$

$$D_{ip} = K_{id} D_{iw}^0 (T, c_{inside}) \quad (5.13)$$

$$c_{inside} = \sum_{i=1}^{N} \frac{c_i(0^+) + c_i(\delta^-)}{2} \quad (5.14)$$

$$\Delta\pi = \pi(0^-) - \pi(\delta^+)$$

ϕ_i, K_{ic}, K_{id} from Table 3.2; γ_i from Table 3.6; ΔW_{DE} from Table 3.3

TABLE 3.6 Activity coefficients of ions in electrolyte aqueous solutions.

	Activity coefficients	References
Debye–Hückel limiting law	$\gamma_i = \exp(-z_i^2(r_B \kappa))$	Bandini and Vezzani (2003), Szymczyk and Fievet (2005)
Debye–Hückel law	$\gamma_i = \exp\left(-z_i^2\left(r_B \frac{\kappa}{1+\kappa r_i}\right)\right)$	Szymczyk and Fievet (2005)
Davies equation	$\gamma_i = \exp\left[-z_i^2\left(\frac{r_B \kappa}{1+\sqrt{I}} - 0.3I\right)\right]$	Geraldes and Brites Alves (2008)
	$r_B = \frac{F^2}{8\pi\varepsilon_0\varepsilon_r RTN_A}$; $\kappa^{-1} = \frac{1}{F}\sqrt{\frac{\varepsilon_0\varepsilon_r RT}{2I}}$	

confined narrow pores. Contrary to the characterization procedure used for neutral solutes, it is necessary to introduce a parameter to characterize the mean pore dimension and the ionic species size, in order to quantify the excess energies contributions (Table 3.3).

Generally speaking, it is possible to evidence the following list:

1. membrane morphological parameters: pore geometry, mean pore radius (r_p), effective membrane thickness (δ);
2. membrane transport parameters: hydraulic permeability (L_{pw}) and/or the membrane permeability (L_p) in case of swelling phenomena;
3. membrane electrical parameter: the volume membrane charge (X), which is a function of the electrolytes types, of the concentration of each ionic species and of the pH in the feed side;
4. solute parameters: the Stokes radius (r_i) or the hydrodynamic coefficients (λ_i), in order to describe solute/membrane interactions, which can be a function of temperature;
5. solvent parameters: the dielectric constant of the solution inside the membrane (ε_{rp}), which is, in principle, a function of the electrolyte type and of the concentration of each ionic species inside the membrane.

Finally, by considering the dielectric phenomena described by the equations reported in Table 3.3, the complexity of the problem is evident, even in the simple case of aqueous solutions containing a single strong binary electrolyte. The complexity is not only mathematical, rather physical: several many phenomena are interdependent each other and it is thus difficult to elaborate a protocol of independent measurements to calculate the model parameters. In some cases, the mathematical complexity has been solved by developing approximated versions of the general model, based on the complete coion exclusion (Bandini & Vezzani, 2003), or on the hypothesis of constant potential gradient along the membrane (Bandini & Bruni, 2010; Bandini & Vezzani, 2003), or on the assumption of finite difference linearization of the concentration gradient (Bandini & Bruni, 2010; Bowen, Cassey, Jones, & Oatley, 2004; Bowen, Welfoot, & Williams, 2008). In addition, a rigorous analytical equation has been also obtained to calculate rejection of single salts in neutral membranes (Bandini & Bruni, 2010; Bandini & Vezzani, 2003).

Nowadays, the ability of the DSPM-DE model to predict NF performances of multicomponent aqueous electrolyte solution is well recognized: numerous examples are reported in literature in which negative rejections of various ions have been correctly simulated (Bandini & Vezzani, 2003; Bowen & Mukhtar, 1996; Cavaco Morão et al., 2008; Deon et al., 2009; Escoda et al., 2011). The great interest in this model is related to its facile extension to dilute multicomponent solutions, which made the DSPM-DE model one of the most studied in the last 20 years. Finally, the model allows the investigation of the role of each transport mechanism across the membrane in determining the ion flux. Bowen and Mohammad (1998) and other authors (Fievet et al., 2002; Szymczyk et al., 2003) obtained that for NaCl–water mixtures in negatively charged membranes, at high membrane charge values, the transport is controlled mainly by ordinary diffusion, although convection and electromigration contributions are not negligible. For membranes close to their isoelectric point, on the contrary, all contributions are remarkable.

3.6.2.1.1 Parameters calculation: discussion

The problem is still open. The most sensitive points are the definition of the experimental measurement protocol and the data processing criteria, with the objective to independently determine each parameter, basing on different sets of experimental data.

At present, most authors agree on the identification of the following steps for the calculations of the morphological and transport parameters of the membrane:

1. perform experiments with pure water and calculate the hydraulic permeability (L_{pw}) by Eq. (5.8)–Table 3.5;
2. choose the pore geometry: experience shows that in case of ionic species the slit-like geometry seems to be more suitable, whereas both the cylindrical and the slit shape can be used for neutral solutes;
3. perform experiments with neutral solutes to obtain a fairly complete rejection versus flux curve; recommendations are to select a rather spherical compound, such as glycerol or glucose, so that the Stokes radius is representative of the solute geometry;
   a. calculate the mean pore radius (r_p) and the effective membrane thickness (δ) by fitting the rejection versus flux data according to Eq. (4.6)–Table 3.4.

Conversely, for the calculation of the volume membrane charge (X) and of the dielectric constant of the solution inside the membrane (ε_{rp}), various protocols have been proposed, introducing different levels of approximations.

Bowen and Welfoot (2002a) followed by Oatley et al. (2012) estimated the water confinement inside the pore as the occurrence of a single oriented layer of water molecules, (as depicted in Fig. 3.1C). They proposed to calculate (ε_{rp}) by fitting the rejection data of single solutes at the isoelectric point of the membrane, by assuming the Born partitioning only, as the DE phenomenon. The membrane charge was then calculated as a function of the total concentration of the feed bulk at a fixed pH value. No evaluation of X as a function of pH was performed, nor tests with mixtures were proposed.

Bandini and Vezzani (2003) and in the following papers (Bandini et al., 2005; Mazzoni & Bandini, 2006; Mazzoni et al., 2007) proposed to calculate the membrane charge by fitting rejection versus flux data of single electrolytes as a function of concentration and pH, by assuming the image force as the unique DE phenomenon. Empirical relationships of X versus total concentration, according to a Freundlich isotherm, were used to simulate multicomponent mixtures with good success, following the original ideas of Tsuru, Nakao, and Kimura (1991) and Bowen and Mukhtar (1996). No well-defined mixing rules were proposed to predict the membrane charge.

Szymczyk and Fievet (2005) proposed to measure the membrane charge by streaming potential measurements and then to calculate the (ε_{rp}) as a function of solute type and of the feed concentration by fitting rejection versus flux data of single electrolytes, by accounting both for Born partitioning and for image force exclusion. No mixing rules for the prediction of membrane charge, nor the effect of pH were considered. Later, in 2010, the same authors proposed a new method for determining the dielectric constant inside the pores using membrane potential measurements (Escoda et al., 2010).

Finally, Deon et al. (2009) suggested to fit the membrane charge and the dielectric constant (ε_{rp}) with experimental data of ternary mixtures as to obtain the dependence with solute type and with the relative composition of the mixture in the feed side, by neglecting the contribution of image force. The same authors improved their method in the following years (Déon et al., 2013; Escoda et al., 2011) by including membrane potential measurements at high salt concentration to estimate the dielectric constant (ε_{rp}) in ternary mixtures as a function of the relative composition of the ionic species in the feed side and by estimating the volume membrane charge from zeta-potential measurements. No pH effect nor details on mixing rules for membrane charge were presented; however, the results reported in Déon et al. (2013) are very encouraging.

Apparently, all the authors agree with the need to estimate the effect of composition on the dielectric constant and on the membrane charge and with the need to develop protocols for membrane characterization independent of the rejections fitting by the partitioning-transport model proposed. However, they do not completely agree with the role of the DE phenomena. Finally, the remarkable effect of pH on the charge of the amphoteric membrane is not typically regarded as relevant to date, as it is evidenced on the contrary by the experience.

3.7 Conclusions and future trends

An overview on the most used models for RO and NF processes has been carried out, with a particular emphasis on the study of transport and partitioning phenomena in the past 30 years.

As it has been documented, most of the known models can be obtained by a rigorous derivation from the statistical–mechanical model introduced by Mason and Lonsdale in 1990, which derives from the general equations of nonequilibrium thermodynamics to the membrane case. The use of that approach is recommended to develop a structural model, where the physical meaning of the parameters is desired. A simple reformulation of the solution–diffusion with no need for approximations has been also performed.

The basic elements of the DSPM-DE, which is typically used for NF, have been introduced. Hindered mass transfer across the membrane is described according to a porous vision of the membrane and all the solute partitioning phenomena have been considered. It is easy to demonstrate that the typical RO models for binary salt–water mixtures are particular cases of the above-mentioned model.

The DSPM-DE model is quite general and the extension of the transport equations to multicomponent mixtures is rather easy. However, some points have not been solved yet.

The phenomena that rule the process are known, as the effects of the solution type and of its concentration on the membrane charge, together with the role of the nanoscale of the pore dimensions in the determination of the solvent properties that produce a decrease in the dielectric constant inside the pore with respect to the bulk value. However, neither the introduced protocols of measure nor the protocols of data processing seems valid to enable the prediction of the behavior of multicomponent mixtures from the information obtained on single electrolytes.

Moreover, the validation on large scale of the models developed to describe the mechanism of the membrane charge formation is still missing. There are no mixing rules of general validity to express the membrane charge and the dielectric constant in mixture as a function of the single species concentration and on the solution pH, starting from the available data of charge obtained with single salts.

Finally, owing to the kind of phenomena involved in transport and in partitioning mechanisms, which for some aspects approach the molecular scale (such as the phenomena in the electric double layer at the feed–membrane interface, the modifications of the water dipole by the pore walls, and so on), it could be possible to describe the phenomena involved starting from a microscopic level, following a molecular dynamics approach.

Appendix. Reformulation of the solution–diffusion model

Under the following hypotheses:

1. the transport mechanism is only diffusive (the convective contributions are neglected)
2. solute-solvent interdiffusion is neglected
3. the membrane is electrically neutral
4. the membrane is a high rejection membrane and the permeating stream is highly diluted

5. isothermal conditions

In the case of binary systems, Eq. (3.1) becomes:

$$\underline{n}_w = -c_w \frac{D_{wM}}{RT} \underline{\nabla}\mu_w\Big|_T$$
$$\underline{n}_s = -c_s \frac{D_{sM}}{RT} \underline{\nabla}\mu_s|_T \tag{A.1}$$

(1) Accounting of the Gibbs–Duhem Eq. (A.2) and of the osmotic pressure definition (A.3):

$$c_w \underline{\nabla}\mu_w|_T + c_s \underline{\nabla}\mu_s|_T = \underline{\nabla}p \tag{A.2}$$

$$\underline{\nabla}\pi = -\frac{RT\underline{\nabla}\ln a_w\Big|_T}{\tilde{V}^o_w} = -c^o_w \underline{\nabla}\mu_w|_T + \underline{\nabla}p \tag{A.3}$$

(2) Assuming the hypothesis of dilute solutions, Van't Hoff equation ($\underline{\nabla}\pi \simeq RT\underline{\nabla}c_s$) can be used, and the simplification ($c_w \tilde{V}^o_w = 1$) can be applied.

Eq. (A.1) can be elaborated as reported in the following:

$$\begin{aligned}\underline{n}_w &= -c_w \frac{D_{wM}}{RT} \underline{\nabla}\mu_w\Big|_T \\ &= -c_w \frac{D_{wM}}{RT}\left[\tilde{V}^o_w \underline{\nabla}p - \tilde{V}^o_w \underline{\nabla}\pi\right] \simeq -\frac{D_{wM}}{RT}\left[\underline{\nabla}p - \underline{\nabla}\pi\right]\end{aligned}$$

$$\begin{aligned}\underline{n}_s &= -c_s \frac{D_{sM}}{RT} \underline{\nabla}\mu_s\Big|_T \\ &= -\frac{D_{sM}}{RT}\left[\underline{\nabla}p - c_w \underline{\nabla}\mu_w\Big|_T\right] \\ &= -\frac{D_{sM}}{RT}\left[\underline{\nabla}p - c_w \tilde{V}^o_w \underline{\nabla}p + c_w \tilde{V}^o_w \underline{\nabla}\pi\right] \\ &\simeq -\frac{D_{sM}}{RT}\underline{\nabla}\pi \simeq -\frac{D_{sM}}{RT} RT\underline{\nabla}c_s \\ &\simeq -D_{sM}\underline{\nabla}c_s\end{aligned}$$

References

Afonso, M. D. (2006). Surface charge on loose nanofiltration membranes. *Desalination, 191*, 262–272.

Afonso, M. D., & de Pinho, M. N. (2000). Transport of $MgSO_4$, $MgCl_2$ and Na_2SO_4 across an amphoteric nanofiltration membrane. *Journal of Membrane Science, 179*, 137–154.

Afonso, M. D., Hagmeyer, G., & Gimbel, R. (2001). Streaming potential measurements to assess the variation of nanofiltration membranes surface charge with the concentration of salt solutions. *Separation and Purification Technology, 22–23*, 529–541.

Ahmad, A. L., Chong, M. F., & Bhatia, S. (2005). Mathematical modeling and simulation of the multiple solutes system for the nanofiltration process. *Journal Membrane Science, 253*, 103–115.

Anderson, J. L., & Quinn, J. A. (1974). Restricted transport in small pores. A model for steric exclusion and hindered particle motion. *Biophysics Journal, 14*, 130–150.

Ariza, M. J., & Benavente, J. (2001). Streaming potential along the surface of polysulfone membranes: a comparative study between two different experimental systems and determination of electrokinetic and adsorption parameters. *Journal of Membrane Science, 190*, 119–132.

Atra, R., Vatai, G., Bekassy-Molnar, E., & Balint, A. (2005). Investigation of ultra- and nanofiltration for utilization of whey protein and lactose. *Journal of Food Engineering, 67*, 325–332.

Bandini, S., & Bruni, L. (2010). Transport phenomena in nanofiltration membranes. In E. Drioli, & L. Giorno (Eds.), *Comprehensive membrane science and engineering* (vol. 2). Oxford: Elsevier.

Bandini, S., Drei, J., & Vezzani, D. (2005). The role of pH and concentration on the ion rejection in polyamide nanofiltration membrane. *Journal of Membrane Science, 264*, 65–74.

Bandini, S., & Morelli, V. (2017). Effect of temperature, pH and composition on nanofiltration of mono/disaccharides: Experiments and modelling assessment. *Journal Membrane Science, 533*, 57–74.

Bandini, S., & Morelli, V. (2018). Mass transfer in 1812 spiral wound modules: Experimental study in dextrose-water nanofiltration. *Separation and Purification Technology, 199*, 84–96.

Bandini, S., & Nataloni, L. (2015). Nanofiltration for dextrose recovery from crystallization mother liquors: A feasibility study. *Separation and Purification Technology, 139*, 53–62.

Bandini, S., & Vezzani, D. (2003). Nanofiltration modeling: the role of dielectric exclusion in membrane characterization. *Chemical Engineering Science, 58*, 3303–3326.

Bargeman, G., Vollenbroek, J. M., Straatsma, J., & Schroen, C. G. P. H. (2005). Nanofiltration of multi-component feeds. Interactions between neutral and charged components and their effect on retention. *Journal of Membrane Science, 247*, 11–20.

Bargeman, G., Westerink, J. B., Guerra Miguez, O., & Wessling, M. (2014). The effect of NaCl and glucose concentration on retentions for nanofiltration membranes processing concentrated solutions. *Separation and Purification Technology, 134*, 46–57.

Bird, R. B., Stewart, W. E., & Lightfoot, E. N. (2007). *Transport phenomena* (2nd ed.). New York, NY: Wiley.

Bouchoux, A., Roux-de Balmann, H., & Lutin, F. (2005). Nanofiltration of glucose and sodium lactate solutions. Variations of retention between single- and mixed-solute solutions. *Journal of Membrane Science, 258*, 123–132.

Bowen, W. R., Cassey, B., Jones, P., & Oatley, D. L. (2004). Modelling the performance of membrane nanofiltration—Application to an industrially relevant separation. *Journal of Membrane Science, 242*, 211–220.

Bowen, W. R., & Doneva, T. A. (2000). Atomic force microscopy studies of nanofiltration membranes: surface morphology, pore size distribution and adhesion. *Desalination, 129*, 163–172.

Bowen, W. R., & Mohammad, A. W. (1998). Characterization and prediction of nanofiltration membrane performance—A general assessment. *Chemical Engineering Research and Design, 76A*, 885–893.

Bowen, W. R., Mohammad, A. W., & Hilal, N. (1997). Characterisation of nanofiltration membranes for predictive purposes — Use of salts, uncharged solutes and atomic force microscopy. *Journal of Membrane Science, 126*, 91–105.

Bowen, W. R., & Mukhtar, H. (1996). Characterization and prediction of separation performance of nanofiltration membranes. *Journal of Membrane Science, 112*, 263–274.

Bowen, W. R., & Sharif, A. O. (1994). Transport through microfiltration membranes—Particle hydrodynamics and flux reduction. *Journal of Colloid interface Science, 168*(2), 414–421.

Bowen, W. R., & Welfoot, J. S. (2002a). Modelling the performance of membrane nanofiltration – Critical assessment and model development. *Chemical Engineering Science, 57*, 1121–1137.

Bowen, W. R., & Welfoot, J. S. (2002b). Modelling the performance of membrane nanofiltration – Pore size distribution effects. *Chemical Engineering Science, 57*, 1393–1407.

Bowen, W. R., Welfoot, J., & Williams, P. M. (2008). Linearized transport model for nanofiltration: Development and assessment. *AIChE Journal, 48*, 760–773.

Boy, V., Roux-de Balmann, H., & Galier, S. (2012). Relationship between volumetric properties and mass transfer through NF membrane for saccharide/electrolyte systems. *Journal of Membrane Science, 390–391*, 254–262.

Brovchenko, I., & Geiger, A. (2002). Water in nanopores in equilibrium with a bulk reservoir — Gibbs ensemble Monte Carlo simulations. *Journal of Molecular Liquids, 96–97*, 195–206.

Bruni, L., & Bandini, S. (2008). The role of the electrolyte on the mechanism of charge formation in polyamide nanofiltration membranes. *Journal Membrane Science, 308*, 136–151.

Bungay, P. M., & Brenner, H. (1973). The motion of a closely-fitting sphere in a fluid-filled tube. *International Journal of Multiphase Flow, 1*, 25–56.

Catarino, I., Minhalma, M., Beal, L. L., Mateus, M., & de Pinho, M. N. (2008). Assessment of saccharides fractionation by ultrafiltration and nanofiltration. *Journal of Membrane Science, 312*, 34–40.

Cavaco Morão, A. I., Szymczyk, A., Fievet, P., & Brites Alves, A. M. (2008). Modelling the separation by nanofiltration of ionic solution relevant to an industrial process. *Journal of Membrane Science, 322*, 320–330.

Chandrapala, J., Duke, M. C., Gray, S. R., Weeks, M., Palmer, M., & Vasiljeciv, T. (2016). Nanofiltration and nanodiafiltration of acid whey as a function of pH and temperature. *Separation and Purification Technology, 160*, 18–27.

Childress, E., & Elimelech, M. (1996). Effect of solution chemistry on the surface charge of polymeric reverse osmosis and nanofiltration membranes. *Journal of Membrane Science, 119*, 253–268.

Condom, S., Larbot, A., Youssi, S. A., & Persin, M. (2004). Use of ultra- and nanofiltration ceramic membranes for desalination. *Desalination, 168*, 207–213.

Cuartas-Uribe, B., Alcaina-Miranda, M. I., Soriano-Costa, E., Mendoza-Roca, J. A., Iborra-Clar, M. I., & Lora-Garcia, J. (2009). A study of the separation of lactose from whey ultrafiltration permeate using nanofiltration. *Desalination, 241*, 244–255.

Dechadilok, P., & Deen, W. M. (2006). Hindrance factors for diffusion and convection in pores. *Industrial Engineering Chemistry Research, 45*, 6953–6959.

Deen, W. M. (1987). Hindered transport of large molecules in liquid-filled pores. *AIChE Journal, 33*, 1409–1425.

Deen, W. M., Satvat, B., & Jamieson, J. M. (1980). Theoretical model for glomerular filtration of charged solutes. *American Journal of Physiology, 238*, F126–F139.

De Lint, W. B. S., Biesheuvel, P. M., & Verweij, H. (2002). Application of the charge regulation model to transport of ions through hydrophilic membranes: One-dimensional transport model for narrow pores (nanofiltration). *Journal of Colloid and Interface Science, 251*, 131–142.

Deon, S., Dutournie, P., Limousy, L., & Bourseau, P. (2009). Transport of salt mixtures through nanofiltration membranes: Numerical identification of electric and dielectric contributions. *Separation and Purification Technology, 69*, 225–233.

Dresner, L. (1972). Some remarks on the integration of extended Nernst-Planck equation in the hyperfiltration of multicomponent solutions. *Desalination, 10*, 27–46.

Déon, S., Escoda, A., Fievet, P., & Salut, R. (2013). Prediction of single salt rejection by NF membranes: An experimental methodology to assess physical parameters from membrane and streaming potentials. *Desalination, 315*, 37–45.

Ernst, M., Bismark, A., Springler, J., & Jekel, M. (2000). Zeta-potential and rejection rates of a polyethersulfone nanofiltration membrane in single salt solutions. *Journal of Membrane Science, 165*, 251–259.

Escoda, A., Deon, S., & Fievet, P. (2011). Assessment of dielectric contribution in the modelling of multi-ionic transport through nanofiltration membranes. *Journal of Membrane Science, 378*, 214–223.

Escoda, A., Lanteri, Y., Fievet, P., Deon, S., & Szymczyk, A. (2010). Determining the dielectric constant inside pores of nanofiltration membranes from membrane potential measurements. *Langmuir: The ACS Journal of Surfaces and Colloids, 26*(18), 14628–14635.

Faxen, H. (1922). The resistance to a motion of a solid sphere in a viscous liquid enclosed between parallel walls. *Annalen der Physik, 68*(10), 89–119.

Feng, Y. M., Chang, X. L., Wang, W. H., & Ma, R. Y. (2009). Separation of galacto-oligosaccharides mixuture by nanofiltration. *Journal of Taiwan Institute of Chemical Engineers, 40*, 326–332.

Ferry, J. D. (1936). Statistical evaluation of sieve constants in ultrafiltration. *Journal of General Physiology, 20*, 95–104.

Fievet, P., Aoubiza, B., Szymczyk, A., & Pagetti, J. (1999). Membrane potential in charged porous membranes. *Journal of Membrane Science, 160*, 267–275.

Fievet, P., Szymczyk, A., Aoubiza, B., & Pagetti, J. (2000). Evaluation of three methods for the characterisation of the membrane–solution interface: Streaming potential, membrane potential and electrolyte conductivity inside pores. *Journal of Membrane Science*, *168*, 87–100.

Fievet, P., Labbez, C., Szymczyk, A., Vidonne, A., Foissy, A., & Pagetti, J. (2002). Electrolyte transport through amphoteric nanofiltration membranes. *Chemical Engineering Science*, *57*, 2921–2931.

Freger, V., Arnot, T. C., & Howell, J. A. (2000). Separation of concentrated organic/inorganic salt mixtures by nanofiltration. *Journal of Membrane Science*, *178*, 185–193.

Fridman-Bishop, N., Tankus, K. A., & Freger, V. (2018). permeation mechanisms and interplay between ions in nanofiltration. *Journal of Membrane Science*, *548*, 449–458.

Garba, Y., Taha, S., Cabon, J., & Dorange, G. (2003). Modeling of cadmium salts rejection through a nanofiltration membrane: Relationships between solute concentration and transport parameters. *Journal Membrane Science*, *211*, 51–58.

Garba, Y., Taha, S., Gondrexon, N., & Dorange, G. (1999). Ion transport modeling through nanofiltration membranes. *Journal Membrane Science*, *160*, 187–200.

Geraldes, V., & Brites Alves, A. M. (2008). Computer program for simulation of mass transport in nanofiltration membranes. *Journal of Membrane Science*, *321*, 172–181.

Goulas, A. K., Kapasakalidis, P. G., Sinclair, H. R., Rastall, R. A., & Gradison, A. S. (2002). Purification of oligosaccharides by nanofiltration. *Journal of Membrane Science*, *209*, 321–335.

Gupta, V. K., Hwang, S. T., Krantz, W. B., & Greenberg, A. R. (2007). Characterization of nanofiltration and reverse osmosis membrane performance for aqueous salt solutions using irreversible thermodynamics. *Desalination*, *208*, 1–18.

Hagmeyer, G., & Gimbel, R. (1998). Modelling the salt rejection of nanofiltration membranes for ternary mixtures and for single salts at different pH values. *Desalination*, *117*, 247–256.

Hagmeyer, G., & Gimbel, R. (1999). Modelling the rejection of nanofiltration membranes using zeta potential measurements. *Separation and Purification Technology*, *15*, 19–30.

Hall, M. S., Starov, V. M., & Lloyd, D. R. (1997). Reverse osmosis of multicomponent electrolyte solutions Part II. Experimental verification. *Journal of Membrane Science*, *128*, 39–53.

Hartnig, C., Witschel, W., & Spohr, E. (1998). Molecular dynamics study of the structure and dynamics of water in cylindrical pores. *Journal of Physical Chemistry B*, *102*, 1241–1249.

Hoffer, E., & Kedem, O. (1967). Hyperfiltration in charged membranes: The fixed charge model. *Desalination*, *2*, 25–39.

<https://www.desalitech.com> (consulted on March 30th, 2021).

<https://www.suezwatertechnologies.com/resources/winflows> (consulted on March 30th, 2021).

Jitsuhara, I., & Kimura, S. (1983). Rejection of inorganic salts by charged ultrafiltration membranes made of sulfonated polysulfone. *Journal of Chemical Engineering of Japan*, *16*, 394–399.

Kiso, Y., Muroshige, K., Oguchi, T., Yamada, T., Hhirose, M., Ohara, T., & Shintani, T. (2010). Effect of molecular shape of uncharged organic componds by nanofiltration membranes and on calculated pore radii. *Journal of Membrane Science*, *358*, 101–113.

Li, W., Li, J., Chen, T., & Chen, C. (2004). Study on Nanofiltration for purifying fructo-oligosaccharides: I. Operation mode. *Journal of Membrane Science*, *245*, 123–129.

Lipp, P., Gimbel, R., & Frimmel, F. H. (1994). Parameters influencing the rejection properties of FT30 membranes. *Journal of Membrane Science*, *95*, 185–197.

Lonsdale, H. K., Merten, U., & Riley, R. L. (1965). Transport properties of cellulose acetate osmotic membranes. *Journal of Applied Polymer Science*, *9*, 1341–1362.

Mason, E. A., & Lonsdale, H. K. (1990). Statistical-mechanical theory of membrane transport. *Journal of Membrane Science*, *51*, 1–81.

Mazzoni, C., & Bandini, S. (2006). On nanofiltration desal-5 DK performances with calcium chloride-water solutions. *Separation and Purification Technology*, *52*, 232–240.

Mazzoni, C., Bruni, L., & Bandini, S. (2007). Nanofiltration: Role of the electrolyte and pH on desal DK performances. *Industrial and Engineering Chemistry Research Journal*, *46*, 2254–2262.

Montesdeoca, V., Janssen, A. E. M., Boom, R. M., & Van der Padt, A. (2019). Fine ultrafiltration of concentrated oligosaccharide solutions – Hydration and pore size distribution effects. *Journal of Membrane Science*, *580*, 161–176.

Mucchetti, G., Zardi, G., Orlandini, F., & Gostoli, C. (2000). The pre-concentration of milk by nanofiltration in the production of Quarg-type fresh cheeses. *Le Lait*, *80*(1), 43–50.

Nakao, S., & Kimura, S. (1982). Models of membrane transport phenomena and their applications for ultrafiltration data. *Journal of Chemical Engineering of Japan, 15*, 200–205.

Nilsson, M., Trägårdh, G., & Östergren, K. (2008). The influence of pH, salt and temperature on nanofiltration performance. *Journal of Membrane Science, 312*, 97–106.

Nyström, M., Kaipia, L., & Luque, S. (1995). Fouling and retention of nanofiltration membranes. *Journal of Membrane Science, 98*, 249–262.

Oatley, D. L., Llenas, L., Perez, R., Williams, P. M., Martinez-Llado, X., & Rovira, M. (2012). Review of the dielectric properties of nanofiltration membranes and verification of the single oriented layer approximation. *Advances in Colloid and Interface Science, 173*, 1–11.

Oatley-Radcliffe, D. L., Williams, S. R., Barrow, M. S., & Williams, P. M. (2014). Critical appraisal of current nanofiltration modelling strategies for seawater desalination and further insights on dielectric exclusion. *Desalination, 343*, 154–161.

Otero, J. A., Lena, G., Colina, J. M., Prádanos, P., Tejerina, F., & Hernández, A. (2006). Characterisation of nanofiltration membranes: Structural analysis by the DSP model and microscopical techiques. *Journal of Membrane Science, 279*, 410–417.

Otero, J. A., Mazarrasa, O., Villasante, J., Silva, V., Prádanos, P., Calvo, J. I., & Hernández, A. (2008). Three independent ways to obtain information on pore size distributions of nanofiltration membranes. *Journal of Membrane Science, 309*, 17–27.

Paine, P. L., & Scherr, P. (1975). Drag coefficients for the movement of rigid spheres through liquid-filled cylindrical pores. *Biophysics Journal, 15*(10), 1087–1091.

Paugam, L., Taha, S., Dorange, G., Jaouen, P., & Quéméneur, F. (2004). Mechanism of nitrate ions transfer in nanofiltration depending on pressure, pH, concentration and medium composition. *Journal of Membrane Science, 231*, 37–46.

Peeters, J. M. M., Mulder, M. H. V., & Strathmann, H. (1999). Streaming potential measurements as a characterization method for nanofiltration membranes. *Colloids and Surfaces A: Physicochemical and Engineering Aspects, 150*, 247–259.

Perry, M., & Linder, C. (1989). Intermediate reverse osmosis ultrafiltration (RO UF) for concentration and desalting of low molecular weight organic solutes. *Desalination, 71*, 233–245.

Pinelo, M., Jonsson, G., & Meyer, A. S. (2009). Membrane technology for purification of enzymatically produced oligosaccharides: Molecular and operational features affecting performance. *Separation and Purification Technology, 70*, 1–11.

Qi, B., Luo, J., Chen, X., Hang, X., & Wan, Y. (2011). Separation of furfural from monosaccharides by nanofiltration. *Bioresource Technology, 102*, 7111–7118.

Quin, J., Oo, M. H., Lee, H., & Coniglio, B. (2004). Effect of feed pH on permeate pH and ion rejection under acidic conditions in NF process. *Journal of Membrane Science, 232*, 153–159.

Rautenbach, R., & Albrecht, R. (1989). *Membrane processes*. John Wiley & Sons.

Rizki, Z., Janssen, A. E. M., Boom, R. M., & Van der Padt, A. (2019). Oligosaccharides fractionation cascades with 3 outlet streams. *Separation and Purification Technology, 221*, 183–194.

Santos, Josè L. C., de Beukelaar, P., Ivo., Venkelcom, F. J., Velizarov, S., & Crespo, G. J. (2006). Effect of solute geometry and orientation on the rejection of uncharged compound by nanofiltration. *Separation Purification Technology, 50*, 122–131.

Schaep, J., & Vandecasteele, C. (2001). Evaluating the charge of nanofiltration membranes. *Journal of Membrane Science, 188*, 129–136.

Seader, D., & Henley, E. J. (2006). *Separation process principles* (2nd ed.). New York, NY: Wiley.

Senapati, S., & Chandra, A. (2001). Dielectric constant of water confined in a nanocavity. *Journal of Physical Chemistry B, 105*, 5106–5109.

Silva, V., Montalvillo, M., Javier Carmona, F., Palacio, L., Hernandez, A., & Pradanos, P. (2016). Prediction of single salt rejection in nanofiltration membranes by independent measurements. *Desalination, 382*, 1–12.

Simons, R., & Kedem, O. (1973). Hyperfiltration in porous fixed charge membranes. *Desalination, 13*, 1–16.

Sjöman, E., Mänttäri, M., Nyström, M., Koivikko, H., & Heikkilä, H. (2007). Separation of xylose from glucose by nanofiltration from concentrated monosaccharide solutions. *Journal of Membrane Science, 292*, 106–115.

Sjöman, E., Mänttäri, M., Nyström, M., Koivikko, H., & Heikkilä, H. (2008). Xylose recovery by nanofiltration from different hemicellulose hydrolyzate feeds. *Journal of Membrane Science, 310*, 268–277.

Soltanieh, M., & Gill, W. N. (1981). Review of reverse osmosis membranes and transport models. *Chemical Engineering Communications*, *12*, 279–363.

Spiegler, K. S., & Kedem, O. (1966). Thermodynamics of Hyperfiltration (Reverse Osmosis): Criteria for efficient membranes. *Desalination*, *1*, 311–326.

Spohr, E., Hartnig, C., Gallo, P., & Rovere, M. (1999). Water in porous glasses. A computer simulation study. *Journal of Molecular Liquids*, *80*, 165–178.

Szoke, S., Patzay, G., & Weiser, L. (2002). Characteristics of thin-film nanofiltration membranes at various pH-values. *Desalination*, *151*, 123–129.

Szymczyk, A., & Fievet, P. (2005). Investigating transport properties of nanofiltration membranes by means of a steric, electric and dielectric exclusion model. *Journal of Membrane Science*, *252*, 77–88.

Szymczyk, A., Fievet, P., Mullet, M., Reggiani, J. C., & Pagetti, J. (1998). Comparison of two electrokinetic methods – Electroosmosis and streaming potential – to determine the zeta-potential of plane ceramic membranes. *Journal of Membrane Science*, *143*, 189–195.

Szymczyk, A., Labbez, C., Fievet, P., Vidonne, A., Foissy, A., & Pagetti, J. (2003). Contribution of convection, diffusion and migration to electrolyte transport through nanofiltration membranes. *Advances in Colloid and Interface Science*, *103*, 77–94.

Szymczyk, A., Fatin-Rouge, N., Fievet, P., Ramseyer, C., & Vidonne, A. (2007). Identification of dielectric effects in nanofiltration of metallic salts. *Journal of Membrane Science*, *287*, 102–110.

Szymczyk, A., Fievet, P., & Bandini, S. (2010). On the amphoteric behavior of desal DK nanofiltration membranes at low salt concentrations. *Journal of Membrane Science*, *355*, 60–68.

Takagi, R., & Nakagaki, M. (1990). Theoretical study of the effect of ion adsorption on membrane potential and its application to colloidon membranes. *Journal of Membrane Science*, *53*, 19–35.

Takagi, R., & Nakagaki, M. (1992). Membrane potential of separation membranes as affected by ion adsorption. *Journal of Membrane Science*, *71*, 189–200.

Takagi, R., & Nakagaki, M. (2003). Ionic dialysis through amphoteric membranes. *Separation and Purification Technology*, *32*, 65–71.

Tanninen, J., & Nystrom, M. (2002). Separation of ions in acidic conditions using NF. *Desalination*, *147*, 295–299.

Tay, J. H., Liu, J., & Sun, D. D. (2002). Effect of solution physico-chemistry on the charge property of nanofiltration membranes. *Water Research*, *36*, 585–598.

Teixeira, M. R., Rosa, M. J., & Nystrom, M. (2005). The role of membrane charge on nanofiltration performance. *Journal of Membrane Science*, *265*, 160–166.

Timmer, J. M. K., van der Horst, H. C., & Robbertsen, T. (1993). Transport of lactic acid through reverse osmosis and nanofiltration membranes. *Journal of Membrane Science*, *85*, 205–216.

Tsuru, T., Nakao, S., & Kimura, S. (1991). Calculation of ion rejection by extended nernst–planck equation with charged reverse osmosis membranes for single and mixed electrolyte solutions. *Journal of Chemical Engineering of Japan*, *24*, 511–517.

Tsuru, T., Urairi, M., Nakao, S., & Kimura, S. (1991). Reverse osmosis of single and mixed electrolytes with charged membranes: Experiments and analysis. *Journal of Chemical Engineering of Japan*, *24*, 518–524.

Van der Bruggen, B., Manttari, M., & Nystrom, M. (2008). Drawbacks of applying nanofiltration and how to avoid them: A review. *Separation and Purification Technology*, *63*, 251–263.

Van der Bruggen, B., Schaep, J., Maes, W., & Vandecasteele, C. (1999). Influence of molecular size, polarity and charge on the retention of organic molecules by nanofiltration. *Journal of Membrane Science*, *156*, 29–41.

van der Host, H. C., Timmers, J. M. K., Robbertsen, T., & Leenders, J. (1995). Use of nanofiltration for concentration and demineralization in the dairy industry, model for mass transport. *Journal of Membrane Science*, *104*, 205–218.

Vellenga, E., & Tragardh, G. (1998). Nanofiltration of combined salt and sugar solutions: Coupling between retentions. *Desalination*, *120*, 211–220.

Wang, R., & Lin, S. (2021). Pore model for nanofiltration: History, theoretical framework, key predictions, limitations, and prospects. *Journal of Membrane Science*, *620*, 118809.

Wang, X. L., Tsuru, T., Nakao, S., & Kimura, S. (1995). Electrolyte transport through nanofiltration membranes by the space-charge model and the comparison with Teorell-Meyer-Sievers model. *Journal of Membrane Science*, *103*, 117–133.

Wang, X., Wang, W., & Wang, D. (2002). Experimental investigation on separation performance of nanofiltration membranes for inorganic electrolyte solutions. *Desalination*, *145*, 115–122.

Xu, Y., & Lebrun, R. E. (1999). Investigation of the solute separation by charged nanofiltration membrane: effect of pH, ionic strength and solute type. *Journal of Membrane Science*, *158*, 93–104.

Yaroshchuk, A. E. (2000). Dielectric exclusion of ions from membranes. *Advances in Colloid and Interface Science*, *85*, 193–230.

Yaroshchuk, A. E. (2001). Non-steric mechanisms of nanofiltration: Superposition of Donnan and dielectric exclusion. *Separation and Purification Technology*, *22–23*, 143–158.

Yaroshchuk, A., Bruening, M. L., & Zholkovskiy, E. (2019). Modelling nanofiltration of electrolyte solutions. *Advances in Colloid and Interface Science*, *268*, 39–63.

Yaroshchuk, A., Martinez-Llado, X., Llenas, L., Rovira, M., & de Pablo, J. (2011). Solution-diffusion-film model for the description of pressure-driven trans-membrane transfer of electrolyte mixtures: One dominant salt and trace ions. *Journal Membrane Science*, *368*, 192–201.

Zemaitis, J. F., Clark, D. M., Rafal, M., Scrivner, N. C. (1986). Handbook of aqueous electrolyte thermodynamics: Theory & application, American Institute of Chemical Engineers, Inc. Available from: https://doi.org/10.1002/9780470938416.

Zhang, L., Davis, H. T., Kroll, D. M., & White, H. S. (1995). Molecular dynamics simulations of water in a spherical cavity. *Journal of Physical Chemistry*, *99*, 2878–2884.

Zhang, Z., Yang, R., Zhang, S., Zhao, H., & Hua, X. (2011). Purification of lactulose syrup by using nanofiltration in a diafiltration mode. *Journal of Food Engineering*, *105*, 112–118.

Zhao, Y., Xing, W., Xu, N., & Wong, F. (2005). Effects of inorganic electrolytes on zeta potentials of ceramic microfiltration membranes. *Separation and Purification Technology*, *42*, 117–121.

CHAPTER

4

Transport phenomena in electrodialysis/reverse electrodialysis processes

R. Zeynali[1], *Kamran Ghasemzadeh*[1] and *Angelo Basile*[2]

[1]Chemical Engineering Faculty, Urmia University of Technology, Urmia, Iran

[2]Hydrogenia, Genoa, Italy

Abbreviations

AD	adsorption desalination
AEM	anion-exchange membrane
AEL	anion-exchange layer
BM	bipolar membrane
CCS	carbon-capture sequestration
CDI	capacitive deionization
CEL	cation-exchange layer
CEM	cation-exchange membrane
CFB	concentration flow battery
DD	Donnan dialysis
ED	electrodialysis
FB	flow battery
FE	flow electrode
FO	forward osmosis
HE	heat engine
HER	hydrogen evolution reaction
HSS	high salinity solution
IEM	ion-exchange membrane
IERB	ion-exchange resin bead
ITO	indium tin oxide coated glass
LSS	low-salinity solution

Current Trends and Future Developments on (Bio-) Membranes
DOI: https://doi.org/10.1016/B978-0-12-822257-7.00001-7

List of symbols and nomenclature

Δ_{Gmix} Gibbs free energy of mixing
V_{LSS} volume of LSS
V_{HSS} volume of HSS
R gas constant (8.314 J/molK)
T absolute temperature (K)
f volume fraction of LSS to the total feed solutions
c_i molar concentration of component i in aqueous solutions (mol/L)
γ activity coefficient
E_{emf} theoretical electromotive force (V)
N number of cell pairs
N_m number of IEMs
α average membrane permselectivity
F Faraday constant (96,485 C/mol)
z ionic valence
I current (A)
S projected area of IEMs (cm^2)
S_m projected area of IEMs (m^2)
R_{stack} the internal resistance (Ω)
R_{CEM} area resistance of CEM (Ωcm^2)
R_{AEM} area resistance of AEM (Ωcm^2)
R_{dbl} diffusion boundary layer resistance (Ω)
R_{edl} electrical double layer resistance (Ω)
σ molar conductivity of solution species ($s \cdot cm^2/mol$)
d intermembrane distance (cm)
c molar concentration of electrolyte (mol/L)
P power generation (W/m^2)
R_L external resistance (Ω)
P_{max} maximum power density (W/m^2)
η energy efficiency
Φ volumetric flow rate of LSS (m^3/s)
δ thickness (μm)

4.1 Introduction

Water is known to be a fundamental resource and asset for humanity. Individuals around the globe are persuaded about the need of identifying the best possible methods for preserving the environment. In spite of its bounty on the planet, the truth of the matter is that just around 0.8% of earth's water is drinkable and fresh. As an outcome, the current surface water assets can never be adequate to meet the future requirements for humankind and water is expected to become rare owing to cluttered utilization, which pollutes our stores and makes them unacceptable for use, either for human utilization, or for modern use. Because of the consideration toward the significance of water throughout everyday life and the water asset deficiency, water desalination has received noticeable importance.

Water desalination is one of the methods of purifying and cleaning water, which makes it conceivable to be utilized. Over the last couple of years, a few wastewater treatment forms have been recommended that could empower water reuse, with uncommon accentuation on waste and squander minimization, in addition to hazardous and perilous waste treatment.

Yet at the same time, decreasing the energy prerequisites and infrastructures and framework expenses of existing desalination innovations continues to be a challenge. In general, desalination advances can be categorized into two diverse component detachments, namely, membrane based and thermal desalination. Membrane-based advancements are receiving increasingly more consideration these days because of their reliable and dependable contaminant production without creation of any deleterious side-effects, particularly in water and wastewater treatment. Electrodialysis (ED) and reverse electrodialysis (RED) are vital techniques for desalination based on membrane technology, which owing to less energy expending is an alluring and less energy consuming strategy in comparison and examination with the other approaches, for example, thermal method. This procedure offers an electrochemical method that expels ionic toxins from a fluid solution, creating two new kinds of solutions: one concentrate of particles and another comprising practically pure and unadulterated water. The first solution can be used in industries and the pure water can be reused. Related to another field of research, for example, innovation of membranes, this strategy can be amazing in the treatment of mechanical and industrial effluents. ED utilizes direct current capacity to evacuate salts and other ionized species through cation and anion particle-specific layers of membrane to a concentrate gathering stream. Practically, all useful procedures, different ED cells are connected in series to develop an ED stack, with rotating anion (AM) and cation (CM) membranes of ED and RED forming the numerous cells. Besides, energy is one of the distinct advantages deciding the general financial turn of events: a powerful energy supply should be deemed essential for a feasible development and an improved expectation for everyday comforts. Energy utilization of the world is expanding at an exponential rate. It is projected that energy utilization of world will ascend by 48% somewhere between 2012 and 2040 (Administration & Office, 2016). Additionally, the foreseen populace increment in the coming decades will create two billion new energy buyers in developing economies by 2050 (Chu & Majumdar, 2012; Tufa et al., 2018). During a similar time interim, worldwide energy-related carbon dioxide emanations are required to increase by 46%, compared to 45 billion metric tons (EIA, 2013).

In this unique situation, there is a dire need for advancements in elective energy assets ready to mitigate the soaring interest for clean energy and related ecological issues (Elimelech & Phillip, 2011). In this chapter, both ED and RED techniques are examined, in addition to a discussion on the difficulties and significant issues associated with the same.

4.2 Electrodialysis process

ED is a process based on membrane technology during which particles (ions) are moved through a semiporous membrane, upon being affected by an electric potential. The layers are cation- or anion-specific and selective, which fundamentally implies that either positive particles or negative particles will move through. Membranes for the selection of cation are polyelectrolytes with oppositely charged material, which dismisses negatively charged particles and permits positively charged particles to move through. By putting multiple rows of membranes, which on the other hand permit positively or negatively charged particles to move through, the particles can be expelled from wastewater. In certain column segments, concentration convergence of particles will occur, and in other segments, particles will be evacuated and removed. The concentrated saltwater stream is circled until it has the ability to facilitate precipitation. Now the stream is released. This procedure can be applied for expelling particles from water. Particles that do not convey an electrical charge are not evacuated. Cation-specific films comprise sulfonated polystyrene, while anion-specific layers comprise polystyrene with quaternary-smelling salts. Sometimes pretreatment is fundamental before ED can occur. Suspended solids with a width that surpasses 10 μm should be expelled, or, in all likelihood, they will plug the film pores. There are additional substances that can kill a layer, for example, huge natural anions, colloids, iron oxides, and manganese oxide. These upset the particular impact of the film. Pretreatment strategies that enable the counteraction of these impacts are filtration procedures flocculation (for colloids), and active carbon filtration (for natural organic materials).

Some studies in the extant literature have investigated the quality of various electrodes, such as activated carbons, alumina, silica and graphene nanocomposites, mesoporous carbons, carbon aerogel, and carbide-derived carbons (Lee et al., 2009; Villar et al., 2011).

Vermaas et al. (2013) evaluated the application of capacitive electrodes instead of conventional ones. It has an active layer of carbon (e.g., wood and coal) to adsorb/desorb ions, and current collector to capture free electrons. Hence without a redox reaction, it converts anionic current in the ED stack into an electric current passing through the external electrical circuit. Furthermore, it does not require toxic components such as Cl_2, but its main disadvantage is carbon layer saturation. Wood-based materials are made to undergo a pyrolysis process to produce activated carbon with high porosity. Then, a polyvinylidene fluoride binds the activated carbon to maintain its structure. If we compare activated carbon with other electrode materials such as carbon aerogel, carbon cloth, carbon felt, carbon paper, it can be found that the activated carbon is more suitable because of its high pore-size distribution and surface area.

On the other hand, the conventional membranes are characterized according to the material and the pore dimension, while ion-exchange membranes (IEMs) are based on interaction of charged molecules. Hence conventional membranes cannot be utilized for ED applications because they are not selective, and they can be considered an ohmic barrier. In recent years, a porous separator known as diaphragm has been used in an electrochemical cell; however, it does not deliver selective separation (Paidar, Fateev, & Bouzek, 2016).

IEMs can be defined as thin polymeric films, such as polysulfone, polyethylene, polystyrene, having charged ion groups. They are mainly categorized on the basis of their charged ion groups into anion-exchange membrane (AEM) and cation-exchange membrane (CEM). The latter possesses negative charges such as phosphonic acid ($-PO_3H^-$), phosphoryl ($-PO_3^{2-}$), carboxylic acid ($-COO^-$), sulfonic acid ($-SO_{3-}$), and $-C_6H_4O^-$ covalently bonded to the backbone of the polymeric membrane, which blocks negative and coins, and permits positive ions to pass (Mei & Tang, 2018). Shortly, the AEM has positive charges such as quaternary amine ($-NR_{3+}$), tertiary amine ($-NR_2H^+$), secondary amine ($-NRH_2{}^+$), ammonium ($-NH_3{}^+$), the opposite of CEM (Galama, Post, Stuart, & Biesheuvel, 2013).

4.2.1 Description of process

Every now and again, ED framework system involves IEMs, assistive materials (gasket seal, electrodes, spacers), power supply, and stacks of ED. The stack of ED is shut using couple end plates and packed by nuts and screws. In the stack, there exists an arrangement of IEMs with series, concentrate compartments, electrodes and feed, gasket gel, and spacers. Additionally, this stack includes couple of electronic compartments, which changes over particles flow going through IEMs (into an electrons), the outer electrical circuit and arrangements. In a lab scale, ED incorporates not many cell sets, while, in a mechanical site, it can reach up to a few several sets of cell (e.g., 500 sets of cell). The ED framework structure system comprising concentrated and diluted compartments is illustrated in Fig. 4.1.

The ED repeating unit, in particular the pair of cell includes diluted and weakened stream spacer, AEM, CEM, and concentrated stream spacer. CEMs and AEMs are orchestrated in an "anode-C-A-C-A-cathode" design between brine (concentrated saltwater) compartments and feed (diluted). The channels encompass spacers made of plastic, introduced with gaskets to direct the arrangement stream and advance the blending of channels while maintaining a fixed intermembrane separation and seal the channels. The spacers make conduits for the water-powered bay, going about as multifold for gathering/appropriating the channels arrangements. In this way, the membranes space is used for feeding channels created by imposing them. As may be required, the applied angle of electric potential changes over particles into electric flow going along ED stack and NaCl utilized for producing the Cl_2 as electrolyte in anodic area (Demircioglu, Kabay, Kurucaovali, & Ersoz, 2003). In this manner, the proposed material is SO_4^{2-}, but it should stay away from the results to the cathode in the anode area and compartment, and 0.1M Na_2SO_4 flush arrangement is eventually confronted with harmful gas creation (Campione et al., 2018). It was discovered that inorganic particles are handily isolated from the feed, in contrast to natural particles. Diluted part mainly includes the natural constituents; while in this case,

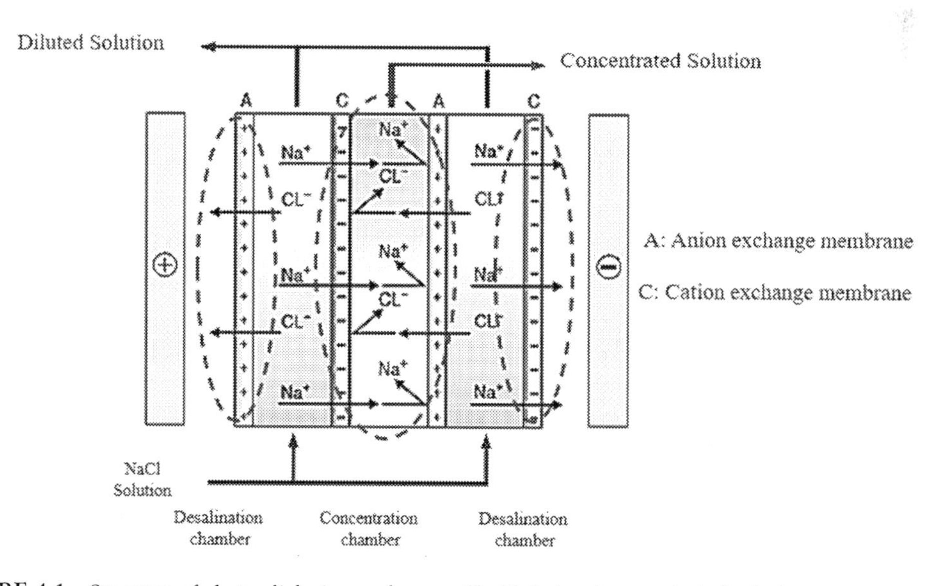

FIGURE 4.1 Structure of electrodialysis membrane with diluted and concentrated solution.

the brackishwater contains inorganic particles. In another study, researchers used a lab-scale ED arrangement comprising couple of chambers and possessing one anion trade and one bipolar film of membrane for the creation of alpha-ketoglutaric corrosive (Szczygiełda & Prochaska, 2017). The successful zone of the films of membranes area was 64 cm^2, and the thickness dispersing was 10 mm. Furthermore, other researchers utilized ED for recouping NaOH from the antacid high arrangement (Merkel, Ashrafi, & Ondrušek, 2017). Afterward, Andrew et al. (2018) performed supplements using wastewater recuperation by means of a pilot-scale ED system (Ward, Arola, Brewster, Mehta, & Batstone, 2018).

4.2.2 Theory of transport phenomena

Accomplishing ED requires a numerical articulation as a crucial asset for investigation and advancement. A large number of studies presented different ED numerical models for salt separation and removal, and they varied in terms of separation system and mass balance (Vickers, 2017), irreversible thermodynamics formalism (Gong, Wang, & Li-Xin, 2005; Koter, 2002), Stefan–Maxwell hypothesis (Kraaijeveld, Sumberova, Kuindersma, & Wesselingh, 1995), Nernst–Planck condition (Boubriak, Urban, & Cui, 2006; Tado, Sakai, Sano, & Nakayama, 2016), and semiexact models; for example, being inclusive of experimental outcomes other than the hypothetical parameters (Tado et al., 2016). Fig. 4.2 outlines ED arrangement of system including the diluted flow, concentrate, anode, cathode, siphons, AEM, CEM, and the development of positive and negative particles, for example, M^+ and X^-, in the stack of ED. Inside the ED stack, convection, movement, and dispersion are the three significant powers that transport and carry a wide scope of negatively and

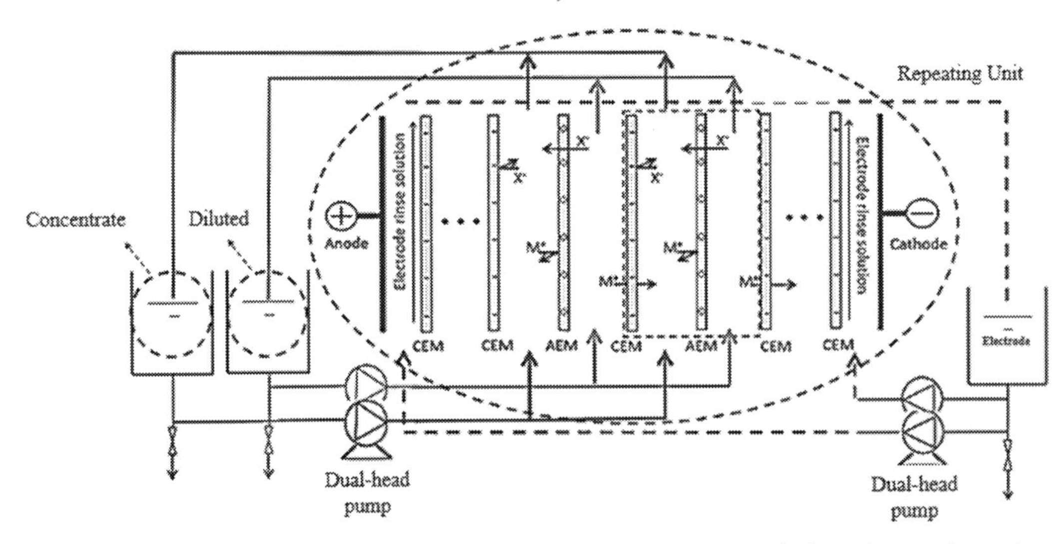

FIGURE 4.2 Structure of electrodialysis diluted flow, concentrate, anode, cathode, siphons, anion-exchange membrane, cation-exchange membrane, and the development of negative and positive ions.

positively charged particles (Jia, Li, Chen, & Wang, 2018). Eq. (4.1) mathematically expresses dispersion and relocation powers:

$$j_i = -D_i \frac{dC_i}{dx} + \frac{t_i I}{ZiF} \quad (4.1)$$

Where the diffusion (dispersion) coefficient is presented by D_i (m^2/s) for component of I, concentration is presented by C_i (M), current is presented by I(A), component transfer through membrane layer is presented by t_i, and Faraday constant is presented by F.

The coefficient of particle diffusion (D_i) can be estimated with the assistance of the equation of Stokes–Einstein (Jia et al., 2018):

$$D_i = \frac{RT\lambda_{i,m}{}^{\infty}}{|Z_i|F^2} \quad (4.2)$$

In this equation, R is the global all-inclusive constant of gas, F is Faraday constant, T is the test temperature, and λ_i^{∞},m based dimension with (S•cm/mol) is the constraining ionic related conductivity. Inorganics and organics leaving the diluted flow can be evaluated using Eq. (4.3):

$$(A^- + X^-) = n_{OH^-} + M^+_{out} - M^+_{in} \quad (4.3)$$

The number of equivalent anions leaving the diluted flow can be represented by A^- and X^- for organic and inorganic ones, respectively, and those entering the same by nOH^-; the number of cations that migrate from diluted solution because of AEM blemished selectivity is represented by $M_{out}{}^+$, the number of cations that are soaked into diluted solution because of inappropriate selectivity of membrane is represented by $M_{in}{}^+$.

The capacity of process (C_F) in ED can be calculated using Eq. (4.4) (Merkel, Ashrafi, & Ečer, 2018):

$$C_F = \frac{m_F}{N.A.t} \quad (4.4)$$

Where:

1. The total time is represented by t
2. Pairs of membrane number are represented by N
3. Membranes active area (surface) is represented by A
4. Dilute stream mass is represented by m_f
5. Flow mean velocity in membranes with dormant spacers is shown with V according to Eq. (4.5) (Yamato, Kimura, Miyoshi, & Watanabe, 2006).

$$V = \frac{Q}{N.\delta.b.\varepsilon_{sp}} \quad (4.5)$$

In this equation, δ presents the thickness of spacers (cm), flow rate in volumetric rate is represented by Q (mL/min), the porosity of spacers is represented by ε_{sp}, and the width of compartment is represented by b. Vickers et al. (2017) defined the linear velocity (u) using the following equation:

$$u = \frac{Q_d}{n.h.W} \quad (4.6)$$

The dimensions of flow channel in feed chamber are represented by h and W, which are related to height and width, respectively; pair number of IEM is represented by n, and the flow rate in volumetric form is presented with Q_d.

Another important factor is hydraulic retention tie, which could be evaluated using Eq. (4.7):

$$\mathrm{HRT} = \frac{n.h.W.L}{Q_d} \tag{4.7}$$

The channel of feed flow length in the stack of ED is presented by L.

There is a relation between final dilute (Q_d) and feed (Q_f), which can be used for estimating the water recovery (%) according to Eq. (4.8):

$$\mathrm{Recovery}(\%) = \frac{Q_d}{Q_f} \times 100 \tag{4.8}$$

4.2.3 Literature on electrodialysis process

Researchers began investigating the dimensionless modeling of ED systems in the late years of the 19th century. Some researchers like Probstein and Sonin (1972) employed a practical equation in order to study the dimensionless density current and used couple of domains for finding some scientific relations in diffusion layers. Couple of researchers in order to investigate the relation between electricity current and solution hydrodynamics in flow employed the revised version of Stanton and Peclet numbers in 1971. They found that there is a weak and unreliable relationship between types of IEM (Kitamoto & Takashima, 1971). Leila et al. (2016) observed that in the subject of ionic mass transfer, the diffusion is a critical factor and stated that Kitamoto & Takashima (1971) results are logical. Probstein et al. (1972) comprehensively investigated polarization and found an exact relation between the current density in both limited and mean condition for minimal salt removal levels.

In the late 1977 a researcher worked on finding the relation between the ionic mass transfer and current density during turbulent and laminar flow condition (Huang, 1977). In another study, Kuroda et al. found that there is no relationship between Sherwood number and types of spacers in ED stack and found its dependence to be $Re^{0.5}$ (Kuroda, Takahashi, & Nomura, 1983). A practical model introduced by Isaacson and Sonin was utilized by Kraaijeveld et al. (1995) for amino acid separation. Moresi and Fidaleo (2005) compared their empirical results with the models of Sonin and Isaacson (1974) and Kuroda et al. (1983) and found that Sherwood number is a function of $Re^{0.5}$ instead of $Re^{0.3}$ (Boubriak et al., 2006). Over the last eight years, a number of equations have been introduced by a group of researchers (Shaposhnik & Grigorchuk, 2010) to find a relationship between Sherwood number and ED systems mass transfer issue and a correlation among the mass transfer in ionic membranes and channels with various spacers design. Most of them used developed Sherwood number and equations in order to understand the current density limitation. Fidaleo and Moresi (2006, 2010) worked on couple of numbers as independent (Re & Sc) and Sherwood number as a dependent variable using ED experimental results with different feeds. Mitko and Turek (2014) used Graetz–Leveque related equations in laminar flow state in stack of ED systems and explored concentration polarization and coefficient of mass transfer. Subsequent to that, (Tadimeti, Chandra, & Chattopadhyay, 2015) employed experimentally obtained Sherwood number for Ca^{2+} and Cl^{-} mass-transfer investigation in a ED system and process with sugar solution as feed.

In recent years, Karimi and Ghassemi (2016) have stated a different point of view regarding recently published papers about ED process and claimed that these studies have focused just in few important determining factors and modeling studies are not complete, for instance, most of them are focused on current density limitation, but there are so many factors such as membrane life, quality of separation, consumption of energy, and there are numerous factors that cause industrial ED systems membrane not to damage.

Some of the studies reported that the accumulation of ions is a result of using high voltage in ED systems membrane and in this regard Karimi and Ghassemi (2016) found that for reducing the risks in ED systems, using low voltage is more acceptable than limiting the density of current. Scarazzato, Buzzi, Bernardes, Tenório, and Espinosa (2015) suggested using 70% or 80% of limited current density as the maximum range of current.

4.3 Overview of reverse electrodialysis process

RED is a clean and nonpolluting practical innovation that changes the free energy of blending of two arrangements of solutions with varying salinity into electrical energy. Regardless of the way that the idea of RED innovation was accounted for some time in 1954 (Pattle, 1954), the most noteworthy pattern of research propels has been done from 2007 onward (Veerman & Vermaas, 2016). Fig. 4.3 shows the numbers of papers on RED distributed yearly over the previous decade, and the key research subjects so far are process investigation, testing, and enhancement or optimization (Turek & Bandura, 2007; Vermaas, Veerman, Saakes, & Nijmeijer, 2014), stack structure (Veerman, Post, Saakes, Metz, & Harmsen, 2008), membrane plan and improvement, fouling simulations with modeling (Tedesco, Cipollina, Tamburini, Van Baak, & Micale, 2012), hybrid applications, and expansions to energy stockpiling as a stream battery. In addition, RED operability has as of late been stretched out from moderately low saline solutions for highly saline mechanical effluents and thermolytic solutions recovered in a shut circle (Hatzell, Ivanov, Cusick, Zhu, & Logan, 2014).

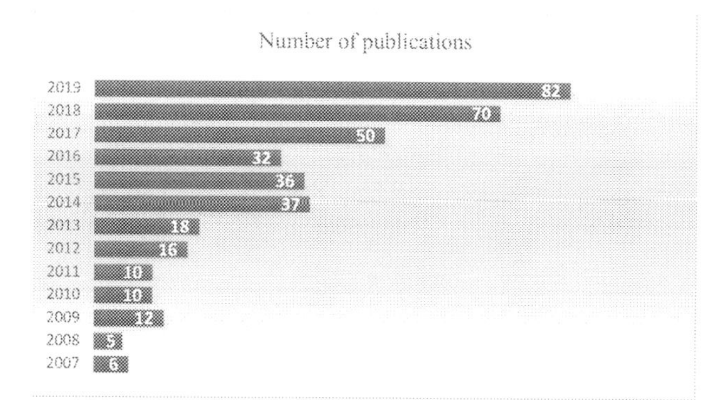

FIGURE 4.3 Progress in reverse electrodialysis research (number of publications): Scopus 2020.

4.3.1 Description of the process

The fundamentals of RED are presented in Fig. 4.1. An RED framework system is made up of membranes for ion exchange and compartments for seawater and waterway water (in substituting request). The membranes for ion exchange are selective for either anions or cations. The salinity contrast between seawater on one side and waterway water on the opposite side of the layer of membrane gives rise to a potential difference. Different cells, each containing a CEM, a seawater section, an AEM and a waterway water compartment, can be accumulated to build the voltage. Electrodes at both ends of the pile facilitate a redox reaction, which generates an electrical current to power an external device. The overall potential for salinity inclination power is very high. Each cubic meter of stream water can create 1.4 MJ of vitality when blended with equivalent measures of seawater and more than 2 MJ when blended in with large amount of seawater (Klaysom, Cath, Depuydt, & Vankelecom, 2013). The overall overflow of stream water into the ocean can possibly produce more than the prospected worldwide power interest for 2012 (Nam, Cusick, Kim, & Logan, 2012). Besides, the force yield from RED could be constrained by managing the water stream, particularly when a lake is accessible for new water stockpiling. Accordingly, salinity slope vitality can be put away and utilized when the force creation from sun and wind is at a low level. Latest advancements have improved the tentatively acquired force thickness (e.g., power per film region) for agent seawater and stream water, to a maximum estimation of 2.2 W/m^2. When considering the vitality spent for siphoning the water, a highest net force thickness of 1.2 W/m^2 was observed (Straub, Deshmukh, & Elimelech, 2016). The expansion in functional net force yield in RED of approximately 1.2 W/m^2 was acquired by advancing the intermembrane separation, forced by spacers. The nonconductive spacers discourage the vehicle of feed water and lessen the particle transport and thus the force yield. To tackle this issue, we propose a without spacer plan that utilizes profiled layers, provided with straight particle conductive edges, to incorporate the film and spacer usefulness. With these profiled membranes, the pumping losses were reduced by a factor of 4 and the absence of the non-conductive spacers reduced the ohmic resistance significantly (Nam et al., 2012). Then again, the limit layer opposition expanded when utilizing profiled films. In any case, the net force thickness was around 10% higher than when a stack with spacers was utilized (Nam et al., 2012). A structure with profiled films not just builds the most extreme estimation of the net force thickness, yet in addition moves the greatest to higher stream rates (Nam et al., 2012) and empowers lower intermembrane separations and distances (Fig. 4.4).

It should be noted that IEMs are at the heart of RED and considerably affect the performance of the overall process. The requirements of the optimal IEMs for RED application comprise high ideal selectivity (above 95%), low electrical resistance (below 1 Ω cm^2), acceptable chemical and mechanical stability, and long lifetime (at least 5 years) (Galama et al., 2013). Besides, these membranes should be of low cost; membrane costs below 4.3 €/m^2 have been observed to be break-even prices for financial feasibility of seawater electricity generation by RED (Daniilidis, Herber, & Vermaas, 2014). High ionic conductivity and low resistance are the key requirements for membranes applied in electrochemical energy systems (Devanathan, 2008; Varcoe et al., 2014). Membrane permselectivity also indicates an important role in maintaining selective transport of counter-ions, which

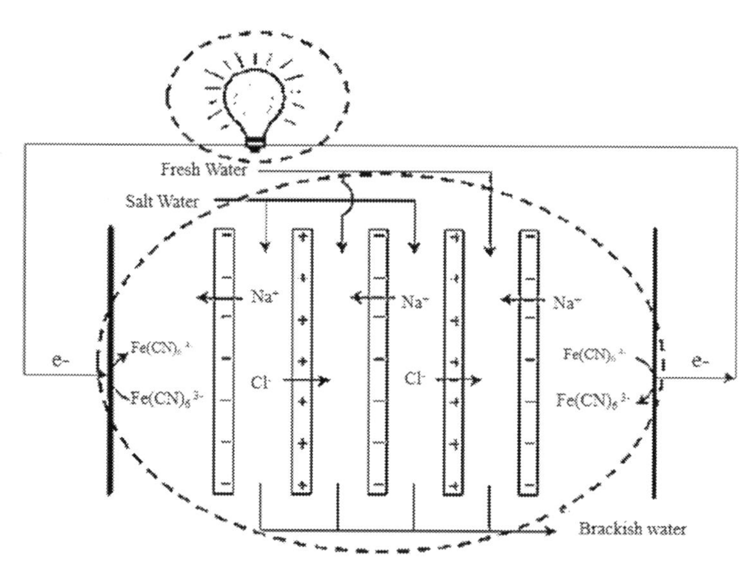

FIGURE 4.4 Principle of reverse electrodialysis.

enhances the OCV. Additionally, both membrane permselectivity and resistance are a function of other electrochemical properties, such as the ion exchange capacity and fixed charge density (Cassady, Cimino, Kumar, & Hickner, 2016; Geise, Cassady, Paul, Logan, & Hickner, 2014). These critical properties are inherently sensitive to the chemical structure of the polymeric materials for IEMs, and there exists a trade-off between the two parameters depending on the structure–property relationships (Varcoe et al., 2014). However, the water transfer coefficients of IEMs need to be minimized as well (Zlotorowicz, Strand, Burheim, Wilhelmsen, & Kjelstrup, 2017).

4.3.2 Theory of transport phenomena

Fig. 4.5 outlines a standard RED unit whose fundamental part is a stack of membrane and cathodes. The membrane film stack is made by repeatedly stacking CEMs and AEMs. A progression of contiguous high-salinity compartments and low-salinity compartments are then shaped and taken care of with a high-salinity solution (HSS) and a low-salinity solution (LSS), separately. AEMs are specific for anions, while CEMs permit the vehicle of cations. The distinction in fixation among HSS and LSS creates a movement of particles from the HSS toward LSS, which is constrained by membrane layers. Cations moved toward the CEM side and anions toward the AEM side, making an ionic transition that can be changed over at the cathodes by utilizing a redox response to control an electrical circuit.

The extractable salinity gradient power from blending VLSS (volume of LSS) and VHSS (volume of HSS) at steady weight and temperature can be determined as the Gibbs

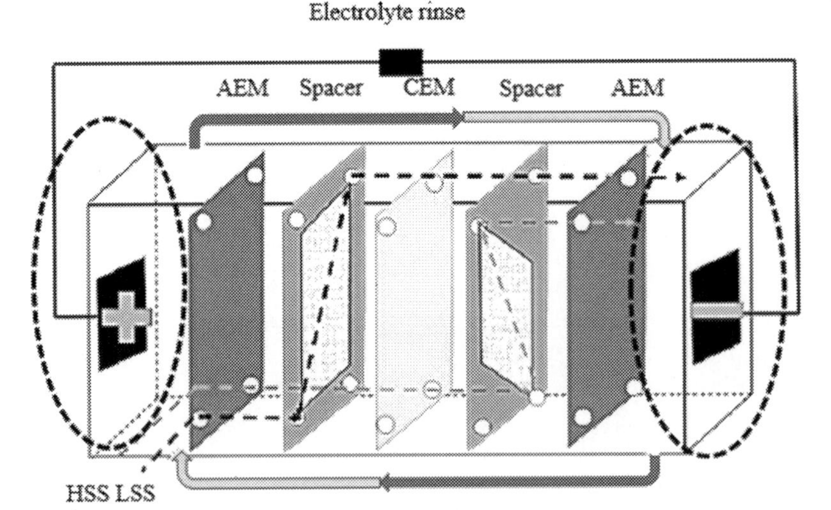

FIGURE 4.5 Principle of reverse electrodialysis.

free-mixing energy ΔG_{mix}, which is expressed mathematically using Eq. (4.9) (Post, Hamelers, & Buisman, 2008; Vermaas et al., 2013):

$$\Delta G_{\text{mix}} = RT\left\{\left[\sum_i c_i \ln(\gamma_i.c_i)\right]_{\text{LSS}} + \frac{(1-f)}{f}\left[\sum_i c_i \ln(\gamma_{i.}c_i)\right]_{\text{HSS}} - \frac{1}{f}\left[\sum_i c_i \ln(\gamma_i.c_i)\right]_{\text{mix}}\right\} \tag{4.9}$$

where R is the universal gas consistent (8.314 J/mol • K), T is the total temperature, f is the volume division of LSS to the absolute feed arrangements (i.e., f = VLSS/V_{mix}), c_i is the molar grouping of part I in watery arrangements (I = Na^+, Cl^-, and so on.), γ is the action coefficient, the addendums blend, HSS and LSS show blended gushing, HSS and LSS, separately.

The hypothetical electromotive power E_{emf} (V) can be determined on the basis of the Nernst condition equation Eq. (4.10) (Post et al., 2008):

$$E_{\text{emf}} = \frac{N_m \alpha \text{RT}}{zF} \ln\left\{\frac{\gamma_{\text{HSS}}.C_{\text{HSS}}}{\gamma_{\text{LSS}}.C_{\text{LSS}}}\right\} \tag{4.10}$$

Where N_m is the quantity of IEMs, α is the normal membrane film permselectivity, F is the Faraday constant (96,485 C/mol), and z is the ionic valence. At the point when RED is associated with an outer burden, the voltage yield U can be determined as the contrast between the electromotive power E_{emf} and the voltage drop over the inward obstruction R_{stack}, which is represented using Eq.(4.11) (Vermaas et al., 2013):

$$U = E_{\text{emf}} - IR_{\text{stack}} \tag{4.11}$$

Where I (An) is flow in the electrical circuit. R_{stack} is the entirety of ohmic opposition and resistance of stack segments and nonohmic obstruction. The opposition of the cathode framework is attributed to the Nernst voltage of the redox response, to overpotential and to ohmic part in the arrangement. The obstruction of the terminals framework can be dismissed here considering the operability of RED for a greater scope.

$$R_{stack} = R_{ohmic} + R_{nonohmic} \tag{4.12}$$

The ohmic opposition and resistance of the stack brought about by ionic vehicle constraints through the stack segments is chiefly represented by the obstruction of the IEMs and resistance of the feed compartments. The force disseminated in the outer resistance RL, which is the force thickness P (W/m^2) produced by RED can be determined as follows [Eq. (4.13)]:

$$P = I^2 R_L = \left(\frac{E_{emf}}{R_{stack} + R_L}\right)^2 R_L \tag{4.13}$$

Where R_L is the outside opposition resistance (Ω). The force thickness density (P_{max} in W/m^2 of film of membrane) can be acquired when the resistance of outside burden rises to the inside stack opposition based on Eq. (4.14). ($R_L = R_{stack}$).

$$P_{max} = \frac{E_{emf}^2}{4N_m S_m R_{stack}} \tag{4.14}$$

The energy of productivity η is dictated by the caught electric energy contrasted with the hypothetical measure of Gibbs free vitality discharged during the blending procedure based on Eq. (4.15) (Vermaas, Saakes, & Nijmeijer, 2011).

$$\eta = \frac{P.N_m.S_m}{\Delta G_{mix}.\varphi} \tag{4.15}$$

Where Φ is the volumetric flow rate of LSS (m^3/s).

4.3.3 Literature on reverse electrodialysis process

Huge advances in RED execution have been accomplished over a decade ago with massive advances in IEMs and spacer structures. In 2012 the main customized IEM explicitly intended for RED demonstrated a force thickness of 1.27 W/m^2 (Guler, Zhang, Saakes, & Nijmeijer, 2012). This has been trailed by a rush of RED film improvement work (Cho et al., 2017; Hong & Chen, 2014; Kim et al., 2015). Most of these customized RED layers are homogeneous; the greater part of them have a lot more slender film thickness (26–91 μm) contrasted with regular IEMs utilized for ED so as to accomplish lower electric opposition (0.28–2.26 Ω · cm^2). Nano/microfluidic RED utilizing particle particular nanochannels show further sensational increment in power thickness (e.g., a force thickness of 20–2600 W/m^2 was normal for a permeable film) (Guo et al., 2010). In the meantime, creative spacer plans (e.g., customized IEMs with edges/column structures (Güler, Elizen, Saakes, & Nijmeijer, 2014; Post et al., 2008), particle conductive spacers, and the utilization of particle trade pitch as space separators lead to enhanced power density by up to 4 times. With the enhancements of RED force age, extensive endeavors have focused on creative utilizations of RED, for example, osmotic heat engine for changing over second-rate heat to power (Luo et al., 2012), fixation battery for energy stockpiling, microbial RED cell for synergistic energy reaping, and for RED-based unmanageable toxin reduction. During the most recent decades, we have likewise seen the authorization of RED pilot plants in the Netherlands (Mei & Tang, 2018) and Italy (Tamburini et al., 2017), which is a basic advance to its viable usage on a global scale. These pilot tests have been enhanced by lab-scale fouling examinations. Notwithstanding, so far, there is no full-scale RED plants.

The transport and movement of ions through IEM depends on a few factors, such as electric potential slope (movement), synthetic/fixation potential angle (dissemination), and weight inclination (convection). The bearing and speed of particles are not a conspicuous issue; the focus slope is only one of the potential factors, and particles could be moved even against their own fixation inclination (large motion) from diluted solution to the compartments with high concentration up to valence-reliant of counter-particles redistribution and Donnan balance are accomplished. Such large movements are distinguished in the system of RED for the instance of covalent particles at first step of present in the diluted flow (Pawlowski, Crespo, & Velizarov, 2016; Vermaas et al., 2014). Vermaas et al. (2014) researched the impact of SO_4^{2-} and Mg^{2+} in stacks of RED outfitted with multiple-layer models. It was seen that the expansion in $MgSO_4$ prompted an expansion in important Ohmic opposition. In the case of blend of seawater and waterway water (including 10% of SO_4^{2-} and Mg^{2+} particles in both of them), a 29–half decrease in power thickness was obtained regarding unadulterated NaCl arrangements. A similar pattern (approximately 15% − 43%) was obtained by Hong et al. In the case of RED worked including counterfeit multiparticle arrangements imitating common waterway and seawater. Covalent particles transport from the LCC solution for HCC arrangement decreased the interior transport of charge, along these lines prompting a few electron numbers that moved from anode to cathode and, eventually, to bring down created power and current. An increase in cost of pumping energy was observed, in comparison to a fixed channel thickness, when breathing cell concept was implemented, nevertheless, since the pumping losses were relatively low (due to a high channel thickness - 480 μm), high values of net power density could have been obtained in a wide range of flow rates. As the pumping losses are relatively low, high values of net power density can be obtained in a wide range of flow rates. As stated by Pawlowski et al., 2016 the advantages of applying this concept may be limited by internal leakages caused by membrane movements, and by mechanical stress on the membranes during oscillations.

Spacer geometry altogether impacts the fixation and stream dispersion along the channels and, thus RED execution. In a similar report utilizing diverse spacer geometries, (Tamburini, La Barbera, Cipollina, Micale, & Ciofalo, 2015) discovered that spacers made by weaving gives a superior trade-off among the polarization and pressure drop contrasted with spacers that are used with covering fibers (not woven). The spacer made by weaving technique configuration is accepted to advance greater blending in RED system (Gurreri, Tamburini, Cipollina, Micale, & Ciofalo, 2016). The net materials of spacer may likewise impact the liquid elements of channel by influencing the no-slip and slip conditions. Materials with hydrophobic characteristic diminishes grating among the streaming arrangement and the wires of spacer, along these lines coming about near to a 40% decrease of the weight drops contrasting with the materials with hydrophobic characteristic. Spacers with no conductivity mostly spread the surface of the film and make longer ways for transport of particles in arrangements. This purported "shadow impact of spacer" could be answerable for nearly multiplying the Ohmic resistance (Pawlowski, Sistat, Crespo, & Velizarov, 2014). In spite of the fact that the shadow impact of spacer can be diminished by expanding the spacer open region (Vermaas, Saakes, & Nijmeijer, 2014) or by utilizing a spacer work more slender than the encompassing gasket, its total expulsion is ideal. Appropriation of various stream bearings can possibly expand RED execution (Hatzell & Logan, 2013).

The system and framework of RED can be divided to three essential designs: counter-, co-, and cross-stream. In many procedures, the counter-stream activity is considerably productive, since it advances increasingly uniform main impetus along the gadget. Contingent upon stream rates, costream activities may be beneficial now and again: (Veerman, Saakes, Metz, & Harmsen, 2010) announced marginally higher force thickness of about 0.05 W/m^2 by costream activity contrasted with counter-stream activity in RED. This impact is presumably brought about by less spillage among LCC and HCC because of the low nearby weight contrasts between the two compartments during costream activity. Veerman et al. (2010) embraced a triple division of anodes in RED, announcing higher force yield up to 11% contrasted with a solitary cathode. Vermaas et al. (2013) thought about the extraction process of energy proficiency when utilizing the three stream modes both with single and numerous anodes sets (division). It was observed that the effectiveness of counter-stream systems are better than cross-stream, though the coflow mode was seen as less proficient (two-overlay lower). It was estimated that up to 95% of the theoretically available energy can be extracted in counter-flow mode (saltwater fraction of 0.13) using a single electrode segment.

4.4 Conclusion and future trends

In last decades, the research in the field of water treatment technologies has led to great achievements such as:

1. High amount of loading rate
2. Separations with great selectivity
3. Less amount of fouling
4. High level of water recovery

All of the mentioned advances require a decrease in freshwater producing cost and consumption of energy. Besides, most of these technologies require using chemicals for pre- and posttreatment steps, but in this method, ED omitted the using of chemicals during treatment process for desalination process.

ED process because of its capabilities in vast range of treatment process has satisfied the requirements for treatment in the industrial, brackish, municipal, and chemical and food industries.

In fact, the recent advancements in the domain of IEMs has led to improvement of ED technology in many areas. In this regard, using modeling studies in this area using mathematical studies and techniques will cause to pave the way for optimization and examining the ED systems.

Future works will highlight advance ED coupling with different innovations to build its supportability, execution, and limit its downsides. Additional pilot-scale examinations are needed to approve the possibility of ED applications and assume a significant job in connecting lab-scale ED application to full-scale execution. Other future papers will introduce more examinations to reproduce ED applications and measure the procedure's financial aspects. Ultimately, advancement in ionic, steady, specific, and conductive AEMs can be helpful for overcoming difficult issues, improving film life expectancy, and anticipating another skyline of uses and new markets. Having a good membrane with some characteristics like such as permselectivity and low cost of materials for membrane preparation is the main challenge for improvements in RED process during the last years. In this regard, innovations in membrane material and reducing the cost of IEM will enable RED technology's commercial implementation. Utilization of natural saline streams as a source for RED resulted in positive net power generation , but to a lesser extent when compared to artificial NaCl solutions. The presence of organic compounds and divalent ions in natural streams reduces (in some cases by more than - 50%) the obtained power due to overlapping electrochemical and fouling phenomena. The negative impact of divalent ions on the performance of RED is one of the major challenges when working with natural streams. Specifically, the generally high wealth of magnesium particles in feed arrangements prompts a radical decrease in Pd, stifling the Nernst potential drop across IEMs for reasons identified with the adjustment of electrochemical layer properties and transport wonders inside the SGP-RED framework. In this system, pretreatment techniques dependent on substance mellowing as well as layer treatment could be visualized. Weight-driven layer activities, for example, nanofiltration, today widely utilized as pretreatment to invert assimilation, can undoubtedly be adjusted to RED. Lastly, the use of RED can possibly be reached out to a few different situations, since brackishwater arrangements (in the end at high temperatures) can be acquired from various modern exercises. A few different applications are emerging, for example, the treatment of wastewater with concurrent age of power. Overall, this represents a promising perspective for RED as a reliable power source once the technological challenges have been overcome.

References

Administration, E. I., & Office, G. P. (2016). *International energy outlook 2016, with projections to 2040*. Government Printing Office.

Boubriak, O., Urban, J., & Cui, Z. (2006). Monitoring of lactate and glucose levels in engineered cartilage construct by microdialysis. *Journal of Membrane Science, 273*, 77–83.

Campione, A., Gurreri, L., Ciofalo, M., Micale, G., Tamburini, A., & Cipollina, A. (2018). Electrodialysis for water desalination: A critical assessment of recent developments on process fundamentals, models and applications. *Desalination, 434*, 121–160.

Cassady, H. J., Cimino, E. C., Kumar, M., & Hickner, M. A. (2016). Specific ion effects on the permselectivity of sulfonated poly (ether sulfone) cation exchange membranes. *Journal of Membrane Science, 508*, 146–152.

Cho, D. H., Lee, K. H., Kim, Y. M., Park, S. H., Lee, W. H., Lee, S. M., & Lee, Y. M. (2017). Effect of cationic groups in poly (arylene ether sulfone) membranes on reverse electrodialysis performance. *Chemical Communications, 53*, 2323–2326.

Chu, S., & Majumdar, A. (2012). Opportunities and challenges for a sustainable energy future. *Nature, 488*, 294–303.

Daniilidis, A., Herber, R., & Vermaas, D. A. (2014). Upscale potential and financial feasibility of a reverse electrodialysis power plant. *Applied Energy, 119*, 257–265.
Demircioglu, M., Kabay, N., Kurucaovali, I., & Ersoz, E. (2003). Demineralization by electrodialysis (ED)—Separation performance and cost comparison for monovalent salts. *Desalination, 153*, 329–333.
Devanathan, R. (2008). Recent developments in proton exchange membranes for fuel cells. *Energy & Environmental Science, 1*, 101–119.
EIA, U. (2013). *Annual energy outlook 2013, United States Energy Information Administration*. DOE/EIA-0383.
Elimelech, M., & Phillip, W. A. (2011). The future of seawater desalination: Energy, technology, and the environment. *Science (New York, N.Y.), 333*, 712–717.
Fidaleo, M., & Moresi, M. (2006). Assessment of the main engineering parameters controlling the electrodialytic recovery of sodium propionate from aqueous solutions. *Journal of Food Engineering, 76*, 218–231.
Fidaleo, M., & Moresi, M. (2010). Application of the Nernst–Planck approach to model the electrodialytic recovery of disodium itaconate. *Journal of Membrane Science, 349*, 393–404.
Galama, A., Post, J., Stuart, M. C., & Biesheuvel, P. (2013). Validity of the Boltzmann equation to describe Donnan equilibrium at the membrane–solution interface. *Journal of Membrane Science, 442*, 131–139.
Geise, G. M., Cassady, H. J., Paul, D. R., Logan, B. E., & Hickner, M. A. (2014). Specific ion effects on membrane potential and the permselectivity of ion exchange membranes. *Physical Chemistry Chemical Physics, 16*, 21673–21681.
Gong, Y., Wang, X.-L., & Li-Xin, Y. (2005). Process simulation of desalination by electrodialysis of an aqueous solution containing a neutral solute. *Desalination, 172*, 157–172.
Güler, E., Elizen, R., Saakes, M., & Nijmeijer, K. (2014). Micro-structured membranes for electricity generation by reverse electrodialysis. *Journal of Membrane Science, 458*, 136–148.
Guler, E., Zhang, Y., Saakes, M., & Nijmeijer, K. (2012). Tailor-made anion-exchange membranes for salinity gradient power generation using reverse electrodialysis. *ChemSusChem, 5*, 2262–2270.
Guo, W., Cao, L., Xia, J., Nie, F. Q., Ma, W., Xue, J., … Jiang, L. (2010). Energy harvesting with single-ion-selective nanopores: A concentration-gradient-driven nanofluidic power source. *Advanced Functional Materials, 20*, 1339–1344.
Gurreri, L., Tamburini, A., Cipollina, A., Micale, G., & Ciofalo, M. (2016). Flow and mass transfer in spacer-filled channels for reverse electrodialysis: A CFD parametrical study. *Journal of Membrane Science, 497*, 300–317.
Hatzell, M. C., Ivanov, I., Cusick, R. D., Zhu, X., & Logan, B. E. (2014). Comparison of hydrogen production and electrical power generation for energy capture in closed-loop ammonium bicarbonate reverse electrodialysis systems. *Physical Chemistry Chemical Physics, 16*, 1632–1638.
Hatzell, M. C., & Logan, B. E. (2013). Evaluation of flow fields on bubble removal and system performance in an ammonium bicarbonate reverse electrodialysis stack. *Journal of Membrane Science, 446*, 449–455.
Hong, J. G., & Chen, Y. (2014). Nanocomposite reverse electrodialysis (RED) ion-exchange membranes for salinity gradient power generation. *Journal of Membrane Science, 460*, 139–147.
Huang, T.-C. (1977). Correlations of ionic mass transfer rate in ion exchange membrane electrodialysis. *Journal of Chemical and Engineering Data, 22*, 422–426.
Jia, Y.-X., Li, F.-J., Chen, X., & Wang, M. (2018). Model analysis on electrodialysis for inorganic acid recovery and its experimental validation. *Separation and Purification Technology, 190*, 261–267.
Karimi, L., & Ghassemi, A. (2016). An empirical/theoretical model with dimensionless numbers to predict the performance of electrodialysis systems on the basis of operating conditions. *Water Research, 98*, 270–279.
Kim, H.-K., Lee, M.-S., Lee, S.-Y., Choi, Y.-W., Jeong, N.-J., & Kim, C.-S. (2015). High power density of reverse electrodialysis with pore-filling ion exchange membranes and a high-open-area spacer. *Journal of Materials Chemistry A, 3*, 16302–16306.
Kitamoto, A., & Takashima, Y. (1971). Transfer rates in electrodialysis with ion exchange membranes. *Desalination, 9*, 51–87.
Klaysom, C., Cath, T. Y., Depuydt, T., & Vankelecom, I. F. (2013). Forward and pressure retarded osmosis: Potential solutions for global challenges in energy and water supply. *Chemical Society Reviews, 42*, 6959–6989.
Koter, S. (2002). Transport of simple electrolyte solutions through ion-exchange membranes—the capillary model. *Journal of Membrane Science, 206*, 201–215.
Kraaijeveld, G., Sumberova, V., Kuindersma, S., & Wesselingh, H. (1995). Modelling electrodialysis using the Maxwell-Stefan description. *The Chemical Engineering Journal and the Biochemical Engineering Journal, 57*, 163–176.

Kuroda, O., Takahashi, S., & Nomura, M. (1983). Characteristics of flow and mass transfer rate in an electrodialyzer compartment including spacer. *Desalination, 46*, 225–232.

Lee, J.-B., Park, K.-K., Yoon, S.-W., Park, P.-Y., Park, K.-I., & Lee, C.-W. (2009). Desalination performance of a carbon-based composite electrode. *Desalination, 237*, 155–161.

Luo, X., Cao, X., Mo, Y., Xiao, K., Zhang, X., Liang, P., & Huang, X. (2012). Power generation by coupling reverse electrodialysis and ammonium bicarbonate: Implication for recovery of waste heat. *Electrochemistry Communications, 19*, 25–28.

Mei, Y., & Tang, C. Y. (2018). Recent developments and future perspectives of reverse electrodialysis technology: A review. *Desalination, 425*, 156–174.

Merkel, A., Ashrafi, A. M., & Ečer, J. (2018). Bipolar membrane electrodialysis assisted pH correction of milk whey. *Journal of Membrane Science, 555*, 185–196.

Merkel, A., Ashrafi, A. M., & Ondrušek, M. (2017). The use of electrodialysis for recovery of sodium hydroxide from the high alkaline solution as a model of mercerization wastewater. *Journal of Water Process Engineering, 20*, 123–129.

Mitko, K., & Turek, M. (2014). Concentration distribution along the electrodialyzer. *Desalination, 341*, 94–100.

Nam, J.-Y., Cusick, R. D., Kim, Y., & Logan, B. E. (2012). Hydrogen generation in microbial reverse-electrodialysis electrolysis cells using a heat-regenerated salt solution. *Environmental Science & Technology, 46*, 5240–5246.

Paidar, M., Fateev, V., & Bouzek, K. (2016). Membrane electrolysis—History, current status and perspective. *Electrochimica Acta, 209*, 737–756.

Pattle, R. E. (1954). Production of electric power by mixing fresh and salt water in the hydroelectric pile. *Nature, 174*, 660. Available from https://doi.org/10.1038/174660a0.

Pawlowski, S., Crespo, J. & Velizarov, S. (2016). Sustainable power generation from salinity gradient energy by reverse electrodialysis. In: Ribeiro, A., Mateus, E., Couto, N. (Eds.), *Electrokinetics across disciplines and continents*. Springer, Cham (pp. 57–80). Available from https://doi.org/10.1007/978-3-319-20179-5_4.

Pawlowski, S., Sistat, P., Crespo, J. G., & Velizarov, S. (2014). Mass transfer in reverse electrodialysis: Flow entrance effects and diffusion boundary layer thickness. *Journal of Membrane Science, 471*, 72–83.

Post, J. W., Hamelers, H. V., & Buisman, C. J. (2008). Energy recovery from controlled mixing salt and fresh water with a reverse electrodialysis system. *Environmental Science & Technology, 42*, 5785–5790.

Scarazzato, T., Buzzi, D. C., Bernardes, A. M., Tenório, J. A. S., & Espinosa, D. C. R. (2015). Current-Voltage curves for treating effluent containing HEDP: Determination of the limiting current. *Brazilian Journal of Chemical Engineering, 32*, 831–836.

Shaposhnik, V., & Grigorchuk, O. (2010). Mathematical model of electrodialysis with ion-exchange membranes and inert spacers. *Russian Journal of Electrochemistry, 46*, 1182–1188.

Straub, A. P., Deshmukh, A., & Elimelech, M. (2016). Pressure-retarded osmosis for power generation from salinity gradients: Is it viable? *Energy & Environmental Science, 9*, 31–48.

Szczygiełda, M., & Prochaska, K. (2017). Alpha-ketoglutaric acid production using electrodialysis with bipolar membrane. *Journal of Membrane Science, 536*, 37–43.

Tadimeti, J., Chandra, A., & Chattopadhyay, S. (2015). Optimum concentrate stream concentration in $CaCl_2$ removal from sugar solution using electrodialysis. *Chemical Engineering and Process Technology, 6*(1). Available from https://doi.org/10.4172/2157-7048.1000216.

Tado, K., Sakai, F., Sano, Y., & Nakayama, A. (2016). An analysis on ion transport process in electrodialysis desalination. *Desalination, 378*, 60–66.

Tamburini, A., La Barbera, G., Cipollina, A., Micale, G., & Ciofalo, M. (2015). CFD prediction of scalar transport in thin channels for reverse electrodialysis. *Desalination and Water Treatment, 55*, 3424–3445.

Tamburini, A., Tedesco, M., Cipollina, A., Micale, G., Ciofalo, M., Papapetrou, M., ... Piacentino, A. (2017). Reverse electrodialysis heat engine for sustainable power production. *Applied Energy, 206*, 1334–1353.

Tedesco, M., Cipollina, A., Tamburini, A., Van Baak, W., & Micale, G. (2012). Modelling the reverse electrodialysis process with seawater and concentrated brines. *Desalination and Water Treatment, 49*, 404–424.

Tufa, R. A., Pawlowski, S., Veerman, J., Bouzek, K., Fontananova, E., Di Profio, G., ... Curcio, E. (2018). Progress and prospects in reverse electrodialysis for salinity gradient energy conversion and storage. *Applied Energy, 225*, 290–331.

Turek, M., & Bandura, B. (2007). Renewable energy by reverse electrodialysis. *Desalination, 205*, 67–74.

Varcoe, J. R., Atanassov, P., Dekel, D. R., Herring, A. M., Hickner, M. A., Kohl, P. A., ... Scott, K. (2014). Anion-exchange membranes in electrochemical energy systems. *Energy & Environmental Science, 7*, 3135–3191.

Veerman, J., Post, J., Saakes, M., Metz, S., & Harmsen, G. (2008). Reducing power losses caused by ionic shortcut currents in reverse electrodialysis stacks by a validated model. *Journal of Membrane Science, 310*, 418–430.

Veerman, J., Saakes, M., Metz, S. J., & Harmsen, G. J. (2010). Electrical power from sea and river water by reverse electrodialysis: A first step from the laboratory to a real power plant. *Environmental Science & Technology, 44*, 9207–9212.

Veerman, J. & Vermaas, D. A. (2016). Reverse electrodialysis: Fundamentals. In: Cipollina, Andrea, Micale, Giorgio (Eds.), Sustainable energy from salinity gradients. Woodhead Publishing (pp. 77–133). Available from https://doi.org/10.1016/B978-0-08-100312-1.00004-3.

Vermaas, D. A., Saakes, M., & Nijmeijer, K. (2011). Doubled power density from salinity gradients at reduced intermembrane distance. *Environmental Science & Technology, 45*, 7089–7095.

Vermaas, D. A., Saakes, M., & Nijmeijer, K. (2014). Enhanced mixing in the diffusive boundary layer for energy generation in reverse electrodialysis. *Journal of Membrane Science, 453*, 312–319.

Vermaas, D. A., Veerman, J., Saakes, M., & Nijmeijer, K. (2014). Influence of multivalent ions on renewable energy generation in reverse electrodialysis. *Energy & Environmental Science, 7*, 1434–1445.

Vermaas, D. A., Veerman, J., Yip, N. Y., Elimelech, M., Saakes, M., & Nijmeijer, K. (2013). High efficiency in energy generation from salinity gradients with reverse electrodialysis. *ACS Sustainable Chemistry & Engineering, 1*, 1295–1302.

Vickers, N. J. (2017). Animal communication: When I'm calling you, will you answer too? *Current Biology, 27*, R713–R715.

Villar, I., Suarez-De La Calle, D. J., González, Z., Granda, M., Blanco, C., Menéndez, R., & Santamaría, R. (2011). Carbon materials as electrodes for electrosorption of NaCl in aqueous solutions. *Adsorption, 17*, 467–471.

Ward, A. J., Arola, K., Brewster, E. T., Mehta, C. M., & Batstone, D. J. (2018). Nutrient recovery from wastewater through pilot scale electrodialysis. *Water Research, 135*, 57–65.

Yamato, N., Kimura, K., Miyoshi, T., & Watanabe, Y. (2006). Difference in membrane fouling in membrane bioreactors (MBRs) caused by membrane polymer materials. *Journal of Membrane Science, 280*, 911–919.

Zlotorowicz, A., Strand, R. V., Burheim, O. S., Wilhelmsen, Ø., & Kjelstrup, S. (2017). The permselectivity and water transference number of ion exchange membranes in reverse electrodialysis. *Journal of Membrane Science, 523*, 402–408.

CHAPTER

5

Transport phenomena in membrane distillation processes

Jianhua Zhang[1], *Jun-De Li*[2], *Zongli Xie*[3], *Xiaodong Dai*[4] *and Stephen Gray*[1]

[1]ISILC, Victoria University, Melbourne, VIC, Australia [2]College of Engineering and Science, Victoria University, Melbourne, VIC, Australia [3]CSIRO Manufacturing, Private Bag 10, Clayton South MDC, VIC, Australia [4]Shengli College, China University of Petroleum, Dongying, P.R.China

Abbreviations

AGMD air gap membrane distillation
DCMD direct contact membrane distillation
LEP liquid entry pressure
MD membrane distillation
PP polypropylene
PTFE polytetrafluoroethylene
PVDF polyvinylidene fluoride
RO reverse osmosis
SGMG sweeping gas membrane distillation
TPC temperature polarization coefficient
VMD vacuum membrane distillation

Symbol

γ_l surface tension
θ contact angle
ΔP total pressure difference between the membrane surfaces
$\Delta P(T)$ vapour pressure difference as a function of T_1 and T_2
ΔP_A vapour pressure difference across the pores
a exponent coefficient
b membrane thickness
B geometric factor

Current Trends and Future Developments on (Bio-) Membranes
DOI: https://doi.org/10.1016/B978-0-12-822257-7.00012-1

C_1 and C_2	constants
d	pore diameter
D_{AB}	relative diffusivity of the vapour (A) to air (B)
E	thermal efficiency
H_{latent}	latent heat of vaporization
J	flux
J_m and J_k	vapour fluxes contributed respectively by molecular and Knudsen diffusions
k_B	Boltzman constant
Kn	Knudsen number
l	mean free path
m_1 and m_a	molecular weights of vapour and air
P_m	mean pressure in the pore
P_{max}	maximum operational pressure
P_{pore}	pressure in the pore
Q	total heat transferred across the membrane
Q_c	heat transfers arising from thermal conduction
Q_l	latent heat associated with mass transfer
R	universal gas constant
R_m and R_k	mass transfer resistances to molecular and Knudsen diffusions
r_{max}	largest pore size
$R_{membrane}$	mass transfer resistance of membrane
t	pore tortuosity
T	mean temperature in the pores,
T_1 and T_2	temperatures at the membrane surfaces of the feed and permeate sides
x_A	mole fraction of vapour
ε	porosity of membrane
η	gas viscosity
λ, λ_{air} and λ_{solid}	thermal conductivities of membrane, air and the membrane material
σ_1 and σ_a	collision diameters of vapour and air

5.1 Introduction

Membrane distillation (MD) is a process driven by the partial vapor pressure difference across a hydrophobic membrane. Although MD is defined as a thermal process due to the involvement of water evaporation, depending on the purpose of separation, elevating the temperature of the feed side might not be necessary or desirable for the separation (Yang, Duke, Zhang, & Li, 2019; Zhang et al., 2019). This special driving force makes MD processes distinctive from other membrane processes in which an absolute hydraulic pressure difference, a concentration gradient, or an electrical potential gradient drives mass transfer through a membrane.

MD was patented in the 1960s for water demineralization purposes (Ahmed, Lalia, Hashaikeh, & Hilal, 2020; Macedonio & Drioli, 2019). Since adequate membranes suitable for MD were not readily available and MD systems were not optimized, the economics of MD were poor and related research ceased at that time (Alklaibi & Lior, 2005; Hanbury & Hodgkiess, 1985). The temperature polarization coefficient of a prototype MD configuration was roughly estimated as 0.32 by Schofield, Fane, and Fell (1987b). Thus for this system, the actual temperature difference driving mass transfer across the membrane was only 3.2°C when the average temperature difference between the hot and cold channels was 10°C.

The emergence of new membrane materials since the1980s has garnered massive interest from many researchers conducting MD studies (Ardeshiri, Akbari, Peyravi, &

Jahanshahi, 2019; Hubadillah et al., 2019). In combination with the reformation of membrane materials, novel module designs that can significantly improve the thermal efficiency or freshwater productivity were also developed on the basis of improved understanding of heat and mass transfer principles (Chung, Swaminathan, & Warsinger, 2016; Dow et al., 2016, 2017; Gao, Zhang, Gray, & Li, 2017). These days, membranes and modules have been tailored to different types of applications, such as environmental protection, low-grade energy recovery, and wastewater treatment areas (Dow et al., 2016; Yang, Pang, Zhang, Liubinas, & Duke, 2017; Zhang et al., 2019).

The terminology for MD was formally settled in the "Round Table" at the "Workshop on Membrane Distillation" in Rome on May 5, 1986 (Cassano, Conidi, & Drioli, 2020). According to the established definitions, the porous membrane is not required to be hydrophobic, but it should not be wetted by the process liquids during operations. Mass transfer across the membrane is driven by the partial pressure gradient of the component in the gas phase, which should not condense in the pores. The membrane should directly contact the liquid with one or both sides, but it must not affect the vapor equilibrium of the different components. Therefore the micropores in the MD membrane act as passages, which only allow components in the gas phase to pass through. The driving force for mass transfer is the vapor pressure gradient that can arise from concentration difference if volatiles are involved, temperature difference, and/or reduced pressure across the membrane.

For most applications, the mass transfer is associated with a phase change, which requires thermal energy input to maintain MD operation. In desalination processes where water is separated from the nonvolatile salts, the mass transfer in MD requires the latent heat to convert the water liquid to vapor, and a vapor pressure difference to drive the vapor across the membrane. Since the vapor pressure variation is in general not very sensitive to feed concentration changes (Al-Salmi et al., 2020; Shao et al., 2019), MD can deal with relatively high concentration feed without a change in operating condition in comparison with other desalination processes.

As per these characteristics, the distinctive advantages of MD include a theoretically 100% rejection of nonvolatile components, no requirement for hydraulic pressure difference for mass transfer, a compact vapor space compared to conventional distillation (multistage flash distillation, MSF), and a mild operating temperature (40°C–80°C) (Hitsov et al., 2017; Kebria & Rahimpour, 2020). Thus theoretically, MD has a relative smaller footprint than MSF, is able to produce more pure water than reverse osmosis (RO) if only nonvolatile components are in the feed, and can be coupled with low-grade heat sources. Furthermore, since the operating conditions of MD are suitable for most of the commonly available plastics, MD can be used in handling highly corrosive wastewater, which could require considerable investment if metal equipment was used.

As shown in Fig. 5.1, there are four major configurations of MD modules that differ according to their configuration on the permeate side of the membrane and are used for different purposes. Both sides of direct contact membrane distillation (DCMD) membrane are in contact with liquid phases. An air gap is inserted between the membrane and the coolant in air gap membrane distillation (AGMD), in which the permeate condenses on the cooling plate. The permeate evaporates across the membrane under reduced pressure in a vacuum chamber in the vacuum membrane distillation (VMD) configuration. For

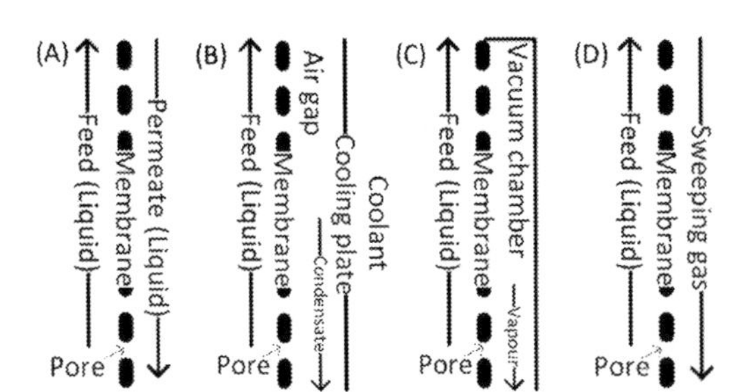

FIGURE 5.1 Schematics of MD configurations (A: DCMD, B: AGMD C: VMD, D: SGMD).

sweeping gas membrane distillation (SGMD), a gas is used to strip the vapor across the membrane. Condensation or adsorption is required for both VMD and SGMD outside the MD module, if the permeated components need to be collected.

The cost estimations vary greatly for MD desalination, in accordance with the availability of heat and heat recovery designed into the process. Estimated by Al-Obaidani et al. (2008), the cost for DCMD was \$1.17 m^{-3} with heat recovery, which was comparable to the cost of other distillation related desalination process, such as multiple effect distillation ($\approx$ \$1.00 m^{-3}) and MSF ($\approx$\$1.40 m^{-3}). If low-grade thermal energy sources are available, the expected cost of MD will be reduced to about \$0.50 m^{-3}, equivalent to the cost of water produced by RO (Fritzmann, Löwenberg, Wintgens, & Melin, 2007). Tavakkoli et al. estimated that the cost would be \$5.70 m^{-3} for desalination of the high salinity water (10 wt.% salt) produced from shale gas at a total recovery of 67% (Tavakkoli, Lokare, Vidic, & Khanna, 2017), which could be reduced dramatically to \$0.74 m^{-3} with integration of waste heat. Walton et al. (El-Bourawi, Ding, Ma, & Khayet, 2006; Khan & Nordberg, 2019) estimated that the cost of a full-scale MD with productivity of 3800 m^3/day was \$0.782 m^{-3} if 30% heat could be recovered.

5.2 Mass and heat transfers in the membrane distillation process

In MD, mass transfer and heat transfer are coupled. Mass transfer in MD is determined by the membrane permeability and the availability of the latent heat at the membrane surface where the evaporation occurs. The heat transfer rate from the bulky feed to the membrane surface could be a key factor limits the mass transfer.

5.2.1 Mass transfer through the membrane

Microporous membrane in MD rejects nonvolatile components in the feed by allowing vapor to pass though. Permeability of the membrane is a key factor to assess whether the membrane is able to achieve reasonable flux.

5.2.1.1 Influence of the membrane characteristics on mass transfer

The membrane in MD is a physical barrier that separates the feed from the permeate and components that permeate through the membrane. Although the existence of the membrane makes the MD module more compact than the conventional distillation technology (Kiss & Kattan Readi, 2018; Tomaszewska, 2000), it also reduces the effective evaporation area of the process liquid, since the porosity of the membrane MD is always less than 100%. Since the pore channels generally do not go straight through the membrane, the path for vapor transport in the membrane will be greater than the membrane thickness. Furthermore, the collision between the pore walls and the vapor molecules decreases the momentum of the molecules and increases the mass transfer resistance. Therefore the existence of the membrane in the distillation process increases the mass transfer resistance. The resistance $R_{membrane}$ of membrane to the mass transfer can be estimated by (Zhang et al., 2010):

$$R_{membrane} \propto \frac{tb}{d^{a}\varepsilon} \tag{5.1}$$

where t is the pore tortuosity, which is defined as the ratio of the pore channel length to the membrane thickness, b is the membrane thickness, d is the diameter of the pore, ε is the porosity of the membrane, and a is an exponent coefficient in the range of 1−2.

By using Eq. (5.1), it can be found that the resistance of a membrane to mass transfer could be mitigated by reducing the membrane thickness (b), enlarging the pore size (d), straightening the pore channel (t) and increasing the porosity (ε). However, the membrane needs mechanical strength to maintain its structure and performance. Thus it is impossible to reduce the thickness and/or increase the porosity infinitely, which could make the membrane unusable. Depending on the materials and fabrication methods (Ho & Zydney, 2000; Schofield, Fane, & Fell, 1990), the tortuosity of the MD membrane varies greatly from close to 1 up to about 27, and there is no acknowledged adverse effect by making the channel straighter. The mass transfer resistance would reduce exponentially as the pore diameter increases. However, in the terminology of MD (Thomas, Mavukkandy, Loutatidou, & Arafat, 2017), it is required that the membrane should not be wetted by the process liquid, which relies on the minimum liquid entry pressure (LEP) calculated using Eq. (5.2) based on the maximum pore size of the membrane (Alklaibi & Lior, 2005; Yazgan-Birgi, Ali, & Arafat, 2018).

$$LEP = \frac{-2B\gamma_l}{r_{max}} cos\theta \tag{5.2}$$

where B is a geometric factor, γ_l is the surface tension of the solution, θ is the contact angle between the solution and the membrane surface which depends on the hydrophobicity of the membrane, and r_{max} is the largest pore size.

The maximum operational pressure (P_{max}) in a MD module can be calculated by Eq. (5.3).

$$P_{max} = LEP + P_{pore} \tag{5.3}$$

where P_{pore} is the pressure in the pore. Therefore for a given system, the maximum pore size of membrane can used in the MD process is:

$$r_{max} = \frac{-2B\gamma_l}{P_{max} - P_{pore}} cos\theta \tag{5.4}$$

Since most process liquids treated by MD are aqueous solutions, based on Eq. (5.4), it will potentially facilitate mass transfer:

- if the pressure difference between the pore and inside the module is low, since large pore membranes can be used with a given membrane material, and
- if the hydrophobicity/contact angle of the membrane material is great, since the membrane used can have larger pore size under the same operation conditions without membrane wetting.

Since the absolute maximum value of $cos\theta$ is 1 in Eq. (5.4), the increment of applicable pore size by increasing hydrophobicity is limited. However, if the difference between the maximum pressure in the module and pressure in the pore is approaching zero, theoretically the pore size could be infinite based on Eq. (5.4) so long as the membrane is hydrophobic ($\theta > 90$ degrees). Thus lowering the process pressure of the operation is also significant in increasing the overall mass transfer in MD process by employment of large pore size membranes.

Hydrophobic materials, such as polyvinylidene fluoride (PVDF), polytetrafluoroethylene (PTFE), and polypropylene (PP) are commonly used for the fabrication of the MD membranes (Li, Rana, Matsuura, & Lan, 2019; Zhang et al., 2010). In Table 5.1, the surface energy of those materials and the typical porosity of membranes fabricated from the different materials are listed (Alklaibi & Lior, 2005; Curcio & Drioli, 2005; Huang, Huang, Xiao, You, & Zhang, 2017; Lloyd, Kinzer, & Tseng, 1990; Mulder, 1996; Zhang et al., 2010). It can be found that the PTFE membrane prepared by the stretching method shows the most suitable characteristics for high mass transfer in MD. The pore size of the MD membrane is normally greater than 0.1 μm but less than 1 μm, which falls into the range of microfiltration membrane.

5.2.1.2 Mathematical approaches for mass transfer across the membrane in different configuration

Knudsen diffusion (K), Poiseuille flow (P), and molecular diffusion (M) are the three fundamental mass transfer mechanisms for the vapor diffusing through the membrane pores (Lawson & Lloyd, 1997; Lei, Chen, & Ding, 2005; Schofield, Fane, & Fell, 1987a; Zhang et al., 2010). The dominant mass transfer mechanism of gas molecule across a given membrane under certain conditions is determined by the Knudsen number (Kn).

$$Kn = \frac{l}{d} \tag{5.5}$$

TABLE 5.1 Pore size and porosity of membrane made from different materials.

Material	Surface Energy ($\times 10^{-3}$ N/m)	Preparation	Porosity (%)
PTFE	9.1	Sintering	10–40
		Stretching	>90
PP	30.0	Stretching	~90
PVDF	30.3	Phase inversion	~80

where d is the mean pore size of the membrane and l is the mean free path of the molecules defined by (Albert & Silbey, 1997; Kuhn & Forstering, 2000):

$$l = \frac{k_B T}{\pi\left((\sigma_1 + \sigma_a)/2\right)^2 P_{pore}} \frac{1}{\sqrt{1 + (m_1/m_a)}} \tag{5.6}$$

where k_B is the Boltzmann constant (1.381×10^{-23} J/K), σ_1 and σ_a are the collision diameters of the vapor and air, T is the mean temperature in Kelvin in the pores, and m_1 and m_a are the molecular weights of the vapor and air. For water vapor molecules transferring through the membrane pores at 60°C, the mean free path is 0.1 μm (Li & Sirkar, 2017).

If it is a gas mixture (vapor + air) and there is no total pressure gradient in the pores, the dominating mass transfer mechanism based on the Kn is shown in Table 5.2 (Lei et al., 2005). The mass transfers in DCMD and AGMD are under the same conditions (Ding, Ma, & Fane, 2003). Since the pore sizes of the membranes used for membrane distillation are in the range of 0.1 to 1.0 μm, Kn will be in the range of 0.1 to 1. Therefore the dominant mass transfer mechanism is Knudsen–molecular diffusion transition or Knudsen, depending on the pore size. Although the pore size distribution of the polymeric membrane will affect the mass transfer mechanism, it has been demonstrated that the transition region is the major mass transfer mechanism for the majority of membranes (Phattaranawik, Jiraratananon, & Fane, 2003a).

In Fig. 5.2, electrical circuit analogs are used to show mass transport models for vapor diffusion of a gas mixture through a membrane (Ahadi, Karimi-Sabet, Shariaty-Niassar, &

TABLE 5.2 Dominant mass transfer mechanism of a gas mixture in membrane pore without a pressure gradient.

$Kn < 0.01$	$0.01 < Kn < 1$	$Kn > 1$
Molecular diffusion	Knudsen–molecular diffusion transition mechanism	Knudsen mechanism

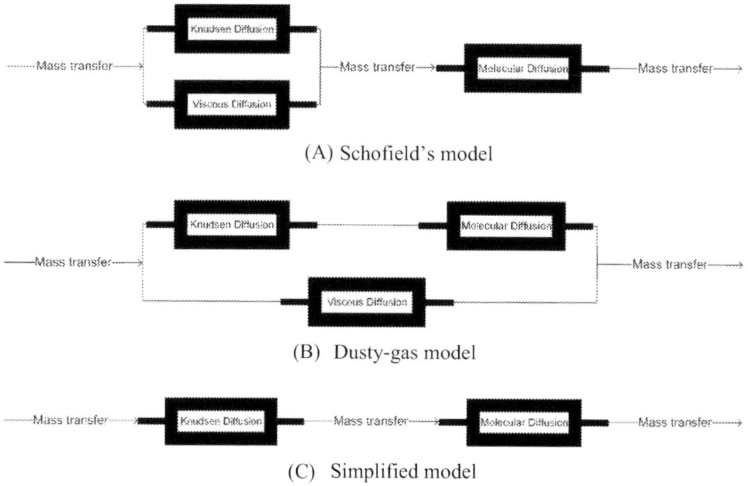

(A) Schofield's model

(B) Dusty-gas model

(C) Simplified model

FIGURE 5.2 Electrical circuit analogs for different transport models.

Matsuura, 2018; Alklaibi & Lior, 2005; Damtie et al., 2019; Phattaranawik et al., 2003a; Schofield et al., 1990). Since in DCMD, AGMD, and SGMD, the total pressure gradient in the pores can be ignored, and viscous diffusion in the Schofield's model (Fig. 5.2A) and Dusty-Gas model (Fig. 5.2B) can be neglected (Li & Sirkar, 2017), both models can be simplified as the shown in Fig. 5.2C, in which only Knudsen and molecular diffusions are involved. Therefore the mass transfer resistance through the membrane pore can be expressed as (Phattaranawik et al., 2003a):

$R_{membrane} = R_k + R_m = \frac{1}{J} = \frac{1}{J_m} + \frac{1}{J_k}$, in which

$$J_k = \frac{4}{3} d \frac{\varepsilon}{bt} \sqrt{\frac{M}{2\pi RT}} \Delta P_A, \text{ and } J_m = \frac{m_1}{1 - x_A} \frac{\varepsilon D_{AB}}{btRT} \Delta P_A \tag{5.7}$$

where R_m and R_k are the mass transfer resistances to molecular and Knudsen diffusions, J is the membrane flux, J_m and J_k are the vapor fluxes contributed respectively by molecular and Knudsen diffusions, R (= 8314 Pa.m^3/mol/K) is the universal gas constant, x_A is the mole fraction of the vapor, ΔP_A is the vapor pressure difference across the pores, and D_{AB} (m^2/s) is relative diffusivity of the vapor (A) to air (B), which can be estimated using Eq. (5.8) in the temperature range of 273–373K (Phattaranawik, Jiraratananon, & Fane, 2003b).

$$D_{AB} = \frac{1.895 \times 10^{-5} T^{2.072}}{P} \tag{5.8}$$

In VMD, the air in the pores will be replaced by the component(s) evaporates from the feed continuously. Therefore it can be assumed that there is no air stagnant in the pores. If water vapor is the only component passing through the membrane pores, it should be a single gas in the pore, which transfers through the membrane under total pressure gradient. The dominant mass transfer mechanism in VMD is shown in Table 5.3.

Since the Kn in VMD is in range of 0.11 to 1.1, the dominant mass transfer mechanisms are Poiseuille flow–Knudsen diffusion transition mechanism and Knudsen diffusion. In Fig. 5.3, the mass transfer is shown as an electrical circuit analog.

TABLE 5.3 Thermal conductivities of common membrane materials.

Membrane Material	PTFE	PP	PVDF
Thermal conductivity (W.m^{-1}K^{-1})	~0.25	~0.17	~0.19

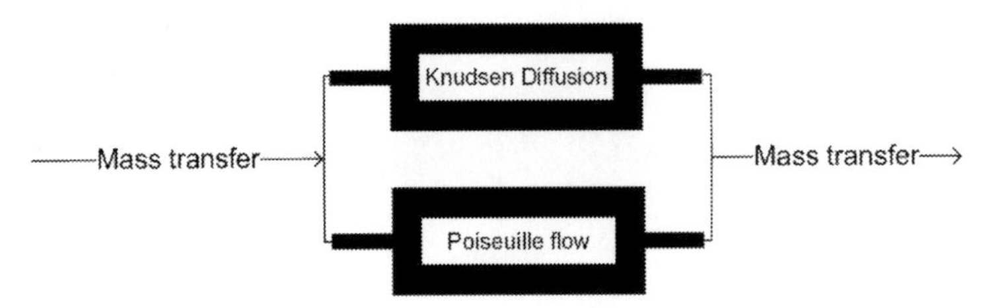

FIGURE 5.3 Electrical circuit analogs for VMD.

Therefore the mass transfer through the membrane pore can be expressed as (Zhang et al., 2013):

$$\frac{1}{R_{membrane}} = \frac{1}{R_p} + \frac{1}{R_k} = J = J_p + J_k = \left(\frac{8}{3}\frac{r\varepsilon}{tb}\sqrt{\frac{1}{2\pi R m_1 T}} + \frac{\varepsilon r^2}{tb}\frac{1}{8\eta}\frac{P_m}{RT}\right)\Delta P \tag{5.9}$$

where η is the viscosity of the gas transferred across the membrane pores, P_m is the mean pressure in the pore and ΔP is the total pressure difference between the membrane surfaces.

5.2.2 Heat transfers through the membrane

Heat transfer is in the same direction as the mass transfer and is in forms of latent heat and sensible heat.

5.2.2.1 Influence of the membrane characteristics on heat transfer

Heat transfers through the membrane via thermal conduction and latent heat, which can be calculate by:

$Q = Q_c + Q_l = \frac{\lambda}{b} A(T_1 - T_2) + JAH_{latent}$, in which

$$\lambda = \lambda_{air}\varepsilon + \lambda_{solid}(1 - \varepsilon) \tag{5.10}$$

where Q *is* the total heat transferred across the membrane; Q_c and Q_l are the heat transfers arising from thermal conduction and the latent heat associated with mass transfer; λ, λ_{air}, and λ_{solid} are the thermal conductivities of the membrane, air and the membrane material; A is the membrane surface area; T_1 and T_2 are the temperatures at the membrane surfaces of the feed and permeate sides, respectively, and H_{latent} is the latent heat of vaporization.

It can be found that the latent heat transfer is associate with the flux (mass transfer) in Eq. (5.10), which is the effective energy driving mass transfer. The thermal efficiency (E) can be calculated by:

$$E = \frac{Q_l}{Q} \tag{5.11}$$

Therefore to increase the thermal efficiency, thermal conduction losses need to be reduced by lowering λ/b. The thermal conductivity of the solid material used for membrane fabrication is generally higher than that of gas. Therefore increasing the porosity and/or using membrane materials with low thermal conductivity would reduce the thermal conduction loss. The thermal conductivities of the most common membrane materials are shown in Table 5.3. In combination with membrane porosity shown in Table 5.1, it can be seen that the PTFE/PP membranes fabricated by the stretching method would possess the lowest heat loss due to thermal conduction.

Increasing the membrane thickness will also lower the heat loss. However, it would also increase the mass transfer resistance as shown in Eq. (5.1). Since the mass transfer is linear to the driving force, vapor pressure difference, Eq. (5.10) can be converted to:

$$Q = \frac{C_1(T_1 - T_2)}{b} + \frac{C_2 \Delta P(T)}{b} \tag{5.12}$$

where C_1 and C_2 are constants, and $\Delta P(T)$ is the vapor pressure difference as a function of T_1 and T_2.

When the membrane thickness increases, it can be speculated that the temperature on the cold side of the membrane will decrease. Therefore the vapor pressure on the cold side will decrease exponentially and the driving force will increase, if the membrane surface temperature on the hot side is assumed approximately constant. From Fig. 5.4, it can be found that when temperature at the cold side of membrane is in range of 20°C to 40°C, a 1°C decline leads to 237 Pa vapor pressure drop (or mass transfer driving force increase) depending on the slope. However, when the temperature at the cold side of membrane surface is in range of 0°C to 20°C, a 1°C decline only leads to about 82 Pa vapor pressure drop. If the temperature at the cold side of the membrane is high, a thicker membrane might increase the thermal efficiency by increasing the Q_l, since a greater vapor pressure difference increase is achievable owing to the temperature decline on the cold side of the membrane. However, when the temperature on the cold surface side of the membrane is low, a thicker membrane might reduce the thermal efficiency by reducing the Q_l, since the contribution from the increasing vapor pressure difference could be less than the resistance increase to the mass transfer. Thus the optimum membrane thickness is related to the operating temperatures. Furthermore, this conclusion is not valid for VMD, in which mass transfer driving force is related to the vapor pressure difference between the vapor pressure on the hot side of the membrane and the vacuum degree on the cold side of the membrane, and is not affected by temperature at the cold side of the membrane. Thus a thinner membrane would enhance mass transfer for VMD.

5.2.2.2 *Heat transfer through the membrane in different configurations*

In all configurations, DCMD has the lowest thermal efficiency in general, since the cold side of the membrane contacts directly with liquid that conducts heat faster than air and vacuum, and this enhances the heat conduction loss with a given membrane. In VMD, the

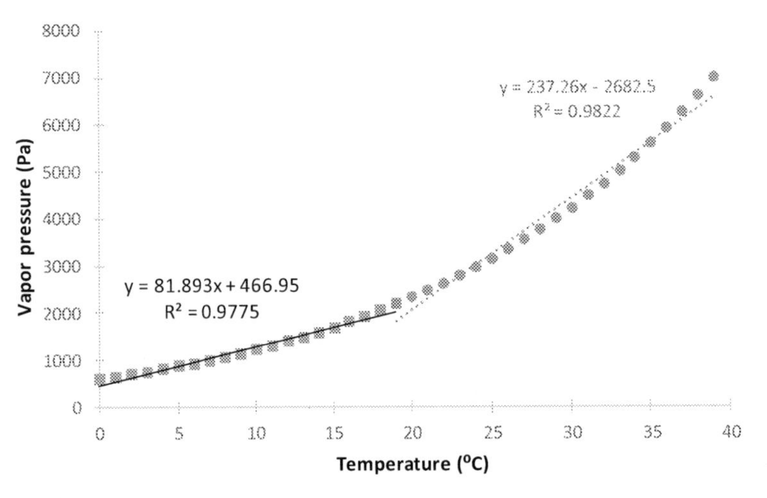

FIGURE 5.4 Vapor pressure change with the temperature.

vapor in the pores and in the chamber contacting the membrane is under the reduced pressure, and will have much lower conductivity than air under atmospheric pressure (Mo & Ban, 2017). Therefore VMD should have the highest thermal efficiency because the thermal conduction loss can be ignored (Zhang et al., 2013). In SGMG and AGMD, the cold side of the membrane contacts a gas phase with lower thermal conductivity than that of a liquid phase. Although they have relatively low fluxes due to the extra resistance to mass transfer in the gas phase, their thermal efficiencies are higher than that of the DCMD since there is less heat loss from thermal conduction due to the higher thermal conduction resistance of the gas layer.

5.2.3 Temperature polarization

In MD, the apparent driving force – the difference between bulk flow temperatures—is different from the actual driving force that is the temperature difference between the membrane surfaces. The lower temperature difference across the membrane surfaces compared to that between the bulk liquids is due to a phenomenon called "temperature polarization." This is caused by the combination of heat transfer between boundary layers on both sides of the membrane and slow heat exchange between the bulk flows and the boundary layers.

For MD, mass transfer across the membrane is associated with evaporation from the boundary layer on the hot side and condensation into the boundary layer on the cold side. The heat transfer is the combination of latent heat transfer and conductive heat transfer as shown in Eq. (5.10). Since the latent heat for evaporation is much greater than the specific heat of water (Bhatt et al., 2018; Daily, 2018; Dudorova & Belan, 2019), significant temperature decrease and increase occur in the boundary layers. This creates a thermal boundary layer on each side of the membrane. Therefore the boundary layers (both hydraulic and thermal) on the hot and cold sides of the membrane will respectively have lower and higher temperatures than that of bulk flows. Hence the temperature difference between both sides of the membrane is less than that between the bulk flows.

Temperature polarization lowers the driving force (i.e., the temperature difference) for both heat transfer and mass transfer. However, since the relationships of temperature to conductive heat transfer and mass transfer are linear and exponential respectively, optimization is required to achieve high thermal efficiency as defined in Eq. (5.11).

5.2.4 Influence of module and membrane configurations on mass and heat transfers

The configurations of membrane and module will affect the resistance to mass transfer and heat transfer. In accordance with the purpose of the project, optimization is required to make both membrane and module suitable for the given conditions and minimize the negative impact.

5.2.4.1 *Mass and heat transfers affected by membrane configurations*

Flat sheet and hollow fiber membranes are the two most common configurations of MD membrane as shown in Fig. 5.5. Flat sheet membranes consist of a thick porous support

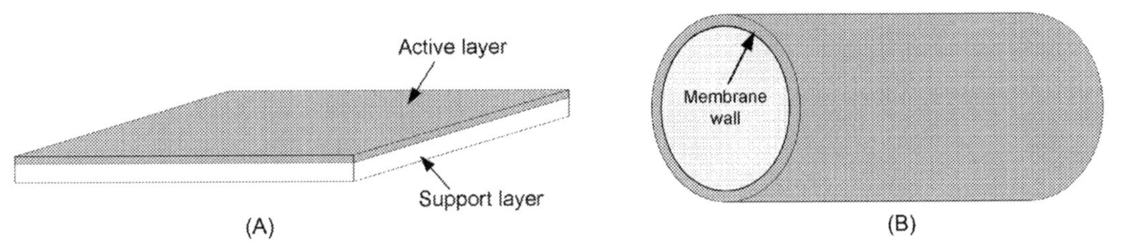

FIGURE 5.5 Schematics of flat sheet membrane (A) and hollow fiber membrane (B).

layer and a thin active layer. This structure is able to provide enough mechanic strength for the membrane, and enables the active layer to be manufactured as thin as possible. The thin active layer reduces the mass transfer resistance based on Eq. (5.1) and flat sheet membranes typically have flux of 20–30 Lm^{-2}/h (Alklaibi & Lior, 2005).

Compared to flat sheet membrane, the hollow fiber membrane is self-supported and homogenous. Since the membrane thickness is much greater than that of the active layer of the flat sheet membrane, the hollow fiber membrane typically has low flux (generally 1–4 $L.m^{-2}/h$) (Bonyadi & Chung, 2007, 2009; Cheng, Wu, & Chen, 2008). Many studies have been conducted to improve the flux of hollow fiber membrane by structure modification. To minimize the mass transfer resistance of the active layer, a dual-layer hydrophilic–hydrophobic hollow fiber membrane was fabricated with a very thin effective hydrophobic PVDF layer (50 μm) (Bonyadi & Chung, 2007). Hollow fiber membranes with asymmetric structure similar to the flat sheet membrane were also developed, which consists of a sponge-like support layer and a thin skin active layer (Bonyadi, Chung, & Rajagopalan, 2009; Guclu et al., 2019; Ravi et al., 2020).

Many novel membranes have been developed in recent years to increase possible MD application areas. Fouling and wetting resistant membranes have been widely researched (Fan et al., 2016; Wang, Tang, & Li, 2017; Wang & Lin, 2017), besides flux-enhanced membranes. However, most of these studies still remain at the laboratory research stage. A commercially available dual-layer membrane has been introduced by (Cheng et al., 2018; Zhang et al., 2019). The membrane could process water containing volatile organics and sodium dodecyl sulfate, but the membrane flux was compromised significantly.

The flux of MD membrane is quite different from the pressure driven membrane processes, and is also related to the module length (Zhang, Gray, & Li, 2012). The difference is caused by the driving force distribution along the membrane. Mass transfer in MD will cause significant loss in driving force due to the decline/increase in feed/permeate temperature from the inlet to outlet along its flow direction. The loss increases with the duration of the mass and heat transfers, which is determined by the membrane length in the flow direction at a given flowrate.

The mean vapor pressure difference between the membrane surfaces calculated according to Zhang et al. (2012) varied with membrane length as shown in Fig. 5.6. The mean driving (mean vapor pressure difference) decreases from 3.03 to 1.12 kPa, as the membrane length increased from 0.1 to 0.5 m. Therefore for the same type of membrane, the shorter membrane (0.1 m) has a flux about three times that of the longer membrane (0.5 m) under the same operating conditions.

Hence under the same conditions, the membrane flux is low when a membrane is long in the flow direction at low flow velocity. Therefore flux is not a perfect performance indicator for the comparison of MD membrane permeability from different works where operating conditions and module design vary.

5.2.4.2 *Mass and heat transfers affected by module configurations*

Hollow fiber and flat sheet membranes are two common types of module configuration as shown in Fig. 5.7. Hollow fiber membranes do not need a separate support to maintain its flow channels owing to their cylindrical structure. The flexible flat sheet polymeric membranes can be deformed under pressure and block the flow channel (Duke et al., 2009). Therefore in the flat sheet module, as shown in Fig. 5.7A, spacers are generally used on both sides of the membrane to maintain the clearance of flow channels. Since the spacers have thicknesses similar to or even greater than that of the membrane, compared

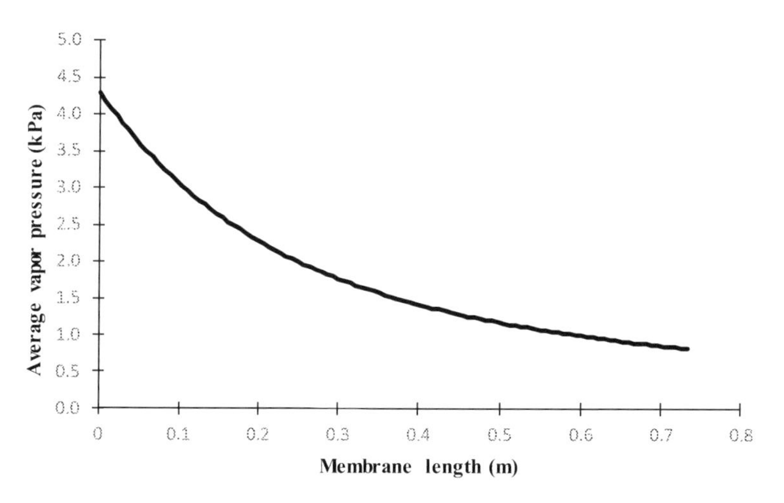

FIGURE 5.6 Mean vapor pressure difference varies with membrane length (Feed inlet temperature = 60°C, permeate inlet temperature = 20°C, and stream velocities = 0.114 m/s).

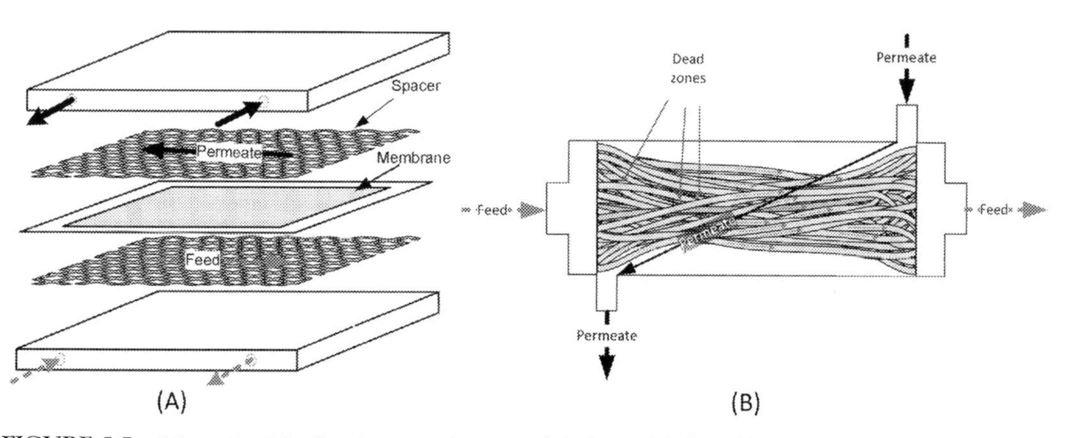

FIGURE 5.7 Schematic of the flat sheet membrane module (A) and hollow fiber membrane module (B).

with hollow fiber membranes, less membrane area could be packed per unit volume of module (Wu et al., 2017). However, the spacers also act as turbulence promoters, and reduce temperature polarization.

Since the existence of spacers reduce temperature polarization, the temperature difference ($T_1 - T_2$) across the membrane will increase. Hence both the flux and heat conduction loss increase according to Eq. (5.12). However, the linear increase of heat conduction loss is generally less than the exponential increase of heat transfer from evaporation, when temperature difference ($T_1 - T_2$) across the membrane becomes large. From Fig. 5.4, it can be found that 1°C variation in temperature on both sides of the membrane will lead to 319 Pa (82 Pa decrease on permeate side and 237 Pa increase on feed side) increase of mass transfer driving force (vapor pressure change) in the MD operating temperature range. Therefore the thermal efficiency calculated by Eq. (5.11) may increase with reduced temperature polarization.

Generally, hollow fiber membranes are randomly packed as a bundle into the module without turbulence promoters as shown in Fig. 5.7B. Since the membranes could contact mutually in the bundle, dead zones may be formed, where the fluid is almost stagnant. Furthermore, the tight membrane bundle could show high resistance to the stream on the shell side of the module, and reduce stream flowing into the fiber bundle. Hence the membranes inside the bundle are not able to contact the stream effectively. The lack of contact between the stream and fiber will lead to the significantly less effective mass transfer area than the nominal membrane area and serious temperature polarization. However, the productivity per unit volume of the hollow fiber module is still generally higher than that of the flat sheet membrane module, since the packing density of the hollow fiber module could be as high as 90% (Ali, Aimar, & Drioli, 2015).

For a symmetric hollow fiber membrane, the hot feed is normally arranged in the lumen side of the hollow fiber membrane, and is able to achieve reasonable turbulence (Calabrò, 2013; Teoh, Bonyadi, & Chung, 2008) that suppresses temperature polarization of the hot feed. Serious temperature polarization occurs on the cold permeate side and will cause high T_2, which will reduce the heat conduction loss from the hot feed to cold permeate based on Eq. (5.12). As T_2 increases, a higher T_1 is expected. As shown in Fig. 5.4, compared with the temperature variation in the high temperature range, the same variation in low temperature range will lead to much less change of vapor pressure. Although the increase of T_2 would be greater than the increase of T_1, the vapor pressure difference could maintain similar, since T_1 and T_2 are respectively in the high and low temperature ranges. Therefore the flux of the hollow fiber membrane would not be reduced significantly owing to the temperature polarization on the permeate side. Furthermore, if the feed is recirculated, the overall thermal efficiency (Eq. (5.11)) will not be compromised either, since the high feed outlet temperature will lower the requirement of the thermal energy to heat the feed up to the set inlet temperature.

In VMD, the permeate side is under low absolute pressure. If the permeate side is arranged on the shell side of the hollow fiber module, the resistance of hollow fiber bundle to the vapor is negligible, since the viscosity of the vapor is low under vacuum pressure (Crifo, 1989). Therefore the dead zones would not exist on the shell side. However, if the feed has to be arranged on the shell side of the module, the advantages of hollow fiber in VMD employment will be comprised.

5.3 Conclusion and future trends

MD is a distinctive membrane process, in which phase change is involved in mass transfer. Since heat transfer is mainly associated with mass transfer in MD, optimization is required to achieve high thermal efficiency, based on the linear and exponential relationship of driving force for mass transfer and heat conduction.

Membranes with high porosity, low thickness and large pore size are preferred in MD operation, as far as it provides enough durability and wetting resistance. Materials with high hydrophobicity have the potential to enable use of membranes with large pore size, although it is also related to the pore size distribution.

Knudsen diffusion (*K*), Poiseuille flow (*P*), and molecular diffusion (*M*) are the fundamental mass transfer mechanisms for the vapor diffusing through the membrane pores. In DCMD, AGMD, and SGMD, Knudsen and molecular diffusions are the dominant mass transfer mechanism. In VDM, the mass transfer is dominated by Poiseuille flow–Knudsen diffusion transition mechanism and Knudsen diffusion.

In the four major MD configurations, DCMD and VMD, respectively have the lowest and highest thermal efficiency. Temperature polarization is caused by the combination of heat transfer between the thermal boundary layers on both sides of membrane and slow heat exchange between the bulk flows and the boundary layers.

Flat sheet membranes and modules show better flux and lower temperature polarization than that of the hollow fiber membranes and modules. However, the unit volume productivity of the hollow fiber module can be higher than that of the flat sheet module.

References

Ahadi, H., Karimi-Sabet, J., Shariaty-Niassar, M., & Matsuura, T. (2018). Experimental and numerical evaluation of membrane distillation module for oxygen-18 separation. *Chemical Engineering Research and Design, 132*, 492–504.

Ahmed, F. E., Lalia, B. S., Hashaikeh, R., & Hilal, N. (2020). Alternative heating techniques in membrane distillation: A review. *Desalination, 496*, 114713.

Al-Obaidani, S., Curcio, E., Macedonio, F., Di Profio, G., Al-Hinai, H., & Drioli, E. (2008). Potential of membrane distillation in seawater desalination: Thermal efficiency, sensitivity study and cost estimation. *Journal of Membrane Science, 323*, 85–98.

Al-Salmi, M., Laqbaqbi, M., Al-Obaidani, S., Al-Maamari, R. S., Khayet, M., & Al-Abri, M. (2020). Application of membrane distillation for the treatment of oil field produced water. *Desalination, 494*, 114678.

Albert, R. A., & Silbey, R. J. (1997). *Physical Chemistry* (2nd ed.). New York: Wiley.

Ali, A., Aimar, P., & Drioli, E. (2015). Effect of module design and flow patterns on performance of membrane distillation process. *Chemical Engineering Journal, 277*, 368–377.

Alklaibi, A. M., & Lior, N. (2005). Membrane-distillation desalination: Status and potential. *Desalination, 171*, 111–131.

Ardeshiri, F., Akbari, A., Peyravi, M., & Jahanshahi, M. (2019). A hydrophilic-oleophobic chitosan/SiO 2 composite membrane to enhance oil fouling resistance in membrane distillation. *Korean Journal of Chemical Engineering, 36*, 255–264.

Bhatt, N., Pati, A., Das, L., Panda, A., Varshney, P., Kumar, A., ... Mohapatra, S. (2018). The diminishment of specific heat and surface tension of coolant droplet in a dropwise evaporation process: A novel methodology to enhance the heat transfer rate. *Experimental Heat Transfer, 31*, 355–372.

Bonyadi, S., & Chung, T. S. (2007). Flux enhancement in membrane distillation by fabrication of dual layer hydrophilic-hydrophobic hollow fiber membranes. *Journal of Membrane Science, 306*, 134–146.

Bonyadi, S., & Chung, T.-S. (2009). Highly porous and macrovoid-free PVDF hollow fiber membranes for membrane distillation by a solvent-dope solution co-extrusion approach. *Journal of Membrane Science, 331*, 66–74.

Bonyadi, S., Chung, T. S., & Rajagopalan, R. (2009). A novel approach to fabricate macrovoid-free and highly permeable PVDF hollow fiber membranes for membrane distillation. *AIChE Journal, 55*, 828–833.

Calabrò, V. (2013). 1 - Engineering aspects of membrane bioreactors. In A. Basile (Ed.), *Handbook of Membrane Reactors* (pp. 3–53). Woodhead Publishing.

Cassano, A., Conidi, C., & Drioli, E. (2020). A Comprehensive Review of Membrane Distillation and Osmotic Distillation in Agro-Food Applications. *Journal of Membrane Science and Research.*

Cheng, D., Zhang, J., Li, N., Ng, D., Gray, S. R., & Xie, Z. (2018). Antiwettability and Performance Stability of a Composite Hydrophobic/Hydrophilic Dual-Layer Membrane in Wastewater Treatment by Membrane Distillation. *Industrial & Engineering Chemistry Research, 57*, 9313–9322.

Cheng, L.-H., Wu, P.-C., & Chen, J. (2008). Modeling and optimization of hollow fiber DCMD module for desalination. *Journal of Membrane Science, 318*, 154–166.

Chung, H. W., Swaminathan, J., & Warsinger, D. M. (2016). Multistage vacuum membrane distillation (MSVMD) systems for high salinity applications. *Journal of Membrane Science, 497*, 128–141.

Crifo, J. (1989). Inferences concerning water vapour viscosity and mean free path at low temperatures. *Astronomy and Astrophysics, 223*, 365.

Curcio, E., & Drioli, E. (2005). Membrane Distillation and Related Operations: A Review. *Separation and Purification Reviews, 34*, 35–86.

Daily, J. W. (2018). *Statistical Thermodynamics: An Engineering Approach*. Cambridge University Press.

Damtie, M. M., Woo, Y. C., Kim, B., Park, K.-D., Hailemariam, R. H., Shon, H. K., & Choi, J.-S. (2019). Analysis of mass transfer behavior in membrane distillation: Mathematical modeling under various conditions. *Chemosphere, 236*, 124289.

Ding, Z., Ma, R., & Fane, A. G. (2003). A new model for mass transfer in direct contact membrane distillation. *Desalination, 151*, 217–227.

Dow, N., García, J. V., Niadoo, L., Milne, N., Zhang, J., Gray, S., & Duke, M. (2017). Demonstration of membrane distillation on textile waste water: assessment of long term performance, membrane cleaning and waste heat integration. *Environmental Science: Water Research & Technology, 3*, 433–449.

Dow, N., Gray, S., Zhang, J., Ostarcevic, E., Liubinas, A., Atherton, P., ... Duke, M. (2016). Pilot trial of membrane distillation driven by low grade waste heat: Membrane fouling and energy assessment. *Desalination, 391*, 30–42.

Dudorova, N. V., & Belan, B.D. (2019). The role of evaporation and condensation of water in the formation of the urban heat island. In *25th international symposium on atmospheric and ocean optics: Atmospheric physics* (pp. 112088J). International Society for Optics and Photonics.

Duke, M., Dow, N., Zhang, J., Li, J.-D., Gray, S., & Osteracevic, E. (2009). Enhancing water recovery of reverse osmosis brine reject by membrane distillation. *AWA Ozwater, 09*, 16–18.

El-Bourawi, M. S., Ding, Z., Ma, R., & Khayet, M. (2006). A framework for better understanding membrane distillation separation process. *Journal of Membrane Science, 285*, 4–29.

Fan, X., Liu, Y., Quan, X., Zhao, H., Chen, S., Yi, G., & Du, L. (2016). High desalination permeability, wetting and fouling resistance on superhydrophobic carbon nanotube hollow fiber membrane under self-powered electrochemical assistance. *Journal of Membrane Science, 514*, 501–509.

Fritzmann, C., Löwenberg, J., Wintgens, T., & Melin, T. (2007). State-of-the-art of reverse osmosis desalination. *Desalination, 216*, 1–76.

Gao, L., Zhang, J., Gray, S., & Li, J.-D. (2017). Experimental study of hollow fiber permeate gap membrane distillation and its performance comparison with DCMD and SGMD. *Separation and Purification Technology, 188*, 11–23.

Guclu, S., Erkoc-Ilter, S., Koseoglu-Imer, D. Y., Unal, S., Menceloglu, Y. Z., Ozturk, I., & Koyuncu, I. (2019). Interfacially polymerized thin-film composite membranes: Impact of support layer pore size on active layer polymerization and seawater desalination performance. *Separation and Purification Technology, 212*, 438–448.

Hanbury, W. T., & Hodgkiess, T. (1985). Membrane distillation - an assessment. *Desalination, 56*, 287–297.

Hitsov, I., Eykens, L., Schepper, W. D., Sitter, K. D., Dotremont, C., & Nopens, I. (2017). Full-scale direct contact membrane distillation (DCMD) model including membrane compaction effects. *Journal of Membrane Science, 524*, 245–256.

Ho, C.-C., & Zydney, A. L. (2000). Measurement of membrane pore interconnectivity. *Journal of Membrane Science, 170*, 101–112.

Huang, Q.-L., Huang, Y., Xiao, C.-F., You, Y.-W., & Zhang, C.-X. (2017). Electrospun ultrafine fibrous PTFE-supported ZnO porous membrane with self-cleaning function for vacuum membrane distillation. *Journal of Membrane Science, 534*, 73–82.

Hubadillah, S. K., Tai, Z. S., Othman, M. H. D., Harun, Z., Jamalludin, M. R., Rahman, M. A., ... Ismail, A. F. (2019). Hydrophobic ceramic membrane for membrane distillation: A mini review on preparation, characterization, and applications. *Separation and Purification Technology*.

Kebria, M. R. S., & Rahimpour, A. (2020). Membrane distillation: Basics, advances, and applications. *Advances in Membrane Technologies*. IntechOpen.

Khan, E. U., & Nordberg, Å. (2019). Thermal integration of membrane distillation in an anaerobic digestion biogas plant—A techno-economic assessment. *Applied Energy, 239*, 1163–1174.

Kiss, A. A., & Kattan Readi, O. M. (2018). An industrial perspective on membrane distillation processes. *Journal of Chemical Technology & Biotechnology, 93*, 2047–2055.

Kuhn, H., & Forstering, H.-D. (2000). *Principles of physical chemistry*. New York: Wiley.

Lawson, K. W., & Lloyd, D. R. (1997). Membrane distillation. *Journal of Membrane Science, 124*, 1–25.

Lei, Z., Chen, B., & Ding, Z. (2005). *Special distillation processes*. Elsevier.

Li, L., & Sirkar, K. K. (2017). Studies in vacuum membrane distillation with flat membranes. *Journal of Membrane Science, 523*, 225–234.

Li, Z., Rana, D., Matsuura, T., & Lan, C. Q. (2019). The performance of polyvinylidene fluoride-polytetrafluoroethylene nanocomposite distillation membranes: An experimental and numerical study. *Separation and Purification Technology, 226*, 192–208.

Lloyd, D. R., Kinzer, K. E., & Tseng, H. S. (1990). Microporous membrane formation via thermally induced phase separation. I. Solid-liquid phase separation. *Journal of Membrane Science, 52*, 239–261.

Macedonio, F., & Drioli, E. (2019). Chapter 5 - Membrane distillation development. In C. M. Galanakis, & E. Agrafioti (Eds.), *Sustainable Water and Wastewater Processing* (pp. 133–159). Elsevier.

Mo, J., & Ban, H. (2017). Measurements and theoretical modeling of effective thermal conductivity of particle beds under compression in air and vacuum. *Case Studies in Thermal Engineering, 10*, 423–433.

Mulder, M. (1996). *Basic Principles of Membrane Technology* (2nd ed.). Dordrecht: Kluwer.

Phattaranawik, J., Jiraratananon, R., & Fane, A. G. (2003a). Effect of pore size distribution and air flux on mass transport in direct contact membrane distillation. *Journal of Membrane Science, 215*, 75–85.

Phattaranawik, J., Jiraratananon, R., & Fane, A. G. (2003b). Effects of net-type spacers on heat and mass transfer in direct contact membrane distillation and comparison with ultrafiltration studies. *Journal of Membrane Science, 217*, 193–206.

Ravi, J., Othman, M. H. D., Matsuura, T., Bilad, M. Ri, El-badawy, T., Aziz, F., ... Jaafar, J. (2020). Polymeric membranes for desalination using membrane distillation: A review. *Desalination, 490*, 114530.

Schofield, R., Fane, A., & Fell, C. (1987a). Heat and mass transfer in membrane distillation. *Journal of Membrane Science, 33*, 299–313.

Schofield, R. W., Fane, A. G., & Fell, C. J. D. (1987b). Heat and mass transfer in membrane distillation. *Journal of Membrane Science, 33*, 299–313.

Schofield, R. W., Fane, A. G., & Fell, C. J. D. (1990). Gas and vapour transport through microporous membranes. II. Membrane distillation. *Journal of Membrane Science, 53*, 173–185.

Shao, Y., Han, M., Wang, Y., Li, G., Xiao, W., Li, X., ... He, G. (2019). Superhydrophobic polypropylene membrane with fabricated antifouling interface for vacuum membrane distillation treating high concentration sodium/magnesium saline water. *Journal of Membrane Science, 579*, 240–252.

Tavakkoli, S., Lokare, O. R., Vidic, R. D., & Khanna, V. (2017). A techno-economic assessment of membrane distillation for treatment of Marcellus shale produced water. *Desalination, 416*, 24–34.

Teoh, M. M., Bonyadi, S., & Chung, T.-S. (2008). Investigation of different hollow fiber module designs for flux enhancement in the membrane distillation process. *Journal of Membrane Science, 311*, 371–379.

Thomas, N., Mavukkandy, M. O., Loutatidou, S., & Arafat, H. A. (2017). Membrane distillation research & implementation: Lessons from the past five decades. *Separation and Purification Technology, 189*, 108–127.

Tomaszewska, M. (2000). Membrane distillation-examples of applications in technology and environmental protection. *Environmental Studies, 9*, 27–36.

Wang, Z., & Lin, S. (2017). Membrane fouling and wetting in membrane distillation and their mitigation by novel membranes with special wettability. *Water Research, 112*, 38–47.

Wang, Z., Tang, Y., & Li, B. (2017). Excellent wetting resistance and anti-fouling performance of PVDF membrane modified with superhydrophobic papillae-like surfaces. *Journal of Membrane Science*, *540*, 401–410.

Wu, B., Christen, T., Tan, H. S., Hochstrasser, F., Suwarno, S. R., Liu, X., ... Fane, A. G. (2017). Improved performance of gravity-driven membrane filtration for seawater pretreatment: Implications of membrane module configuration. *Water Research*, *114*, 59–68.

Yang, X., Duke, M., Zhang, J., & Li, J.-D. (2019). Modeling of heat and mass transfer in vacuum membrane distillation for ammonia separation. *Separation and Purification Technology*, *224*, 121–131.

Yang, X., Pang, H., Zhang, J., Liubinas, A., & Duke, M. (2017). Sustainable waste water deammonification by vacuum membrane distillation without pH adjustment: Role of water chemistry. *Chemical Engineering Journal*, *328*, 884–893.

Yazgan-Birgi, P., Ali, M. I. H., & Arafat, H. A. (2018). Estimation of liquid entry pressure in hydrophobic membranes using CFD tools. *Journal of Membrane Science*, *552*, 68–76.

Zhang, J., Dow, N., Duke, M., Ostarcevic, E., Li, J.-d, & Gray, S. (2010). Identification of material and physical features of membrane distillation membranes for high performance desalination. *Journal of Membrane Science*, *349*, 295–303.

Zhang, J., Gray, S., & Li, J.-d (2012). Modelling heat and mass transfers in DCMD using compressible membranes. *Journal of Membrane Science*, *387*, 7–16.

Zhang, J., Li, J.-d, Duke, M., Hoang, M., Xie, Z., Groth, A., ... Gray, S. (2013). Modelling of vacuum membrane distillation. *Journal of Membrane Science*, *434*, 1–9.

Zhang, J., Li, N., Ng, D., Ike, I. A., Xie, Z., & Gray, S. (2019). Depletion of VOC in wastewater by vacuum membrane distillation using a dual-layer membrane: mechanism of mass transfer and selectivity. *Environmental Science: Water Research & Technology*, *5*, 119–130.

CHAPTER

6

Transport phenomena in dialysis processes

Marco Cocchi, Leone Mazzeo and Vincenzo Piemonte

Faculty of Engineering, University Campus Biomedico of Rome, Rome, Italy

Abbreviations

AKI acute kidney injuries
CKD chronic kidney disease
DRU dialysate regeneration unit
EO electro-oxidation
ESKD end-stage kidney disease
HD hemodialysis
HDF hemodiafiltration
HF hemofiltration
MW molecular weight
OL-HDF online hemodiafiltration
PD peritoneal dialysis
RRTs renal replacement therapies
SC sieving coefficient
TR transport resistance

Nomenclature

A membrane surface area
AR adsorption ratio
c_i interface concentration
$c_i^{'}$ bulk concentration
$\overline{C_i}$ membrane solute concentration
CL clearance
C_s solute concentration
D diffusivity
DL dialysance
D_m membrane diffusivity

Current Trends and Future Developments on (Bio-) Membranes
DOI: https://doi.org/10.1016/B978-0-12-822257-7.00009-1

DR desorption ratio
G solute generation rate
GFR glomerular ultrafiltration rate
J_c solute convective flux
J_d solute diffusive flux through a semipermeable membrane
J_{IC} solute flux due to passive diffusion across the membrane from the intracellular to the extracellular environment
J_{IC}^{ACT} intracellular to extracellular target solute flux due to the active pumping across the membrane
J_V volume flux through the membrane
J^Z dialysate target solute flux due to diffusion across the dialyzer
K_B transport coefficient in blood side
K_D transport coefficient in dialysate side
K_{GF} hydraulic permeability
K_r endogenous clearance
K_0 overall mass transport coefficient
K_{UF} ultrafiltration coefficient
L dialyzer length
L_P membrane hydraulic permeability
— membrane permeability
Pe Peclet number
P_B blood pressure
P_D dialysate pressure
Q_B blood flow rate
Q_F ultrafiltration rate
S_{EC} extracellular solute mass concentration
S_m membrane surface
S^z dialyzer membrane cross area
T absolute temperature
V extracellular volume
Z, R dimensionless parameters for CL evaluation
α countercurrent/cocurrent factor
δ_m membrane thickness
Δc local solute concentration difference between blood and dialysate
ΔP difference in the hydrostatic pressure between the blood capillaries
$\Delta \pi$ difference in the osmotic/oncotic pressure
ε membrane hindrance
$\varnothing_A$ partition coefficient
Φ^Z dialysate target solute flux due to convection across the dialyzer
$\mu_{i,reg}$ regeneration effectiveness of solute i
π_B blood osmotic pressure
σ and σ_S Staverman reflection coefficient

6.1 Introduction

Chronic kidney disease (CKD) has been reported to be an increasing public health problem, and nowadays more than 2 million of patients in the world suffer from end-stage kidney disease (ESKD), in which it is necessary to intervene with certain renal replacement therapies (RRTs). The global estimated prevalence of CKD is 13.4% and through its effect on cardiovascular system, it directly affects burden of morbidity, mortality, and quality of life of patients worldwide, although patient survival has gradually increased during the past two decades (Salani, Roy, and Fissell, 2018). Current modalities of RRTs include in-center hemodialysis (HD), peritoneal dialysis (PD), home HD, and kidney transplantation, but research and developments in sorbent technology, nanotechnology, and cell culture techniques may provide promise for new interesting innovations in ESKD management. In this scenario, new approaches including artificial kidneys (AKs) and implantable AKs, in the field of miniaturization and tissue engineering, could really represent the future step for health patient conditions' improvement.

The kidneys are two bean-shaped organs located in the retroperitoneal space, one on each side of the spine, as shown in Fig. 6.1A. Blood irroration is fed by renal arteries, while the excretion of urine to the bladder is assured by the paired ureters Fig. 6.1B. In the human organisms' balance, a substantial contribution is made by kidneys, which play a fundamental role during the excretion of waste products, the reabsorption of essential compounds, and the regulation of endocrine and metabolic activities. All these functions are regulated by the kidney's functional unit: the nephron Fig. 6.1C. Each kidney is characterized by more than 1 million of nephrons; each one is made up of two subunits: the glomerulus, which is aimed at performing blood ultrafiltration (UF) and removing the metabolic products, and the tubules, which provide fine regulation of the excretion of water and solutes in the urine (di Elisabetta, Francesco, and Francesco, 2018).

The glomerulus is a network of blood nestled capillaries inside a cup, known as Bowman's capsule, along which the pressure drop permits blood filtration through its moderate selective membrane, primarily using the phenomenon of convection. The glomerular barrier consists of three distinct but closely interacting layers: glomerular endothelial cell, the basement membrane and visceral epithelial cell (podocyte) (Menon, Chuang,

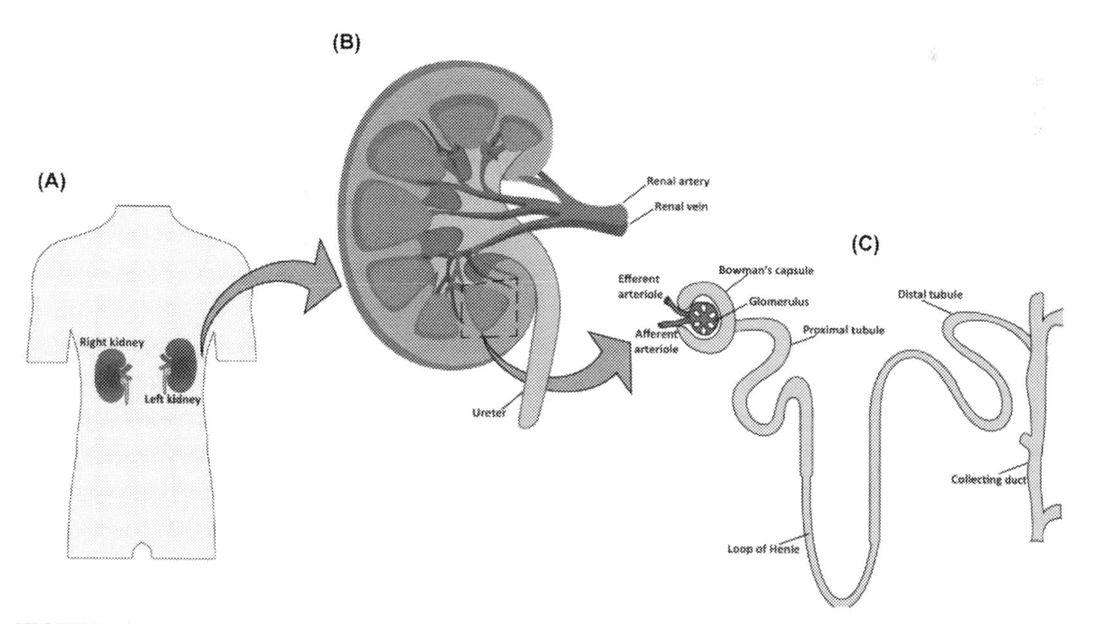

FIGURE 6.1 Kidney's anatomy. Location (A), structure (B), and nephron (C) of the kidney (Faria, Ahmed, Gerritsen, Mihaila, and Masereeuw, 2019).

TABLE 6.1 Glomerular filtration of different molecules.

Molecule	Molecular weight (g/mol)	Sieving coefficient
Water	18	1.0
Sodium	23	1.0
Glucose	180	1.0
Insulin	5000	1.0
Inulin	5500	1.0
Myoglobin	17,000	0.75
Hemoglobin	68,000	0.03
Albumin	69,000	0.005

and He, 2012), which together prevent the passage for the blood cells and almost all plasma proteins, while electrolytes and small solutes passage (glucose, amino acids, urea, creatinine, etc.) are allowed in the glomerular filtrate. The selectivity and the filterability of the membrane are based on solute molecular size and electrical charge. Table 6.1 enumerates the effects of molecular weight (MW) on filterability of different molecules, in which the sieving coefficient (SC), representing the ratio between the filtrate to the plasma concentration, is often used as an evaluation parameter.

An SC value of 1.0 means that the substance is filtered as freely as water, while larger molecules, connected to lower SC values, experience a progressively higher restriction to passage in the filtrate, and molecules larger than 100,000 kDa are completely rejected (Seifter, Ratner, and Sloane, 2005). Even though the molecular diameter of albumin is only about 6 nm, it is almost completely retained from filtration, because of its negative charge at physiological pH, which counteracts the same nature glomerular membrane's charge (Kasztan and Pollock, 2018).

Despite a wide variety of causes including metabolic abnormalities, aging, obesity, diabetes mellitus, hypertension, autoimmune diseases, drug-induced nephrotoxicity and genetic background, common pathologic hallmarks of CKD are decreased glomerular filtration and loss of glomerulus functionalities (Metcalfe, 2007). Indeed, glomerular UF rate (GFR), obtained by Eq. (6.1).

$$\mathrm{GFR} = K_{\mathrm{GF}}(\Delta P - \Delta\pi) \tag{6.1}$$

where K_{GF} is the hydraulic permeability, ΔP the difference in the hydrostatic pressure between the blood capillaries (about 55 mmHg) and the filtrate (about 15 mmHg) and $\Delta\pi$ the difference in the osmotic/oncotic pressure, indicates the formation velocity of the filtrate in the glomerulus.

An estimation of GFR, referred to 1.73 m^2 of body surface area for at least three months, can provide a good idea of normal or stages of pathological kidneys' functionality, as shown in Fig. 6.2, in which is also reported a level evaluation of albuminuria.

Renal diseases can be clinically distinguished in acute kidney injuries (AKIs), which encompasses prerenal (inadequate blood flow), intrarenal (nephrons' damages) or postrenal (obstructions and structural defects) pathologies, and CKDs. In AKIs, characterized by an abrupt

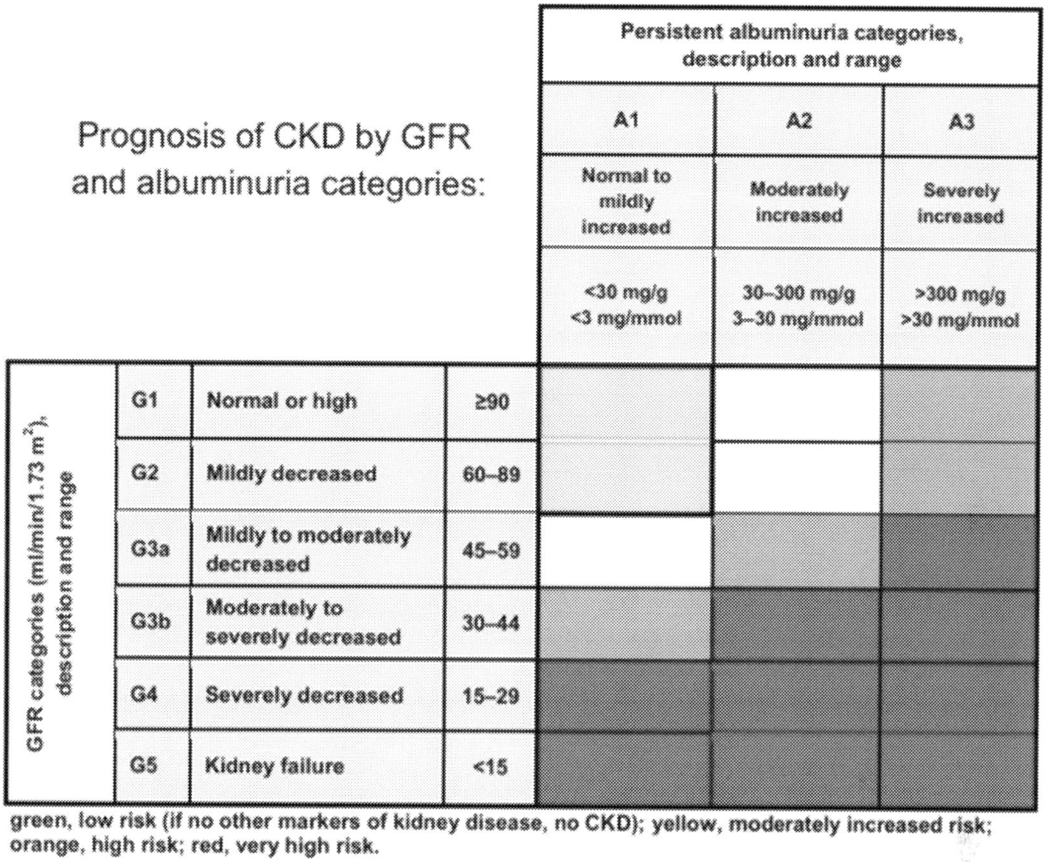

Prognosis of CKD by GFR and albuminuria categories:

GFR categories (ml/min/1.73 m^2), description and range			Persistent albuminuria categories, description and range		
			A1	A2	A3
			Normal to mildly increased	Moderately increased	Severely increased
			<30 mg/g <3 mg/mmol	30–300 mg/g 3–30 mg/mmol	>300 mg/g >30 mg/mmol
G1	Normal or high	≥90			
G2	Mildly decreased	60–89			
G3a	Mildly to moderately decreased	45–59			
G3b	Moderately to severely decreased	30–44			
G4	Severely decreased	15–29			
G5	Kidney failure	<15			

green, low risk (if no other markers of kidney disease, no CKD); yellow, moderately increased risk; orange, high risk; red, very high risk.

FIGURE 6.2 Glomerular ultrafiltration rate categories (G1–G5), and albuminuria categories (A1–A3) for the classification of normal or pathological kidneys' conditions (Ketteler et al., 2018).

(within hours) deterioration of renal functionalities and retention of toxins, fluids, and end metabolic products, the diseases are usually reversible, but if not correctly treated, they could drastically evolve into CKDs. In such chronic nephron-pathologies, instead, resulting from gradual and progressive loss of renal roles, the excessive retention and accumulation of a variety of endogenous metabolites require dialysis or kidney transplantation as the unique solution for patient's survival. In particular, six categories of CKD are identified depending on the GFR: values of GFR greater than 60 mL/min indicate normal renal activities (G1–G2), in which patient symptoms are not able to indicate if the kidneys are damaged; a progressive decrease of GFR (>15 mL/min) implies moderate and severe abnormalities (G3a/b-G4); moreover, when GFR <15 mL/min, an effective kidney failure is verified, which need RRTs management (G5). In addition, the description of albuminuria levels (A1–A3) and its range is useful to better identify the potential risk in the prognosis of CKDs (Fig. 6.2).

Instead, renal tubules are formed by proximal tubule, Henle loop, distal tubule, and collector duct, and they provide, in different ways, for the reabsorption of most part of water

TABLE 6.2 Tubular reabsorption of different substances.

Substance	Reabsorption (%)
Water	>99
Glucose, amino acids, phosphate, lactate, citrate	~100
Ions (Na^+, K^+, Ca^{2+}, Cl^-)	~99.5
Uric acid	~91
Urea	~50
Creatinine	0

(more than 99%), and solutes filtered in the glomerulus, as well as to the secretion of endogenous metabolites and drugs, as shown in Table 6.2 (Basicmedical Key, 2020).

The renal tubules' functions are very difficult to replicate in an artificial way because contrary to a moderate selective role of glomerular membrane, transport in tubular system is very selective and is primarily driven by the phenomenon of diffusion, during which the drop pressure (opposite in relation to the glomerular case) further increases the passage of molecules from tubules to the bloodstream. Moreover, each characteristic renal tract is responsible of a different reabsorption percentage of different substances (the first half of the proximal tubule, for example, reabsorbs around the 70% of the glomerular filtrate) (di Elisabetta et al., 2018).

In case of kidney failure, the patient requires RRT, which can include extracorporeal dialysis – in its different forms as HD, hemofiltration (HF), or hemodiafiltration (HDF) – or intracorporeal PD. Although these therapies have incredibly changed the prognosis of renal diseases, they cannot be considered as a complete replacement treatment because they do not provide the homeostatic, regulatory, metabolic and endocrine functions of the kidney, but they only replace the majority of physiological filtration activity. This is the reason why the patients, especially who are suffering from ESKD, continue to have a wide variety of inter- and posttherapy problems (Rocha, Sousa, Teles, Coelho, and Xavier, 2015). In this scenario, AK development means realizing an innovative device which significantly mimic physiological renal functions, simulate its behavior and eventually substitute the role of kidneys in pathologic conditions.

Nowadays, the impact of tissue engineering for the manufacturing of an implantable bioartificial kidney, composed of both biologic and synthetic components, could really represents the future solution, with less risk of infection, reduced costs and substantial benefits for patients, in terms of life expectancy, mobility and quality of life (Humes, Buffington, Westover, Roy, and Fissell, 2014).

6.1.1 Brief history of dialysis

In clinical use, the term dialysis, derived from Greek words *"dia"* (i.e., "through") and *"lysis"* (i.e., "to separate"), is referred to the removing process of excess water, solutes, and toxins from the blood in patients whose kidneys can no longer perform their physiological functions naturally. Actually, it can be performed in its various modalities, which include intracorporeal or extracorporeal (traditional dialysis) treatment. The latter therapy involves blood circulation into an external artificial device that operates by means of a selective membrane for blood detoxification, the first one method, exploits the human biological filtration properties in order to exchange water and solutes with a washing solution and it doesn't require the use of any artificial mass transfer devices.

The idea to use dialysis in the clinical practice originates from the "father of dialysis" Thomas Graham, when in the middle of the 19th century, first observed that colloid and crystalloid substances in liquids can be separated by vegetable parchments, which serve as semipermeable membranes, and decided to apply this principle in medicine (Stewart Cameron, 2012). Gradually, it was at the beginning of the 20th century that Abel and coworkers conducted the first experiments using a "vividiffusion" device (for which they coined the name of "artificial kidney") to treat the blood of dogs in extracorporeal circulation (Eknoyan, 2009). Only in 1924, the German physician Georg Haas performed the first human HD treatment in the history of medicine, using a dialyzer consisting of U-shaped collodion tubes, immersed into a glass cylinder with a bath solution, and introducing heparin, as anticoagulant substance, during the procedure (Paskalev, 2001). Another milestone in the development of modern dialysis techniques was the "rotating drum" dialysis machine invented by Willem Johan Kolff in the Nazi-occupied Netherlands in the early 1940s, and its modification to form a twin-coil dialyzer by 1956, which made him one of the pioneers in the field of artificial organs (Heiney, 2003). A breakthrough in dialysis history was achieved by Nils Alwall's device, which represented an innovative AK able not only to purify the blood, but also to remove excess water after the dialysis procedure (Kurkus and Ostrowski, 2019). In these early years, despite not totally reliable performances of the devices, the use of dialysis was mainly limited by the difficulties in having repeated access to the patient's circulation without compromising the blood vessels. The first solution became available in 1960, thanks to the Scribner's invention, which allowed to perform frequent treatments, on severe ill patients, with a semipermanent arteriovenous cannulation access. Further improvements occurred in 1966 by Brescia, Cimino and coworkers, with a new surgical procedure to connect an artery directly to a vein, not requiring exteriorized shunts, capable of delivering optimal flow for HD, which currently become a widely used therapeutic method (Agarwal, Haddad, Vachharajani, and Asif, 2019). A schematic representation of a generic current dialytic circuit is presented in Fig. 6.3.

The dirty blood is collected from the patient artery and it is directed to the dialyzer using opportune roller peristaltic pumps. Once the blood purification from waste compounds occurred, it is restored in patient body through its vein. A precise pressure sensors' system and air detector are equipped in order to monitor the physiological blood pressure and the eventual bubbles' presence, respectively (Hojs, Fissell, and Roy, 2020). Moving from these attempts, technological developments were aimed at increasing efficiency and safety of dialysis, trying to achieve a more comfortable treatment reducing the dialysis system dimensions and costs. In this scenario, the shape and the dimensions of the dialyzer and the materials from which the membranes were made were also improved constantly. Nowadays, special polymeric and synthetic membranes and dialyzers formed by bundle of hollow fibers are the most frequently used devices in clinical applications (Thijssen, Raimann, and Levin, 2012), but future innovative developments are focusing continuous attention by researchers and engineers in order to unravel new frontiers in RRT panorama of both intra- and extracorporeal dialysis.

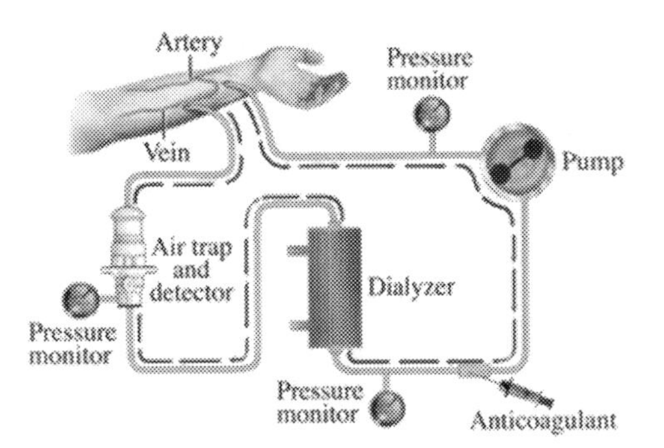

FIGURE 6.3 Schematic circuit of a convectional hemodialysis treatment.

6.2 Background

In the following subsections an introductory classification of different HD operations and their application is reported. For instance, low-permeability membranes with high selectivity are employed when liquid retention is required and a purely diffusive flux occurs, while, on the other hand, high-permeability membranes allow moderate water flux and are employed in HF systems.

6.2.1 Dialysis

There are two main types of RRT, which can be distinguished:

1. PD(intracorporeal)
2. Traditional dialysis (extracorporeal)

PD is a suggestive RRT, which includes the various applications of intracorporeal dialysis treatment. In particular, it exploits the physiological properties of the peritoneal membrane in order to detoxicate the blood from its waste compounds. During PD, a cleansing fluid flows through a tube (catheter) into part of the abdomen. The lining of the abdomen (peritoneum) acts as a filter and removes waste products from the blood. After a set period of time, the fluid with the filtered waste products flows out of your abdomen and is discarded Fig. 6.4.

These treatments can be done at home, at work or while traveling. But PD is not an option for everyone with kidney failure.

With regard to extracorporeal dialysis, it is appropriate to distinguish among three different modalities:

1. HD
2. HF
3. HDF.

In all cases, it involves membrane-based processes used in clinical practice for RRTs, whose operating principles can be based on convection or diffusion phenomena. While PD takes advantage of permeable peritoneal membrane, extracorporeal dialysis involves the

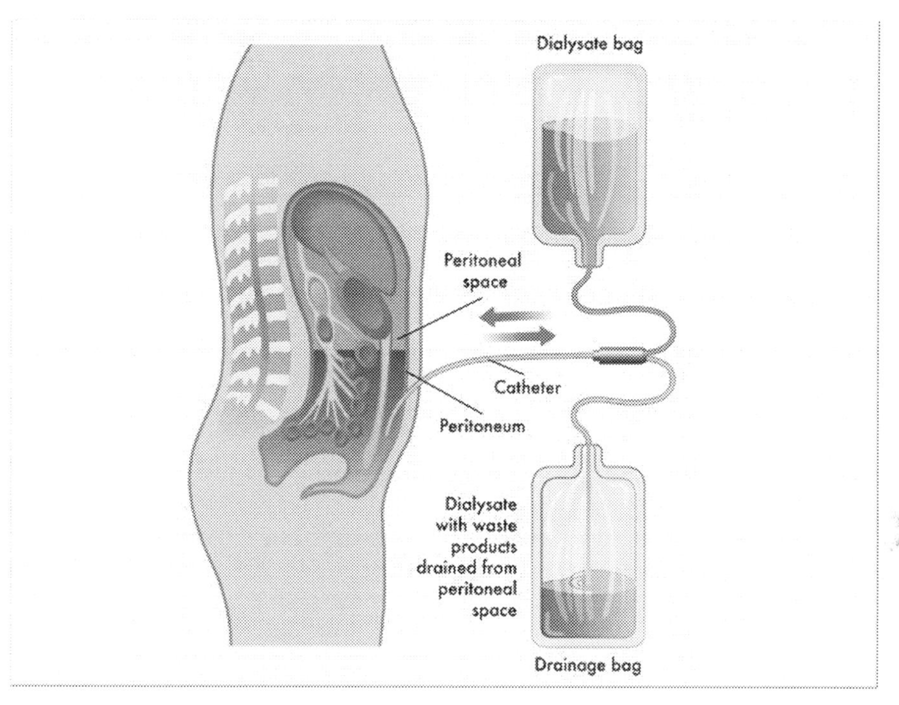

FIGURE 6.4 Representation of peritoneal dialysis

construction of a circuit, external to the body, which performs as an AK, in order to achieve patient blood detoxification, as above mentioned.

6.2.2 Hemodialysis

HD is the most common RRT used in case of kidney failure. It represents a remarkable membrane-based process during which the toxin removal mainly depends on toxin diffusion through a specific membrane. In detail, the driving force is given by toxin concentration's difference between the blood, which flows in the upstream of the membrane, and a washing solution, which, instead, flows downwards. The rinsing solution, also known as dialysate, is typically obtained by diluting a concentrated solution of electrolytes and glucose; chloride, sodium and magnesium ions form the solution that results to be at the same plasma concentration. Furthermore, acetate or bicarbonate is added to buffer the solution pH. A fine monitoring of chemical composition of dialysate is customized in accordance with the patient needs. Therefore HD is primarily a diffusion-based transfer of small molecules (<500 Da), because of larger substances are usually rejected (Misra, 2008), as shown in Fig. 6.5.

The diffusion is a chemical–physical phenomenon whereby the substances move in all directions in a random way and it statistically results in the solutes' passage from a more concentrated to a less concentrated area (Mazzeo, Bianchi, Cocchi, and Piemonte, 2019).

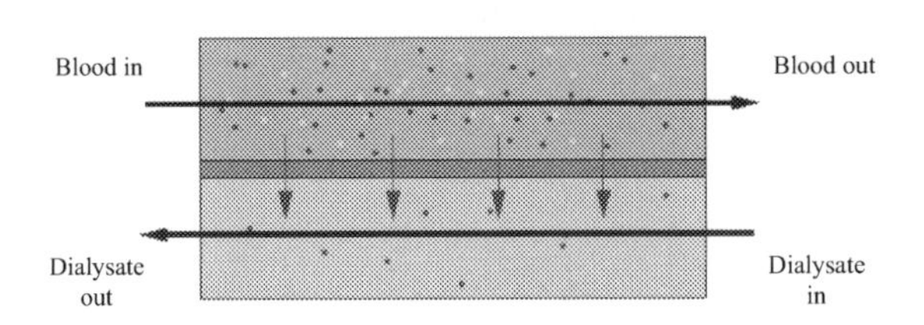

FIGURE 6.5 A schematic circuit of hemodialysis. The process is mainly driven by the phenomenon of diffusion.

The solute diffusive flux J_d through a semipermeable membrane is given by Fick's law Eq. (6.2):

$$J_d = -D \times \frac{\partial C}{\partial x} \tag{6.2}$$

It is directly proportional to the diffusivity coefficient (D) and the toxin concentration difference (C), while it is inversely proportional to the membrane thickness (x). The negative sign of the equation is because of the flux moving from a higher concentration site to a lower concentration site.

In typical clinical applications, during a hemodialytic treatment, a blood flow, variable from 200 to 400 mL/min, is usually sent to the dialyzer, with a ratio of dialysate to blood flow rate set in the range 1÷1.5. In this way, it is possible to treat the whole patient's blood every 15 minutes. Even if HD is typically effective for the removal of small toxins, the convective transport might eliminate some middle-MW molecules. Indeed, the pressure difference between blood and dialysate stream (100÷500 mmHg) results in an evident water flux and allows the excessive water removal from the body (2÷4 L per session) (Ikizler and Schulman, 2005).

6.2.3 Hemofiltration

The convection is a physical phenomenon based on a fluid movement due to a transmembrane pressure gradient (Kotanko, Kuhlmann, and Levin, 2010). Therefore the solute convective flux J_c is obtained by Eq. (6.3):

$$J_c = J_V \times C_s \times SC \tag{6.3}$$

and it depends on the volume flux through the membrane (J_V), the solute concentration (C_s) and the Sieving Coefficient (SC). In literature, SC is also expressed, under ideal conditions, in the form $SC = 1 - \sigma$, where σ represents the Staverman reflection coefficient, which can assume values from 0 for solutes smaller than the membrane pores to 1, indicating solutes larger than membrane pores, which are completely rejected. In other words, the efficacy of convection is strongly dependent on the dialyzer membranes pores' size. The dialytic process mainly based on convective transport is typically named "hemofiltration," which essentially represents a UF process Fig. 6.6.

During HF, the water removal rate is higher than that required to remove the excessive water in the organism and, in order to keep volume homeostasis, a sterile, nonpyrogenic, and endotoxin-free replacement fluid is reinfused to the patient (dialysate is not used).

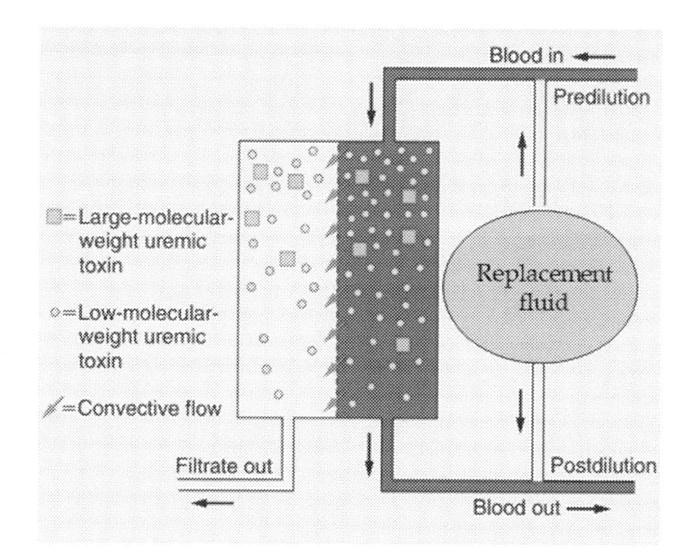

FIGURE 6.6 A schematic circuit of hemofiltration. This process is mainly driven by the phenomenon of convection. Substitution fluid is usually given either pre- or postdilution, but not both.

Infusion of the substitution fluid can occur by a predilution (upstream of the membrane) or a postdilution technique (downstream of the membrane). The predilution method allows to achieve very high UF rates, with significant clearance of middle-MW substances (Forni and Hilton, 1997). The major disadvantage of HF is related to the cost, because of the ultrapure replacement fluid, which is administered intravenously, is greater than the typical dialytic one. In conclusion, if HD is more effective for small-MW molecules removal, on the other hand, HF is more useful in clinical conditions requiring the clearance of middle-MW molecules. However, nowadays, there are no powered randomized trials demonstrating the superiority of one therapy over the other, in terms of morbidity and mortality (Friedrich, Wald, Bagshaw, Burns, and Adhikari, 2012).

6.2.4 Hemodiafiltration

Between these two limiting conditions, a wide variety of intermediate renal treatments, generally referred to as "hemodiafiltration," were proposed for clinical use. In particular, HDF combines diffusive and convective solute transport in a single therapy, in order to obtain a final result, which is the union of the best HD (high clearance of small molecules) and HF (high clearance of middle molecules) features. As shown in Fig. 6.7, it basically consists in a HD combined with a controlled UF, during which UF rate is higher than that in HD but lower than that in HF.

In order to implement HDF as a daily treatment, online preparation of substitution fluid has become the most relevant solution. This revolutionary method consists of three additive components to conventional HD machine: a pump for delivering substitution fluid, a package of specific ultrafilters to produce sterile substitution fluid from dialysate, and a controlling system to monitor the fluid exchange and removal (Canaud, Vienken, Ash, and Ward, 2018). Currently, online HDF (OL-HDF) represents a practical RRT and recent

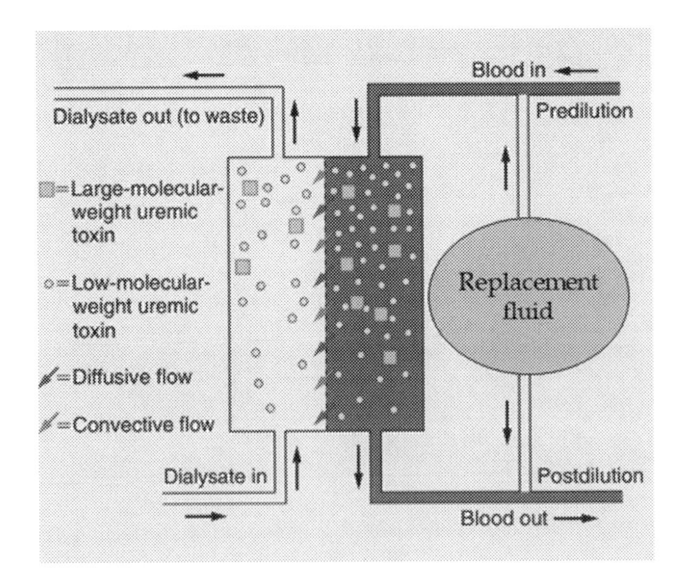

FIGURE 6.7 A schematic circuit of hemodiafiltration. The process is driven by diffusive and convective phenomena. Substitution fluid is usually given either pre- or postdilution but not both.

randomized controlled studies have demonstrated not only its safety and sustainability, but also a higher efficiency in terms of outcomes achieved compared with standard HD, when a large substitution volume is delivered (Canaud et al., 2017). Moreover, OL-HDF has reached its full maturity phase, but further clinical trials are necessary to better show the superiority of HDF compared with high-flux HD and the optimal convective dose in different clinical settings. Nevertheless, the choice of the more appropriate treatment is strongly depending on specific pathological patient conditions and this is the reason why, nowadays, all four different modalities (including PD) are used in clinical practice.

6.3 Role of semipermeable membrane in artificial kidney

Each AK is constituted by a fundamental operating element: the dialyzer. This structured system comprises of three main components:

1. the blood compartment,
2. the semipermeable membrane,
3. the dialysate compartment.

These three units are strictly linked into a final configuration which permits the optimal performance of the dialyzer as a solute and water exchanger. Different chemical–physical parameters and processes, including convection and diffusion phenomena, chemical composition of patient blood and infused dialysate, or also shape and dimensions of the device, could influence its behavior and reliability. In order to realize a more efficient extracorporeal dialysis, the knowledge of the basic mechanisms on solute transportation and the proper choice of materials definitely play a key role. Therefore the practical development of an effective AK is mainly because of the better semipermeable membrane

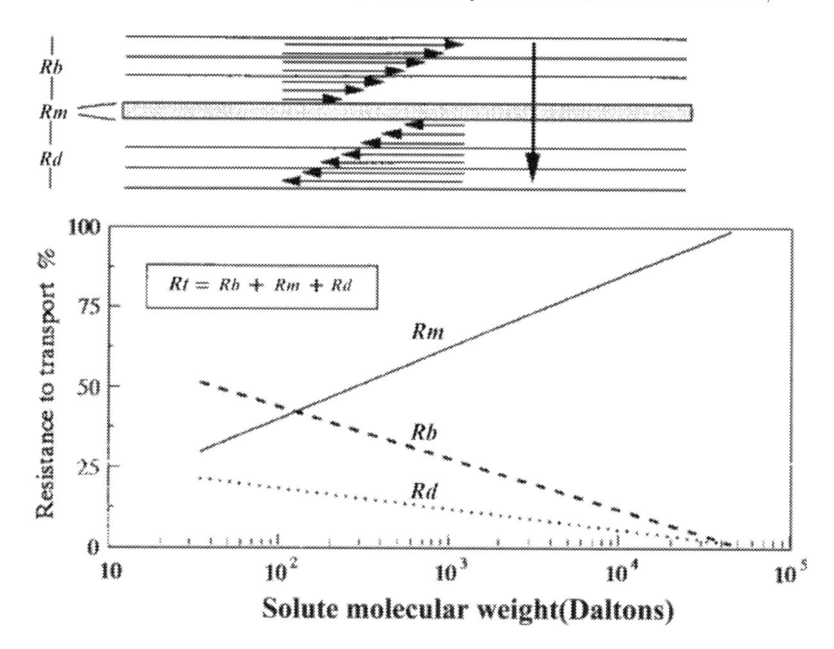

FIGURE 6.8 The overall resistance to solute transport (R_t) is the sum of blood resistance (R_b), membrane resistance (R_m) and dialysate resistance (R_d). R_m progressly decreases, in terms of percentage, for smaller solutes, while R_b and R_d become more and more important (Ronco, Ghezzi, Brendolan, Crepaldi, and La Greca, 1998).

application, as well as the accurate blood and dialysate flow rates moving in the dialyzer. While impermeable membranes act as real barriers for any molecules, avoiding the passage through it, the membranes used in dialysis applications base their behavior on two fundamental features: biocompatibility and selectivity. Dense membranes generally present high selective properties (diffusive transport), due to their high transport resistance (TR), with low mass flow rates. On the other hand, porous membranes (characterized by specific pores' distribution of different sizes) are more permeable, because of their lower TR and larger mass flow rates (convective transport). A profile evolution of TR in relation to solute MW is shown in Fig. 6.8.

Intermediate and more efficient performances are given by asymmetric membranes, formed by a very thin perm-selective layer (which determines the membrane's selectivity) and a highly porous layer (which provides the mechanical strength to withstand pressure differences) (Annesini, Marrelli, Piemonte, and Turchetti, 2017). Several kinds of dialysis membranes have been developed over the history of RRT. The first prototypes in collodion, cuprophane and then cellophane are now totally substituted by the current forms of modified cellulosic and synthetic membranes, which guarantee reliable properties in terms of biocompatibility. In detail, the extreme hydrophilic behavior of cellulosic materials, connected to the high water adsorption, permits the development of very thin membranes (thickness of 5–15 μm), with good efficiency for the removal of small-MW molecules, but poor permeability for middle-MW substances. Even if the use of modified cellulosic components partially mitigates the complement activation phenomenon, synthetic membranes represent a necessary step for middle-MW substances compliance. Thus different polymers, with hydrophobic, hydrophilic and excellent biocompatible properties, have been introduced for clinical use. In particular, the development of a higher membrane thickness

(>20 μm) enhances the mechanical resistance and the permeability to larger molecules (Mineshima, 2015). A comparation of these two different typologies, in terms of diffusivity, is presented in Fig. 6.9, in which it is evident that further improvements are fundamental to better mimic renal physiological functions.

In the proper dialyzer planning, also blood and dialysate strongly influence the resulting performance of the device. In blood compartment, different blood flow conditions influence its purification through the membrane.

In Fig. 6.10 is clearly demonstrated that, the SC and the wall shear rate, defined as the ratio between the speed variation of the fluid in the dialyzer and the variation of the distance from the center of the dialyzer, significantly modify the expressions of diffusive and convective solute transport. In particular, the permeability of the membrane increases with high shear rates and high blood flows, that results in higher UF rates and solute clearance. As a consequence, the sum of diffusive and convective clearances is minor than what it would have been if the two processes had been independent of each other (Ronco et al., 1998).

While several tests have been performed for the optimization of blood compartment, in terms of geometry and flow distribution, *minor attention is kept for dialysate* compartment. A nonhomogeneous dialysate distribution flow, inside the dialyzer, can cause the phenomenon of channeling, in which the stream flows externally to fibers without reaching the

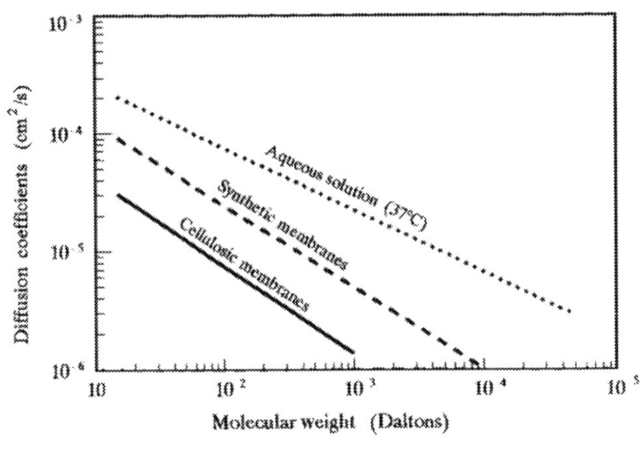

FIGURE 6.9 Diffusion coefficient profiles in relation to different molecular weight molecules. Synthetic membranes present higher performances than those of cellulosic membranes, but lower efficiency compared with a general aqueous solution (Ronco et al., 1998).

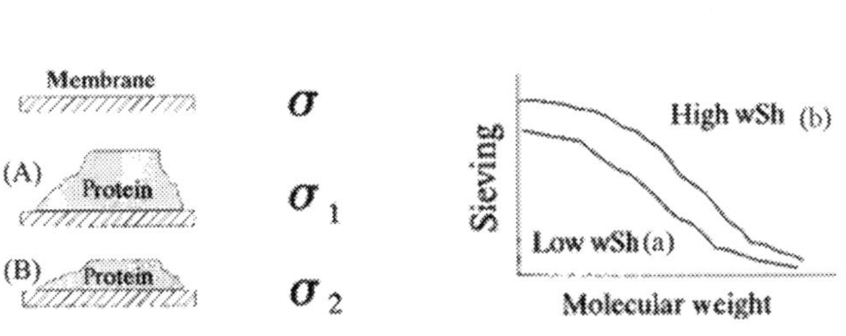

FIGURE 6.10 The impact of wall shear rate (wSh) on convective transport. σ_1 and $\sigma_2 = f(\text{wSh})$ represent the different Staverman reflection coefficients of the functional membrane (membrane + protein layer) in the presence of low shear rates (A) and high shear rates (B) (Ronco et al., 1998).

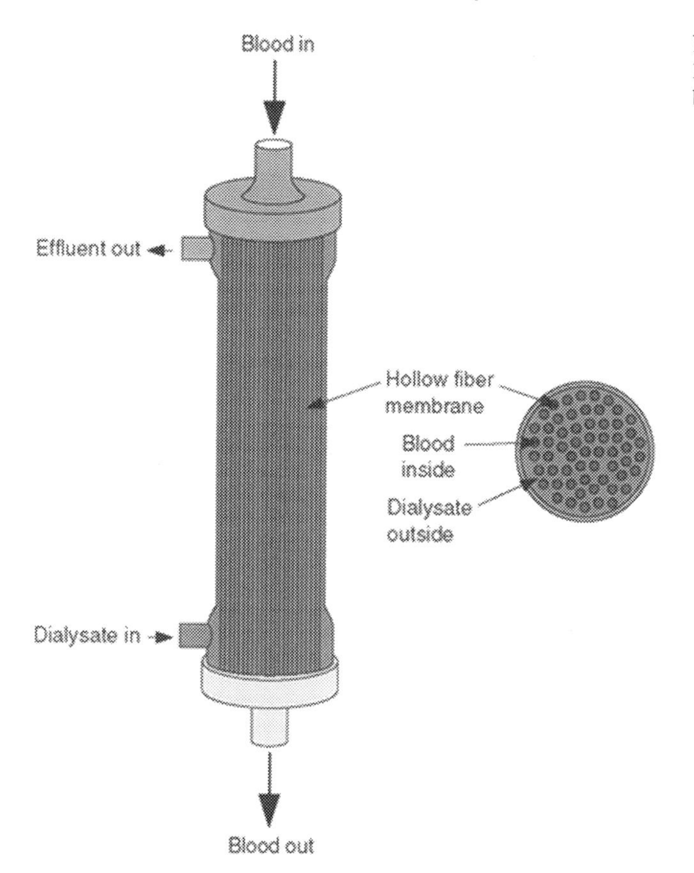

FIGURE 6.11 Representation of a capillary dialyzer for hemodialysis constituted by hollow fiber membranes.

space available within the fibers in the central region of the bundle. Over the years, different dialyzers' designs have been developed: from the first rotating drums, spiral, parallel flow and electrodialyzers to the plate-flate, pseudocapillary and coil ones (Twardowski, 2008). Nowadays, in order to prevent channeling problem and to increase the final device performance, an innovative approach is currently in clinical use and provides the adequate distance between the various adjacent fibers in the bundle in the so called capillary dialyzer configuration.

As shown in Fig. 6.11, the blood passes through a bundle of hollow fibers made from a microporous membrane, while the dialysate is forced to stream in polycarbonate shells. Even if, the major refinement of a cocurrent flow is essentially used on pediatric patients and some first dialysis patients, the countercorrent flow design is the most frequent used because it improves general device's performance (Yehl, Jabra, and Zydney, 2019). In order to avoid the contact between the blood and the dialysate, the fibers are covered with a polymeric material (usually polyurethane) and to meet medical requirements, dialyzers of different membrane areas ranging from 0.2 to about 2 m^2, with up to 15,000 fibers, and a priming volume of 50–150 mL are manufactured (Annesini et al., 2017).

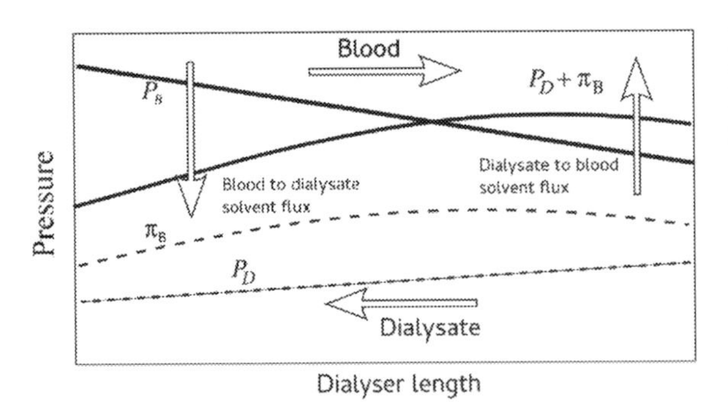

FIGURE 6.12 Blood and dialysate pressure profile along a hemodialysis dialyzer length. In the distal section can occur the phenomenon of back-filtration (Annesini et al., 2017).

The dialyzer is also a significant hydraulic unit in which water kinetics plays a crucial role. Especially, using high-flux membranes, it is possible to mimic the physiological tubular reabsorption, through the so called "back-filtration" phenomenon. As shown in Fig. 6.12, during an HD treatment, different pressure profiles along the dialyzer length are responsible of this process. Both blood and dialysate pressures decrease in their respective flow directions, while the blood osmotic pressure (π_B), owing to the larger molecules rejected from the membrane, increases in the direction of blood flow, at least until water removal from blood. In detail, in the dialyzer's proximal section (blood inlet), the blood pressure (P_B) drop is higher than the dialysate one because blood section is lower than that of dialysate.

In absence of π_B, dialysate pressure (P_D) will never overcome P_B leading to a continuos passage of substances from blood to dialysate. While π_B presence provides the profile switch and the consequent opposite molecules passage from dialysate to blood (back-filtration) in the dialyzer's distal section (blood outlet) (Misra, 2008). Only using unidirectional membranes and not allowing transfer of impure solutes to the blood, it is possible to circumvent this phenomenon.

6.4 Mathematical models of kidney transport phenomena

Currently, the best way to predict the effects of RRTs and minimize inter- and postdialytic disorders is using mathematical models. They provide a helpful tool for physicians and nurses in order to achieve the desired solute clearance, while avoiding wasteful excesses in kidneys, in according to specific patient needs.

As briefly discussed in the previous chapter, the efficiency of a dialysis treatment is strongly dependent on operating parameters, which influence the dialyzer's performance (Hirano et al., 2011).

Especially, in biomedical field, different devices are classified in according to:

1. $K_0 \times A$ value; where K_0 is the overall mass transport coefficient and A is the membrane surface area. In particular, K_0 can be calculated by Eq. (6.4) (the subscript "i" indicates that the variable is referred to the specific solute or molecule "i"):

$$\frac{1}{K_{i0}} = \frac{1}{K_{iB}} + \frac{1}{K_{iD}} + \frac{\delta_m}{?_i} \tag{6.4}$$

where K_B and K_D are the transport coefficient for blood and dialysate, respectively, δ_m is the membrane thickness and ? represents the membrane permeability. High-performing dialytic sessions are usually related to high $K_0 \times A$ values.

1. UF coefficient K_{UF}; defined as the ratio between UF rate (Q_F) and the transmembrane pressure difference between blood (P_B) and dialysate (P_D):

$$K_{UF} = \frac{Q_F}{P_B - P_D} \tag{6.5}$$

FDA defines low- and high-permeability dialyzers if the K_{UF} values are lower or higher than 8 mL/h × mmHg, respectively (Annesini et al., 2017).

Moreover, in order to evaluate the dialyzer's efficiency, it is possible to define two main variables:

1. Clearance (CL): it is indicated as the ratio of the toxin removal rate from blood to toxin concentration in the inlet blood, assuming that its concentration in inflowing dialysate (c_{iD}) is zero:

$$CL = Q_B \frac{c_{iB,in} - c_{iB,out}}{c_{iB,in}} \tag{6.6}$$

It represents a crucial factor, evaluating which is possible to describe the impact of the dialysis process on the patient and the blood purification profile over time.

1. Dialysance (DL): it is expressed by the ratio of the toxin removal rate from blood to difference toxin concentration:

$$DL = Q_B \frac{c_{iB,in} - c_{iB,out}}{c_{iB,in} - c_{iD,in}} \tag{6.7}$$

Only if the dialysate fluid is free from the solute considered (like in the case of single-pass dialysis), CL and DL are equal. Moreover, both the dialysance and clearance depend on the dialyzer geometry, membrane properties and both blood and dialysate operating flow conditions.

The final role of dialysis mathematical models is to describe solute and solvent kinetics across the AK's membrane, using and controlling all these key parameters. Indeed, they really represent preventive instruments able to verify the proper choice for the dialytic session in order to overcome different pathological conditions of the patients.

6.4.1 Dialysis model

During a conventional dialysis treatment, blood purification occurs through solutes' passage to the dialysate, using a combination of both diffusive and convective phenomena (HDF therapy). In order to derive a general model of a dialytic process, it is necessary to study the balance of all quantities moving across the dialyzer unit. Referring to Fig. 6.13, the blood inlet and dialysate outlet occur in $z = 0$, while blood outlet and dialysate inlet

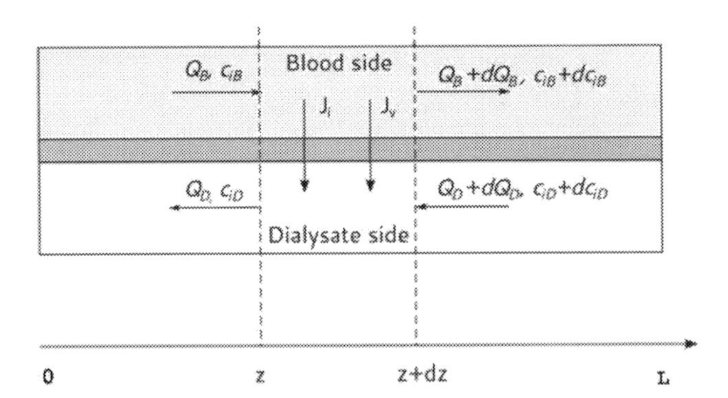

FIGURE 6.13 Mass balance during hemodiafiltration treatment (Annesini et al., 2017).

occur at $z = L$. In this configuration the countercurrent flow is provided and the mathematical behavior is constructed by combining three main elements:

1. The differential mass balance equations for solvent and toxins:

$$\frac{dQ_B}{dz} = \alpha\frac{dQ_D}{dz} = -J_V\frac{A}{L} \tag{6.8}$$

$$\frac{d(Q_B c_{iB})}{dz} = \frac{d(Q_D c_{iD})}{dz} = -J_i\frac{A}{L} \tag{6.9}$$

where $\alpha = 1$ or -1 for countercurrent or cocurrent, respectively, J_V is the solvent flux, J_i is the flux of the solute (toxin) "*I*" and L indicates the dialyzer length.

2. The expressions for the solvent flux J_V and the toxin flux J_i.

The solvent flux is mainly due to the phenomenon of convection and so its driven force is the local transmembrane pressure difference. According to the Starling equation, J_V is expressed by Eq. (6.10):

$$J_V = L_P[(\Delta P - \Delta\pi)] \tag{6.10}$$

where L_P is the membrane hydraulic permeability, while the pressure term assumes the profile just discussed in Fig. 6.12.

The osmotic pressure difference is given by the Van't Hoff formula, Eq. (6.11):

$$\Delta\pi = \sigma RT\Delta c \tag{6.11}$$

where R is the universal gas constant, T the absolute temperature and Δc represents the local solute concentration difference between blood (c_{iB}) and dialysate (c_{iD}).

On the other hand, the toxin flux J_i needs a more detailed formulation because it is function of both diffusive and convective phenomena. More especially, during a conventional HDF treatment, the expression of J_i is given by Eq. (6.12):

$$J_i = -D\frac{\partial C}{\partial z} + J_V(1-\sigma)c_i \tag{6.12}$$

where the first and second terms indicate the diffusive and the convective solute transport, respectively, and c_i is the solute concentration. Although, the latter formula clearly expresses the verifying processes, many coefficients are still too generic. Indeed, in transport through porous membranes, the diffusivity coefficient (D) has to include geometry factors, which link the number and the dimensions of the membrane's pores. Assuming that the solute flows only through the pores, it is important to define the partition coefficient $\varnothing_A$, as the ratio to total pores' area and the whole membrane's area, and the membrane hindrance ε, which consideres the effective solute passagge across the single pore, through the expression (Dechadilok and Deen, 2006) by Eq. (6.13):

$$\varepsilon = \left(1 - \frac{R}{r}\right)^2 \left[2.1\left(\frac{R}{r}\right) + 2.09\left(\frac{R}{r}\right)^3 - 0.95\left(\frac{R}{r}\right)^5\right] \tag{6.13}$$

with r and R are the pore radius and the effective radius of the solute passed over the membrane, respectively. Thus a more complex formulation of membrane diffusivity D_m is given by Eq. (6.14):

$$D_m = \varnothing_A \times \varepsilon \times D \tag{6.14}$$

Obviously, the convective term is also influenced by geometry factors and the Staverman coefficient can be rewritten as (Liao et al., 2005):

$$\sigma_S = \left(1 - \left(1 - \frac{R}{r}\right)^2\right)^2 \tag{6.15}$$

Integrating Eq. (6.9), it is possible to obtain the solute flux as developed by Villarroel, Klein, and Holland (1977) and then elaborated by Waniewski et al. (2008):

$$J_i = -D_m \Delta c + J_V(1 - \sigma_S)\overline{C_i} \tag{6.16}$$

where $\overline{C_i}$ is the mean intramembrane concentration of the solute, defined as:

$$\overline{C_i} = c_{iB}(1 - \beta) + c_{iD} \tag{6.17}$$

in which c_{iB} and c_{iD} are the boundary values of the solute concentration in blood and dialysate, respectively, with:

$$\beta = \frac{1}{\mathrm{Pe}} - \frac{1}{\exp(\mathrm{Pe}) - 1} \tag{6.18}$$

where Pe represents the Peclet number, which may be thought as a dimensionless coefficient showing the ratio of convective to diffusive transport:

$$\mathrm{Pe} = \frac{J_V(1 - \sigma_S)}{D_m} \tag{6.19}$$

Thus in Eq. (6.16), solute flux is a nonlinear function of solvent flux. This nonlinearity can be approximated in a linear form by substituting parameter β with a constant F, which is function of Pe. F value varies in according to the specific conditions; the most widely

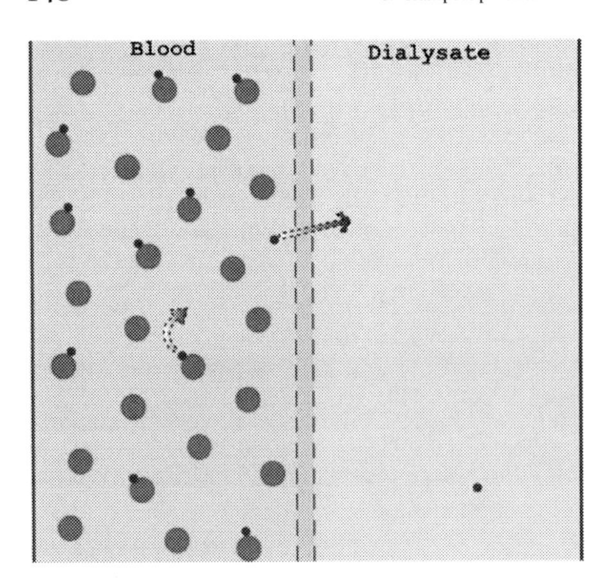

FIGURE 6.14 Only the solutes no protein-bound can pass from blood to dialysate.

used approximation is $F = 0.5$ for low Pe (Pe < 1), while approximations of $F = 0$ are usually indicated when Pe $>> 1$ (Waniewski, 1994).

Moreover, in blood, the solute protein-binding is a very common phenomenon, which rents the blood purification more difficult because only the free solute portion will be available for diffusive or convective transport, as shown in Fig. 6.14. Indeed, the value of C_{iB} must be replaced by the free concentration C_{iBf} with:

$$C_{iBf} = f_i \times C_{iB} \tag{6.20}$$

where f_i is the fraction of solute that is not bound to protein (Meyer, 2004).

An alternative interesting form of J_i, obtained from the integration of Eq. (6.16) through the membrane thickness, is the following Eq. (6.21) (Annesini et al., 2017):

$$J_i = \text{Pe} \times D_m \left(\frac{\exp(\text{Pe})}{\exp(\text{Pe}) - 1} c_{iB} - \frac{1}{\exp(\text{Pe}) - 1} c_{iD} \right) \tag{6.21}$$

In according to thin layer theory, the boundary layers concentrations are different from those in the bulk of the fluid phase, due to the mass TRs. The relations between the interface (c_i) and the bulk (c_i') concentrations in blood and dialysate, respectively, are expressed by Annesini et al. (2017):

$$c_{iB} = c_{iB}' \left[e^{J_V / k_{iB}} + \frac{J_i}{J_V c_{iB}'} \left(1 - e^{J_V / k_{iB}} \right) \right] \tag{6.22}$$

$$c_{iD} = c_{iD}' \left[e^{J_V / k_{iD}} + \frac{J_i}{J_V c_{iD}'} \left(1 - e^{J_V / k_{iD}} \right) \right] \tag{6.23}$$

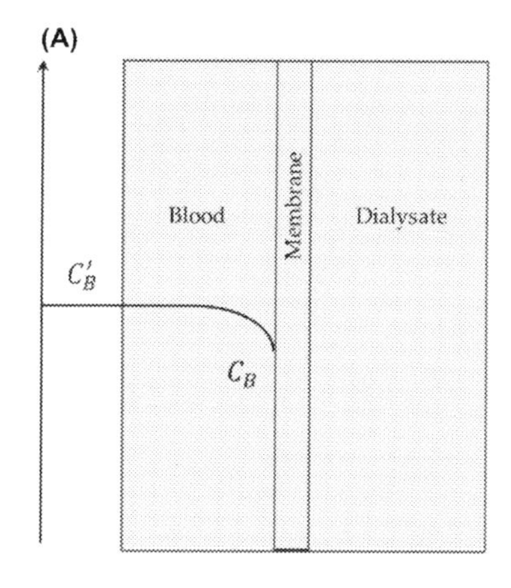

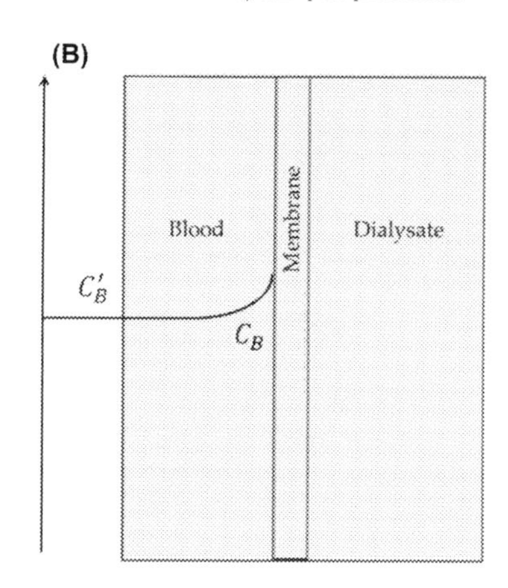

FIGURE 6.15 Solute concentration profile in blood in the case of pure dialysis (A) and polarization phenomenon (B).

In the case of pure dialysis (no UF$->J_V=0$), the Eq. (6.22) becomes:

$$c_{\mathrm{iB}}=c_{\mathrm{iB}}^{'}\left[1-\frac{J_i}{k_{\mathrm{iB}}c_{\mathrm{iB}}^{'}}\right] \tag{6.24}$$

and the solute concentration at the membrane interface is smaller than that in the bulk (Fig. 6.15A). In the case of UF ($J_i=0$) with a solute completely rejected by the membrane ($\sigma=1$), Eq. (6.23) becomes:

$$c_{\mathrm{iB}}=c_{\mathrm{iB}}^{'}e^{J_V/k_{\mathrm{iB}}} \tag{6.25}$$

and the boundary solute concentration is higher than that in the bulk, as shown in Fig. 6.15B (membrane polarization).

3. The momentum balance equation; useful to evaluate the local pressure in both blood and dialysate. Hagen–Poiseuille equation holds in intralumen flow in order to evaluated the pressured drop for hollow fiber membrane, while for the shell a modification is required. In particular, the dialysate momentum effect can be neglected compared to that of the blood due to lower viscosity of the dialysate and its low flow rate during a common RRT (Kim et al., 2013).

In order to complete the system of differential equations listed above (continuity equation and momentum balance equation) and obtain a solution, it is necessary to add boundary conditions, for the flow rate, composition and pressure of both blood and dialysate streams. Finally, a simplified example of dialytic mathematical model is constructed and it can be used with the aim to better evaluate patient solutes' kinetics and clearances. In detail, in the assumption of a pure dialysis therapy, during which diffusive solute transport completely exceeds the convective one, the solute flux is simply directly proportional to D_m and concentration's difference Eq. (6.10), where $J_V=0$). In these conditions, an alternative expression of CL is obtained with the introduction of the two dimensionless

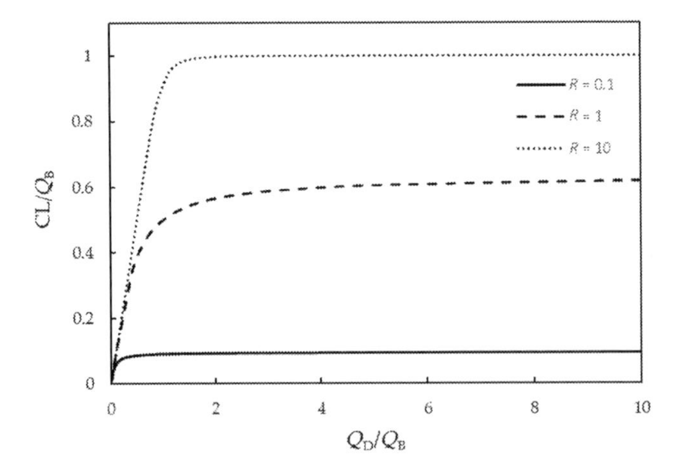

FIGURE 6.16 Device efficiency profiles in relation to the ratio dialysate to blood flow rate, for different values of *R*. Increasing dialyzer clearance means to reduce the mass transport resistance and use modules with high membrane area.

parameters Z (depending only on operating conditions) and R (including also dialyzer's characteristics):

$$R = \frac{K_0 A}{Q_B}; Z = \frac{Q_B}{Q_D}$$

$$\mathrm{CL} = Q_B \frac{1 - \exp[R(1 - Z)]}{Z - \exp[R(1 - Z)]} \tag{6.26}$$

where the dimensionless ratio clearance to blood flow rate is a device efficiency index ($\frac{\mathrm{CL}}{Q_B} = 1$ indicates the maximum performance).

Fig. 6.16 shows clearance profile in relation to the ratio between dialysate and blood flow rate, for different values of R. In particular, the dialyzer's efficiency increases with increasing dialysate flow rate, maintaining both R and Q_B constant, because of the reduction of toxin concentration provides a growth in the driving force for the toxin transfer. Although an excessive increase of Q_D ($\frac{Q_D}{Q_B} > 1$) does not imply an evident performance improvement, clearance also increases by increasing R (depending on K_0 and A) values (Annesini et al., 2017).

Finally, in recent RRTs, in which sorbent cartridges play a key role, it is significant to model the regenerative dialysis unit's behavior. Indeed, the regeneration effectiveness of the solute i ($\mu_{i,\mathrm{reg}}$), defined as:

$$\mu_{i,\mathrm{reg}} = \frac{c_{\mathrm{iD,out}} - c_{\mathrm{iD,in}}}{c_{\mathrm{iD,out}}} \tag{6.27}$$

assumes relevance in the final solute clearance expression:

$$\frac{Q_B}{\mathrm{CL}_i} = \frac{Q_B}{\mathrm{DL}_i} + \frac{Q_B}{Q_D}\frac{1 - \mu_{i,\mathrm{reg}}}{\mu_{i,\mathrm{reg}}} \tag{6.28}$$

The clearance is strongly dependent on $\mu_{i,\mathrm{reg}}$ values. In detail, for high regeneration performances ($\mu_{i,\mathrm{reg}} \sim 1$), clearance and dialysance can be approximated as equal, while for scarce regeneration efficiencies ($\mu_{i,\mathrm{reg}} \sim 0$), CL_i is directly proportional to dialysate flow rate, according to the relation $\mathrm{CL}_i \cong Q_D \times \mu_{i,\mathrm{reg}}$ (Annesini et al., 2017).

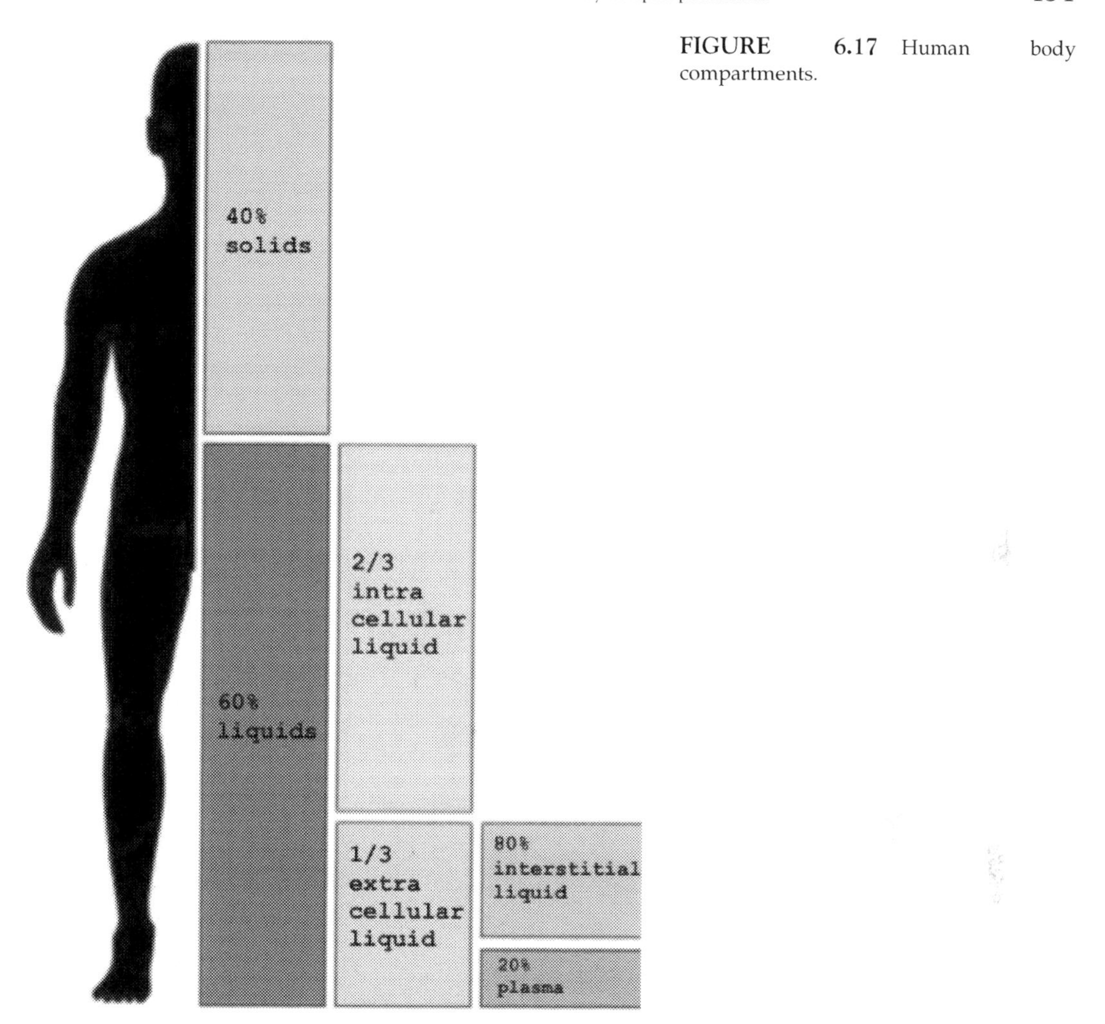

FIGURE 6.17 Human body compartments.

6.4.2 Patient device models

In order to better mimic physiological conditions during a dialysis treatment, is necessary to extend the mathematical concept to patient point of view. Indeed, the proper modeling is obtained only if different human solutes' kinetics are considered. In literature, a wide variety of patient device models have been reported with the aim to suggest solutes' profiles interacting between patient and dialysis machine. All such models are based on compartmental theory, according to which human body can be approximated to a set of different well-mixed compartments, through which mass transports are simulated. As shown in Fig. 6.17, up to 60% of the human adult organism is composed by liquids, of which two-third is intracellular liquid, while the remaining one-third indicates the extracellular one.

In this latter compartment, two additional regions can be distinguished: the interstitial liquid and plasma, representing a percentage of 80%–20%, respectively (Hall, 2015). In each compartment, the mathematical modeling of target solutes' mass balances provides the concentration evolution during the dialysis treatment as well as in the interdialytic period. Moreover, the rates of solute generation and consumption by metabolic processes and exchanges with neighboring compartments are analyzed. Mathematically, according to the compartmental approach, the specific solute kinetics can be described as a progressive sequence of different contributes:

$$\frac{d(V \quad S_{\mathrm{EC}})}{dt} = J_{\mathrm{IC}} \quad S_m + J_{\mathrm{IC}}{}^{\mathrm{ACT}} \quad S_m + J^Z \quad S^z + \Phi^Z S^z \tag{6.29}$$

Where $V\ S_{\mathrm{EC}}$ represents the extracellular target solute mass; J_{IC} the target solute flux due to passive diffusion across the membrane from the intracellular to the extracellular environment; S_m the membrane surface; $J_{\mathrm{IC}}{}^{\mathrm{ACT}}$ the eventual intracellular to extracellular target solute flux due to the active pumping across the membrane (not negligible contribute in the case of sodium or potassium); J^Z dialysate target solute flux due to diffusion across the dialyzer; S^z the dialyzer membrane cross area and Φ^Z the dialysate target solute flux due to convection across the dialyzer.

The general form of expression (6.29) is extremely important because it distinguishes between cellular transport, from EC to IC (the first two terms), and dialysis transport, from EC to dialysate fluid (the latter two terms) and it clarifies the concept that only the final portion of blood is in effective contact with the dialyzer.

Although the mathematical model is more accurate with a decreasing level of human compartments' approximations, a more detailed phenomena's description provides a general system complexity increase (Schneditz and Daugirdas, 2001). In order to preserve the model's predictive capacity, probably the compromise between simplifications and intricacy may be the best solution. Over the years, several mathematical approaches have been proposed in the literature to describe solute transports and assist clinicians in individualizing the proper prescriptions during the dialytic therapies. Firstly, the single-pool or pseudo-one compartment models had been applied to simulate the kinetics of small molecules (mainly urea). Then, multicompartment models have been introduced, allowing the better description of simultaneous contribution of different solutes and the kinetic simulation of small molecules not freely permeable (such as phosphate and creatinine) as well as of larger substances (β_2-microglobulin, for example).

6.4.3 Single-compartment model

The simplest approach is based on single compartment models, in which the patient body is described as a simple one homogeneous compartment of volume V. As shown in Fig. 6.18, solute is generated in the unique pool (with a generation rate G) and removed by an endogenous clearance (K_r), due to residual renal functions. The connection with the dialyzer, providing a specific solute clearance during dialysis treatment, is also illustrated.

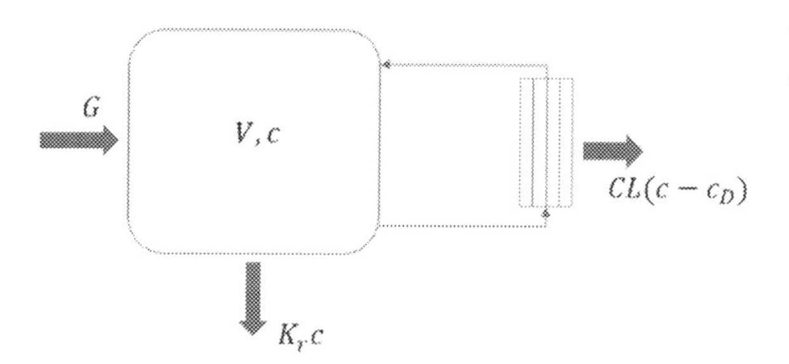

FIGURE 6.18 Schematics of a single-compartment model.

The governing differential equation for the solute mass balance is given by:

$$\frac{d(V\quad c)}{dt} = G - K_r \quad c - \mathrm{CL}(c - c_D) \tag{6.30}$$

where c is the solute concentration in the volume (equal to the concentration in blood) and c_D is the solute concentration in the dialysate. Integration of Eq. (6.30), in the case of toxins like urea or creatinine, generally characterized by a negligible concentration in dialysis fluid ($c_D = 0$), yields:

$$c(t) = c_0 e^{-K_d t/V} + \frac{G}{K_d}\left(1 - e^{-K_d t/V}\right) \tag{6.31}$$

where $c(t = 0) = c_0$ is the initial solute concentration and $K_d = K_r + \mathrm{CL}$. The ratio $\frac{G}{K_d}$ is small during dialysis and the total solute concentration profile basically depends on $\frac{K_d t}{V}$ parameter. The change of volume has been neglected, during the dialysis treatment, because of its little influence on the modeling efficiency (Ziółko, Pietrzyk, and Grabska-Chrząstowska, 2000). Since the specific patient information (V, G, and K_r) are included, the equations' model can be used to analyze clinical data and establish the proper specific dialytic dose. As a result, it is possible to estimate the therapy's duration (T) as:

$$T = \frac{V}{K_d}\ln\frac{c_0}{c_T} \tag{6.32}$$

where c_T is the theoretical solute concentration value at the end of the treatment.

A wide variety of single- and pseudo-one-compartment models have been elaborated by several authors, in which different solutes' kinetics, such as urea, creatinine, β_2-microglobulin, potassium, phosphate, phosphorus or calcium, have been simulated (Agar, Culleton, Fluck, and Leypoldt, 2015; Debowska, Poleszczuk, Wojcik-Zaluska, Ksiazek, and Zaluska, 2015; di Filippo et al., 2018; Leypoldt, Cheung, and Deeter, 1997; Sanfelippo, Walker, Hall, and Swenson, 1978; Yashiro et al., 2015).

Different concentration's time profile could be predicted, in according to the specific dialytic regimens, as reported in Table 6.3. In detail, Fig. 6.19 shows that an urea concentration reduction is attained with both an increase in the dialysis clearance and the duration of the session.

Although each model presents acceptable results, none of these is able to predict the phenomenon of postdialysis solute rebound, during which the solute releases from a not

TABLE 6.3 Operating conditions for urea concentration profile reported in Figure 6.19. The obtained parameters are all referred to a thrice-weekly dialysis treatment with the assumptions of $V = 40$ L and $G = 6.25$ mg/min (Annesini et al., 2017).

Operating conditions	T (h)	CL (mL/min)	K_r (mL/min)	$\frac{K_d t}{V}$
A	3	200	0	0.9
B	3	266	0	1.2
C	4	200	0	1.2
D	4	250	0	1.5

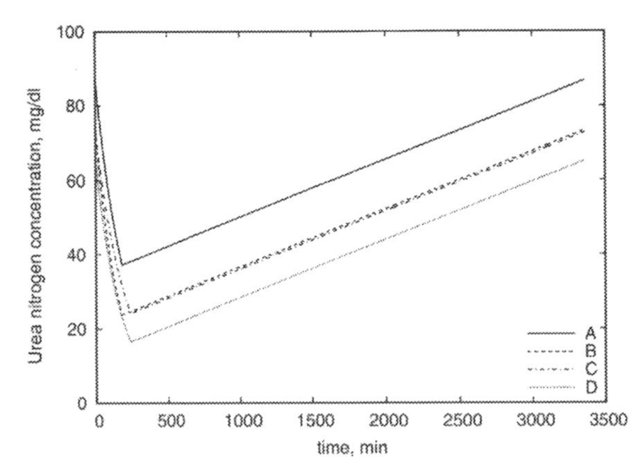

FIGURE 6.19 Urea concentration's profiles predicted with a single-compartment model. All the curves suffer from the phenomenon of solute rebound (Annesini et al., 2017).

perfused plasma region (not in direct contact with the dialyzer) to blood. As a result, blood toxins' levels sharply increase in the early postdialytic period. In order to overcome this limitation, a more complex processes' formulation becomes necessary.

6.4.4 Multicompartment model

Actually, a better description of the solute removal in RRTs requires to consider different compartments in human body solute distribution. Thus the idea on the basis of the multipool models is that fluids' exchanges can occur through the various compartments, but only the latter one perfuses the dialyzer. As it regards two-compartment models, patient volume is divided in the intra- and extracellular regions. Fig. 6.20 shows that the solute transport from the first pool is proportional to the intracompartmental mass transport coefficient (K_c) and only the second pool, characterized by a generation rate G and a residual endogenous clearance K_r, is in effectively contact with the dialyzer

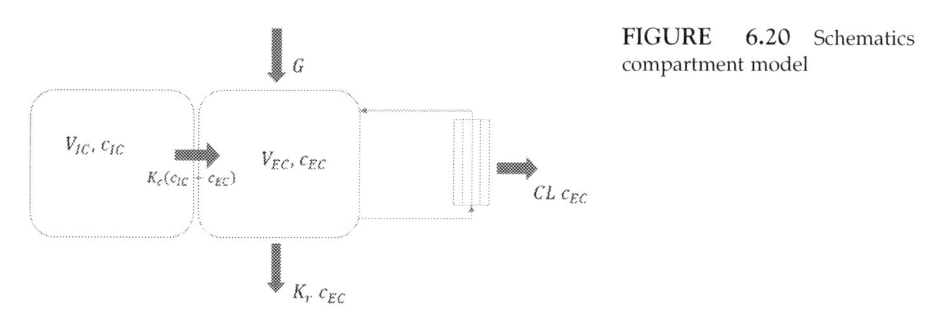

FIGURE 6.20 Schematics of a two-compartment model

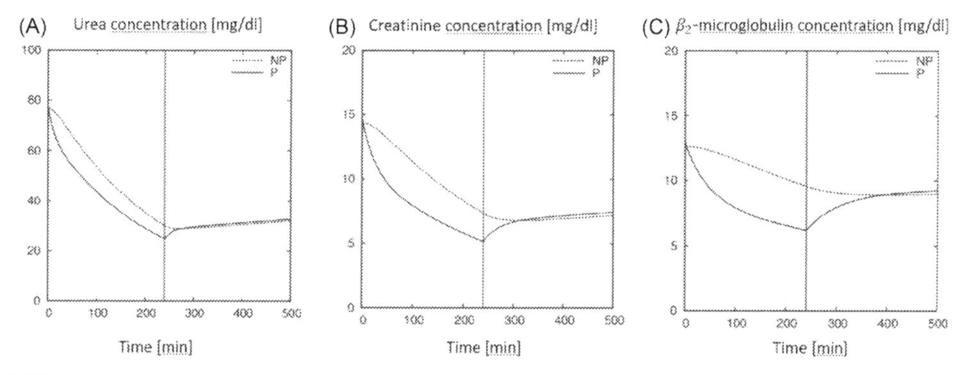

FIGURE 6.21 Concentration profiles of urea (A), creatinine (B) and β_2-microglobulin (C), predicted by a two-compartment model (Clark et al., 1999; Leypoldt, Jaber, Lysaght, McCarthy, and Moran, 2003).

With the additional assumption of constant volumes (Ziółko et al., 2000), the mass balance can be expressed by the following two equations:

$$\frac{dc_{\mathrm{IC}}}{dt} = -\frac{K_c}{V_{\mathrm{IC}}}(c_{\mathrm{IC}} - c_{\mathrm{EC}}) \tag{6.33}$$

$$\frac{dc_{\mathrm{EC}}}{dt} = \frac{K_c}{V_{\mathrm{EC}}}(c_{\mathrm{IC}} - c_{\mathrm{EC}}) + \frac{G}{V_{\mathrm{EC}}} - \frac{(K_r + \mathrm{CL})}{V_{\mathrm{EC}}}c_{\mathrm{EC}} \tag{6.34}$$

where IC and EC are related to the intra- and extracellular compartment, respectively, and c_{EC} is equal to the solute concentration in blood. Solving Eqs. (6.33) and (6.34), the model is able to describe not only a nonexponential decay of solute concentration, but also the phenomenon of solute rebound. As a result, it is possible to predict, for example, the solute concentration profiles in intradialytic and early postdialytic periods in both perfused and not perfused compartments. Fig. 6.21 shows a small intercompartmental concentration difference in the case of urea, which assumes more evidence for creatinine and even more for β_2-microglobulin.

Moreover, the possibility to personalize the dialytic therapy, also in terms of treatment's frequency, strongly influences the solutes' kinetics. In a general way, a more uniform concentration profile is obtained applying frequent short dialysis sessions.

In a similar manner, the development of multicompartment models are all based on the idea to approximate human body in different pools, of which only the latter one can be considered as perfused by the dialyzer. In particular, the emerging need for treatment customization, due to individual variable tolerance when similar therapies are involved, has been revealed by using models capable to support clinical observations. Moreover, the specific dialytic procedure induces substantial changes in osmotic balances, with rapid variations in fluid volumes and electrolytic concentrations across the patient's body compartments. In this scenario, one of the most interesting mathematical approach derives from Ursino and coworkers, who have elaborated different models, including many solutes' kinetics and the realization of acid−base equilibrium through different buffer systems. Their previous studies were simply able to predict small toxins' transport, such as urea or creatinine (Ursino et al., 1997), while more recent works have demonstrated the effective solute osmolarity achieved, in comparation to clinical data. More complex three-compartment models (where the extracellular environment is divided into the interstitial and plasma regions) (Ursino et al., 1999, 2000) are also included with the aim to better evaluate potassium's profile (Ursino and Donati, 2017), also in special HDF treatment with online regeneration of ultrafiltrate (Ursino et al., 2006). Waniewski et al. are also committed in the development of a more accurate and predictable method in order to reduce postdialysis patient morbidities. Their contribution was focalized on sodium kinetics during HD (Waniewski, 2006), but especially, on PD, mathematically designed through a three pore model, in which the transport phenomena occur across large, small and ultra-small pores (Waniewski, 2013). Additional significant improvements are actually represented by recent models in which phosphate kinetics is simulated for the prevention of hyperphosphatemia and related complications (Laursen, Vestergaard, and Hejlesen, 2018). During the last years, not only the physiological role of sodium−potassium pump has been modeled (Pietribiasi, Waniewski, Wójcik-Załuska, Załuska, and Lindholm, 2018) but also the simulation of AK's conditions has been implemented (Kim et al., 2016; Rambod, Beizai, and Rosenfeld, 2010) in order to obtain a final model that could really minimize patients' morbidity and eventually substitute the conventional dialytic therapy.

6.4.5 Modeling of regenerative dialysis

The possibility to transfer the dialytic machine into a wearable device represents a concrete challenge in biomedical field. The current dialysis treatment paradigms can exploit sophisticated smaller and lighter devices, intended for use outside the clinic. As deeply discussed in the previous chapter, the capability to perform a regenerative dialysis could really revolutionize the RRT panorama, even if more clinical trials and studies are still necessary. The mathematical modeling is totally equivalent to those obtained for the conventional dialytic procedures (still based on compartmental theory), with the only exception due to the contribute of the dialysate regeneration unit (DRU). As shown in Fig. 6.22, the dialyzer is perfused by an amount of dialysate, which periodically undergoes regeneration in according to the specific patient needs.

In according to the specific patient conditions, it is possible to estimate the proper dialysis dose with the adequate AK treatment. As an example, referring to the

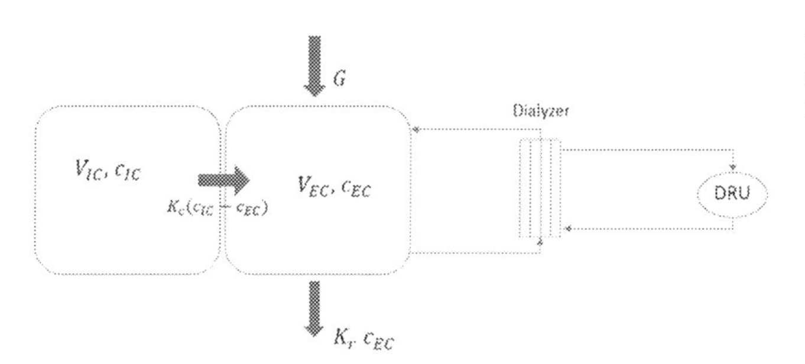

FIGURE 6.22 Schematics of a two-compartment model with a dialysate regeneration unit.

TABLE 6.4 Operating conditions for creatinine concentration profile reported in Fig. 6.23 (Kim et al., 2016).

Operating conditions	Body weight (kg)	Daily water intake volume (L/day)	Creatinine dialysis clearance (mL/min)	Daily dialysis pause time (h/day)
A	100	1	25	12
B	70	1	20	6
C	70	1	30	4
D	70	1	60	2

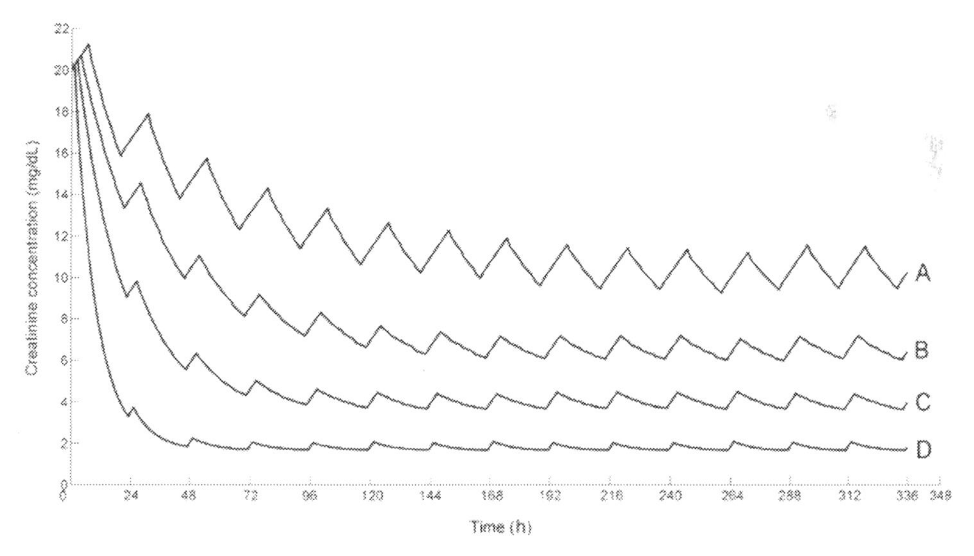

FIGURE 6.23 Creatinine concentration profiles of the perfused compartment predicted with a two-compartment model, in according to the operating conditions reported in Table 6.4 (Kim et al., 2016).

operating conditions of Table 6.4, Fig. 6.23 shows the creatinine concentration profiles of the perfused compartment, for a period of two week, by using a two-compartment model.

In regenerative dialysis treatments, a key role is played by sorbent cartridges because they are able to purify the exhausted dialysate from a variety of chemical substances maintaining the appropriate electrolytes' concentration. Actually, the proper sorbent material choice is the activated carbon, due to its capability to remove all organic uremic toxins, with the notable exception of urea (affinity for activated carbon ~0.1 mmol/g) (Ash, 2009). Although urea is often considered a relatively harmless solute, it remains the main nitrogenous waste product, with the highest daily production (Singh, Dar, and Ahmad, 2009). As a consequence, the greatest challenge for successful dialysate regeneration is urea removal. The currently widely used DRU is formed by a resin bed including at least a four-layer sorbent cartridge, in order to convert urea to ammonium and carbonate using urease as catalyst, and remove ammonium ions by a zirconium phosphate exchange layer (Karoor et al., 2011). From the kinetic point of view, adsorption can be described as a bonding reaction strongly depending on the sorbate solutes (S_i) and the adsorption centers on sorbent surface (C^*):

$$S_i + C^* \leftrightarrow MC_i$$

where MC_i is the complex formed on the surface of the activated carbon layer. The chemical reaction is reversible and it depends on adsorption constant k_1 (in the right direction) and desorption constant k_2 (in the left direction).

The sorbate concentration's profile (C) can be expressed by Bazaev, Bizyukov, and Streltsov (2017):

$$\frac{dC}{dt} = -k_1 C[A_m - (C_0 - C)V] + k_2(C_0 - C)V \tag{6.35}$$

where A_m is the total number of adsorption centers, C_0 is the initial sorbate concentration and V is the volume of the solution considered. The time dependence of sorbate concentration, obtained by solving Eq. (6.35), can be used to evaluate the dynamics of changing of the sorption columns in an AK in according to patient need. Kim et al. (2009) have elaborated a model for a cold dialysate regeneration system (CDRS) with activated carbon at low temperature (5°C) in order to enhance the toxins' removal (Lim, Kim, and Shim, 2010). The evaluation of adsorption ratio (AR) and desorption ratio (DR) given by:

$$\text{AR} = \frac{C_0 - C}{C_0}; \text{DR} = \frac{\text{amount of desorption}}{\text{amount of adsorption}}$$

and their final mathematical model have shown promising results with short-term dialytic therapy with CDRS, comparable to those obtained through a three-weekly HD treatment. Although some studies report side effects of long-term use of activated carbon cartridges, the in-vivo experiments of CDRS do not register atypical symptoms, demonstrating the possibility to improve the quality of life of chronic renal patients. Another interesting approach is characterized by urease immobilization achieved by special physical or chemical bindings in order to increase urease activity (Krajewska, 2009). Both urea removal for adsorption and urea hydrolysis present several disadvantages including large amounts of cations exchange materials (>0.5 kg) required to remove the generated ammonium, or the ammonium partially exchanged for sodium, resulting in high sodium release, which complicates body fluid and blood pressure management. In addition, urease cartridges are expensive and they show a not negligible aluminum release (Ash, 2009). An alternative strategy is the electro-oxidation (EO), which represents an attractive technique for RRT because it converts urea into gaseous products (nitrogen, hydrogen, and carbon dioxide), that can be easily removed from dialysate by a bubble trap. The process can be performed by using small and light modules, which are also relatively cheap and potentially reusable. The bigger problem of EO is related to the lack of control of formed products during urea oxidation, such as active chlorine species and chloramines, which need to be removed. Nowadays, the previous prototypes in platinum and ruthenium are completely substituted by graphite electrodes combined with activated carbon, owing to their better capability to overcome EO-drawbacks (Wester et al., 2014). Wester et al. (2018) have elaborated a mathematical modeling of a AK with urea EO on goats, but future researches are still warranted to elucidate if the phenomenon can be safely applied in vivo. In conclusion, an innovative method provides the use of two-dimensional titanium carbide sorbent cartridge, which exploiting its high selectivity for urea adsorption from dialysate and its relevant biocompatibility, can open a new opportunity in the field of regenerative dialysis (Meng et al., 2018).

6.5 Conclusion and future trends

ESKD is a common social and economic burden and more than 3 million patients receive dialysis treatments, worldwide. They are typically forced to perform in-center dialytic therapies, characterized by long duration sessions. Novel engineering devices represent a challenge for the modern biomedical research with the aim to improve patients' quality of life and possibly reduce their morbidities. Efforts in understanding and modeling transport phenomena in AKs are crucial and this chapter gives an overview of the engineering aspects of the design of a dialysis systems, on the basis of the fundamentals equations in transport phenomena. Medical requirements and characteristics for artificial devices were defined after a short overview of functions and pathologies of the renal system together with a summary of the historical development of the AK. Afterwards, the different ways in which membrane separation is achieved in RRT, as well as the characteristics of membrane modules used, are presented. On the basis of this information, mathematical models are then developed to describe the behavior of these systems and optimize their design. Finally, patient-device models were derived in order to predict the evolution in time of toxin levels in patient blood and optimized the dialysis protocol.

References

Agar, B. U., Culleton, B. F., Fluck, R., & Leypoldt, J. K. (2015). Potassium kinetics during hemodialysis. *Hemodialysis International, 19*(1), 23–32. Available from https://doi.org/10.1111/hdi.12195.

Agarwal, A. K., Haddad, N. J., Vachharajani, T. J., & Asif, A. (2019). Innovations in vascular access for hemodialysis. *Kidney International, 95*, 1053–1063. Available from https://doi.org/10.1016/j.kint.2018.11.046.

Annesini, M. C., Marrelli, L., Piemonte, V., & Turchetti, L. (2017). *Artificial organ engineering*. London: Springer London, ISBN 978-1-4471-6442-5.

Ash, S. R. (2009). Sorbents in treatment of uremia: A short history and a great future. *Seminars in Dialysis, 22*, 615–622. Available from https://doi.org/10.1111/j.1525-139X.2009.00657.x.

Basicmedical Key. (2020). *Solute and water transport along the nephron: Tubular function*. <https://basicmedicalkey.com/solute-and-water-transport-along-the-nephron-tubular-function/#cetable1>. Accessed 20.04.20.

Bazaev, N. A., Bizyukov, I. O., & Streltsov, E. V. (2017). Mathematical modeling of sorption in a wearable artificial kidney with dialysate regeneration. *BioMedical Engineering, 50*, 318–320. Available from https://doi.org/10.1007/s10527-017-9646-2.

Canaud, B., Chénine, L., Leray-Moraguès, H., Patrier, L., Rodriguez, A., Gontier-Picard, A., & Moréna, M. (2017). Hémodiafiltration en ligne: Modalités pratiques, sécurité et efficacité de la méthode. *Néphrologie & Thérapeutique, 13*, 189–201. Available from https://doi.org/10.1016/j.nephro.2017.02.007.

Canaud, B., Vienken, J., Ash, S., & Ward, R. A. (2018). Hemodiafiltration to address unmet medical needs ESKD patients. *Clinical Journal of the American Society of Nephrology: CJASN, 13*, 1435–1443. Available from https://doi.org/10.2215/CJN.12631117.

Clark, W. R., Leypoldt, J. K., Henderson, L. W., Mueller, B. A., Scott, M. K., & Vonesh, E. F. (1999). Quantifying the effect of changes in the hemodialysis prescription on effective solute removal with a mathematical model. *Journal of the American Society of Nephrology: JASN, 10*, 601–609.

Debowska, M., Poleszczuk, J., Wojcik-Zaluska, A., Ksiazek, A., & Zaluska, W. (2015). Phosphate kinetics during weekly cycle of hemodialysis sessions: Application of mathematical modeling. *Artificial Organs, 39*, 1005–1014. Available from https://doi.org/10.1111/aor.12489.

Dechadilok, P., & Deen, W. M. (2006). Hindrance factors for diffusion and convection in pores. *Industrial & Engineering Chemistry Research, 45*, 6953–6959. Available from https://doi.org/10.1021/ie051387n.

di Elisabetta, A.; Francesco Saverio, A.-I.; Francesco, C. (2018). *Le basi cellulari e molecolari delle malattie*; Idelson-Gnocchi, (Ed.). ISBN 8879476718.

di Filippo, S., Carfagna, F., la Milia, V., Bellasi, A., Casagrande, G., Bianchi, C., et al. (2018). Assessment of intra-dialysis calcium mass balance by a single pool variable-volume calcium kinetic model. *Hemodialysis International. International Symposium on Home Hemodialysis, 22*, 126–135. Available from https://doi.org/10.1111/hdi.12531.

Eknoyan, G. (2009). The wonderful apparatus of John Jacob Abel called the "Artificial Kidney. *Seminars in Dialysis, 22*(3), 287–296. Available from https://doi.org/10.1111/j.1525-139X.2009.00527.x.

Faria, J., Ahmed, S., Gerritsen, K. G. F., Mihaila, S. M., & Masereeuw, R. (2019). Kidney-based in vitro models for drug-induced toxicity testing. *Archives of Toxicology, 93*, 3397–3418. Available from https://doi.org/10.1007/s00204-019-02598-0.

Forni, L. G., & Hilton, P. J. (1997). Continuous hemofiltration in the treatment of acute renal failure. *The New England Journal of Medicine, 336*, 1303–1309. Available from https://doi.org/10.1056/NEJM199705013361807.

Friedrich, J. O., Wald, R., Bagshaw, S. M., Burns, K. E., & Adhikari, N. K. (2012). Hemofiltration compared to hemodialysis for acute kidney injury: Systematic review and *meta*-analysis. *Critical Care (London, England), 16*, R146. Available from https://doi.org/10.1186/cc11458.

Hall, J. E. (2015). *Guyton and hall textbook of medical physiology; Guyton physiology*. Elsevier Health Sciences, ISBN 9781455770052.

Heiney, P. (2003). *The nuts and bolts of life: Willem Kolff and the invention of the kidney machine*. Sutton, ISBN 9780750928960.

Hirano, A., Yamamoto, K., Matsuda, M., Ogawa, T., Yakushiji, T., Miyasaka, T., & Sakai, K. (2011). Evaluation of dialyzer jacket structure and hollow-fiber dialysis membranes to achieve high dialysis performance. *Therapeutic Apheresis and Dialysis: Official Peer-reviewed Journal of the International Society for Apheresis, the Japanese Society for Apheresis, the Japanese Society for Dialysis Therapy, 15*, 66–74. Available from https://doi.org/10.1111/j.1744-9987.2010.00869.x.

Hojs, N., Fissell, W. H., & Roy, S. (2020). Ambulatory hemodialysis-technology landscape and potential for patient-centered treatment. *Clinical Journal of the American Society of Nephrology: CJASN, 15*, 152–159. Available from https://doi.org/10.2215/CJN.01970219.

Humes, H. D., Buffington, D., Westover, A. J., Roy, S., & Fissell, W. H. (2014). The bioartificial kidney: Current status and future promise. *Pediatric Nephrology (Berlin, Germany), 29*, 343–351. Available from https://doi.org/10.1007/s00467-013-2467-y.

Ikizler, T. A., & Schulman, G. (2005). Hemodialysis: Techniques and prescription. *American Journal of Kidney Diseases: The Official Journal of the National Kidney Foundation, 46*, 976–981. Available from https://doi.org/10.1053/j.ajkd.2005.07.037.

Karoor, S., Donovan, B., Hai, T.T., Katada, M., Lu, L., Martis, L. et al. (2011). *Method and composition for removing uremic toxins in dialysis processes*. US7241272B. Baxter International Inc.

Kasztan, M., & Pollock, D. M. (2018). A more direct way to measure glomerular albumin permeability—Even in human glomeruli!. *Kidney International, 93*, 1035–1047. Available from https://doi.org/10.1016/j.kint.2018.01.016.

Ketteler, M., Block, G. A., Evenepoel, P., Fukagawa, M., Herzog, C. A., McCann, L., et al. (2018). Diagnosis, evaluation, prevention, and treatment of chronic kidney disease–mineral and bone disorder: Synopsis of the kidney disease: improving global outcomes 2017 clinical practice guideline update. *Annals of Internal Medicine, 168*, 422. Available from https://doi.org/10.7326/M17-2640.

Kim, D. K., Lee, J. C., Lee, H., Joo, K. W., Oh, K., Kim, Y. S., ... Kim, H. C. (2016). Calculation of the clearance requirements for the development of a hemodialysis-based wearable artificial kidney. *Hemodialysis International. International Symposium on Home Hemodialysis, 20*, 226–234. Available from https://doi.org/10.1111/hdi.12343.

Kim, J. C., Cruz, D., Garzotto, F., Kaushik, M., Teixeria, C., Baldwin, M., et al. (2013). Effects of dialysate flow configurations in continuous renal replacement therapy on solute removal: Computational modeling. *Blood Purification, 35*, 106–111. Available from https://doi.org/10.1159/000346093.

Kim, J. H., Kim, J. C., Moon, J.-H., Park, J. Y., Lee, K. K., Kang, E., ... Ronco, C. (2009). Development of a cold dialysate regeneration system for home hemodialysis. *Blood Purification, 28*, 84–92. Available from https://doi.org/10.1159/000218088.

Kotanko, P.; Kuhlmann, M.K.; Levin, N.W. (2010). Chapter 89 – Hemodialysis: Principles and techniques. In: Floege, J, Johnson, RJ, Feehally, J, editors, Comprehensive clinical nephrology (4th ed.); Mosby: Philadelphia, pp. 1053–1059. <https://doi.org/10.1016/B978-0-323-05876-6.00089-7>.

Krajewska, B. (2009). Ureases. II. Properties and their customizing by enzyme immobilizations: A review. *Journal of Molecular Catalysis B: Enzymatic, 59*, 22–40. Available from https://doi.org/10.1016/j.molcatb.2009.01.004.

Kurkus, J., & Ostrowski, J. (2019). Nils Alwall and his artificial kidneys: Seventieth anniversary of the start of serial production. *Artificial Organs, 43*, 713–718. Available from https://doi.org/10.1111/aor.13545.

Laursen, S. H., Vestergaard, P., & Hejlesen, O. K. (2018). Phosphate kinetic models in hemodialysis: A systematic review. *American Journal of Kidney Diseases: The Official Journal of the National Kidney Foundation, 71*, 75–90. Available from https://doi.org/10.1053/j.ajkd.2017.07.016.

Leypoldt, J. K., Cheung, A. K., & Deeter, R. B. (1997). Single compartment models for evaluating β2-microglobulin clearance during hemodialysis. *ASAIO Journal (American Society for Artificial Internal Organs: 1992), 43*, 904–909. Available from https://doi.org/10.1097/00002480-199711000-00011.

Leypoldt, J. K., Jaber, B. L., Lysaght, M. J., McCarthy, J. T., & Moran, J. (2003). Kinetics and dosing predictions for daily haemofiltration. *Nephrology, Dialysis, Transplantation: Official Publication of the European Dialysis and Transplant Association - European Renal Association, 18*, 769–776. Available from https://doi.org/10.1093/ndt/gfg019.

Liao, Z., Klein, E., Poh, C. K., Huang, Z., Lu, J., Hardy, P. A., & Gao, D. (2005). Measurement of hollow fiber membrane transport properties in hemodialyzers. *Journal of Membrane Science, 256*(1-2), 176–183. Available from https://doi.org/10.1016/j.memsci.2005.02.032.

Lim, K. M., Kim, J. H., & Shim, E. B. (2010). Mathematical analysis of the long-term efficacy of daily home hemodialysis therapy with a cold dialysate regeneration system. *Blood Purification, 29*, 27–34. Available from https://doi.org/10.1159/000245044.

Mazzeo, L., Bianchi, M., Cocchi, M., Piemonte, V. (2019). Chapter 10 - Drug delivery with membranes systems. In: Basile, A, Charcosset, C, editors. Current trends and future developments on (bio-) membranes; Elsevier, pp. 291–309. <https://doi.org/10.1016/B978-0-12-813606-5.00010-5>.

Meng, F., Seredych, M., Chen, C., Gura, V., Mikhalovsky, S., Sandeman, S., Ingavle, G., Ozulumba, T., Miao, L., Anasori, B., et al. (2018). MXene sorbents for removal of urea from dialysate: A step toward the wearable artificial kidney. *ACS Nano, 12*(10), 10518–10528. Available from https://doi.org/10.1021/acsnano.8b06494.

Menon, M. C., Chuang, P. Y., & He, C. J. (2012). The glomerular filtration barrier: Components and crosstalk. *International Journal of Nephrology, 2012*, 749010. Available from https://doi.org/10.1155/2012/749010.

Metcalfe, W. (2007). How does early chronic kidney disease progress? A background paper prepared for the UK consensus conference on early chronic kidney disease. *Nephrology, Dialysis, Transplantation: Official Publication of the European Dialysis and Transplant Association - European Renal Association, 22*, ix26–ix30. Available from https://doi.org/10.1093/ndt/gfm446.

Meyer, T. W. (2004). Increasing dialysate flow and dialyzer mass transfer area coefficient to increase the clearance of protein-bound solutes. *Journal of the American Society of Nephrology: JASN, 15*, 1927–1935. Available from https://doi.org/10.1097/01.ASN.0000131521.62256.F0.

Mineshima, M. (2015). The past, present and future of the dialyzer. *Contributions to Nephrology, 185*, 8–14. Available from https://doi.org/10.1159/000380965.

Misra, M. (2008). Basic mechanisms governing solute and fluid transport in hemodialysis. *Hemodialysis International. International Symposium on Home Hemodialysis., 12*, S25–S28. Available from https://doi.org/10.1111/j.1542-4758.2008.00320.x.

Paskalev, D. (2001). Georg Haas (1886-1971): The forgotten hemodialysis pioneer. *Dialysis & Transplantation, 30*, 828–832.

Pietribiasi, M., Waniewski, J., Wójcik-Załuska, A., Załuska, W., & Lindholm, B. (2018). Model of fluid and solute shifts during hemodialysis with active transport of sodium and potassium. *PLoS One, 13*, e0209553. Available from https://doi.org/10.1371/journal.pone.0209553.

Rambod, E., Beizai, M., & Rosenfeld, M. (2010). An experimental and numerical study of the flow and mass transfer in a model of the wearable artificial kidney dialyzer. *Biomedical Engineering Online, 9*, 21. Available from https://doi.org/10.1186/1475-925X-9-21.

Rocha, A., Sousa, C., Teles, P., Coelho, A., & Xavier, E. (2015). Frequency of intradialytic hypotensive episodes: Old problem, new insights. *Journal of the American Society of Hypertension: JASH, 9*(10), 763–768. Available from https://doi.org/10.1016/j.jash.2015.07.007.

Ronco, C., Ghezzi, P. M., Brendolan, A., Crepaldi, C., & La Greca, G. (1998). The haemodialysis system: Basic mechanisms of water and solute transport in extracorporeal renal replacement therapies. *Nephrology Dialysis Transplantation Official Publication of European Dialysis and Transplantation Association. - Eur. Ren. Assoc, 13* (Suppl 6), 3–9. Available from https://doi.org/10.1093/ndt/13.suppl_6.3.

Salani, M., Roy, S., & Fissell, W. H. (2018). Innovations in wearable and implantable artificial kidneys. *American Journal of Kidney Diseases: The Official Journal of the National Kidney Foundation, 72*, 745–751. Available from https://doi.org/10.1053/j.ajkd.2018.06.005.

Sanfelippo, M. L., Walker, W. E., Hall, D. A., & Swenson, R. S. (1978). Clinical application of a single compartment model to urea and creatinine kinetics in dialysis therapy. *Computer Programs in Biomedicine, 8*(1), 44–50. Available from https://doi.org/10.1016/0010-468X(78)90058-2.

Schneditz, D., & Daugirdas, J. T. (2001). Compartment effects in hemodialysis. *Seminars in Dialysis, 14*, 271–277. Available from https://doi.org/10.1046/j.1525-139X.2001.00066.x.

Seifter, J., Ratner, A., & Sloane, D. (2005). *Concepts in medical physiology* (1st ed.). Philadelphia, PA: Lippincott Williams & Wilkins.

Singh, L. R., Dar, T. A., & Ahmad, F. (2009). Living with urea stress. *Journal of Biosciences, 34*, 321–331. Available from https://doi.org/10.1007/s12038-009-0036-0.

Stewart Cameron, J. (2012). Thomas Graham (1805–1869) — The "Father" of dialysis. In: TS Ing, editor. Dialysis; World Scientific, pp. 19–25. <https://doi.org/10.1142/9789814289764_0003>.

Thijssen, S., Raimann, J. G., & Levin, N. W. (2012). The evolution of dialysis. In T. S. Ing (Ed.), . Dialysis (pp. 233–243). World Scientific. Available from https://doi.org/10.1142/9789814289764_0027.

Twardowski, Z. J. (2008). History of hemodialyzers' designs. *Hemodialysis International. International Symposium on Home Hemodialysis, 12*, 173–210. Available from https://doi.org/10.1111/j.1542-4758.2008.00253.x.

Ursino, M., Colí, L., Brighenti, C., Chiari, L., De Pascalis, A., & Avanzolini, G. (2000). Prediction of solute kinetics, acid-base status, and blood volume changes during profiled hemodialysis. *Annals of Biomedical Engineering, 28*, 204–216. Available from https://doi.org/10.1114/1.245.

Ursino, M., Colì, L., Brighenti, C., De Pascalis, A., Chiari, L., Dalmastri, V., … Stefoni, S. (1999). Mathematical modeling of solute kinetics and body fluid changes during profiled hemodialysis. *The International Journal of Artificial Organs, 22*, 94–107. Available from https://doi.org/10.1177/039139889902200207.

Ursino, M., Colì, L., Dalmastri, V., Volpe, F., La Manna, G., Avanzolini, G., … Bonomini, V. (1997). An algorithm for the rational choice of sodium profile during hemodialysis. *The International Journal of Artificial Organs, 20*, 659–672.

Ursino, M., Colì, L., Magosso, E., Capriotti, P., Fiorenzi, A., Baroni, P., & Stefoni, S. A. (2006). Mathematical model for the prediction of solute kinetics, osmolarity and fluid volume changes during hemodiafiltration with on-line regeneration of ultrafiltrate (HFR). *The International Journal of Artificial Organs, 29*, 1031–1041. Available from https://doi.org/10.1177/039139880602901103.

Ursino, M., & Donati, G. (2017). Mathematical model of potassium profiling in chronic dialysis. *Contributions to Nephrology*, *190*, 134–145. Available from https://doi.org/10.1159/000468960.

Villarroel, F., Klein, E., & Holland, F. (1977). Solute flux in hemodialysis and hemofiltration membranes. *The American Society for Artificial Internal Organs*, *23*, 225–232. Available from https://doi.org/10.1097/00002480-197700230-00061.

Waniewski, J. (1994). Linear approximations for the description of solute flux through permselective membranes. *Journal of Membrane Science*, *95*, 179–184. Available from https://doi.org/10.1016/0376-7388(94)00110-3.

Waniewski, J. (2006). Mathematical modeling of fluid and solute transport in hemodialysis and peritoneal dialysis. *Journal of Membrane Science*, *274*, 24–37. Available from https://doi.org/10.1016/j.memsci.2005.11.038.

Waniewski, J. (2013). Peritoneal fluid transport: Mechanisms, pathways, methods of assessment. *Archives of Medical Research*, *44*, 576–583. Available from https://doi.org/10.1016/j.arcmed.2013.10.010.

Waniewski, J., Werynski, A., Ahrenholz, P., Lucjanek, P., Judycki, W., & Esther, G. (2008). Theoretical basis and experimental verification of the impact of ultrafiltration on dialyzer clearance. *Artificial Organs*, *15*, 70–77. Available from https://doi.org/10.1111/j.1525-1594.1991.tb00763.x.

Wester, M., Simonis, F., Lachkar, N., Wodzig, W. K., Meuwissen, F. J., Kooman, J. P., … Gerritsen, K. G. (2014). Removal of urea in a wearable dialysis device: A reappraisal of electro-oxidation. *Artificial Organs*, *38*, 998–1006. Available from https://doi.org/10.1111/aor.12309.

Wester, M., van Gelder, M. K., Joles, J. A., Simonis, F., Hazenbrink, D. H. M., van Berkel, T. W. M., … Gerritsen, K. G. F. (2018). Removal of urea by electro-oxidation in a miniature dialysis device: A study in awake goats. *American Journal of Physiology*, *315*, F1385–F1397. Available from https://doi.org/10.1152/ajprenal.00094.2018.

Yashiro, M., Kotera, H., Matsukawa, M., Kita, T., Tanaka, H., & Sakai, R. (2015). Evaluation of a pseudo-one-compartment model for phosphorus kinetics by later-phase dialysate collection during blood purification. *The International Journal of Artificial Organs*, *38*, 126–132. Available from https://doi.org/10.5301/ijao.5000384.

Yehl, C. J., Jabra, M. G., & Zydney, A. L. (2019). Hollow fiber countercurrent dialysis for continuous buffer exchange of high-value biotherapeutics. *Biotechnology Progress*, *35*, e2763. Available from https://doi.org/10.1002/btpr.2763.

Ziółko, M., Pietrzyk, J. A., & Grabska-Chrząstowska, J. (2000). Accuracy of hemodialysis modeling. *Kidney International*, *57*, 1152–1163. Available from https://doi.org/10.1046/j.1523-1755.2000.00942.x.

CHAPTER

7

Transport phenomena in pervaporation

Axel Schmidt and Jochen Strube

Institute for Separation and Process Technology, Clausthal University of Technology, Clausthal-Zellerfeld, Germany

Nomenclature

A	m^2 (area)
A_i	−(permeance coefficient)
a	−(activity)
a_1 *to* a_4	−(Sh/Nu correlation parameters)
B_i	−(permeance coefficient)
c	$\frac{mol}{m^3}$ or $\frac{kg}{m^3}$ (concentration)
$\tilde{c}_P$	$\frac{kJ}{kg\ K}$ or $\frac{kJ}{kmol\ K}$ (specific heat capacity)
D	$\frac{m^2}{s}$ (diffusion coefficient)
d	m (diameter)
E_i	$\frac{kJ}{kg}$ (activation energy)
$\dot{H}$	$\frac{kJ}{s}$ (enthalpy flow)
$\tilde{h}$	$\frac{kJ}{kg}$ (specific enthalpy)
h	m (height)
J	$\frac{kg}{m^2 h}$ or $\frac{mol}{m^2 h}$ (permeate flux)
L	− (proportionality factor)
l	m (length)
K	bar (sorption coefficient)
k	m/h (mass transfer coefficient)
M	$\frac{kg}{kmol}$ (molar mass)
m	kg (mass)
$\dot{m}$	$\frac{kg}{s}$ (mass flow)
$\dot{n}$	$\frac{kmol}{s}$ (mole flow)
Nu	− (Nusselt number)
P	$\frac{kg\ m}{m^2 h\ bar}$ (permeability)
p	bar (pressure)
Pr	− (Prandtl number)

DOI: https://doi.org/10.1016/B978-0-12-822257-7.00002-9

Q $\frac{kg}{m^2 h\ bar}$ (permeance)
Q_i^0 $\frac{kg}{m^2 h\ bar}$ (permeance coefficient)
R $\frac{kJ}{mol\ K}$ (gas constant)
Re – (Reynolds number)
Sc – (Schmidt number)
Sh – (Sherwood number)
T K (temperature)
u $\frac{m}{s}$ (velocity)
U_{wetted} m (wetted perimeter)
$\dot{V}$ $\frac{m^3}{s}$ (volume flow)
v $\frac{m^3}{mol}$ (molar volume)
w_i $\frac{kg}{kg}$ (mass fraction)
w m (width)
x $\frac{kmol}{kmol}$ (molar fraction (liquid)
y $\frac{kmol}{kmol}$ (molar fraction (gaseous)

Greek letters

α $\frac{J}{h\ K}$ (heat transfer coefficient)
γ – (activity coefficient)
δ m (boundary layer thickness)
ζ – (drag coefficient)
λ $\frac{W}{m\ K}$ (thermal conductivity)
η Pa • s (dynamic viscosity)
μ $\frac{kJ}{mol}$ (chemical potential)
ν $\frac{m^2}{s}$ (kinematic viscosity)
ρ $\frac{kg}{m^3}$ (density)

Subscripts

b bulk
ch channel
Diff diffusion
eff effective
f feed
fm membrane–feed side
G gas
h hydraulic
i, j components
L liquid
l length
m membrane
p permeate
pm membrane–permeate side
Q cross-section
r retentate
sat saturated
tot total
TM transmembrane
vap vaporization
0 reference

7.1 Introduction

Energy efficiency of new as well as already established processes plays an increasingly important role in a globalized market characterized by rising energy costs and growing international competition (Hessel, Gürsel, Wang, Noël, & Lang, 2012). In addition to classic pinch analyses for the optimal recycling of various forms of energy in the process and process intensification as an overall concept, consideration must be given to replacing or at least expanding energy-inefficient process strategies. Among other things, this reasoning has generated a massive research interest in pervaporation (Bausa & Marquardt, 2001).

Pervaporation is a thermal separation process wherein a liquid solution flows across a membrane and the components dissolved in the membrane are withdrawn in vapor form on the opposite side (Rautenbach & Albrecht, 1982). It is mainly used for solvent dehydration and separation of organic components in water treatment as stand-alone unit operation or in combination with distillation (Zhu et al., 2021). Because of the separation principle, it exhibits significantly more efficient utilization of process energy than classical distillation/rectification, saving up to 28% of energy costs. In addition, it allows bypassing azeotropic points in the equilibrium diagram, thus eliminating the use of expensive and difficult-to-separate entraining agents (Del Pozo Gómez et al., 2008) (Fig. 7.1).

7.2 Fundamentals

Pervaporation in general is the partial evaporation of a liquid mixture through a selectively acting membrane layer. It is therefore a process of thermal-separation technology. The principle of this process is illustrated schematically in Fig. 7.2. A distinguishing

Applications

- For challenging separations
- Solvent dehydration
- Solvent separation
- Fruit juice concentration
- Others

Benefits

- Energy efficient
- Suited for hybrid processes
- Azeotropic mixtures

Materials

- Polymers
- Zeolite
- Ceramics

FIGURE 7.1 Main application, benefits, and materials in pervaporation.

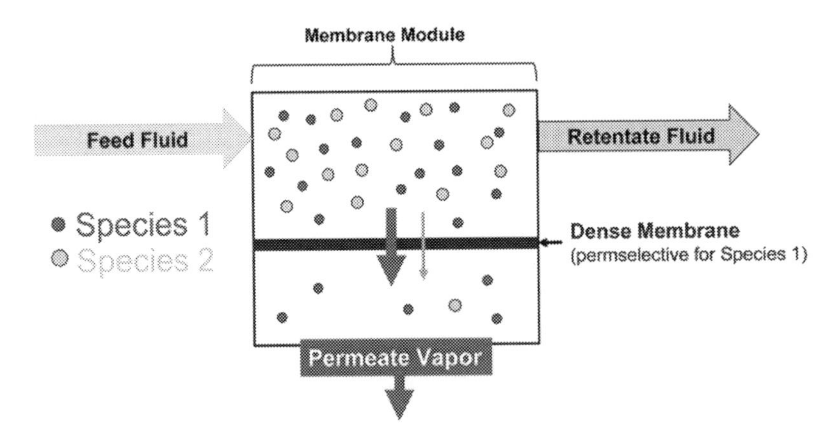

FIGURE 7.2 The general pervaporation process. Source: *Simplified representation slightly modified from Vane, L.M. (2019). Review: Membrane materials for the removal of water from industrial solvents by pervaporation and vapor permeation. Journal of Chemical Technology and Biotechnology (Oxford, Oxfordshire: 1986), 94, 343–365. https://doi.org/10.1002/jctb.5839 (Vane, 2019).*

feature of pervaporation from the other density membrane processes is that it entails phase change from liquid to vapor from the feed to the permeate side (Rautenbach & Albrecht, 1982).

In particular, those membrane materials are extensively used that enable a selective transport of the components through the membrane. In this way, the more permeable component can be preferentially evaporated on the permeate side. The advantages of such a process compared to classical distillation/rectification are, on the one hand, the bypassing (not displacement) of azeotropic points and, on the other hand, the lower energy consumption due to the exclusive evaporation of the permeate instead of the entire feed (Schiffman, 2013).

The most common subdivision of the various pervaporation processes is usually made on the basis of the solvation tendency of the membrane used. The simplest subdivision is therefore that into

1. hydrophilic pervaporation and
2. organophilic pervaporation.

Hydrophilic pervaporation was the first large-scale implementation of the pervaporation process with the aim of ethanol isolation. However, this category also includes all other processes in which solvents are to be dewatered. More rarely, it is also used for separating mixtures of two or more organic solvents according to their degree of polarity (Baker, 2012).

Organophilic pervaporation, on the other hand, is typically used in wastewater treatment. It is often used to remove the last aromatic constituents from aqueous solutions that have already been treated. In rare cases, it can also be used to separate solvent mixtures according to their degree of nonpolarity. The combination of distillation, hydrophilic, and organophilic pervaporation has been reported for the separation of alcohol–water mixtures by Thi, Mizsey, and Toth (2020). For a comprehensive overview of membrane materials and processes in the separation of polar and nonpolar organic solvents, reference is made to the review of Yushkin et al. (2020) and Li, Estager, Monbaliu, Debecker, and Luis (2020) (Fig. 7.3).

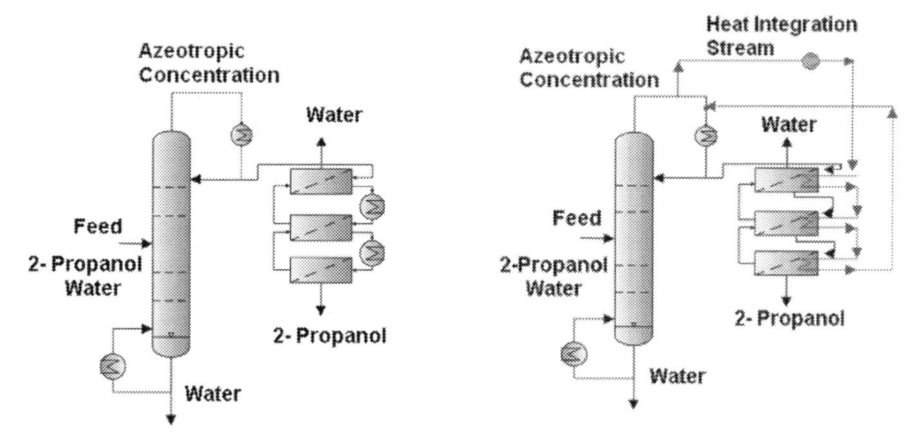

FIGURE 7.3 Hybrid process for the dehydration of 2-propanol. Without heat integration (left) and with heat integration (right) (Del Pozo Gómez et al., 2008).

In addition, pervaporation is occasionally classified according to the source of the driving force and the use of methods that support the driving force. A possible classification in this sense would be:

1. Vacuum pervaporation,
2. thermal pervaporation, and
3. Sweep gas pervaporation (Lipnizki, Hausmanns, Ten, Field, & Laufenberg, 1999).

For the sake of completeness, the individual outlines are briefly explained with reference to Fig. 7.4. The driving force of vacuum pervaporation is the negative pressure applied to the permeate side. This allows the liquid dissolved in the membrane to evaporate, so that more feed-side solution can be taken up by the membrane, and thus the process continues steadily. In thermo-pervaporation, the driving force is generated by a significantly higher feed temperature compared to the permeate side. Both the diffusion through the membrane and the vapor pressure of the components increase with the temperature and thus also the permeate flux.

In sweep gas pervaporation, an inert gas is fed into the permeate chamber for improved removal of the permeate. By heating the inert gas, the necessary evaporation enthalpy can be provided and, in the best case, completely replace heating of the feed (Lipnizki et al., 1999).

The most common process in the literature is the combination of vacuum and thermal pervaporation, as it allows the highest possible flows by increasing the driving force on both the feed and permeate sides (Fig. 7.5).

If the membrane/module geometry or the high permeate fluxes make the vacuum too expensive, all three methods can be combined. Where possible, however, the combination of vacuum and sweep gas should be avoided, since the lowest energy costs are usually always achieved by using only one of the two (Vallieres & Favre, 2004).

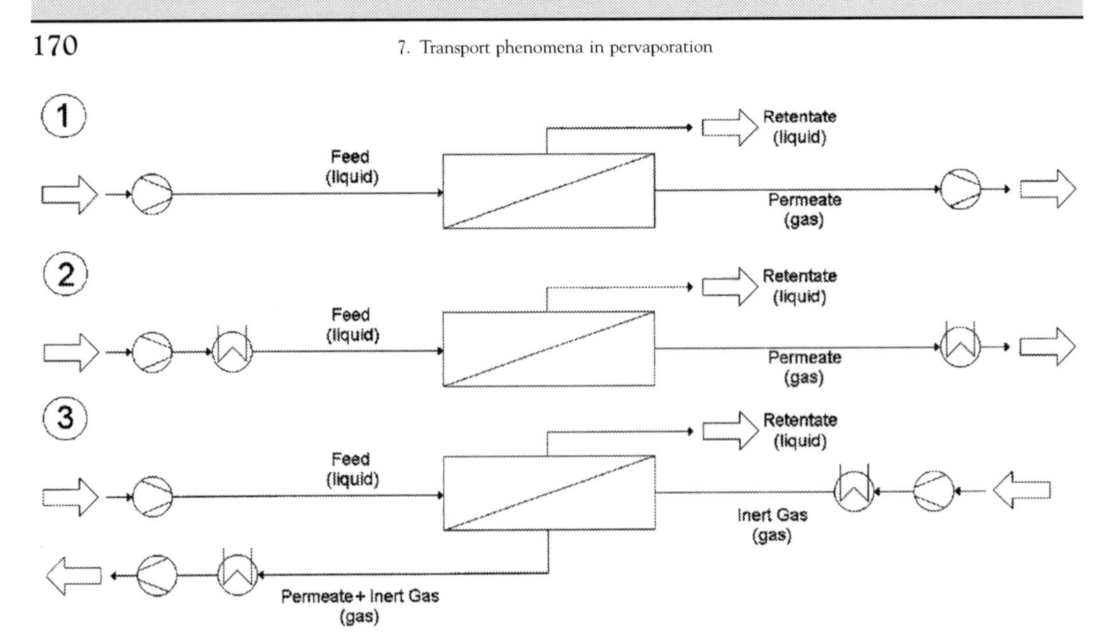

FIGURE 7.4 Possible categorization of pervaporation. (1) Vacuum pervaporation, (2) thermal pervaporation, and (3) sweep gas pervaporation (Lipnizki et al., 1999).

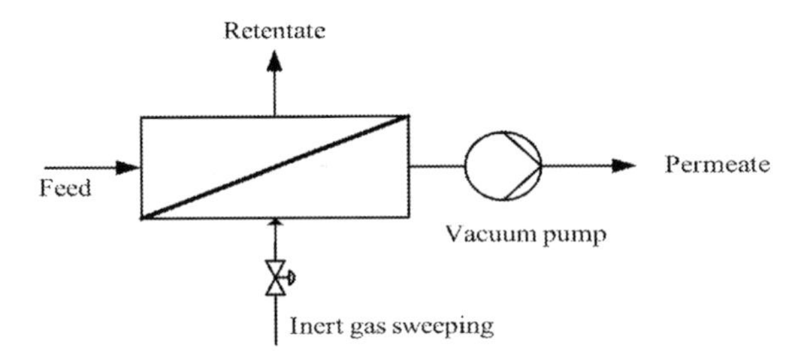

FIGURE 7.5 Combination of vacuum and sweep gas pervaporation. Additional heating of the feed (not shown here) would be a combination of all three basic processes. Source: *Presentation slightly modified taken from Vallieres, C. & Favre, E. (2004). Vacuum vs sweeping gas operation for binary mixtures separation by dense membrane processes. Journal of Membrane Science, 244, 17–23. https://doi.org/10.1016/j.memsci.2004.04.023.*

7.3 Transport phenomena

The main transfer processes in pervaporation are mass transfer, heat transfer, and momentum transfer. In the following, first an approach to describe the momentum transport in terms of pressure drop is described. Subsequently, the derivation of the solution-diffusion model (SDM) is presented, which is the basis of most scientific approaches to simulate mass transfer in pervaporation processes. Finally, the essential enthalpy balances for the representation of the heat transport along the membrane are shown. Mass transfer and driving force reducing effects, such as concentration and temperature polarization, are also briefly discussed.

7.3.1 Pressure drop

The pressure drop plays a decisive role on both sides of the membrane. While on the feed side, a drop below the vapor pressure of the components would lead to evaporation in the feed channel, the pressure on the permeate side must be kept low in order to maintain the greatest possible driving force for pervaporation. If the partial pressure of a component on the permeate side becomes greater than the partial pressure of the corresponding component on the feed side, pervaporation will stop. The pressure on the feed side can usually be set sufficiently high, thereby avoiding the problem of evaporation. Maintaining the vacuum on the permeate side, in turn, becomes more difficult as the permeate flux increases and the distance between the vacuum pump and the membrane increases (Vallieres & Favre, 2004; Wijmans & Baker, 1995).

The derivation of the pressure loss is carried out here as an example for the permeate side, but it can also be set up in an analogous manner for the feed side. For a better understanding of the distinctions between the feed and permeate sides, the essential geometry of the membrane module used in the experiment and in the simulation is shown in Fig. 7.6.

The pressure drop in a duct over a length l is generally derived using Eq. (7.1):

$$\Delta p = \zeta \quad a \quad \frac{\rho \quad v^2}{2} \tag{7.1}$$

Here, the factor "a" signifies the ratio of duct length l and the hydraulic diameter $d_{h,P}$ VDI (2006):

$$a = f \quad \left(\frac{l}{d_h}\right) \tag{7.2}$$

Converted according to the differential change of the pressure over the path length change dz, the following equation follows for the pressure loss on the permeate side:

$$\frac{dp_P}{dz} = \zeta_P \quad \frac{\rho_P \quad v_P^2}{2d_{h,P}} \tag{7.3}$$

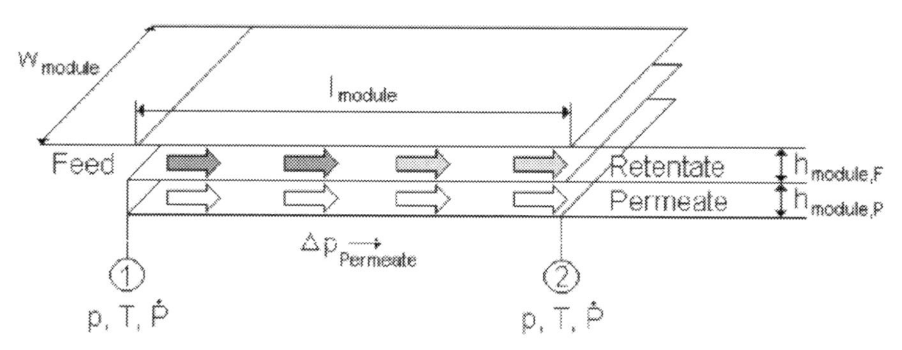

FIGURE 7.6 Basic geometry of the plate module under consideration. The experiments as well as the simulation were carried out in "cocurrent flow" as shown here. However, a counter-current flow mode is also conceivable. Source: *Slightly modified from Lipnizki, F. & Field, R.W. (1999). Simulation and process design of pervaporation plate-and-frame modules to recover organic compounds from waste water.* Chemical Engineering Research and Design, *77, 231–240. https://doi.org/10.1205/026387699526142.*

For the pressure loss coefficient, ζ, different correlations can be found in the literature; these correlations mainly take into account the type of flow regime and the deviation of the present channel geometry from the ideal circular cross-section. In this work, common correlations for plate moduli from the literature are used to model pervaporation. Thus Eq. (7.4) holds for the laminar region ($Re < 2320$) (Lipnizki & Field, 1999):

$$\zeta_P = \frac{38}{Re_P} \tag{7.4}$$

In the case of turbulent flow ($Re > 2320$), on the other hand, Eq. (7.5) is used (Lipnizki & Field, 1999):

$$\zeta_P = \frac{1.22}{Re_P^{0.252}} \tag{7.5}$$

The dimensionless Reynolds number Re is a parameter of fluid mechanics and gives the ratio of inertial and viscous force. The corresponding definition is shown in Eq. (7.6):

$$Re_P = \frac{\rho_P \quad v_P \quad d_{h,P}}{\eta_P} = \frac{v_P \quad d_{h,P}}{\nu_P} \tag{7.6}$$

The velocity v_P of the permeate flow $\dot{n}_P$ is derived via the ideal-gas relationship, since the permeate pressure p_P is considered sufficiently small. This gives the relationship shown in Eq. (7.7):

$$v_P = \frac{R \quad T_P \quad \dot{n}_P}{p_P \quad A_Q} = \frac{R \quad T_P \quad \dot{n}_P}{p_P \quad b_M \quad h_{K,P}} \tag{7.7}$$

Thus in the case of a laminar flow regime, substituting Eqs. (7.4), (7.6), and (7.7) into Eq. (7.3) yields the relationship shown in Eq. (7.8) for the differential pressure drop across the path-length change dz:

$$\frac{dp_P}{dz} = 19\eta_P \quad \frac{R \quad T_P \quad \dot{n}_P}{p_P \quad b_M \quad h_{K,P} \quad d_{h,P}^2} \tag{7.8}$$

For a turbulent flow regime, analogously, by substituting Eqs. (7.5), (7.6), and (7.7) into Eq. (7.3), the relationship shown in Eq. (7.9) is obtained for the differential pressure drop versus path-length change dz:

$$\frac{dp_P}{dz} = 0.61\eta_P^{0.252} \quad \rho_P^{0.748} \quad \frac{\left(\frac{R}{p_P} \quad \frac{T_P}{b_M} \quad \frac{\dot{n}_P}{h_{K,P}}\right)^{1.748}}{d_{h,P}^{1.252}} \tag{7.9}$$

7.3.2 Mass transfer

For the description of the mass transfer on the feed side, the dispersion model is used, which is extended by a mass transfer term to account for the decrease in feed concentration due to the pervaporation process. The simple dispersion model describes the change in concentration of a component i over time due to diffusion superimposed convection.

Accordingly, convection and axial dispersion are considered separately in the balance (Levenspiel, 1999) (Fig. 7.7).

According to the dispersion model extended by the pervaporation process, the following general relationship can be established for mass transport:

$$\frac{\partial w_i}{\partial t} = \frac{\partial w_{i,\mathrm{konv}}}{\partial t} + \frac{\partial w_{i,\mathrm{disp}}}{\partial t} + \frac{\partial w_{i,PV}}{\partial t} \tag{7.10}$$

For a better understanding, the balancing is first carried out on the basis of mass conservation. The temporal mass-related change in the component i due to convection is shown in Eq. (7.11).

$$\frac{\partial m_{i,\mathrm{konv}}}{\partial t} = \frac{\partial \dot{m}_{i,\mathrm{konv}}}{\partial z} \cdot dz = -\dot{V} \cdot \frac{\partial c_i}{\partial z} \cdot dz \tag{7.11}$$

The temporal mass-related change in the component i due to axial dispersion is shown in Eq. (7.12):

$$\frac{\partial m_{i,\mathrm{disp}}}{\partial t} = \frac{\partial \dot{m}_{i,\mathrm{disp}}}{\partial z} \cdot dz = \frac{\partial}{\partial z}\left(A_Q \cdot D_{ax} \cdot \frac{\partial c_i}{\partial z}\right) \cdot dz = A_Q \cdot D_{ax} \cdot \frac{\partial^2 c_i}{\partial z^2} \cdot dz \tag{7.12}$$

Finally, the temporal mass-related change due to the pervaporation process, shown in Eq. (7.13), must be taken into account:

$$\frac{\partial m_{i,PV}}{\partial t} = \frac{\partial \dot{m}_{i,PV}}{\partial z} \cdot dz = -J_i \cdot dA = -J_i \cdot b_{\mathrm{mem}} \cdot dz \tag{7.13}$$

The description of mass transfer in terms of mass fractions as shown in Eq. (7.10) is achieved by dividing Eqs. (7.11) to (7.13) by the total mass m_{ges}. The total mass m_{ges} can be expressed as follows:

$$m_{\mathrm{ges}} = \rho \cdot dV_K = \rho \cdot A_Q \cdot dz = \rho \cdot (b_M \quad h_K) \cdot dz \tag{7.14}$$

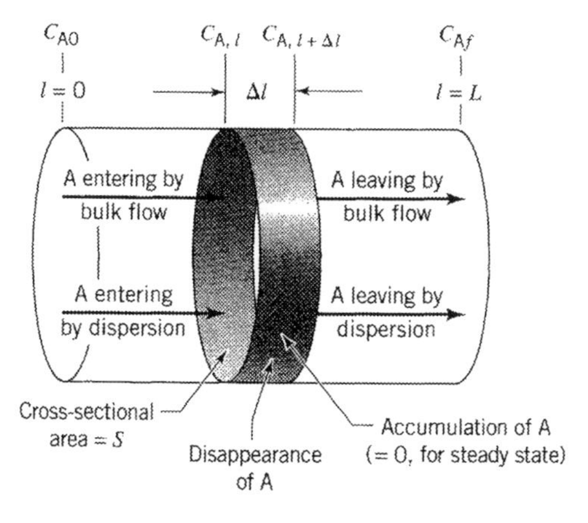

FIGURE 7.7 Schematic representation of the dispersion model (Levenspiel, 1999).

Eqs. (7.11) to (7.13) divided by Eq. (7.14) now give the respective mass fraction–related changes in the component i:

$$\frac{\partial w_{i,\mathrm{konv}}}{\partial t} = -v \cdot \frac{\partial w_i}{\partial z} \tag{7.15}$$

$$\frac{\partial w_{i,\mathrm{disp}}}{\partial t} = D_{ax} \cdot \frac{\partial^2 w_i}{\partial z^2} \tag{7.16}$$

$$\frac{\partial w_{i,\mathrm{PV}}}{\partial t} = -J_i \cdot \frac{1}{\rho \cdot h_K} \tag{7.17}$$

If the Eqs. (7.15) to (7.17) are now substituted into Eq. (7.10), the relationship shown in Eq. (7.18) is obtained for mass transfer in the feed channel according to a modified dispersion model:

$$\frac{\partial w_i}{\partial t} = -v \cdot \frac{\partial w_i}{\partial z} + D_{ax} \cdot \frac{\partial^2 w_i}{\partial z^2} - J_i \cdot \frac{1}{\rho \cdot h_K} \tag{7.18}$$

7.3.3 Solution–diffusion model

The starting point of the mathematical formulation of permeate flux in the SDM is the thermodynamic description of the driving force in terms of a gradient of the chemical potential of species i across the differential path length dz orthogonal to the membrane area (Wijmans & Baker, 1995). Here we follow a derivation given by Wijmans and Baker (1995), other derivations, for example, for a pore-diffusion model (PDM) were also given by Okada and Matsuura (1991) (Fig. 7.8).

Using a proportionality factor L_i, which is not necessarily linear, the permeate flux of species i through the membrane can thus be formulated as follows:

$$J_i = -L_i \frac{d\mu_i}{dz} \tag{7.19}$$

Restricted to pressure and molar concentration gradients as driving forces, the expression results for the differential chemical potential:

$$d\mu_i = RT \quad d\ln(\gamma_i x_i) + v_i \quad dp \tag{7.20}$$

For incompressible fluids, the volume does not change with pressure. Therefore after integration over the pressure, the following expression results for the chemical potential.

$$\mu_i = \mu_i^0 + RT \quad \ln(\gamma_i x_i) + v_i(p - p_i^0) \tag{7.21}$$

The molar volume of compressible fluids such as gases, on the other hand, is pressure dependent. The integration over the pressure with the ideal gas law converted to the molar volume thus yields the expression for the chemical potential:

$$\mu_i = \mu_i^0 + RT \quad \ln(\gamma_i x_i) + RT \quad \ln\left(\frac{p}{p_i^0}\right) \tag{7.22}$$

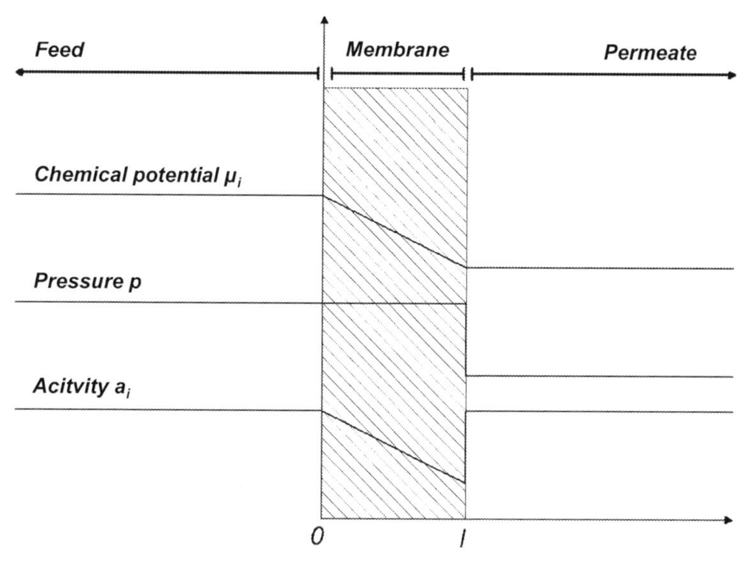

FIGURE 7.8 Qualitative trajectories of μ, p, and a in solution-diffusion model. Source: *Adapted from Wijmans, J. G. & Baker, R.W. (1995). The solution-diffusion model: A review. Journal of Membrane Science, 107, 1–21. https://doi.org/10.1016/0376-7388(95)00102-I.*

If the saturated vapor pressure is used as the reference pressure, the chemical potential for an incompressible fluid is:

$$\mu_i = \mu_i^0 + RT \quad \ln(\gamma_i x_i) + v_i(p - p_{i_{sat}}) \tag{7.23}$$

Correspondingly for a compressible fluid:

$$\mu_i = \mu_i^0 + RT \quad \ln(\gamma_i x_i) + RT \quad \ln\left(\frac{p}{p_{i_{sat}}}\right) \tag{7.24}$$

For the derivation of the permeate flux through the membrane as a function of a pressure or molar concentration gradient, the following simplifications have to be made (Wijmans & Baker, 1995):

1. Adsorption and desorption processes at the membrane occur at significantly higher velocity rates than diffusion through the membrane. Consequently, the latter is a rate-determining step.
2. The phases on both sides of the membrane are in a thermodynamic equilibrium. This results in a continuous gradient of the chemical potential across the membrane thickness and the listed equilibrium conditions can be used in further detail:

$$\mu_{i_0} = \mu_{i_{0(m)}} \tag{7.25}$$

$$\mu_{i_{l(m)}} = \mu_{i_l} \tag{7.26}$$

In the SDM, the sorption and diffusion processes are approximated by considering the membrane as a liquid barrier. Accordingly, the pressure is constant across the membrane thickness. The driving force of the permeate flux J_i within the membrane can therefore also be expressed by the molar concentration gradient. First, Eq. (7.19) is combined with Eq. (7.20):

$$J_i = -L_i \frac{d}{dz}\left(\mu_i^0 + RT \ln(\gamma_i x_i) + v_i \left(p - p_{i_{sat}}^0\right)\right) \tag{7.27}$$

This expression can be further simplified. The considerations necessary for this are summarized below:

1. The standard chemical potential μ_i^0 is independent of location, and thus:

$$\frac{d\mu_i^0}{dz} = 0 \tag{7.28}$$

1. The saturated vapor pressure of the species and the total pressure are of the same order of magnitude. Since the molar volume of the species is small compared to the molar volume of the polymer, the expression as a whole can be neglected and the following equation holds:

$$\frac{d \; v_i \left(p - p_{i_{sat}}^0\right)}{dz} = 0 \tag{7.29}$$

Eq. (7.27) inserted into Eq. (7.9) gives the following expression with the above simplifications and by truncating the activity coefficient γ_i:

$$J_i = -L_i \; RT \frac{d\ln(\gamma_i x_i)}{dz} = -L_i \frac{RT}{x_i} \frac{dx_i}{dz} \tag{7.30}$$

The following relationship can now be established by analogy with Fick's first law:

$$D_i = L_i \frac{RT}{x_i} \tag{7.31}$$

With this expression for the diffusion coefficient and integration over the membrane thickness l, it finally follows:

$$J_i = D_i \frac{x_{i_{0(m)}} - x_{i_{l(m)}}}{l} \tag{7.32}$$

In the following, the principles worked out up to this point for describing the flux through the membrane as a function of the gradient of the chemical potential will now be applied to pervaporation. Thermodynamic equilibrium applies to the transfer of species i from the feed to the membrane surface, so the equilibrium condition from Eq. (7.25) can be used:

$$\mu_i^0 + RT \ln(\gamma_{i_0} x_{i_0}) + v_i(p_0 - p_{i_{sat}}) = \mu_i^0 + RT \ln\left(\gamma_{i_{0(m)}} x_{i_{0(m)}}\right) + v_i(p_0 - p_{i_{sat}}) \tag{7.33}$$

By shortening and rearranging Eq. (7.33), the concentration at the feed-side membrane surface can be expressed in terms of the ratio of the activity coefficients and the concentration in the feed:

$$x_{i_{0(m)}} = \frac{\gamma_{i_0} x_{i_0}}{\gamma_{i_{0(m)}}} = K_i \cdot x_{i_0} \tag{7.34}$$

K_i represents the sorption coefficient for the liquid–liquid transition of species i from the feed to the membrane surface. Using the same procedure, the concentration on the membrane can be expressed for the transition of species i from the membrane surface on the permeate side to the gaseous permeate stream via the equilibrium condition in Eq. (7.26):

$$\mu_i^0 + \mathrm{RTln}\left(\gamma_{i_l} x_{i_l}\right) + \mathrm{RTln}\left(\frac{p_l}{p_{i_{sat}}}\right) = \mu_i^0 + \mathrm{RTln}\left(\gamma_{i_{l(m)}} x_{i_{l(m)}}\right) + v_i(p_0 - p_{i_{sat}}) \tag{7.35}$$

Shortening and transforming results in:

$$x_{i_{l(m)}} = \frac{\gamma_{i_l}}{\gamma_{i_{l(m)}}} \cdot \frac{p_l}{p_{i_{sat}}} \cdot x_{i_l} \cdot \exp\left(\frac{-v_i \quad (p_0 - p_{i_{sat}})}{\mathrm{RT}}\right) \tag{7.36}$$

The simplification made in Eq. (7.29) applies again, because of which the exponential term in Eq. (7.36) can be neglected as a factor. Therefore the following applies to the concentration on the membrane:

$$x_{i_{l(m)}} = \frac{\gamma_{i_l}}{\gamma_{i_{l(m)}}} \cdot x_{i_l} \cdot \frac{p_l}{p_{i_{sat}}} \tag{7.37}$$

The product $x_{i_l} \cdot p_l$ is equal to the partial pressure p_{i_l} of species i. Therefore:

$$x_{i_{l(m)}} = \frac{\gamma_{i_l}}{\gamma_{i_{l(m)}}} \cdot \frac{p_{i_l}}{p_{i_{sat}}} = K_i^G \cdot p_{i_l} \tag{7.38}$$

K_i^G represents the sorption coefficient for the gas-phase transition of species i. The expressions for the membrane concentrations in Eqs. (7.34) and (7.38) could now be integrated into the Fick's approach from Eq. (7.32). However, a simpler formulation can be obtained by transforming the respective sorption coefficients. For this purpose, species i is assumed to be a hypothetical gas at the feed-side membrane surface, which is in equilibrium with the feed solution (Wijmans & Baker, 1993) (Fig. 7.9).

It should be noted at this point that in reality no such gas phase exists between feed and membrane and this assumption serves the sole purpose of finally being able to represent the driving force as a difference of partial pressures. The equilibrium condition from Eq. (7.25) is used again:

$$\mu_i^0 + \mathrm{RT}\ ln\left(\gamma_{i_0}^L \quad x_{i_0}^L\right) + v_i \quad \left(p_0 - p_{i_{sat}}\right) = \mu_i^0 + \mathrm{RT}ln\left(\gamma_{i_0}^G \quad x_{i_0}^G\right) + \mathrm{RT}\ ln\left(\frac{p_0}{P_{i_{sat}}}\right) \tag{7.39}$$

Transforming analogously the steps of Eqs. (7.35) to (7.38) results in:

$$x_{i_0} = \frac{\gamma_{i_0}^G}{\gamma_{i_0}^L \cdot p_{i_{sat}}} p_{i_0} \tag{7.40}$$

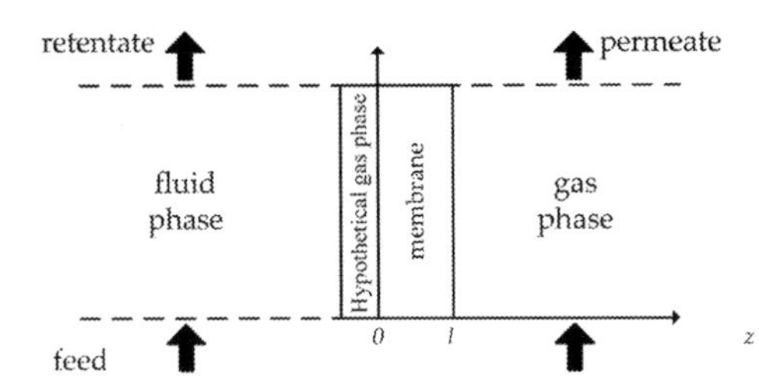

FIGURE 7.9 Hypothetical gas phase between feed solution and membrane. Source: *Simplified representation adapted from Wijmans, J.G. & Baker, R.W. (1993). A simple predictive treatment of the permeation process in pervaporation. Journal of Membrane Science, 79(1), 101–113. https://doi.org/10.1016/0376-7388(93)85021-N.*

By substituting Eq. (7.40) into Eq. (7.34), the expression for the concentration of species i on the feed-side membrane surface follows:

$$x_{i_{0(m)}} = \frac{\gamma_{i_0}^G}{\gamma_{i_{0(m)}}} \cdot \frac{p_{i_0}}{p_{i_{sat}}} = K_i^G \cdot p_{i_0} \tag{7.41}$$

Eqs. (7.38) and (7.41) can now be integrated into the Fick's approach from Eq. (7.32):

$$J_i = D_i \quad K_i^G \cdot \frac{p_{i_0} - p_{i_l}}{l} = P_i^G \cdot \frac{p_{i_0} - p_{i_l}}{l} \tag{7.42}$$

The product $D_i \quad K_i^G$ thus represents the permeability P_i^G. The partial pressures on the feed and permeate side can be expressed via Raoult's law extended with activity coefficients:

$$p_{i_0} = p_i^0 \cdot x_{i_{0(m)}} \cdot \gamma_{i_0} \tag{7.43}$$

$$p_{i_l} = P_p \cdot y_{i_l} \tag{7.44}$$

Eqs. (7.43) and (7.44) inserted into Eq. (7.42) provide for the permeate flux through the membrane:

$$J_i = P_i^G \cdot \frac{\left(p_i^0 \cdot x_{i_{0(m)}} \cdot \gamma_{i_0} - P_p \cdot y_{i_l}\right)}{l} \tag{7.45}$$

With permeability P_i as a product of the form $D_i \cdot K_i$, Eq. (7.42) can also be expressed as follows:

$$J_i = \frac{P_i}{l} \cdot \left(x_{i_0} - \frac{p_{i_0}}{H_i}\right) \tag{7.46}$$

Where H_i represents the Henry coefficient and is defined as:

$$H_i = \frac{\gamma_{i_0}^L \cdot p_{i_{sat}}}{\gamma_{i_0}^G} \tag{7.47}$$

Up to this point, Eqs. (7.32), (7.42), (7.45), as well as Eq. (7.46) provide four equations describing permeation through the membrane according to the SDM and an approach analogous to Fick's first law. The latter three are based in their driving force on the difference of the partial pressures of species i on the feed and permeate side. The first equation, on the other hand, traces the flux through the membrane back to the concentration difference between the feed-side and permeate-facing membrane surfaces.

7.3.3.1 Permeance

Since the exact dimensions of the separation-active polymer layer are usually not accessible or can only be determined imprecisely experimentally, in practice the ratio of permeability P_i of the membrane and the thickness of the separation-active layer l is expressed by the permeance Q_i (Kreis & Górak, 2006):

$$Q_i = \frac{P_i}{l} \tag{7.48}$$

Permeance, like permeability, thus describes the sorption and diffusion behavior of the individual components in the membrane under consideration. Both processes show a strong dependence on the composition of the feed solution and the temperature. Different semiempirical equations have been established in the literature for describing the respective membrane substance system.

On the one hand, a constant value for the permeance can be assumed as a shortcut model. For small variations in feed composition and constant temperature, a meaningful prediction of the magnitude of flux and purification can be obtained in this way by means of only a few experiments (Kreis & Górak, 2006).

$$Q_i = \text{constant} \tag{7.49}$$

This simplified model is not sufficient for more detailed observations and for all temporally transient processes in which neither feed composition nor temperature can be assumed to be constant.

In order to take the kinetics of sorption and diffusion into account, a representation based on the Arrhenius approach has therefore become established as the most common basic equation for describing permeance (Schiffmann & Repke, 2011):

$$Q_i = Q_i^0 \cdot \exp\left(-\frac{E_i}{R} \left(\frac{1}{T_0} - \frac{1}{T} \right) \right) \tag{7.50}$$

This representation contains two pseudo-physical parameters, which have to be determined experimentally. One is the reference permeance at reference temperature T_0 and the other is the activation energy for sorption and diffusion E_i. The exponential representation, which can be approximated to experimental data via two parameters, often already yields sufficiently accurate results.

However, especially in hydrophilic pervaporation with polymer membranes, the composition of the feed plays a nonnegligible role. In contrast to inert membrane materials, such as silica or zeolite membranes, polymers are altered in their sorption, diffusion, as well as structural properties by the presence of a plasticizing agent, often the preferred permeating component. To account for the influence of feed composition on permeance, Eq. (7.50) can be modified as follows (Koch & Górak, 2014):

$$Q_i = Q_i^0 \cdot w_i^{A_i} \cdot \exp\left(-\frac{E_i}{R} \left(\frac{1}{T_0} - \frac{1}{T} \right) \right) \tag{7.51}$$

Here, A_i corresponds to an optional third parameter via which the model can be further approximated to the experimentally found courses.

In some cases, it can be observed that, contrary to expectation, the permeance either does not change with temperature or changes only very slightly. This is not a contradiction to the temperature dependence of sorption and diffusion but indicates a trade-off between the shift in equilibrium of the exothermic sorption process and the acceleration of diffusion. In these cases, it may be more appropriate to describe the permeance in a temperature-independent manner (Koch, Sudhoff, Kreiß, Górak, & Kreis, 2013):

$$Q_i = Q_i^0 \cdot \exp \quad (A_i \cdot w_i) \tag{7.52}$$

Developing improved permeability and permeance functions that describe component flux more accurately is still part of today's research (Szilagyi & Toth, 2020).

7.3.4 Concentration polarization

In the following, the effect of concentration polarization will be explained. According to the so-called film model, in the case of a transversely overflowed plane, in this case the membrane, a standing film is formed because of friction near the surface, at which the overflow velocity v is zero (Baker et al., 1997).

Because of the transport of the preferentially permeating component through the membrane, the retained component is concentrated in the area of the laminar boundary film between feed solution and membrane. An equilibrium results from the convective uptake, the flux through the membrane, and the diffusion through the film along the present concentration gradient between feed solution and membrane surface (Figs. 7.10 and 7.11). Although this boundary film does not move in the direction of overflow, it becomes subject to the concentration polarization effect because of the flux orthogonal to the membrane

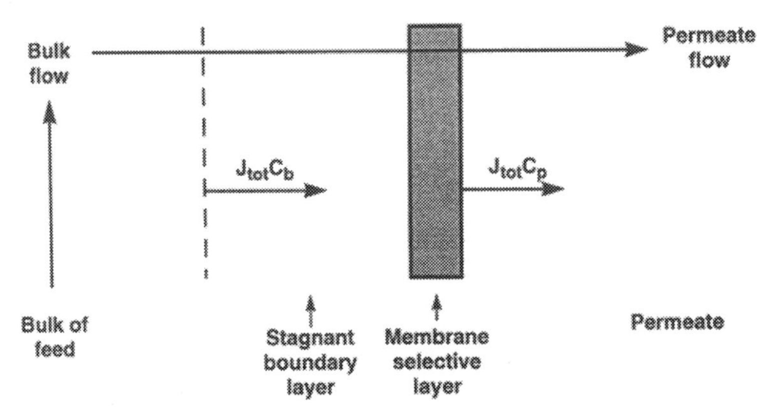

FIGURE 7.10 Representation of the transport processes in the boundary film. The convective transport, the flux through the membrane, and the diffusion through the boundary film are shown. Source: *Illustration taken from Wijmans, J.G., Athayde, A.L., Daniels, R., Ly, J.H., Kamaruddin, H.D., & Pinnau, I. (1996). The role of boundary layers in the removal of volatile organic compounds from water by pervaporation. Journal of Membrane Science, 109, 135–146. https://doi.org/10.1016/0376-7388(95)00194-8 (Wijmans et al., 1996).*

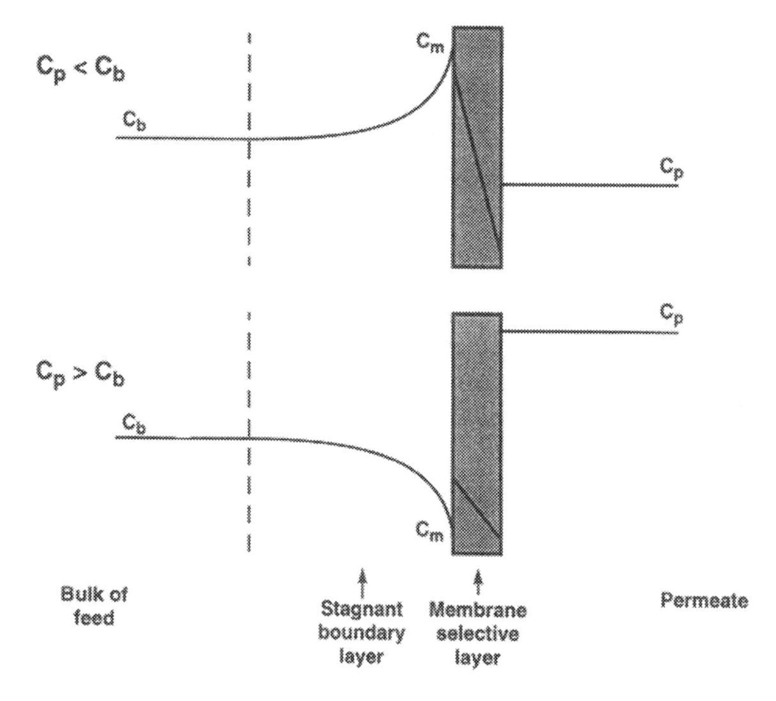

FIGURE 7.11 Representation of the film model and concentration polarization. While the retained component is concentrated at the membrane, the concentration of the preferentially permeating component decreases. Source: *Illustration taken from Wijmans, J.G., Athayde, A.L., Daniels, R., Ly, J.H., Kamaruddin, H.D., & Pinnau, I. (1996). The role of boundary layers in the removal of volatile organic compounds from water by pervaporation.* Journal of Membrane Science, 109, *135–146. https://doi.org/10.1016/0376-7388(95)00194-8.*

surface (Sherwood, Brian, Fisher, & Dresner, 1965). The general balance equation for mass transfer through the boundary film is thus:

$$J_{i,\mathrm{Konv}} + J_{i,\mathrm{Diff}} = J_{i,TM} \tag{7.53}$$

The convective mass transport of component i toward the membrane surface is composed of the total flux through the membrane and the respective mass fraction of component i:

$$J_{i,\mathrm{Konv}} = J_{\mathrm{ges}} \cdot w_i \tag{7.54}$$

The diffusive mass transfer from the feed solution to the membrane in the case of the preferentially permeating component, or from the membrane to the feed solution for the retained component, results from Fick's first law:

$$J_{i,\mathrm{Diff}} = \rho \cdot D \ \frac{dw_i}{dz} \tag{7.55}$$

The flux of the component i through the membrane results from the total flux and the mass fraction in the permeate.

$$J_{i,\mathrm{TM}} = J_{\mathrm{ges}} \cdot w_{i,P} \tag{7.56}$$

Eqs. (7.54) to (7.56) can now be implemented in the general mass transfer accounting in Eq. (7.53):

$$J_{\mathrm{ges}} \cdot w_i + \rho \cdot D \ \frac{dw_i}{dz} = J_{\mathrm{ges}} \cdot w_{i,P} \tag{7.57}$$

This expression can now be integrated over the thickness δ of the laminar boundary layer and over the mass fractions at the corresponding boundaries to determine the mass fractions at the membrane surface (Sherwood et al., 1965).

$$w_i = w_{i,FM}; z = 0 \tag{7.58}$$

$$w_i = w_{i,F}; z = \delta \tag{7.59}$$

By rearranging Eqs. (7.57) and (7.65) and defining the integration limits defined in Eqs. (7.58), (7.59), and (7.55), we obtain:

$$\int_{w_i=w_{i,FM}}^{w_i=w_{i,F}} \frac{dw_i}{w_i - w_{i,P}} = -\int_{z=0}^{z=\delta} \frac{J_{ges}}{\rho \cdot D} dz \tag{7.60}$$

Solving Eq. (7.60) finally leads to the relation first found by Sherwood et al. (1965):

$$\frac{w_{i,FM} - w_{i,P}}{w_{i,F} - w_{i,P}} = \exp\left(\frac{J_{ges}}{\rho}\frac{\delta}{D}\right) = \exp\left(\frac{J_{ges}}{\rho \quad k_{eff,i}}\right) \tag{7.61}$$

Since the thickness of the laminar boundary layer is very difficult to determine experimentally, it is combined with the diffusion coefficient to form an effective mass transfer coefficient:

$$k_{eff,i} = \frac{D}{\delta} \tag{7.62}$$

The mass transfer coefficient is usually determined using Sherwood correlations. In this work, the correlation equations and parameters for plate moduli commonly found in the literature on pervaporation are used again:

$$Sh = \frac{k_{eff} \quad d_h}{D} = a \cdot Re^b \cdot Sc^c \cdot \left(\frac{d_h}{l}\right)^d \tag{7.63}$$

Parameters a, b, c, and d depend on the flow regime, the type of phase of the flow, and the separation process considered. For the modeling and simulation of plate-and-frame module, the following values can be used:

The Schmidt number Sc gives the ratio of diffusive momentum to mass transport and is defined as:

$$Sc = \frac{\eta}{\rho \cdot D} \tag{7.64}$$

7.3.5 Heat transfer

In the following, the fundamentals for modeling heat transport in pervaporation are explained. First, the general enthalpy balance for determining the temperature drop in the feed over the membrane length is discussed and then the phenomenon of temperature polarization due to boundary film effects is described.

During the pervaporation process, the heated feed solution transports an enthalpy current $\dot{H}_F$ into the plate module. Because of pervaporation, the enthalpy of the retentate decreases over the membrane length by the amount of the enthalpy current $\dot{H}_P$ of the permeate (Fig. 7.12).

The general balance equation for enthalpy flows is:

$$\dot{H}_F = \dot{H}_R + \dot{H}_P \tag{7.65}$$

The enthalpy flow through the feed solution can be determined via the temperature T_F, the averaged specific heat capacity $\sim c_{p,F}$, and the mass flow $\dot{m}_F$:

$$\dot{H}_F = \dot{m}_F \cdot \sim c_{p,F} \quad (T_F - T_0) \tag{7.66}$$

For simplicity, the same temperature should prevail on the permeate side as on the feed side. The enthalpy flow of the permeate is composed of the flux J_{tot} through the membrane, the membrane area A_M, and the averaged specific enthalpy $\sim h_P^G$ of the vaporous permeate:

$$\dot{H}_P = \sim h_P^G \cdot J_{\text{tot}} \cdot A_M = \sim h_P^G \cdot J_{\text{ges}} \cdot b_M \cdot l_M \tag{7.67}$$

If now the change of the enthalpy flux over the discrete membrane length Δz is considered, the following relation can be established:

$$\dot{n}_F \cdot \sim c_{p,F} \quad (T_F - T_0)\big|_z = \dot{n}_F \cdot \sim c_{p,F} \quad (T_F - T_0)\big|_{z+\Delta z} + \sim h_P^G \cdot J_{\text{tot}} \cdot b_M \cdot \Delta z\big|_z \tag{7.68}$$

Thus the differential change in enthalpy current on the feed side is:

$$\frac{d\dot{H}_R(z)}{dz} = -J_{\text{ges}}(z) \cdot b_M \cdot \sim h_P^G \tag{7.69}$$

The differential change of the enthalpy current on the permeate side is accordingly:

$$\frac{d\dot{H}_P(z)}{dz} = J_{tot}(z) \cdot b_M \cdot \sim h_P^G \tag{7.70}$$

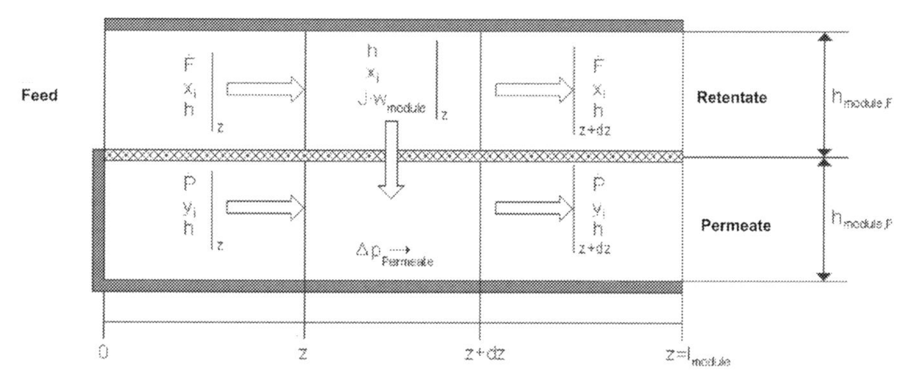

FIGURE 7.12 Balance areas in the plate module under consideration. Source: *Illustration taken from Lipnizki, F., Olsson, J., & Trägårdh, G. (2002). Scale-up of pervaporation for the recovery of natural aroma compounds in the food industry. Part 1: Simulation and performance. Journal of Food Engineering, 54, 183–195. https://doi.org/10.1016/S0260-8774(01)00200-X (Lipnizki, Olsson, & Trägårdh, 2002).*

7.3.6 Temperature polarization

During the pervaporation process, desorption and evaporation of the components dissolved in the membrane take place continuously on the membrane surface on the permeate side. The necessary enthalpy of vaporization Δh_i^V is extracted from the feed and corresponds to the difference in enthalpy between the gaseous and liquid phases:

$$\Delta \sim h_{\text{tot}}^V = \sim h_P^G - \sim h_P^L \tag{7.71}$$

The heat extracted from the feed leads to a temperature gradient between feed and membrane surface because of the slower heat transfer through the laminar boundary layer (Fig. 7.13).

This effect, like concentration polarization, has a negative impact on the total flux and must therefore be taken into account in the modeling. The heat flux extracted from the feed consists of the mass flow rate of the permeate and the enthalpy of vaporization defined in Eq. (7.71) (Lipnizki & Field, 1999):

$$\dot{H}_V = J_{\text{tot}} \cdot A \cdot \Delta \sim h_{\text{tot}}^V = J_{\text{tot}} \cdot A \cdot \left(\sim h_P^G - \sim h_P^L \right) \tag{7.72}$$

The resulting temperature difference is in turn dependent on the extracted heat flux as well as the heat conduction properties in the laminar boundary layer. The contribution of the temperature difference due to this polarization results from a general heat transport approach:

$$(T_F - T_{\text{FM}}) = \frac{\dot{H}_V}{\lambda \cdot A_M} \cdot \delta \tag{7.73}$$

Analogous to concentration polarization, the thickness of the laminar boundary layer is difficult to determine. Therefore the heat transfer coefficient is introduced as a semiempirically describable quantity:

$$\alpha = \frac{\lambda}{\delta} \tag{7.74}$$

The heat transfer coefficient is usually determined using Nußelt correlations.

$$Nu = \frac{\alpha \cdot d_h}{\lambda} = a \cdot Re^b \cdot Pr^c \cdot \left(\frac{d_h}{l_M}\right)^d \tag{7.75}$$

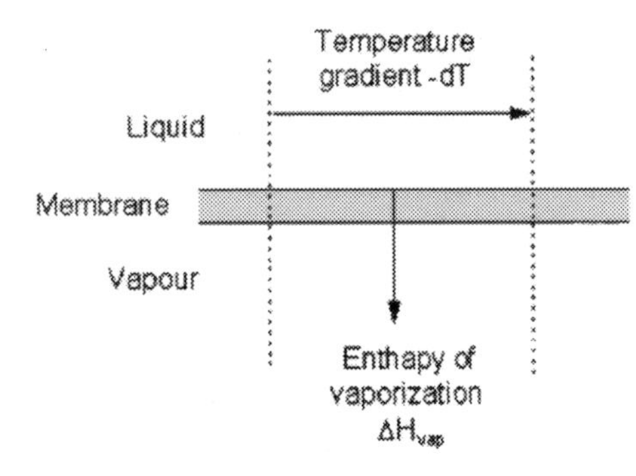

FIGURE 7.13 Representation of heat transport by evaporation. Source: *Illustration slightly modified and taken from Lipnizki, F., Hausmanns, S., Ten, P.-K., Field, R.W., Laufenberg, G. (1999). Organophilic pervaporation: Prospects and performance. Chemical Engineering Journal, 73, 113–129. https://doi.org/10.1016/S1385-8947(99)00024-8.*

The values for the parameters a, b, c, and d are usually taken from the same Sherwood correlations as in the case of concentration polarization (see Table 7.1). The Prandtl number Pr describes the ratio of diffusive momentum and heat transport and is defined as:

$$Pr = \frac{\eta \cdot c_p}{\lambda} \tag{7.76}$$

7.4 Application to process simulation as scaleup tool

Applying physico-chemical models to simulate the pervaporation process of multitude of industrially relevant organics is still subject of today's research as evidenced by the magnitude of simulation studies published in recent years (Ashraf, Schmidt, Kujawa, Kujawski, & Arafat, 2017; Azimi, Thibault, & Tezel, 2019; Dashti et al., 2018; Dong et al., 2020; Dudek & Borys, 2019; Ebneyamini, Azimi, Thibault, & Tezel, 2018; Farhadi, Pazuki, & Raisi, 2018; Gómez-García, Dobrosz-Gómez, & Osorio Viana, 2017; Haaz & Toth, 2018; Haáz et al., 2019; Hassankhan & Raisi, 2020; Krishna, 2019; León & Fontalvo, 2018; Luis, 2018; Qiu et al., 2019; Raoufi, Asadollahzadeh, & Shirazian, 2018; Shan et al., 2021; Sharma & Jain, 2020; Song, Pan, Li, Quan, & Jiang, 2019; Valentinyi et al., 2020; Yong & Zhang, 2021; Zhang, Hou, Ma, Yuan, & Zeng, 2021; Zhang, Peng, Jiang, & Gu, 2017). As for other types of processes, validated process models are necessary, especially for the correct design and optimization of hybrid processes (Babaie & Esfahany, 2020; Daviou, Hoch, & Eliceche, 2004; Franke, Górak, & Strube, 2004; Franke, Nowotny, Ndocko, Górak, & Strube, 2008; Lee, Li, & Chen, 2016; León & Fontalvo, 2020; Naidu & Malik, 2011; Sosa & Espinosa, 2011; Verhoef et al., 2008). In the following, we will briefly present the implementation of the basic principles developed in the previous section in a process model as it is published by Thiess, Schmidt, and Strube (2018). In the example discussed here, a total of five balance volumes are considered. The first balance volume represents the feed channel. Here, the axial dispersion and the temperature drop along the length of the channel are considered. In the second balance volume, the mass and heat transport processes determined by the laminar boundary layer are described (Fig. 7.14).

This mainly concerns the transfer of the components from the feed channel into the membrane phase and the associated concentration polarization as well as the temperature polarization resulting from the slowed heat transfer through the boundary layer. In the third balance volume, the rate-determining process step is now described. Sorption into the membrane phase, diffusion through the membrane, and the subsequent desorption

TABLE 7.1 Parameters of Sherwood correlations for plate modules.

Regime	a	b	c	d
Laminar	1.615	0.33	0.33	0.33
Turbulent	0.026	0.80	0.30	0

From Lipnizki, F. & Field, R.W. (1999). Simulation and process design of pervaporation plate-and-frame modules to recover organic compounds from waste water, Chemical Engineering Research and Design, 77, 231–240. https://doi.org/10.1205/026387699526142.

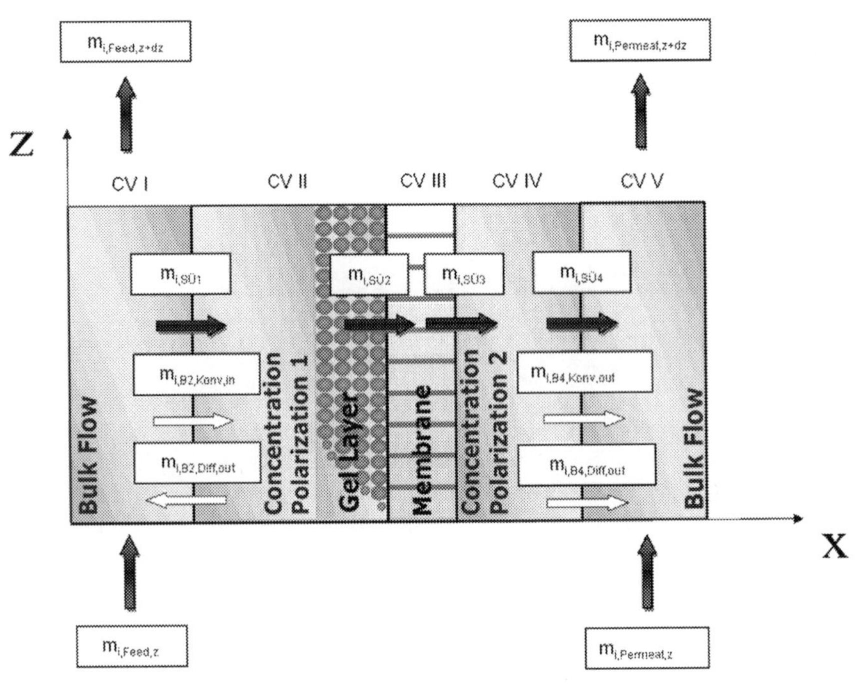

FIGURE 7.14 The five balance volumes in pervaporation modeling and simulation (Grote, Fröhlich, & Strube, 2012).

into the permeate channel (balance volume V) are represented by the SDM. The necessary thermodynamic data can be loaded into commercial simulation software via CAPE-OPEN packages from thermodynamic data libraries. For the description of the permeance Q, often a semiempirically equation is used, which takes into account both the temperature dependence and the concentration dependence. The fourth balance volume is theoretically subject to further concentration polarization. However, the influence of these effects is widely estimated to be small in the literature because of the adjacent vacuum and the associated rapid removal of the components into the permeate space. Accordingly, the fourth balance space is not considered in most of the published work.

In order to apply the developed physico-chemical model as an engineering tool for the design of pervaporation processes, a proof of principle for the scalability of the model is of key importance. Large-scale data of a pervaporation unit with 200 m^2 membrane area was compared to a simulation run with the model in a study by Thiess et al. (2018). The comparison between the recorded experimental profiles and the profiles predicted with the aid of the model shows a high predictive accuracy of the process model used. Owing to the temperature dependence of the pervaporation kinetics, the cooling of the retentate flow along the membrane has the expected strong effect on the mass transfer efficiency. This effect, caused by evaporation and the associated removal of the necessary evaporation enthalpy in the form of heat, is only insignificantly enhanced by temperature polarization. This study has shown that it is important to demonstrate the predictivity of models across different size scales when validating process models. Only then is reliable large-scale process design based on laboratory and/or miniplant tests possible. With the simulation tool used, both the mass fraction of water in the feed (Fig. 7.15) and the mass fraction of water in the permeate (Fig. 7.16) as well as the water flux (Fig. 7.17) can be predicted with high accuracy. The obtained results match with the experimental data of a ternary, as well as a large-scale binary pervaporation.

The results of this study show that when the main transport phenomena in pervaporation are taken into account, simulation tools can be provided that have very good prediction accuracy, even for large-scale plants, in accordance with property data libraries as well as small-scale determined model parameters. Owing to the wide availability of experimentally validated approaches to describe fluid dynamics and pressure drop, a reliable prediction of the process of a 15,000 kg ethanol–water separation based on miniplant experiments has thus been possible. In implementing the effects, a distinction between

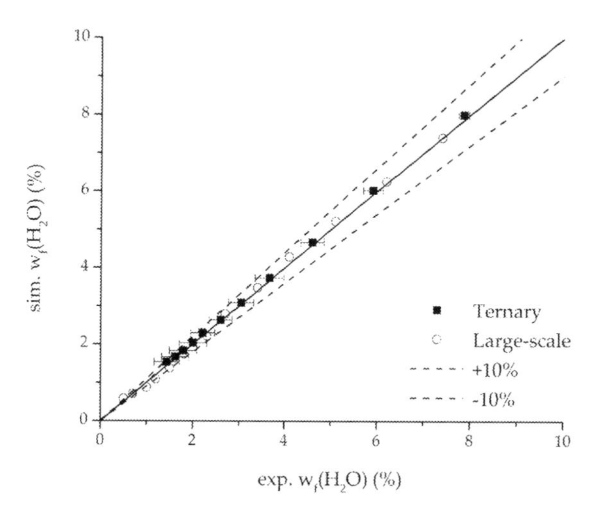

FIGURE 7.15 Experimental and simulated water mass fraction (feed) for a ternary system and a large-scale pervaporation (Thiess et al., 2018).

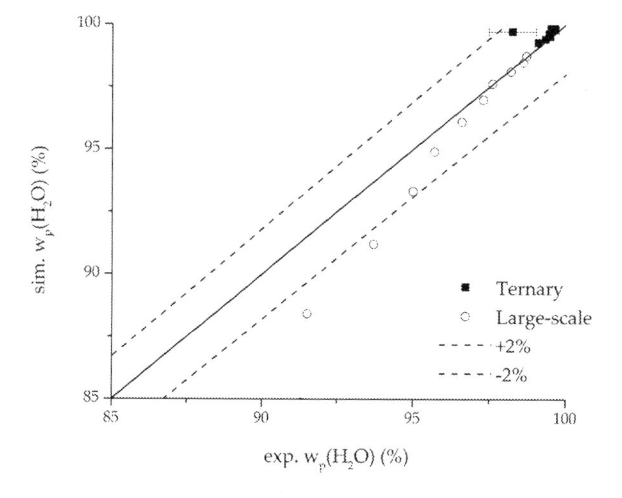

FIGURE 7.16 Experimental and simulated water mass fraction (permeate) for a ternary system and a large-scale pervaporation (Thiess et al., 2018).

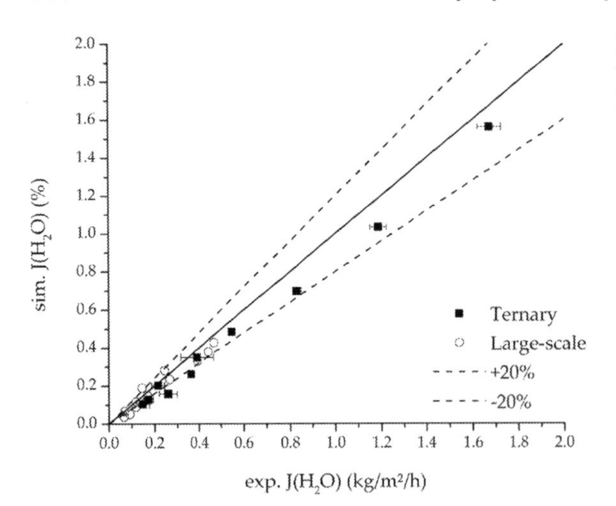

FIGURE 7.17 Experimental and simulated water flux for a ternary system and a large-scale pervaporation (Thiess et al., 2018).

fluid dynamics, equilibrium, and kinetics is necessary. While the thermodynamic equilibrium and mass transfer kinetics should be independent of the size scale of the process under consideration, it is usually the fluid dynamics that change in a scaleup and thus affect the process performance. Special attention should be given here to a correct description of the fluid dynamics when the module geometry is changed. Not only do path length and possible backmixing have an effect on separation but, in particular, the pressure conditions in the permeate channel. If no reliable correlations and/or robust theoretical frameworks are available to describe the geometry used, the correct description of the fluid dynamics and pressure drop should be validated in any case.

7.5 Conclusions and future trends

The pressure drop in pervaporation is significantly influenced by the geometry of the respective module and the flow regime. Common calculation methods are based on the Hagen–Poiseuille law. As a driving force influencing effect, special attention must be paid to the correct description of the pressure loss.

Mass transport in pervaporation can be described according to both the SDM and the PDM. The velocity-determining effect in describing mass transfer is usually concentration polarization, similar to many other membrane processes. Heat transport during pervaporation must be described at least in terms of the cooling retentate flux, due to the removal of the evaporation enthalpy on the permeate side, and in terms of temperature polarization. This chapter gives an overview of the corresponding phenomena and thus provides an introduction to the modeling and simulation of pervaporation processes. The correct description of the phenomena by the process model requires comprehensive validation, which has already been described in the literature for other membrane processes (Huter & Strube, 2019). Physico-chemical models are necessary key technologies of a resource-saving and time-efficient process optimization and design of pervaporation processes.

References

Ashraf, M. T., Schmidt, J. E., Kujawa, J., Kujawski, W., & Arafat, H. A. (2017). One-dimensional modeling of pervaporation systems using a semi-empirical flux model. *Separation and Purification Technology, 174*, 502–512. Available from https://doi.org/10.1016/j.seppur.2016.10.043.

Azimi, H., Thibault, J., & Handan Tezel, F. (2019). Separation of butanol using pervaporation: A review of mass transfer models. *Journal of Fluid Flow, Heat and Mass Transfer (JFFHMT), 6*, 9–38. Available from https://doi.org/10.11159/jffhmt.2019.002.

Babaie, O., & Esfahany, M. N. (2020). Optimization of a new combined approach to reduce energy consumption in the hybrid reactive distillation–pervaporation process. *Chemical Engineering and Processing: Process Intensification, 151*, 107910. Available from https://doi.org/10.1016/j.cep.2020.107910.

Baker, R. W. (2012). *Membrane technology and applications*. Chichester: John Wiley & Sons, Ltd.

Baker, R. W., Wijmans, J. G., Athayde, A. L., Daniels, R., Ly, J. H., & Le, M. (1997). The effect of concentration polarization on the separation of volatile organic compounds from water by pervaporation. *Journal of Membrane Science, 137*, 159–172. Available from https://doi.org/10.1016/S0376-7388(97)00189-0.

Bausa, J., & Marquardt, W. (2001). Detailed modeling of stationary and transient mass transfer across pervaporation membranes. *AIChE Journal. American Institute of Chemical Engineers, 47*, 1318–1332. Available from https://doi.org/10.1002/aic.690470610.

Dashti, A., Asghari, M., Dehghani, M., Rezakazemi, M., Mohammadi, A. H., & Bhatia, S. K. (2018). Molecular dynamics, grand canonical Monte Carlo and expert simulations and modeling of water–acetic acid pervaporation using polyvinyl alcohol/tetraethyl orthosilicates membrane. *Journal of Molecular Liquids, 265*, 53–68. Available from https://doi.org/10.1016/j.molliq.2018.05.078.

Daviou, M. C., Hoch, P. M., & Eliceche, A. M. (2004). Design of membrane modules used in hybrid distillation/pervaporation systems. *Industrial & Engineering Chemistry Research, 43*, 3403–3412. Available from https://doi.org/10.1021/ie034259c.

Del Pozo Gómez, M.T., Carreira, P.R., Repke, J.-U., Klein, A., Brinkmann, T., & Wozny, G. (2008). Study of a novel heat integrated hybrid pervaporation distillation process: Simulation and experiments. In: *18th European Symposium on Computer Aided Process Engineering* (pp. 73–78). Elsevier.

Dong, G., Nagasawa, H., Yu, L., Wang, Q., Yamamoto, K., Ohshita, J., ... Tsuru, T. (2020). Pervaporation removal of methanol from methanol/organic azeotropes using organosilica membranes: Experimental and modeling. *Journal of Membrane Science, 610*, 118284. Available from https://doi.org/10.1016/j.memsci.2020.118284.

Dudek, G., & Borys, P. (2019). A simple methodology to estimate the diffusion coefficient in pervaporation-based purification experiments. *Polymers, 11*(2), 343. Available from https://doi.org/10.3390/polym11020343.

Ebneyamini, A., Azimi, H., Thibault, J., & Tezel, F. H. (2018). Description of butanol aqueous solution transport through commercial PDMS pervaporation membrane using extended Maxwell–Stefan model. *Separation Science and Technology, 53*, 1611–1627. Available from https://doi.org/10.1080/01496395.2018.1441303.

Farhadi, M., Pazuki, G., & Raisi, A. (2018). Modeling of the pervaporation process for isobutanol purification from aqueous solution using intelligent systems. *Separation Science and Technology, 53*, 1383–1396. Available from https://doi.org/10.1080/01496395.2017.1405987.

Franke, M., Górak, A., & Strube, J. (2004). Auslegung und optimierung von hybriden trennverfahren. *Chemie Ingenieur Technik, 76*, 199–210. Available from https://doi.org/10.1002/cite.200406150.

Franke, M. B., Nowotny, N., Ndocko, E. N., Górak, A., & Strube, J. (2008). Design and optimization of a hybrid distillation/melt crystallization process. *AIChE Journal. American Institute of Chemical Engineers, 54*, 2925–2942. Available from https://doi.org/10.1002/aic.11605.

Grote, F., Fröhlich, H., & Strube, J. (2012). Integration of reverse-osmosis unit operations in biotechnology process design. *Chemical Engineering & Technology, 35*, 191–197. Available from https://doi.org/10.1002/ceat.201100182.

Gómez-García, M. Á., Dobrosz-Gómez, I., & Osorio Viana, W. (2017). Experimental assessment and simulation of isoamyl acetate production using a batch pervaporation membrane reactor. *Chemical Engineering and Processing: Process Intensification, 122*, 155–160. Available from https://doi.org/10.1016/j.cep.2017.09.012.

Haaz, E., & Toth, A. J. (2018). Methanol dehydration with pervaporation: Experiments and modelling. *Separation and Purification Technology, 205*, 121–129. Available from https://doi.org/10.1016/j.seppur.2018.04.088.

Haáz, E., Valentinyi, N., Tarjani, A. J., Fózer, D., André, A., Khaled Mohamed, S. A., ... Tóth, A. J. (2019). Platform molecule removal from aqueous mixture with organophilic pervaporation: Experiments and modelling. *Periodica Polytechnica Chemical Engineering, 63*(1), 138–146. Available from https://doi.org/10.3311/PPch.12151.

Hassankhan, B., & Raisi, A. (2020). Separation of isobutanol/water mixtures by hybrid distillation-pervaporation process: Modeling, simulation and economic comparison. *Chemical Engineering and Processing: Process Intensification, 155*, 108071. Available from https://doi.org/10.1016/j.cep.2020.108071.

Hessel, V., Gürsel, I. V., Wang, Q., Noël, T., & Lang, J. (2012). Potenzialanalyse von milli- und mikroprozesstechniken für die verkürzung von prozessentwicklungszeiten – chemie und prozessdesign als intensivierungsfelder: Potential analysis of smart flow processing and micro process technology for fastening process development – use of chemistry and process design as intensification fields. *Chemie Ingenieur Technik, 84*, 660–684. Available from https://doi.org/10.1002/cite.201200007.

Huter, M. J., & Strube, J. (2019). Model-based design and process optimization of continuous single pass tangential flow filtration focusing on continuous bioprocessing. *Processes, 7*, 317. Available from https://doi.org/10.3390/pr7060317.

Koch, K., & Górak, A. (2014). Pervaporation of binary and ternary mixtures of acetone, isopropyl alcohol and water using polymeric membranes: Experimental characterisation and modelling. *Chemical Engineering Science, 115*, 95–114. Available from https://doi.org/10.1016/j.ces.2014.02.009.

Koch, K., Sudhoff, D., Kreiß, S., Górak, A., & Kreis, P. (2013). Optimisation-based design method for membrane-assisted separation processes. *Chemical Engineering and Processing: Process Intensification, 67*, 2–15. Available from https://doi.org/10.1016/j.cep.2012.09.013.

Kreis, P., & Górak, A. (2006). Process analysis of hybrid separation processes. *Chemical Engineering Research and Design, 84*, 595–600. Available from https://doi.org/10.1205/cherd.05211.

Krishna, R. (2019). Highlighting thermodynamic coupling effects in alcohol/water pervaporation across polymeric membranes. *ACS Omega, 4*, 15255–15264. Available from https://doi.org/10.1021/acsomega.9b02255.

Lee, H.-Y., Li, S.-Y., & Chen, C.-L. (2016). Evolutional design and control of the equilibrium-limited ethyl acetate process via reactive distillation – pervaporation hybrid configuration. *Industrial & Engineering Chemistry Research, 55*, 8802–8817. Available from https://doi.org/10.1021/acs.iecr.6b01358.

Levenspiel, O. (1999). *Chemical reaction engineering* (3rd ed.). New York: Wiley.

León, J. A., & Fontalvo, J. (2018). Tools for the design of hybrid distillation–pervaporation columns in a single unit: Hybrid rectifying–pervaporation section. *Industrial & Engineering Chemistry Research, 57*, 11970–11980. Available from https://doi.org/10.1021/acs.iecr.8b02078.

León, J. A., & Fontalvo, J. (2020). Analysis of a hybrid distillation-pervaporation column in a single unit: Intermediate membrane section in the rectifying and stripping section. *The Canadian Journal of Chemical Engineering, 98*(10), 2227–2237. Available from https://doi.org/10.1002/cjce.23765.

Lipnizki, F., & Field, R. W. (1999). Simulation and process design of pervaporation plate-and-frame modules to recover organic compounds from waste water. *Chemical Engineering Research and Design, 77*, 231–240. Available from https://doi.org/10.1205/026387699526142.

Lipnizki, F., Hausmanns, S., Ten, P.-K., Field, R. W., & Laufenberg, G. (1999). Organophilic pervaporation: Prospects and performance. *Chemical Engineering Journal, 73*, 113–129. Available from https://doi.org/10.1016/S1385-8947(99)00024-8.

Lipnizki, F., Olsson, J., & Trägårdh, G. (2002). Scale-up of pervaporation for the recovery of natural aroma compounds in the food industry. Part 1: Simulation and performance. *Journal of Food Engineering, 54*, 183–195. Available from https://doi.org/10.1016/S0260-8774(01)00200-X.

Li, W., Estager, J., Monbaliu, J.-C. M., Debecker, D. P., & Luis, P. (2020). Separation of bio-based chemicals using pervaporation. *Journal of Chemical Technology and Biotechnology (Oxford, Oxfordshire: 1986), 95*, 2311–2334. Available from https://doi.org/10.1002/jctb.6434.

Luis, P. (2018). Pervaporation. In P. Luis (Ed.), *Fundamental modelling of membrane systems* (pp. 71–102). Elsevier. Available from https://doi.org/10.1016/B978-0-12-813483-2.00003-4.

Naidu, Y., & Malik, R. K. (2011). A generalized methodology for optimal configurations of hybrid distillation–pervaporation processes. *Chemical Engineering Research and Design, 89*, 1348–1361. Available from https://doi.org/10.1016/j.cherd.2011.02.025.

Okada, T., & Matsuura, T. (1991). A new transport (model for pervaporation. *Journal of Membrane Science, 59*, 133–149. Available from https://doi.org/10.1016/S0376-7388(00)81179-5.

Qiu, B., Wang, Y., Fan, S., Liu, J., Jian, S., Qin, Y., ... Wang, W. (2019). Ethanol mass transfer during pervaporation with PDMS membrane based on solution-diffusion model considering concentration polarization. *Separation and Purification Technology, 220*, 276–282. Available from https://doi.org/10.1016/j.seppur.2019.03.021.

Raoufi, N., Asadollahzadeh, M., & Shirazian, S. (2018). Investigation into ethanol purification using polymeric membranes and a pervaporation process. *Chemical Engineering Technology., 41*(2), 278–284. Available from https://doi.org/10.1002/ceat.201700303.

Rautenbach, R., & Albrecht, R. (1982). Die trennung engsiedender und azeotroper Gemische durch pervaporation. *Chemie Ingenieur Technik, 54*, 260–261. Available from https://doi.org/10.1002/cite.330540316.

Schiffman, P. (2013). Three step modelling approach for the simulation of industrial scale pervaporation modules. Dissertation, Freiberg.

Schiffmann, P. & Repke, J.-U. (2011). Design of pervaporation modules based on computational process modelling. In: *21st European Symposium on Computer Aided Process Engineering* (pp. 397–401). Elsevier.

Shan, H., Li, S., Zhang, X., Meng, F., Zhuang, Y., Si, Z., … Qin, P. (2021). Molecular dynamics simulation and preparation of vinyl modified polydimethylsiloxane membrane for pervaporation recovery of furfural. *Separation and Purification Technology, 258*, 118006. Available from https://doi.org/10.1016/j.seppur.2020.118006.

Sharma, R., & Jain, M. (2020). Removal of benzothiophenes from model diesel/jet oil fuel by using pervaporation process: Estimation of mass transfer properties of the different membranes and dynamic modeling of a scale-up batch process. *Journal of Membrane Science, 595*, 117500. Available from https://doi.org/10.1016/j.memsci.2019.117500.

Sherwood, T. K., Brian, P. L. T., Fisher, R. E., & Dresner, L. (1965). Salt concentration at phase boundaries in desalination by reverse osmosis: Industrial & engineering chemistry fundamentals. *Industrial & Engineering Chemistry Fundamentals, 4*, 113–118. Available from https://doi.org/10.1021/i160014a001.

Song, Y., Pan, F., Li, Y., Quan, K., & Jiang, Z. (2019). Mass transport mechanisms within pervaporation membranes. *Frontiers of Chemical Science and Engineering, 13*, 458–474. Available from https://doi.org/10.1007/s11705-018-1780-1.

Sosa, M. A., & Espinosa, J. (2011). Feasibility analysis of isopropanol recovery by hybrid distillation/pervaporation process with the aid of conceptual models. *Separation and Purification Technology, 78*, 237–244. Available from https://doi.org/10.1016/j.seppur.2011.02.009.

Szilagyi, B., & Toth, A. J. (2020). Improvement of component flux estimating model for pervaporation processes. *Membranes, 10*(12), 418. Available from https://doi.org/10.3390/membranes10120418.

Thi, H. T. D., Mizsey, P., & Toth, A. J. (2020). Separation of alcohol-water mixtures by a combination of distillation, hydrophilic and organophilic pervaporation processes. *Membranes, 10*(11), 345. Available from https://doi.org/10.3390/membranes10110345.

Thiess, H., Schmidt, A., & Strube, J. (2018). Development of a scale-up tool for pervaporation processes. *Membranes, 8*, 4. Available from https://doi.org/10.3390/membranes8010004.

Valentinyi, N., Andre, A., Haaz, E., Fozer, D., Toth, A. J., Nagy, T., & Mizsey, P. (2020). Experimental investigation and modeling of the separation of ternary mixtures by hydrophilic pervaporation. *Separation Science and Technology, 55*, 601–617. Available from https://doi.org/10.1080/01496395.2019.1569692.

Vallieres, C., & Favre, E. (2004). Vacuum vs sweeping gas operation for binary mixtures separation by dense membrane processes. *Journal of Membrane Science, 244*, 17–23. Available from https://doi.org/10.1016/j.memsci.2004.04.023.

Vane, L. M. (2019). Review: Membrane materials for the removal of water from industrial solvents by pervaporation and vapor permeation. *Journal of Chemical Technology and Biotechnology (Oxford, Oxfordshire: 1986), 94*, 343–365. Available from https://doi.org/10.1002/jctb.5839.

VDI (Ed.), (2006). *VDI-Wärmeatlas*. Berlin, Heidelberg: Springer Berlin Heidelberg.

Verhoef, A., Degrève, J., Huybrechs, B., van Veen, H., Pex, P., & van der Bruggen, B. (2008). Simulation of a hybrid pervaporation–distillation process. *Computers & Chemical Engineering, 32*, 1135–1146. Available from https://doi.org/10.1016/j.compchemeng.2007.04.014.

Wijmans, J. G., Athayde, A. L., Daniels, R., Ly, J. H., Kamaruddin, H. D., & Pinnau, I. (1996). The role of boundary layers in the removal of volatile organic compounds from water by pervaporation. *Journal of Membrane Science, 109*, 135–146. Available from https://doi.org/10.1016/0376-7388(95)00194-8.

Wijmans, J. G., & Baker, R. W. (1995). The solution-diffusion model: A review. *Journal of Membrane Science, 107*, 1–21. Available from https://doi.org/10.1016/0376-7388(95)00102-I.

Wijmans, J. G., & Baker, R. W. (1993). A simple predictive treatment of the permeation process in pervaporation. *Journal of Membrane Science, 79*(1), 101–113. Available from https://doi.org/10.1016/0376-7388(93)85021-N.

Yong, W. F., & Zhang, H. (2021). Recent advances in polymer blend membranes for gas separation and pervaporation. *Progress in Materials Science, 116*, 100713. Available from https://doi.org/10.1016/j.pmatsci.2020.100713.

Yushkin, A. A., Golubev, G. S., Podtynnikov, I. A., Borisov, I. L., Volkov, V. V., & Volkov, A. V. (2020). Separation of mixtures of polar and nonpolar organic liquids by pervaporation and nanofiltration (review). *Petroleum Chemistry*, *60*, 1317–1327. Available from https://doi.org/10.1134/S0965544120110201.

Zhang, C., Peng, L., Jiang, J., & Gu, X. (2017). Mass transfer model, preparation and applications of zeolite membranes for pervaporation dehydration: A review. *Chinese Journal of Chemical Engineering*, *25*, 1627–1638. Available from https://doi.org/10.1016/j.cjche.2017.09.014.

Zhang, Q., Hou, W., Ma, Y., Yuan, X., & Zeng, A. (2021). Dynamic control analysis of eco-efficient double side-stream ternary extractive distillation process. *Computers & Chemical Engineering*, *147*, 107232. Available from https://doi.org/10.1016/j.compchemeng.2021.107232.

Zhu, Z., Li, S., Meng, D., Qi, H., Zhao, F., Li, X., Cui, P., Wang, Y., Xu, D., & Ma, Y. (2021). Energy efficient and environmentally friendly pervaporation-distillation hybrid process for ternary azeotrope purification. *Computers & Chemical Engineering*, *147*, 107236. Available from https://doi.org/10.1016/j.compchemeng.2021.107236.

CHAPTER

8

Transport phenomena in gas membrane separations

Foroogh Mohseni Ghaleh Ghazi, Mitra Abbaspour and Mohammad Reza Rahimpour

Department of Chemical Engineering, Shiraz University, Shiraz, Iran

List of Acronyms

CMS carbon molecular sieve
CNFs carbon nanofibers
CNTs carbon nanotubes
GPU gas permeation unit
MMM mixed matrix membranes
PEBA poly (ether-block-amide)
PEG polyethylene glycol
PIM polymers of intrinsic microporosity
RO reverse osmosis

Nomenclature

A preexponential factor
c concentration
$(\partial c/\partial x)$ concentration gradient
D diffusivity (diffusion) coefficient
E_d diffusion activation energy
ΔH_s sorption enthalpy
J flux
l membrane thickness
Δp_i transmembrane pressure
p pressure
P permeability
R ideal gas constant
S solubility coefficient
T absolute temperature

Current Trends and Future Developments on (Bio-) Membranes
DOI: https://doi.org/10.1016/B978-0-12-822257-7.00005-4

Subscript $_i$ component permeating through the membrane
Subscript $_1$ penetrant at a low concentration
Subscript $_2$ penetrant at a high concentration
Subscript $_n$ nanocomposite
Subscript $_P$ pure polymer matrix
V^* penetrant molecule's minimal free-volume element size
V_f average free volume in the media that penetrants can use for transport
Φ_f: nanofiller volume fraction
γ overlap factor

8.1 Introduction

Membrane technology concentrates on the chemical or physical interactions of certain fluids with the membrane's composition. With their high performance, ease of operation, and low cost, these processes are considered efficient and viable technologies for separating gaseous or liquid mixtures on a global scale (Brunetti, Scura, Barbieri, & Drioli, 2010). As a general idea, a membrane is a selective barrier between two phases. More precisely, a membrane is a thin, observable solid or fluid film with a slight but discernible thickness. The selectivity and flux features offer practical transport through the barrier (Biniaz, Makarem, & Rahimpour, 2020; Biniaz, Roostaie, & Rahimpour, 2021; Biniaz, Torabi Ardekani, Makarem, & Rahimpour, 2019).

Gas separation technology based on membranes refers to different methods for separating gases for purifying a single product or produce several compounds. The gas matches the structure of the container and, even in the presence of gravity and regardless of the material quantity of the container, acquires a uniform density within the container. Gases, also known as vapors, are dispersed in space when not contained in a bottle. In gaseous states, atoms or molecules of matter move freely and are often packed in a loose state compared to the same solid or liquid material molecules. Gases have the following essential properties: volume (V), pressure (P), and temperature (T), which define them. Gas activity in gas mixtures depends on the amount of gas molecules and not on their identity (Hwang, 2011).

Furthermore, for each gas separately and the mixture as a whole, the ideal gas equation would apply. In addition to these characteristics, the same is true for all molecules in the gas mixture, that is, all the gas mixture molecules have precisely the same behavior. Both businesses that develop membrane technology and other developers of gas separation technology are highly competitive with the field of membrane gas separation. The critical events in the construction of membranes for gas separation are summarized in Fig. 8.1.

Membrane separation processes are used in hydrogen separation from gases such as methane and nitrogen; nitrogen or oxygen separated from the air; methane separation from the other biogas components; hydrogen recovery from ammonia plant product streams; hydrogen recovery processes in oil refinery; elimination of water vapor from different gases; CO_2 and H_2S removal from natural gas; oxygen enrichment of the air for medical or metallurgical applications; nitrogen enrichment of ullage in inerting systems to avoid fuel tank explosions; volatile organic liquids elimination from exhaust streams' air (Uragami, 2017). Fig. 8.2 illustrates some commercial applications of membrane gas separation.

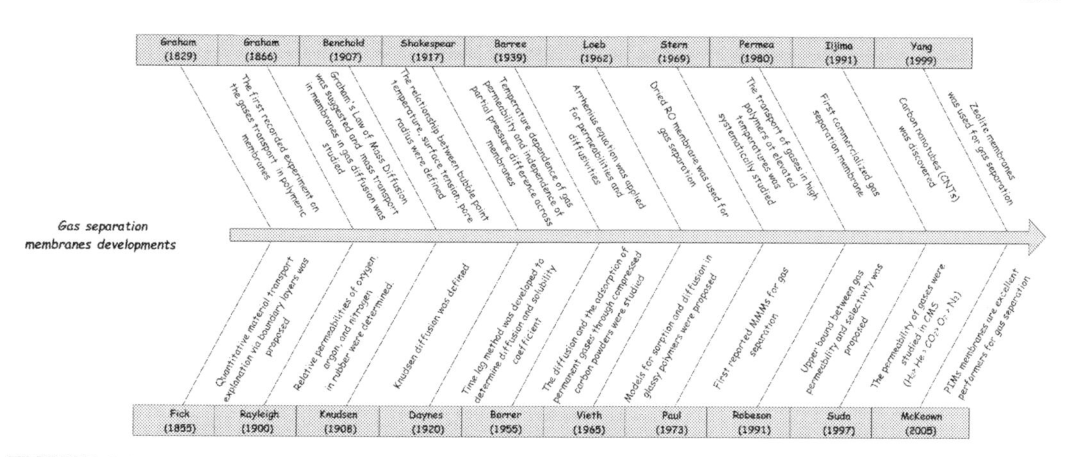

FIGURE 8.1 Membrane-based gas separation development chart (Ismail, Khulbe, & Matsuura, 2015).

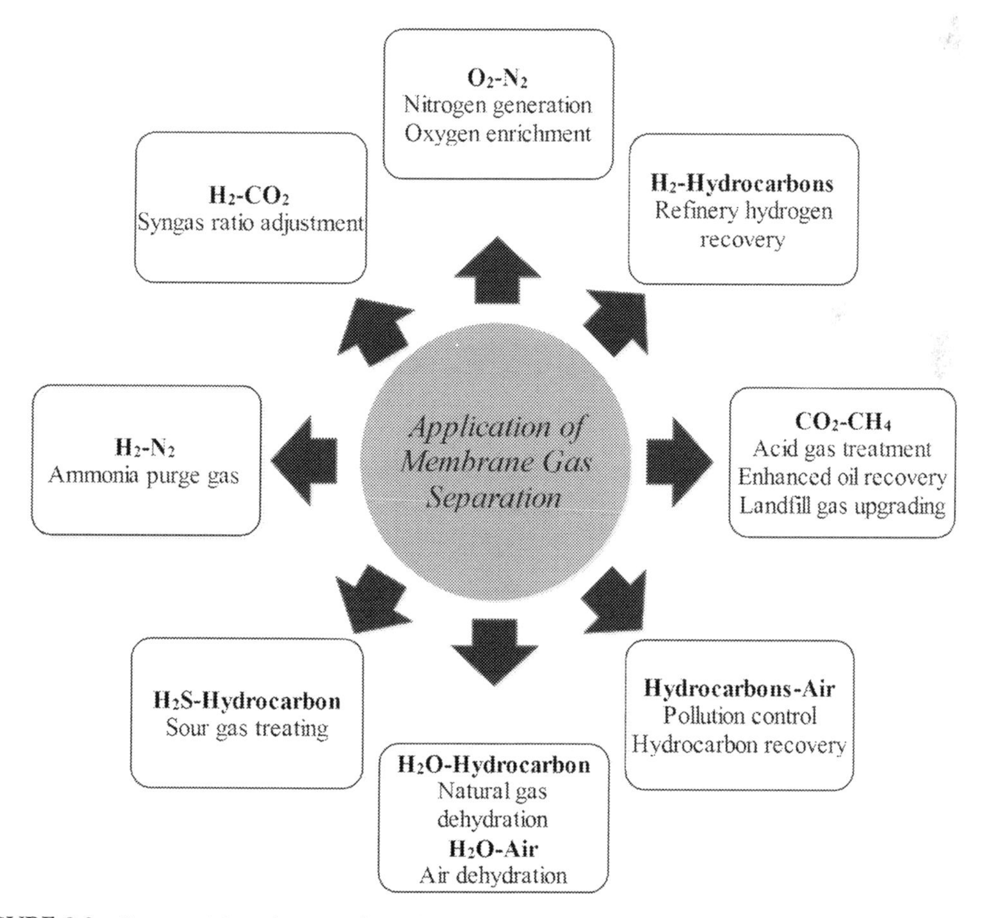

FIGURE 8.2 Commercial application of membrane gas separation (Ismail, Khulbe et al., 2015).

The subject of transport phenomena deals with the movement of various entities in every system and explains the fundamental concepts of transport. Transport phenomena of energy (heat transfer), the transport phenomena of momentum/the volume of motion/(viscous flow/ momentum in fluids/fluid dynamics), and material/mass transport phenomena (diffusion) are typical examples of transport phenomena that take place over a medium such as a solid or fluid under no uniform situations existing inside the medium. For instance, the relative motion of the different chemical species present in a medium is caused by variations in concentration, and this mass transport is referred to as diffusion. Variations in velocity within a fluid cause momentum to be transported, which is known as viscous flow. Temperature changes cause energy to be transported; a phenomenon is known as heat conduction.

8.2 Membrane gas separation

Certain fundamentals terms (laws, processes, or words) of the membrane process for gas separation are summarized in Table 8.1 to help better understand the process of gas membrane–based separation.

TABLE 8.1 Fundamental terms of membrane gas separation (Kumar, 2012).

Term	Definition
Selectivity	• This term indicates that certain substances can pass through the membrane while others remain. • The cell of the membrane, for example, keeps ions out, whereas allowing tiny hydrophobic substances in. • The most effective indicator of the membrane's capacity to separate two gases, namely A and B, is the permeability of gases, $\alpha_{A/B}$, also known as selectivity of the membrane.
Diffusion	• Diffusion is a mechanism through which the high-level concentration molecules transfer to low-level regions. • Space is said to be in "equilibrium" when the molecules are distributed throughout it. The three significant forms of membrane diffusion are: Knudsen diffusion, molecular diffusion/molecular sieving, and solution diffusion.
Knudsen diffusion	• Knudsen diffusion may occur through pinholes in dense polymeric membranes or microporous inorganic membranes. • Since molecules often collide with the pore wall, Knudsen diffusion occurs in a long pore with narrow diameter (2–50 nm). • This transport mode is necessary if the gas mean free path is larger than the pores. • In such cases, molecular collisions are more common with the pore wall than collisions among molecules. • The separation selectivity of these procedure mechanisms is proportionate to the molecular weight reverse square root ratio. • In macroporous and mesoporous membranes, this process is always prominent.
Molecular diffusion	• The mean free path of gas molecules is smaller in molecular diffusion than the pores, and diffusion mainly occurs by molecule collisions. • The driving force is the gradients of the composition in molecular diffusion. • When the pressure gradient in these pore regimes is applied, bulk flow occurs (laminar), as determined by the Poiseuille equation. • This form of transportation is often called Poiseuille or viscous flow.

(*Continued*)

TABLE 8.1 (Continued)

Term	Definition
Surface diffusion/solution diffusion	• When a permeating species has a deep affinity for the membrane surface and adsorbs along the pore walls, this is called surface diffusion. • Differences in the adsorption of permeating species are attributable to this process. • Surface diffusion usually occurs in conjunction with other transport mechanisms like Knudsen diffusion.
Configurational or micropore diffusion	• Such diffusion could be regarded as surface diffusion at the limit where the pores have the same size as the molecular size. • Diffusion is considered an "activated" process in this system, and separation is a strong molecular size and shape function, pores size, and interactions among gas molecules and pore wall. • Microporous zeolite membranes and carbon molecular sieves are mainly employed in this operation.
Graham's law	• The law of Graham notes that the gas diffusion rate is inversely proportional to the square root of its molecular weight's.
Fick's first law	• The first law of Fick concerns the diffusion of flux through the concentration of steady-state assumptions. • It assumes that the flux moves from high-concentration areas to low-concentration areas, with a magnitude proportional to the gradient of concentration.
Fick's second law	• The second law of Fick explains how concentration changes over time because of diffusion.
Henry's law	• According to Henry's Law, the quantity of gas dissolving in a volume of liquid at a constant temperature is directly proportional to the gas's partial pressure in equilibrium with this fluid (liquid).
Capillary condensation	• One surface flow form is capillary condensation, in which one of the gases can be condensed. • The condensed gas fully fills the pores at specific critical pressures, particularly in small macropores and mesopores. • Transport is possible because of the formation of menisci at both ends of the pore, which allows for hydrodynamic flow powered via a capillary pressure gradient. • With menisci formed on both ends of the pores, a capillary pressure differential from both ends allows transportation through a hydrodynamic flow. • Capillary condensation can theoretically be used to achieve very high selectivities, as the fluid layer structure of the condensable gas blocks and prevents noncondensable gas flow.
Molecular sieve	• It is a compound with exact, uniform tiny holes. • These holes are sufficiently small to block large molecules and allow small molecules to pass through. • Silica gel and activated charcoal are typical examples. • Microporous materials have a pore diameter below 2 nm, and macroporously generated materials are above 50 nm (500 Å) in pore diameters. Therefore the mesoporous category lies between 2 and 50 nm (20–500 Å) in the center of the pore category.
Molecular sieve effect	• The surface with pores interacting with the outside space can refer to the internal surfaces concerning porous solids. • Since the sizes of the fluid molecules will affect the accessibility of pores, the size of the molecules that make up the internal surface will influence its size.

(Continued)

TABLE 8.1 (Continued)

Term	Definition
Free volume	• The area in which polymer molecules are not filled is known as the free volume. • The occupied volume is usually included with van der Waals volume multiplied by a factor (generally 2.2) to consider that the packing density is limited even for the perfect crystal at absolute zero.
Mean free path	• The average distance between collisions with another molecule is known as the mean free path. • A molecule's mean free path is proportional to its size; the bigger the molecule, the smaller the mean free path.

8.3 Fundamentals equations of membrane transport

The difference in concentration between the two membrane phases is the driving force behind the phenomenon of transport that involves sorption, penetration, and diffusion. Thus the transport process tries to gradually equalize the difference in concentration or penetration chemical potential during the membrane separation process.

By using the First law diffusion equation of Fick, this mechanism can be explained under steady-state assumptions, according to which the flux J is proportional to the gradient of concentration $(\partial c/\partial x)$ in the flow direction as

$$J = -D\left(\frac{\partial c}{\partial x}\right) \tag{8.1}$$

Here D is the coefficient of diffusion. However, the second law of Fick defines the nonsteady condition, which is determined by the rate of penetrating concentration change $(\partial c/\partial x)$ as follows:

$$\frac{\partial c}{\partial t} = D\left(\frac{\partial c^2}{\partial x^2}\right) \tag{8.2}$$

It is a perfect case of isotropic membranes, and the coefficient of diffusion is independent of time, distance, and concentration. Several solutions are available for Eq. (8.2) according to the boundary conditions. Strong interaction occurs between polymers and penetrating organic molecules, and thus D depends on concentration. Consequently, Eq. (8.2) becomes Eq. (8.3)

$$\frac{\partial c}{\partial t} = D\frac{\partial D(c)\frac{\partial c}{\partial x}}{\partial x} \tag{8.3}$$

Since this equation is difficult to solve analytically, another equation, Eq. (8.4), is widely used.

$$\frac{\partial c}{\partial t} = D(c)\left(\frac{\partial c^2}{\partial x^2}\right) + \frac{\partial D(c)}{\partial c}\left(\frac{\partial c}{\partial x}\right)^2 \tag{8.4}$$

Experiments are generally performed over relatively short c intervals, and the variable $\partial D(c)/\partial c$ is insignificant in comparison to with $D(c)$. And then, a diffusion coefficient integral is obtained according to Eq. (8.5).

$$\overline{D} = \int_{C_1}^{C_2} D(c)\frac{dc}{c_1 - c_2} \tag{8.5}$$

where c_1 and c_2 are the penetrant concentrations at the film's low and high concentration faces, respectively.

Diffusion flow is constant in the steady-state, and the coefficient of diffusion is concentration-independent. Therefore Eq. (8.6) can be derived by integrating Eq. (8.1)

$$J = \frac{D(c_1 - c_2)}{l} \tag{8.6}$$

where l is the thickness of the membrane. The Nernst distribution law describes the distribution of penetrant between the polymer phase and ambient penetrant (Eq. 8.7).

$$c = \text{KC} \tag{8.7}$$

where c represents the sorbed concentration, K depends on c and temperature, and C represents the penetrant concentration of the ambient in contact with the polymer's surface. Instead of ambient penetrant concentration, pressure p is used in the case of transporting gases and vapors. Eq. (8.8) represents Henry's law Eq. (8.8)

$$c = pS \tag{8.8}$$

where S presents the solubility coefficient. The well-recognized permeation Eq. (8.9) is obtained by combining Eqs. (8.6) and (8.8), in which the ambient pressures on both sides of a thickness film (l) are p_1 and p_2.

$$J = \frac{\text{SD}(p_1 - p_2)}{l} \tag{8.9}$$

The permeability coefficient P is defined as the term SD, and thus Eq. (8.10)

$$P = \text{SD} \tag{8.10}$$

When it comes to permeability, the equation of flux (8.9) can be expressed as Eq. (8.11)

$$J = \frac{P(p_1 - p_2)}{l} \tag{8.11}$$

The transport qualities of a specific penetrant differ from one polymer to the next. Transport features are determined by the free volume within the polymer and the segmental mobility of the polymer chains. The extent of unsaturation, degree of crystallinity, degree of cross-linking, and character of substituents influence the polymer chains segmental mobility.

8.4 Permeation of gases through membranes

The ability of barrier materials to permit gases (e.g., O_2, CO_2, and N_2) to pass at a particular time is known as gas permeability. Pressure, temperature, and humidity can affect gas permeability. Depending on the physical properties of the membranes, for example, porous, nonporous, rubbery, and glassy, different gas permeability mechanisms can involve transporting gases across them. A pressure difference across the membrane drives the membrane gas separation mechanism.

The efficiency of gas separation is determined by the permeability and selectivity of the membrane content. A membrane can be broadly divided into two groups of porous and nonporous in accordance with flux and selectivity. The porous membranes are rigid, highly void structures with interconnected pores that are randomly distributed. The permeant characteristics and membrane features, including the polymer molecular size, the pore size of the membrane polymer, and the pore-size distribution, are primarily responsible for separating materials by a porous membrane. Generally, microporous membranes can effectively separate those molecules with a wide range of sizes.

8.4.1 Gas permeation in porous membranes

When it comes to applying porous membranes in gas separation, it is worth noting that although porous membranes have extremely high flux levels, they have poor selectivity values. The membrane porosity, average pore diameter, and membrane tortuosity contribute to the microporous membrane's overall characteristics. The pores of porous membranes used for the separation of gases must have a smaller diameter than the gas molecules' mean free path. The diameter is around 50 nm under normal conditions (100 kPa, 300K). The molecule's velocity determines the gas flux through the pore, and the flux through a porous membrane is three to five orders of magnitude greater than that through a nonporous membrane (Sanders et al., 2013).

Porous membranes can use four different diffusion mechanisms to achieve separation, including (A) diffusion by Knudsen diffusion (free molecule diffusion), (B) surface diffusion, (C) capillary condensation, (D) molecular sieving (see Fig. 8.3). Surface diffusion enhances the permeability of the adsorbent components to the membrane pores more strongly (Fig. 8.3B). Gas molecules adsorb on the membrane's pores and move over the membrane's surface. The efficient diameter of the pore is also reduced. The transport and selectivity of the nonadsorbent components is thus reduced. The positive surface diffusion contribution only works for specific temperature ranges and diameters of the pore. The diffusion of surfaces may occur in parallel to the diffusion of Knudsen. Capillary condensation (Fig. 8.3C) occurs when the membrane pores are filled (partially) by a condensed phase. Only soluble species in the condensed phase will pass through the membrane if the pores are completely filled with the condensed phase. For capillary condensation, selectivities and fluxes are typically high. On the other hand, the capillary condensation appearance is highly dependent on pore size, gas composition, and pore size uniformity. When pore sizes are small enough (3.0–5.2 Å), molecular sieving (Fig. 8.3D) occurs, resulting in the separation of molecules of different kinetic diameters: the pore size becomes so small that only the smaller gas molecules may pass through the membrane.

8.4.2 Gas permeation in nonporous membranes

Gas permeation in dense, nonporous polymer membranes is a method of fractionating gas mixtures with selective gas permeability mainly based on a solution–diffusion

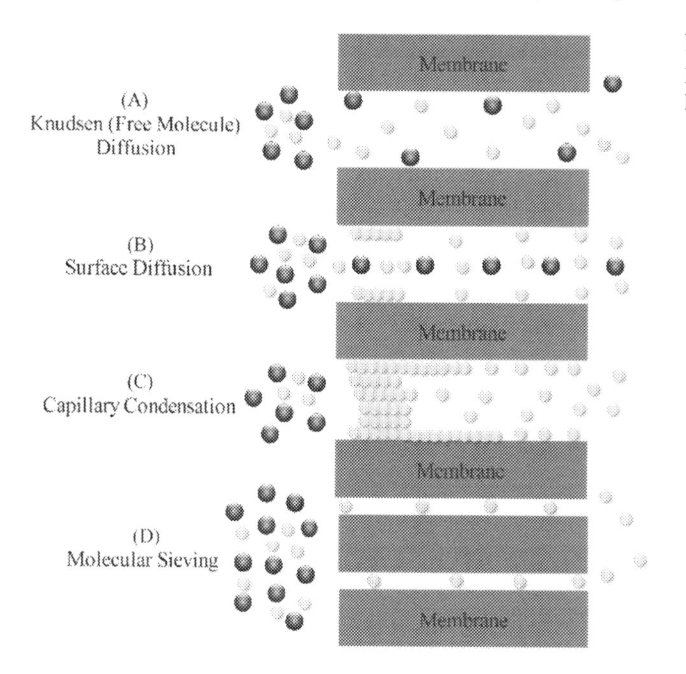

FIGURE 8.3 A diagram of four of the possible gas diffusion mechanisms in porous membranes.

process. Either a flat sheet or hollow fiber can take the shape of the membrane. Hollow fibers are generally recommended because they reach a higher efficient membrane zone within a specific module volume. By using a pressure gradient on either side of the membrane, the gas is transmitted through the membrane (Javaid, 2005). The proposed pressure difference creates a difference in the concentration of dissolved gas between the two membrane sides. The following three stages are involved in gas permeation through the membrane: permeable species absorption into the polymer, diffusion of a species into a polymer, desorption and elimination of species that permeate from the membrane surface.

Furthermore, gas permeation is influenced by solubility and diffusiveness of the small molecule in the membrane, chain packing, and the complexity of the side group, crystallinity, polarity, orientation, humidity, fillers, and plasticization. In traditional materials design for particular permeability applications such as gas barriers, the ability to link polymer molecular structures to gaseous transport features is critical. However, the availability of experimental permeability data was historically limited to commercial/conventional polymers, and this data showed that the rates of gas transport of polymer membranes differ by several orders of magnitude. Therefore the assessment of three fundamental characteristics associated with transport, including permeability, diffusion, and solubility coefficient, is needed to characterize gas separation membranes (George & Thomas, 2001) (Fig. 8.4).

The permeation module is dominated by two primary parameters in the solution–diffusion model: diffusivity and solubility coefficients. As represented in Eq. (8.12) in penetrant permeation, diffusion is the rate-controlling stage. Thus the permeability of the gas through the membrane determines the membrane's productivity (Stern, 1994).

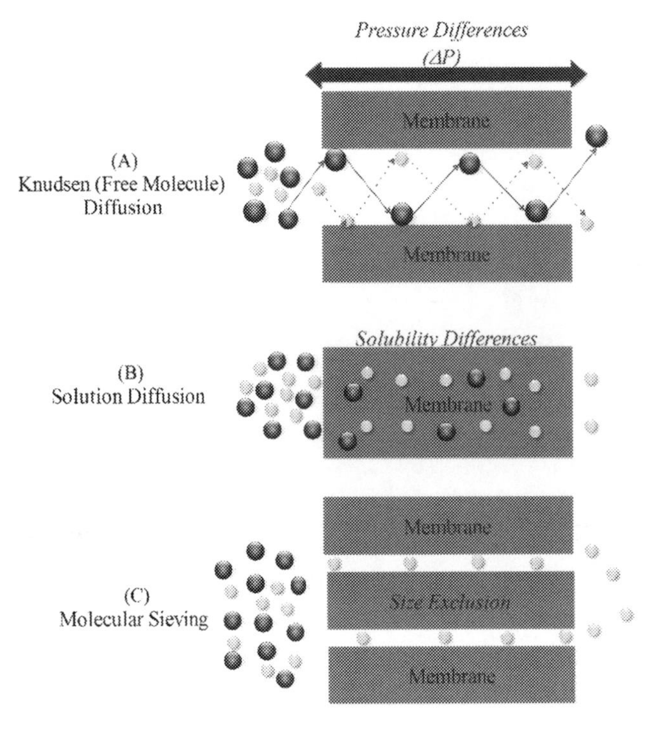

FIGURE 8.4 A diagram of three of the gas diffusion mechanisms in dense, nonporous polymeric membranes.

The gas permeability is defined by:

$$P_i = D_i S_i \tag{8.12}$$

D_i and S_i, respectively, indicate the component's diffusion and solubility coefficients. Permeability may also be represented by Eq. (8.13)

$$P_i = \frac{Jl}{\Delta p_i} \tag{8.13}$$

i is component, J is the flux of permeation, l is the film thickness, and Δp_i is transmembrane pressure. The Barrer is a widely used unit for measuring gas permeability, which is equal to 10^{-10} $\mathrm{cm^3_{STP}cm/cm^2scmHg}$. If, as in asymmetric membranes, the film thickness is hard to define, the gas permeation unit (GPU) is applied, which is determined as $1\mathrm{GPU} = 10^{-6}\mathrm{cm^3_{STP}.cm/cm^2.s.cmHg}$.

The membrane material and the operating conditions affect the permeability value. The solution diffusion model is used for rubbery polymer membranes, and the dual sorption model is applied for glassy polymer membranes. Furthermore, the pore model can be used to assess transport in porous membranes. On the other hand, the resistance model can also be used to assess transport in asymmetric and composite membranes, in which membranes are made up of many barrier layers with different properties. Contrary to asymmetric membranes, the composite membrane is discontinuous in both the chemical structure and morphology of the material at the borders of two adjacent barrier layers (Robeson, 2008).

8.5 Strategies to enhance gas permeation in membranes

Flux reduction can be produced by various factors, including concentration polarization, gel layer development, adsorption, and pore plugging. All these elements increase the resistance to transfer over the membrane on the feed side. The magnitude of these effects is highly dependent on the membrane technique types and feed solution used. The flux flow is proportional to the driving force, with the proportionality constant being the inverse of the total of all resistances. In circumstances where the concentration of polarization is quite severe (ultrafiltration/microfiltration), the flux decrease can be considered to be significant. However, when the concentration polarization is rarely present in processes such as gas separation, the flux is still reasonably constant over time (Firouzjaei, Shamsabadi et al., 2018; Kiadehi, Rahimpour, Jahanshahi, & Ghoreyshi, 2015).

Fouling is a problem with membrane technology, which results in higher operating and membrane replacement costs. Pretreatment is critical for creating the optimal treatment process train in order to ensure higher membrane treatability and acceptable effluent quality. New materials and methods for membrane production are being studied to improve the performance of polymeric membranes. Polymeric membranes feature excellent selectivity and gas flow, as well as outstanding thermal and chemical stability. Inorganic particles (such as carbon molecular sieve, zeolite, and inorganic oxides) and organic fillers (such as fullerenes and carbon nanotubes [CNTs]) have been employed to make mixed matrix membranes (MMMs) throughout the last decade.

Amirkhani, Harami, and Asghari (2020) synthesized PEBA, that is, poly(ether-block-amide), MMMs filled with varying concentrations of nano ZnO (up to 1 wt.%). They evaluated the performance of CO_2 separation from CH_4 and N_2 pure gas and their binary combinations with produced PEBA–ZnO MMMs membrane applying dry-phase inversion method. The MMMs were structurally studied to see how the presence of ZnO affected their morphologies and their thermal and mechanical properties. Compared to simple PEBA membranes, the membrane separation performance and the influence of ZnO loading on membrane parameters were studied using molecular simulation to predict manufactured membranes' structural and transportation characteristics. The results showed that ZnO-filled PEBA membranes could be a promising candidate for a marketable membrane because of their small concentration of 0.5 wt.% nanofillers and the simplified manufacturing process.

Surface modification is another effective way of increasing membrane antifouling. The fundamental purpose of membrane surface modification is to increase permeate flux and improve membrane efficiency. Hydrophobic membranes can be transformed into highly hydrophilic membranes by including inorganic nanoparticles such as TiO_2, Fe_3O_4, Al_3O_4, SiO_2, and ZrO_2 into the membrane (Laohaprapanon, Vanderlipe, Doma, & You, 2017; Yalcinkaya, Boyraz, Maryska, & Kucerova, 2020; Yalcinkaya, Siekierka, & Bryjak, 2017).

The interfacial interaction among nanoparticles and polymer matrix is critical to separation performance. MMMs take advantage of inorganic and organic materials' main transport capabilities; it is also straightforward to make base polymer membranes. Many MMMs (that incorporate nanoparticles such as TiO_2 and SiO_2) have better permeability and improved selectivity when compared to pure polymer membranes. Several researchers (Azizi, Mohammadi, & Behbahani, 2017) have investigated impacts of TiO_2 nanoparticles and polyethylene glycol (PEG) integration with PEBAX-1074 on CO_2 and CH_4 permeability values and optimal CO_2/CH_4 selectivity via the produced membranes. PEBAX is a revolutionary group of usable elastomers with an overall structure of $(A-B)_n$. They feature linear chains of rigid polyamide and soft polyether segments, whereby both segments are relatively short blocks. At constant feed pressure of 2 bar and feed temperature of 25°C, the effect of TiO_2 nanoparticles loading on gas permeation capabilities of composite PEBAX/PEG membranes with 40 wt.% PEG content revealed that both gases' permeability values rise as the TiO_2 loading increases. The creation of voids at the PEBAX-TiO_2 nanoparticles interfaces due to incompatibility among the surface of the TiO_2 nanoparticles and the polymer matrix is one reason for increased permeability. In addition, lower crystallinity caused by the breakdown of interchain hydrogen bonds among PA segments and, as a result, increased polymer matrix's free volume could be another reason why permeability values of nanocomposite membranes are higher than those of pure PEBAX membranes. In their other studies, a group of researchers (Azizi, Mohammadi, & Behbahani, 2017) compared the CO_2/CH_4 permeation capabilities of three types of nano-sized inorganic fillers (Al_2O_3, TiO_2, and SiO_2) incorporated into the PEBAX-1074 matrix. The findings of the gas permeation studies with a constant volume setup revealed that the membrane with Al_2O_3 nanoparticles has a higher ability to pass CO_2 molecules than the other two produced membranes and PEBAX/TiO_2 nanocomposite membranes have greater thermal stability than other membranes.

Carbon nanostructures (such as CNTs) with nanometer-scale diameters and large surface areas, on the other hand, present a promising opportunity to create membranes with excellent selectivity (Bastani, Esmaeili, & Asadollahi, 2013; Pechar, Kim et al., 2006). Carbon nanofibers (CNFs) are another type of carbon nanostructure. CNFs and CNTs have the potential to revolutionize many sectors in material science, as well as to pave the path for nanotechnology.

CNFs are nanostructures in a cylindrical shape with stacked graphene layers structured as cups, cones, or plates. They are discontinuous, highly graphitic, and disseminated in an isotropic or anisotropic mode. They are also very compatible with most polymer processing processes. CNFs offer great mechanical qualities, high electrical and thermal conductivities, and can be used in various matrixes such as thermosets, thermoplastics, ceramics, elastomers, and metals. They resemble CNTs in many ways and are almost more economical than them. As a result, using CNF in MMMs saves more money (Kiadehi et al., 2015).

8.5.1 Gas transport models in nanocomposite membrane modules

Polymeric membranes are currently one of the most used materials for gas separation processes such as landfill gas recovery, natural gas sweetening, flue gas and air separations, hydrogen recovery, and purification. This is because these materials have the best mechanical properties and can be easily turned into various modules.

However, there is a trade-off limitation between selectivity and permeability in polymeric membranes. In order to boost both selectivity and flux stream, researchers strive to address these issues by modifying polymeric membranes physically or chemically. Nanocomposite membrane modules, in particular, have proven to be a great choice for this type of membrane because of several advantages mentioned previously. Rezakazemi, Azarafza, Dashti, and Shirazian (2018) used intelligent methods to predict the permeation of gases, namely, CO_2, H_2, CH_4, and C_3H_8, in nanocomposite membranes fabricated from fumed silica and octatrimethylsiloxy polyhedral oligomeric silsesquioxane nanoparticles integrated into a polydimethylsiloxane polymer matrix. They examined the effects of various critical parameters, including pressure, nanoparticle loading and the kinetic diameter of the gas, on permeability using two innovative, intelligent, rigorous hybrid models, namely, adaptive neuro-fuzzy inference system combined with differential evolution optimization algorithm and coupled simulated annealing-least square support vector machine model. The former has proven to be a precise and reliable model for use in gas separation processes and membrane technology design by forecasting gas permeation more correctly and reliably.

Table 8.2 summarizes the three other possible mechanism processes for the transportation of gases via nanocomposite membranes outlined by Cong et al.

TABLE 8.2 Three different mechanisms of transporting gas across nanocomposite membranes (Cong, Hu, Radosz, & Shen, 2007).

Mechanism	Definition	Equation	Parameters
Maxwell	• The gas permeability of a polymer is reduced when impermeable inorganic nanoparticles are added. • Maxwell's model was designed to examine the diluted sphere suspension's steady-state dielectric characteristics. • The Mimic Maxwell is widely utilized in membranes filled with approximately spherically impermeable particles to model permeability. • Maxwell's model has a weakness in that it ignores interactions between polymer chains and nanofillers, as well as penetrators and nanofillers.	$P_n = P_P\left[\left(1-\phi_f\right)/\left(1+0.5\phi_f\right)\right]$ (8.14)	• P_n: Nanocomposite permeability • P_P: Pure polymer • matrix permeability • Φ_f: Nanofiller volume fraction • The numerator shows the decrease in membrane solubility because of a reduction in the volume of polymer accessible for sorption. • The denominator shows a decrease in diffusivity as the length of the penetrant diffusion path increases. • With increased particle volume fraction, these factors reduce permeability.

(*Continued*)

TABLE 8.2 (Continued)

Mechanism	Definition	Equation	Parameters
Free-volume increase	• Statistical and mechanical diffusion description commonly models the effect of polymer-free volume on penetrating diffusion coefficients. • The mechanism of free volume increase presents an overall comprehension of the interaction of the nanofillers and polymer chain segments • Nanofillers can interrupt the polymer chain's packaging, enhance free volume among the polymer chains, increase gas diffusion, and increase gas permeability. • An enhancement in polymer-free volume should increase penetrant diffusion.	$D = A\exp\left(-\gamma V * / V_f\right)$ (8.15)	• D: Diffusion coefficients • A: Preexponential factor • γ: Introduced overlap factor to prevent double-counting of free volume elements • V^*: A penetrant molecule's minimal free volume element size (and is linked to the size of the penetrant) • V_f: The average free volume in the media that penetrants can use for transport.
Solubility increase	• The mechanism is based on the interaction among nanofillers and the penetrants. • Functional groups on the surface of the inorganic nanofiller phase, hydroxyl for instance, could interact with polar gases like SO_2 and CO_2 enhance the solubility of penetrants in nanocomposite membranes	$P = A\exp\left(-(E_d + \Delta H_s)/RT\right)$ (8.16)	• P: Gas permeability • P_0: Preexponential factor • $E_d + \Delta H_s$: Apparent activation energy • E_d: Diffusion activation energy • ΔH_s: Sorption enthalpy • T: Absolute temperature • R: Ideal gas constant

8.6 Conclusions and future trends

Membrane systems are extremely useful for gas separation. Continued advancements in membrane technology development are an obvious choice for the future. Membrane technology has several advantages over other technologies, including minimal capital investment, low weight, minimal space demand, and high process flexibility. Natural gas purification, hydrogen recovery, and air separation are some of the most common membrane-based gas separations. Moreover, membranes are currently used in numerous applications, such as carbon capture and ethanol/water separations, owing to advancements in materials and a changing industrial environment. The membrane field will continue to evolve and flourish as a result of these separations and material advancements. In recent years, significant progress has been achieved in the creation of novel membranes and their applications. Membrane materials with enhanced selectivity and greater fluxes are made from new inorganic and organic components, supermolecular structures with specialized binding capabilities. Nonetheless, more durable membranes and modules will be required for success in the refinery, petrochemical, and natural gas markets than those currently in use.

Selective permeation is used as a basis for gas separation modules. The technology benefits from the dissolution and diffusion of gases into polymeric materials.

Transport through a permeation (film) will occur if a pressure differential is established on opposing sides of the membrane. The rate of permeation is determined by the product of a solubility coefficient and a diffusion coefficient. Membrane separations based on mixed-matrix materials can address the needs of demanding gas separation in a future where energy resources are limited and environmental concerns are growing. To some extent, any membrane will separate gases. However, selecting the polymeric material that makes up the membrane is critical because it dictates the gas separation module's ultimate performance. As a filler for mixed matrix membranes, high-performance materials can further lead to gas-separating membranes in unexplored areas.

Several critical concerns in membrane operations insert significant effects on membrane separation and transport, impacting the overall process's performance and economics. Therefore for more extensive membrane applications, it is critical to understand the causes of flux drop and be able to estimate flux performance. Mass transfer mechanisms, diffusion processes analyzed in membrane pores, and the transport resistances arising from membrane fouling will lead to a comprehensive understanding of the fouling phenomenon and the appropriate gas permeation flux. The transport processes in gas membrane separations were studied in this chapter, in addition to several analytical models linked to contemporary membrane technology. However, in order for membrane science and technology to remain appealing in the future, both fundamental and applied research to improve today's available membranes and membrane processes are required.

References

Amirkhani, F., Harami, H. R., & Asghari, M. (2020). CO_2/CH_4 mixed gas separation using poly(ether-b-amide)-ZnO nanocomposite membranes: Experimental and molecular dynamics study. *Polymer Testing, 86*, 106464.

Azizi, N., Mohammadi, T., & Mosayebi Behbahani, R. (2017). Comparison of permeability performance of PEBAX-1074/TiO_2, PEBAX-1074/SiO_2 and PEBAX-1074/Al_2O_3 nanocomposite membranes for CO_2/CH_4 separation. *Chemical Engineering Research and Design, 117*, 177–189.

Bastani, D., Esmaeili, N., & Asadollahi, M. (2013). Polymeric mixed matrix membranes containing zeolites as a filler for gas separation applications: A review. *Journal of Industrial and Engineering Chemistry, 19*(2), 375–393.

Biniaz, P., Makarem, M. A., & Rahimpour, M. R. (2020). Membrane reactors. In Maurizio Benaglia, & Alessandra Puglisi (Eds.), *Catalyst immobilization: Methods and applications* (pp. 307–324)). Wiley-VCH Verlag GmbH & Co. KgaA. Available from https://doi.org/10.1002/9783527817290.ch9.

Biniaz, P., Roostaie, T., & Rahimpour, M. R. (2021). Biofuel purification and upgrading: Using novel integrated membrane technology. In M. R. Rahimpour, R. Kamali, M. Amin Makarem, & M. K. D. Manshadi (Eds.), *Advances in bioenergy and microfluidic applications* (pp. 69–86). Elsevier.

Biniaz, P., Torabi Ardekani, N., Makarem, M. A., & Rahimpour, M. R. (2019). Water and wastewater treatment systems by novel integrated membrane distillation (MD). *ChemEngineering, 3*(1), 8.

Brunetti, A., Scura, F., Barbieri, G., & Drioli, E. (2010). Membrane technologies for CO_2 separation. *Journal of Membrane Science*, *359*(1–2), 115–125.

Cong, H., Hu, X., Radosz, M., & Shen, Y. (2007). Brominated poly (2, 6-diphenyl-1, 4-phenylene oxide) and its silica nanocomposite membranes for gas separation. *Industrial & Engineering Chemistry Research*, *46*(8), 2567–2575.

Firouzjaei, M. D., Shamsabadi, A. A., Aktij, S. A., Seyedpour, S. F., Sharifian Gh, M., Rahimpour, A., ... Soroush, M. (2018). Exploiting synergetic effects of graphene oxide and a silver-based metal–organic framework to enhance antifouling and anti-biofouling properties of thin-film nanocomposite membranes. *ACS Applied Materials & Interfaces*, *10*(49), 42967–42978.

George, S. C., & Thomas, S. (2001). Transport phenomena through polymeric systems. *Progress in Polymer Science*, *26*(6), 985–1017.

Hwang, S.-T. (2011). Fundamentals of membrane transport. *Korean Journal of Chemical Engineering*, *28*(1), 1–15.

Ismail, A. F., Khulbe, K. C., & Matsuura, T. (2015). Introduction. In A. F. Ismail, K. C. Khulbe, & T. Matsuura (Eds.), *Gas separation membranes* (pp. 1–10). Switzerland: Springer.

Javaid, A. (2005). Membranes for solubility-based gas separation applications. *Chemical Engineering Journal*, *112*(1–3), 219–226.

Kiadehi, A. D., Rahimpour, A., Jahanshahi, M., & Ghoreyshi, A. A. (2015). Novel carbon nano-fibers (CNF)/polysulfone (PSF) mixed matrix membranes for gas separation. *Journal of Industrial and Engineering Chemistry*, *22*, 199–207.

Kumar, A. (2012). *Fundamentals of membrane processes. Membrane technology and environmental applications* (pp. 75–95). American Society of Civil Engineers. Available from https://doi.org/10.1061/9780784412275.ch03.

Laohaprapanon, S., Vanderlipe, A. D., Doma, B. T., Jr, & You, S.-J. (2017). Self-cleaning and antifouling properties of plasma-grafted poly (vinylidene fluoride) membrane coated with ZnO for water treatment. *Journal of the Taiwan Institute of Chemical Engineers*, *70*, 15–22.

Pechar, T. W., Kim, S., Vaughan, B., Marand, E., Tsapatsis, M., Jeong, H. K., & Cornelius, C. J. (2006). Fabrication and characterization of polyimide–zeolite L mixed matrix membranes for gas separations. *Journal of Membrane Science*, *277*(1–2), 195–202.

Rezakazemi, M., Azarafza, A., Dashti, A., & Shirazian, S. (2018). Development of hybrid models for prediction of gas permeation through FS/POSS/PDMS nanocomposite membranes. *International Journal of Hydrogen Energy*, *43*(36), 17283–17294.

Robeson, L. M. (2008). The upper bound revisited. *Journal of Membrane Science*, *320*(1–2), 390–400.

Sanders, D. F., Smith, Z. P., Guo, R., Robeson, L. M., McGrath, J. E., Paul, D. R., & Freeman, B. D. (2013). Energy-efficient polymeric gas separation membranes for a sustainable future: A review. *Polymer*, *54*(18), 4729–4761.

Stern, S. A. (1994). Polymers for gas separations: The next decade. *Journal of Membrane Science*, *94*(1), 1–65.

Uragami, T. (2017). *Science and technology of separation membranes*. Wiley Online Library.

Yalcinkaya, F., Boyraz, E., Maryska, J., & Kucerova, K. (2020). A review on membrane technology and chemical surface modification for the oily w7astewater treatment. *Materials*, *13*(2), 493.

Yalcinkaya, F., Siekierka, A., & Bryjak, M. (2017). Preparation of fouling-resistant nanofibrous composite membranes for separation of oily wastewater. *Polymers*, *9*(12), 679.

CHAPTER

9

Transport phenomena in membrane contactor systems

Rahim Aghaebrahimian, Parisa Biniaz, Seyed Mohammad Esmaeil Zakeri and Mohammad Reza Rahimpour

Department of Chemical Engineering, Shiraz University, Shiraz, Iran

Abbreviations

CNT	Carbon nanotube
CFD	Computational fluid dynamics
HF	Hollow fiber
MC	Membrane contactors
MD	Membrane distillation
PVDF	Polyvinylidene fluoride
VOC	Volatile organic compound

Nomenclature

C^* **and** C_δ^* **(kg/m³)**	Concentrations on the two membrane sides
c_{Ai}	The concentration of particle A at the phase interface in the liquid part of the system
c_{Ab}	The concentration of particle A at the bulk in the liquid side
C_A^*	The concentration of particle A that exists in the feed at the equilibrium condition with the phase in the membrane–liquid interface
$D_{A,\ F}$	The coefficient of diffusion for particle A that exists in the feed
$D_{A,\ rece}$	The coefficient of diffusion for particle A in the receiving phase
D_i **(m²/s)**	The diffusivity of the solute in membrane pores
$D_{o,i}$ **(m²/s)**	The diffusivity of the solute in bulk solvent
$D_{i,membrane}$	The coefficient of diffusion for particle A in the membrane pores
d_p **(m)**	Pore diameter
F	Feed
Gz_{ext}	Graetz number in the shell
Gz_{int}	Graetz number in the tube side
H	The coefficient of the solubility of the solute inside the immiscible liquid phase
k_1 **(m/s)**	The coefficient of mass transfer in the external feed side boundary layer

Current Trends and Future Developments on (Bio-) Membranes
DOI: https://doi.org/10.1016/B978-0-12-822257-7.00015-7

k_2 (m/s)	The coefficient of mass transfer in the permeate phase boundary layer
k^o (m/s)	The coefficient of mass transport inside the membrane layer
k_{ov} (m/s)	The coefficient of overall mass transport
$k_{g(gas)}$	Mass transport coefficient of the gas
$k_{l(liquid)}$	Mass transport coefficient of liquid
$k_{m(membrane)}$	Mass transport coefficient of membrane
k_m	The coefficient of mass transfer in membrane
L_i (m^2/Pa or Barrer)	The permeability of partial i
M (kg/Kmol)	Molecular weight
m_i	The solubility of particle i in the receiving phase
n	The number of fibers
p_{Ai}	The concentration of particle A at the phase interface in the gas part of the system
p_{Ab}	The concentration of particle A at the bulk in the gas side
P_i (Pa)	The pressure of component i
R (J/kmol K)	Gas constant
$r = 0$	Fiber center
r_i	Inner radius
r_o	Outer radius
r_e	Free surface radius (the distance of Happel's free in the fiber)
r_{cont}	Hollow fiber contactor radius
S	The geometric factor on the basis of the outer radius
Sh_m	Sherwood number
Subscripts b	Bulk phase
Subscripts k	Knudsen
Superscript $_o$	Bulk phase
Superscript *	Interface
Superscript $_1$	Feed phase
Superscript $_2$	Permeate phase
Superscript $_\delta$	Permeate side of the membrane
T (K)	Temperature
$z = 0$	(1) The inlet position of the fiber. (2) Fluid 1 is fed to the shell
$z = L$	(1) The outlet position of the fiber. (2) Fluid 2 transfers through the tube at z = L. (3) Considering that the receiving phase have a negligible amount of particle A at z = L
δ (m)	The thickness of the membrane
ε_p	Pore porosity
τ	Tortuosity
$\xi_{d,i}$	Diffusion hindrance factor
φ	The density of fiber packing
ε	Membrane porosity

9.1 Introduction

Membrane contactors (MCs) are novel models of the phase-contacting method applied in liquid–liquid extraction and gas transfer processes to increase the mass transfer. The membranes' primary role in the process is to act as the interface separating two phases without controlling the permeants' rate. According to the particular purpose, the phase of liquid can be organic or aqueous, while the membrane can be hydrophilic or hydrophobic. The difference in chemical potential on both sides of the membrane is the principal driving force for the system's separation process. In accordance with different solubility coefficients in particles, some species accumulate in one phase simultaneously other species accumulate in the other phase (Li, Fane, Ho, & Matsuura, 2011; Biniaz, Torabi Ardekani, Makarem, & Rahimpour, 2019; Biniaz, Makarem, & Rahimpour, 2020).

Hollow fiber (HF) and flat-sheet membranes are conventionally utilized in MC equipment. Regarding flat-sheet membranes, applying plate, spiral wound, and frame configurations have been offered. Although flat-sheet membranes have wide availability and perfectly effective permeability, HF configurations are generally preferred because of their high packing density. In order to avoid membrane movement and pressure inversion, cocurrent flow is applied in the process. Furthermore, utilizing cocurrent flow guarantees constant pressure difference throughout the membrane. In typical commercial HF membrane modules, the surface area varies in the contactor volume between 1500 and 3000 m^2/m^3 offering high contact area per unit volume, while in traditional contractors such as plate, packed, and bubble column, it is between 100 and 800 m^2/m^3.

In most cases, MCs are made in shell-and-tube configuration, including microporous thin HF membranes with sufficiently small pores such that capillary forces hinder direct mixing of the flows on each side of the membrane.

When it comes to physicochemical and morphological factors, MC systems must have nonwetting membranes with a pore size in the range of 0.02–0.2 μm and high porosity of over 75%–80% to satisfy the essential requirements of an efficient MC system, including large interfacial area per unit volume and high rate of volumetric mass transport. (Gugliuzza & Basile, 2013).

Mastering the MC processes is essential in order to stabilize the optimum operating conditions. Managing the optimum operating conditions that maximize the mass transfer coefficient and minimize the required area of the membrane is not always accessible. This is mainly because operating parameters such as temperature or concentration contribute to a complicated situation.

9.1.1 Advantages and disadvantages of membrane contactors

MC systems have considered well-developed, waste-free, and eco-friendly methods that offer tremendous benefits, such as large contact areas that are approximately 10 times greater than those of a similar-sized tower. Further considerable benefit offered by MCs is that they separate the counter-flowing phases physically. Furthermore, the membrane is entirely independent of flow rates; for instance, high flow ratio differences can be applied without creating operating challenges. One of the most severe problems with MC systems is the membranes' hydrophobicity issue, which means membranes do not absorb absorbent components inside. Furthermore, partial wetting of membrane pores results in a significant increase in the resistance of mass transfer and a considerable reduction in the rate of mass transfer and stability of the system. Therefore, the membrane pores need to be filled with the gas phase.

9.1.2 Application of membrane contactors in different operating units

MCs can be applied in various industrial sectors, including liquid–liquid extraction [e.g., oil extraction of volatile organic compounds (VOCs) from water, phenol extraction, packed columns, centrifugal devices, mixer-settler], drug recovery, gas absorption, desorption, scrubbing, and stripping (e.g., ammonia stripping), wastewater treatment (e.g., membrane distillation), membrane crystallization, advanced separation systems, particularly biopharmaceutical industry, osmotic distillation, high-pressure homogenizers, emulsification and biochemical reactors (Joscelyne & Trägårdh, 2000; Curcio & Drioli, 2005; Gaeta, 2009; Drioli & Giorno, 2010; Enrico & Efrem, 2015; Luis, 2018; Biniaz, Roostaie, & Rahimpour, 2021).

9.2 Transport phenomena

MCs are classified on the basis of phase distribution and mass transport inside the membrane pores. Liquid–liquid and gas–liquid systems are two common types of MC processes. Fig. 9.1 clearly illustrates a component's concentration profile that transfers from one phase to another in the MC system (Nagy, 2019a, 2019b). High mass transfer and consequently the efficient performance of an MC process greatly depends on the following factors: excellent structure characteristics of the membrane, high and appropriate options for operating conditions of the system, low resistances in heat and mass transfer in the external phases, controlled and significantly diverse phases pressure to prevent breakthrough values in the pressure of flowing phases. The low affinity of the membrane to water is a preventing factor for the liquid phase from passing over the pores in operating a hydrophobic membrane unless pressures higher than the "breakthrough pressure" are utilized. According to the Young–Laplace equation, breakthrough pressure is the pressure higher than that for the aqueous phase

9.2.1 Transport phenomena in liquid–liquid membrane contactor systems

Liquid–liquid systems contain two immiscible liquid phases, which have direct contact with each other. In this type of MC system, membrane pores fill with a liquid having the same polarity as the membrane, which results in separating two miscible phases. Mass transport phenomenon takes place through the membrane pores mainly by diffusion mechanism. The difference in concentration is the main driving force, and the diffusion is conducted mostly along with a chemical reaction with the transferred component.

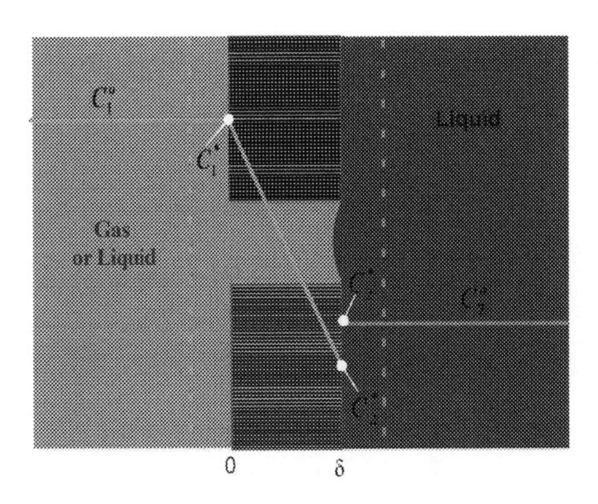

FIGURE 9.1 A schematic diagram of a concentration profile of a component in an membrane contactor that transfers from one phase to another.

Generally, feed, membrane, and permeate diffusive mass transport rates are given as following equations, respectively:

$$J = k_1 \left(C_1^o - C_1^* \right) \tag{9.1}$$

$$J = k^o \left(C^* - C_\delta^* \right) \tag{9.2}$$

$$J = k_2 \left(C_2^* - C_2^o \right). \tag{9.3}$$

Considering Eqs. (9.1)–(9.3), interface concentrations are related by $C_1^* H = C^*$, and $C_\delta^* H = C_2^*$ therefore the overall rate of mass transport is given by:

$$J_{ov} = k_{ov} \left(C_1^o - \frac{C_2^o}{H} \right). \tag{9.4}$$

In which:

$$\frac{1}{k_{ov}} = \frac{1}{k_1} + \frac{1}{k^o} + \frac{1}{k_2 H}. \tag{9.5}$$

The symbols are enumerated in Table 9.1.

9.2.2 Transport phenomena in gas–liquid membrane contactor systems

In gas–liquid MC systems, membrane pores are filled with the gas phase. The mass transfer phenomena take place on the gas–liquid interface at the liquid part of the system. The variance in the concentration of the components is the main driving force in the system. Moreover, the gas's existence on the pores of the membrane leads to a minimum resistance in mass transport due to the high rate of diffusion in the gas phase. Porous

TABLE 9.1 List of symbols used in liquid–liquid membrane contactor mass transfer equations.

Symbol	Name
k_1 (m/s)	The coefficient of mass transfer in the external feed side boundary layer
k_2 (m/s)	The coefficient of mass transfer in the permeate phase boundary layer
H	The coefficient of the solubility of the solute inside the immiscible liquid phase
Superscript$_o$	Bulk phase
Superscript*	Interface
Superscript$_1$	Feed phase
Superscript$_2$	Permeate phase
Superscript$_\delta$	Permeate side of the membrane
k^o (m/s)	The coefficient of mass transport inside the membrane layer
k_{ov} (m/s)	The coefficient of overall mass transport
C^* and C_δ^* (kg/m^3)	Concentrations on the two membrane sides

membranes applied for gas–liquid MC processes are mainly polypropylene, polytetrafluoroethylene, poly (vinylidene fluoride), and poly-4-methyl-1-pentene. These systems are utilized for the following:

(1) A gas–liquid system such as stripping, absorption, scrubber that contains a hydrophobic membrane, a gas phase, and a nonwetting liquid, which moves at the same or higher pressure than that of the gas phase (Drioli & Giorno, 2018).
(2) A gas-liquid system such as membrane distillation systems that contain two liquid phases, and liquid phases separate with a porous membrane, in which pores are filled with vapor phase. Temperature and pressure difference between two membrane sides are the main driving force for transporting evaporated species. Therefore the transport mechanism is conducted partly by convection and partly by diffusion (Drioli, Criscuoli, & Curcio, 2011).

Molecular diffusion, viscous flow, and Knudsen diffusion are three principal flow mechanisms of gas transport through porous membranes. Furthermore, transferring is dependent on the size of pores of the membrane (Nagy, 2019a, 2019b). The following Equation is the general gas transfer through the membranes.

$$J_i = L_i \frac{\Delta P_i}{\delta}. \tag{9.6}$$

The diffusion rate through the layer of membrane is affected by various factors, and the efficient diffusion coefficient is determined as:

$$D_i = D_{o,i} \frac{\varepsilon_p}{\tau} \xi_{d,i}. \tag{9.7}$$

It should be noticed that the coefficient of the membrane in the gas phase is determined by the combination of bulk coefficient and Knudsen diffusion coefficient as follows (Seidel-Morgenstern, 2010):

$$\frac{1}{D_i} = \frac{1}{D_{b,i}} + \frac{1}{D_{k,i}}. \tag{9.8}$$

In which the coefficient of Knudsen diffusion is calculated as (Drioli et al., 2011):

$$D_{k,i} = \frac{1}{3} d_p \sqrt{\frac{8R T}{\pi M}}. \tag{9.9}$$

In the case of gas-filled pores, the diffusion mechanism into the membrane (Knudsen diffusion or bulk diffusion) is defined on the basis of the pores' diameter (Drioli et al., 2011).

The symbols are listed in Table 9.2.

9.2.2.1 *Film theory in gas–liquid membrane contactor systems*

The film theory has been widely used in MC systems to explain the mass transport in fluid phases. The theory takes the mass transfer resistance of turbulent fluid, which exists in both gas and liquid phases, into consideration. Furthermore, the membrane's resistance as the third phase is considered as well (Seader, Henley, & Roper, 1998; Seader, Henley, & Roper, 2011). Fig. 9.2 illustrates a schematic diagram of the interphase of a gas–liquid

TABLE 9.2 List of symbols used in transfer phenomena equations in gas-liquid membrane contactor.

Symbol	Name
L_i (m^2/Pa or Barrer)	The permeability of partial i
P_i (Pa)	The pressure of component i
δ (m)	The thickness of the membrane
ε_p	Pore porosity
τ	Tortuosity (In the range of 1.4–7)
$\xi_{d.i}$	Diffusion hindrance factor
$D_{o,i}$ (m^2/s)	The diffusivity of the solute in bulk solvent
D_i (m^2/s)	The diffusivity of the solute in membrane pores
Subscripts b	Bulk phase
Subscripts k	Knudsen
d_p (m)	Pore diameter
R (J/Kmol K)	Gas constant
T (K)	Temperature
M (kg/Kmol)	Molecular weight

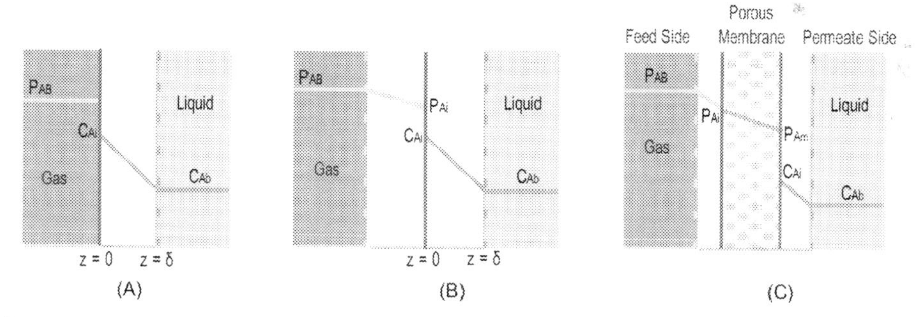

FIGURE 9.2 A schematic diagram of partial pressure and concentration profiles of solute transfer through membranes in an MC system (A) the film theory; (B) the double-film theory; (C) the double-film theory with a porous membrane separating the two phases in an MC system.

system. (1) The particle in phase A disperses into phase B, (2) double-film theory takes place (one film is considered for each phase), (3) MC system in which a porous membrane exists between the two fluid phases and the film layer is next to the fluid phase membrane interface (Luis, 2018).

By taking phase equilibrium at the interface of gas–liquid systems into account, Henry's law $\left(H_A = c_A/p_A\right)$ or partition coefficient $\left(P = c_A/c_{A'}\right)$ can be considered for these types of systems. δ is the film thickness in which molecular diffusion $(c_{Ai} - c_{Ab})$ takes place as a driving force.

According to Fig. 9.2, the average concentration of component A in bulk is c_{Ab}, and therefore by using Fick's first law, the mass transfer equation can be determined as

$$J_A = \frac{D_{AB}}{\delta}(c_{Ai} - c_{Ab}) = \frac{c.D_{AB}}{\delta}(x_{Ai} - x_{Ab}). \tag{9.10}$$

By extending the film theory into two film layers in series, despite the assumption that the concentrations of the fluid phases at the interface are in equilibrium, each film layer shows resistance to mass transport. In steady-state conditions for mass transfer in gas-liquid MC systems, Eq. (9.9) can be applied for component A, which transfers from a gas phase to the liquid phase by passing through the gas–liquid interface. Generally, the ratio D_{AB}/δ is varied considering that the film thickness depends on the flow conditions and alters by the coefficient of mass transfer for each film layer that considers the mass transfer resistance for each film layer (Luis, 2018).

9.2.2.2 Resistance-in-series theory in gas–liquid membrane contactor systems

Film theory is the base of the resistance-in-series model because the membrane and each fluid film layer create distinct resistance described by the inverse of each specific mass transport coefficient. Therefore using the following equations and symbols listed in Table 9.3, the overall coefficient of mass transfer based on the resistance-in-series model is determined by Eq. (9.17).

$$J_A = k_g(p_{Ab} - p_{Ai}). \tag{9.11}$$

$$J_A = k_l(c_{Ai} - c_{Ab}). \tag{9.12}$$

$$c_{Ai} = p_{Ai}H_A. \tag{9.12a}$$

p_{Ai} and c_{Ai} are related by the Henry's law Eq. (9.12a) considering that they are in equilibrium. Substituting Eq. (9.12) in Eq. (9.11) to eliminate c_{Ai} and combining with Eq. (9.10) results in the following equation:

$$J_A = \frac{p_{Ab}H_A - c_{Ab}}{(H_A/k_g) + (1/k_l)}. \tag{9.13}$$

TABLE 9.3 List of symbols used in resistance-in-series theory in gas-liquid membrane contactor systems.

Symbol	Name
p_{Ai}	The concentration of particle A at the phase interface in the gas part of the system
c_{Ai}	The concentration of particle A at the phase interface in the liquid part of the system
p_{Ab}	The concentration of particle A at the bulk in the gas side
c_{Ab}	The concentration of particle A at the bulk in the liquid side
$k_{g(\text{gas})}$	Mass transport coefficient of the gas
$k_{l(\text{liquid})}$	Mass transport coefficient of liquid
$k_{m(\text{membrane})}$	Mass transport coefficient of membrane

Considering the membrane in the system and Henry's law Eq. (9.14) with the equilibrium at the interface of gas and liquid, mass transfer through the membrane in which pores are filled with gas is determined by Eq. (9.15)

$$c_{\mathrm{Ai}} = p_{\mathrm{Am}} H_A. \tag{9.14}$$

$$J_A = k_m \left(p_{\mathrm{Ai}} - p_{\mathrm{Am}} \right). \tag{9.15}$$

Therefore with the same substitution algorithm, Eq. (9.13) would become:

$$J_A = \frac{p_{\mathrm{Ab}} H_A - c_{\mathrm{Ab}}}{\left(H_A/k_g\right) + \left(1/k_m\right) + \left(1/k_l\right)}. \tag{9.16}$$

Finally, the overall mass transport coefficient and mass transport through the membrane can be determined as follows:

$$\frac{1}{k_{\mathrm{Overall}}} = \frac{H_A}{k_{\mathrm{gas}}} + \frac{1}{k_{\mathrm{membrane}}} + \frac{1}{k_{\mathrm{liquid}}}. \tag{9.17}$$

$$J_A = k_{\mathrm{Overall}} \left(p_{\mathrm{Ab}} H_A - c_{\mathrm{Ab}} \right). \tag{9.18}$$

9.3 Mass transfer in shell-and-tube hollow fiber membrane contactor

As mentioned previously, MCs are mainly produced in shell-and-tube configuration with microporous thin HF membranes. Mass transport occurs through the membrane's pores, which are small enough to hinder direct mixing by capillary forces. Direct mixing may occur in phases on each membrane surface. A laminar flow regime model is utilized to define mass transfer on the tube side. Furthermore, Happel's free surface model is applied for mass transport on the shell side (Happel, 1959; Luis, Garea, & Irabien, 2010).

Considering the local equilibrium in the interface of phases and assuming that mass transport occurs only with diffusion mechanism, the mass transfer equations are determined according to Happel's free surface model and investigations by Luis et al. (2010) on the modeling of an HF ceramic contractor for SO_2 capture on radial and axial coordinates, illustrated in Fig. 9.3.

The symbols are listed in Table 9.4.

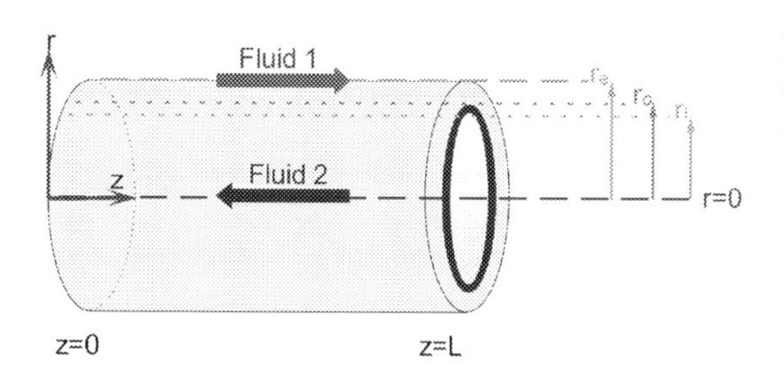

FIGURE 9.3 A schematic diagram of SO_2 capture on radial and axial coordinates in hollow fiber ceramic contactor.

TABLE 9.4 List of symbols used in mass transfer equations of shell-and-tube membrane contactor system with microporous thin hollow fiber membranes and having radial and axial coordinates of the fiber.

Symbol	Name
$r = 0$	Fiber center
r_i	Inner radius
r_o	Outer radius
r_e	Free surface radius (the distance of Happel's free in the fiber)
$z = 0$	• The inlet position of the fiber • Fluid 1 is fed to the shell
$z = L$	• The outlet position of the fiber • Fluid 2 transfers through the tube at $z = L$ • Considering that the receiving phase have a negligible amount of particle A at $z = L$
φ	The density of fiber packing
n	The number of fibers
r_{cont}	Hollow fiber contactor radius
F	Feed
$D_{A, F}$	The coefficient of diffusion for particle A that exists in the feed
S	The geometric factor on the basis of the outer radius
C_A^*	The concentration of particle A that exists in the feed at the equilibrium condition with the phase in the membrane–liquid interface
k_m	The coefficient of mass transfer in membrane
Sh_m	Sherwood number
Gz_{ext}	Graetz number in the shell
$D_{A, rece}$	The coefficient of diffusion for particle A in the receiving phase
Gz_{int}	Graetz number in the tube side
$D_{i,membrane}$	The coefficient of diffusion for particle A in the membrane pores
ε	Membrane porosity
τ	Membrane tortuosity
m_i	The solubility of particle i in the receiving phase

9.3.1 Mass transport in shell side

Considering the isothermal and steady-state condition of the system without axial diffusion and assuming constant physical characteristics of the phase and the shell side's constant pressures, the mass transport in the shell side can be determined.

In cylindrical coordinates, the equation of partial differential for the mass balance based on Fick's law is:

$$u_{z,g}\frac{\partial C_{A,\text{feed}}}{\partial z} = D_{A,\text{feed}}\left[\frac{1}{r}\frac{\partial}{\partial r}\left(r\frac{\partial C_{A,F}}{\partial r}\right)\right]. \tag{9.19}$$

The profile of velocity in the shell side:

$$u_{z,g} = u_{\max,g}.g(r) = 2u_{m,g}.g(r) \tag{9.20}$$

$$g(r) = \left[1-\left(\frac{r_o}{r_e}\right)^2\right].\left[\frac{\left(r/r_e\right)^2-\left(\frac{r_o}{r_e}\right)^2+2.\ln\left(\frac{r_o}{r}\right)}{3+\left(\frac{r_o}{r_e}\right)^4-4.\left(\frac{r_o}{r_e}\right)^2+4.\ln\left(\frac{r_o}{r_e}\right)}\right]. \tag{9.21}$$

$$r_e = \left(\frac{1}{\varphi}\right)^{0.5}.r_o. \tag{9.22}$$

$$\varphi = \frac{n.r_o^2}{r_{\text{cont}}^2}. \tag{9.23}$$

With the following boundary conditions:

$$r = r_e, \frac{\partial C_{A,F}}{\partial r} = 0. \tag{9.24}$$

$$r = r_o, D_{A,F}\frac{\partial C_{A,F}}{\partial r} = k_m.S.\left(C_{A,F} - C_A^*\right). \tag{9.25}$$

$$Z = 0, C_{A,F} = C_{A,\text{in}}. \tag{9.26}$$

By using the following dimensionless parameters, the differential mass balance in the Eq. (9.19) can be transformed into dimensionless.

$$\theta = \frac{r}{r_o}, \overline{z} = \frac{z}{L}, \overline{C} = \frac{C_{A,F}}{C_{A,\text{in}}}. \tag{9.27}$$

$$\text{Sh}_m = \frac{k_m.S.r_o}{D_{A,F}}. \tag{9.28}$$

$$\text{Gz}_{\text{ext}} = \frac{u_{m,g}d_o^2}{D_{A,F}L}, d_o = 2r_o. \tag{9.29}$$

The final dimensionless mass balance equation would be

$$\left(\text{Gz}_{\text{ext}}/2\right).g(\overline{r}).\frac{\partial \overline{C}_A}{\partial \overline{z}} = \frac{1}{\theta}\frac{\partial}{\partial \theta}\left(\theta\frac{\partial \overline{C}_A}{\partial \theta}\right). \tag{9.30a}$$

$$g(\overline{r}) = \left[1-\left(\frac{r_o}{r_e}\right)^2\right].\left[\frac{\left(\frac{\overline{r}.r_o}{r_e}\right)^2-\left(\frac{r_o}{r_e}\right)^2+2.\ln\left(\frac{1}{\overline{r}}\right)}{3+\left(\frac{r_o}{r_e}\right)^4-4.\left(\frac{r_o}{r_e}\right)^2+4.\ln\left(\frac{r_o}{r_e}\right)}\right]. \tag{9.30b}$$

With these boundary conditions:

$$\theta = \frac{r_e}{r_o}, \ \frac{\partial \overline{C}_A}{\partial \theta} = 0. \tag{9.31}$$

$$\theta = 1, \ \frac{\partial \overline{C}_A}{\partial \theta} = \mathrm{Sh}_m.(\overline{C}_A - \overline{C}_{A,\mathrm{rec}}). \tag{9.32}$$

$$z = 0, \ \overline{C}_A = 1. \tag{9.33}$$

9.3.2 Mass transport inside the fiber

Considering the isothermal and steady-state condition of the system without axial diffusion and completely developing a parabolic profile for velocity in the HF as well as assuming constant physical properties of the phase and constant pressures of the tube side, the mass transport in HFs can be determined as follows (Bird, 2002).

$$u_{z,l}\frac{\partial C_{A,\mathrm{rec}}}{\partial z} = D_{A,\mathrm{rec}}\left[\frac{1}{r}\frac{\partial}{\partial r}\left(r\frac{\partial C_{A,\mathrm{rec}}}{\partial r}\right)\right]. \tag{9.34}$$

Considering that velocity is completely achieved inside the laminar flow, the axial direction's velocity can be formulated as

$$u_{z,l} = u_{\mathrm{maximum},l}\left[1-\left(\frac{r}{r_i}\right)^2\right] = 2u_{m,l}\left[1-\left(\frac{r}{r_i}\right)^2\right]. \tag{9.35}$$

$$2u_{m,l}\left[1-\left(\frac{r}{r_i}\right)^2\right]\frac{\partial C_{A,\mathrm{rece}}}{\partial z} = D_{\mathrm{SO}_{2,l}}\left[\frac{1}{r}\frac{\partial}{\partial r}\left(r\frac{\partial C_{A,\mathrm{rece}}}{\partial r}\right)\right]. \tag{9.36}$$

With these boundary conditions:

$$r = 0, \ \frac{\partial C_{A,\mathrm{rece}}}{\partial r} = 0. \tag{9.37}$$

$$r = r_i, D_{\mathrm{SO}_{2,l}}\frac{\partial C_{A,\mathrm{rece}}}{\partial r} = D_{\mathrm{SO}_{2,g}}\frac{\partial C_{A,\mathrm{rece}}}{\partial r}. \tag{9.38}$$

$$Z = L, \ C_{A,\mathrm{rece}} = 0. \tag{9.39}$$

The bulk average values of the concentration of particle A in the receiving phase at z = 0 is:

$$C_{A,z=0} = \frac{\int_0^{r_i} C_{A,\mathrm{rece}}u_{z,l}2\pi r dr}{\int_0^{r_i} 2\pi u_{z,l} r dr} = \frac{4}{r_i^2}\int_0^{r_i} C_{A,\mathrm{rece}}\left[1-\left(\frac{r}{r_i}\right)^2\right] r dr. \tag{9.40}$$

The dimensionless form of the equation would be:

$$\frac{\mathrm{Gz}_{\mathrm{int}}}{2}\left[1-\overline{r}^2\right]\frac{\partial \overline{C}_{A,\mathrm{rece}}}{\partial \overline{z}} = 1\overline{r}\frac{\partial}{\partial \overline{r}}\left(\overline{r}\frac{\partial \overline{C}_{A,\mathrm{rece}}}{\partial \overline{r}}\right). \tag{9.41}$$

With these boundary conditions:

$$\bar{r}=0,\quad \frac{\partial \overline{C}_{A,\mathrm{rece}}}{\partial \bar{r}}=0. \tag{9.42}$$

$$\bar{r}=1,\quad \frac{\partial \overline{C}_{A,\mathrm{rece}}}{\partial \bar{r}}=\frac{\partial \overline{C}_{A}}{\partial \bar{r}}\cdot\frac{D_{A,F}}{D_{A,\mathrm{rece}}}=H. \tag{9.43}$$

$$\overline{Z}=1,\ \overline{C}_{A,\mathrm{rece}}=0. \tag{9.44}$$

By using the following dimensionless parameters

$$\bar{r}=\frac{r}{r_i},\ \bar{z}=\frac{z}{L},\ \overline{C}_A=\frac{C_A}{C_{A,\mathrm{sat}}}. \tag{9.45}$$

$$\mathrm{Gz}_{\mathrm{int}}=\frac{u_{m,l}d_i^2}{D_{A,\mathrm{feed}}L},\ d_i=2r_i. \tag{9.46}$$

The dimensionless equation for bulk average could also be defined as follows:

$$\overline{C}_{A,\mathrm{rece},\bar{z}=0}=\int_0^1 4\overline{C}_{A,\mathrm{rec}}\left[1-\bar{r}^2\right]\bar{r}d\bar{r}. \tag{9.47}$$

9.3.3 Mass transport in membrane

Considering that the diffusion mechanism is the dominant mechanism for mass transfer, the mass transfer in the membrane can be defined as follows (F & Al-Marzouqi, 2009):

$$D_{i,\mathrm{membrane}}\left[\frac{\partial^2 C_{i,\mathrm{membrane}}}{\partial r^2}+\frac{1}{r}\frac{\partial C_{i,\mathrm{membrane}}}{\partial r}+\frac{\partial^2 C_{i,\mathrm{membrane}}}{\partial z^2}\right]=0. \tag{9.48}$$

$$D_{i,\mathrm{membrane}}=\frac{D_i.\varepsilon}{\tau}. \tag{9.49}$$

With these following boundary conditions:

$$\mathrm{At}{:}r=R_1,\ C_{i,\mathrm{membrane}}=\frac{C_{i,\mathrm{lumen}}}{m_i}. \tag{9.50}$$

$$\mathrm{At}{:}r=R_2,\ C_{i,\mathrm{membrane}}=C_{i,\mathrm{shell}}. \tag{9.51}$$

9.4 Membrane wetting and mass transfer resistance

In a typical gas–liquid MC process, the molecules of gas transfer through the membrane pores and are captured by the liquid on the other side of the membrane. The mass transport occurs at the interface of gas and liquid by the mechanism of diffusion. The separated phases provide a controlled independent flow rate in the system without encountering difficulties as those in traditional contacting devices. Furthermore, the liquid

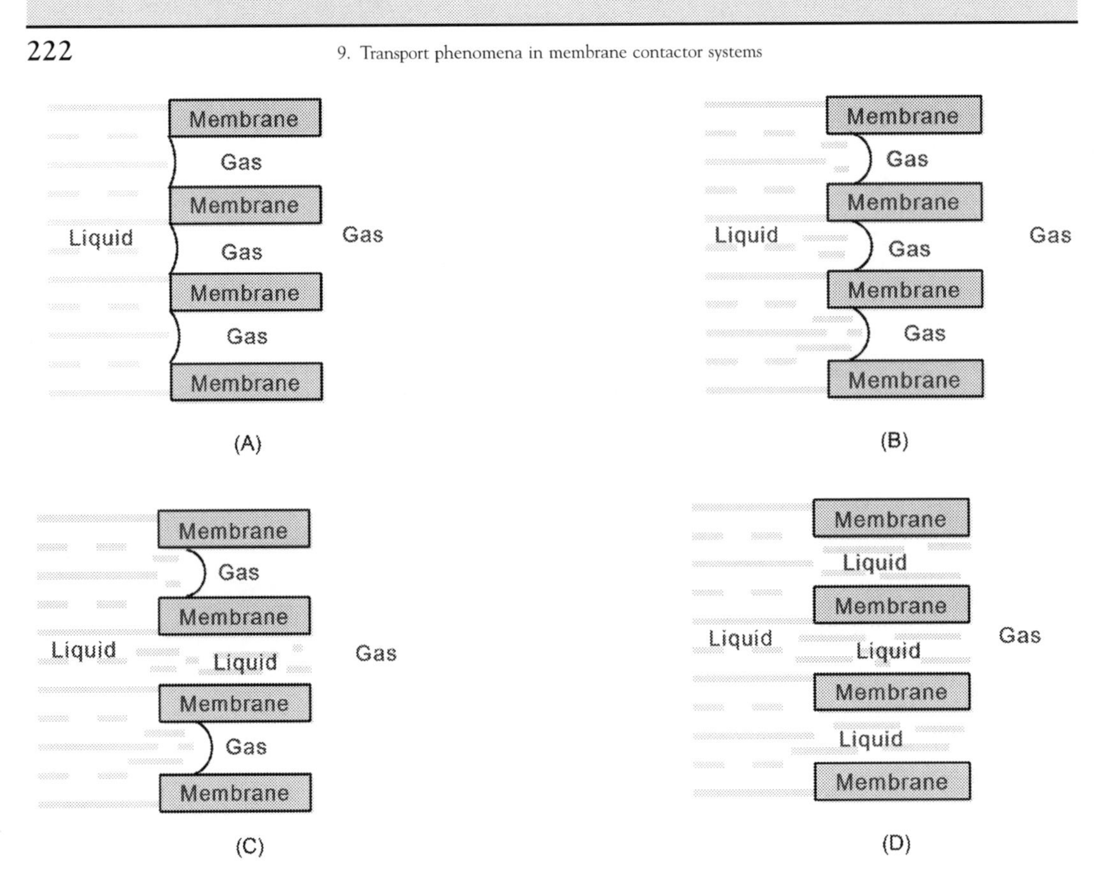

FIGURE 9.4 A Schematic diagram of wettability phenomenon in gas–liquid membrane contactor system. Steps: (A) nonwetted, (B) surface-wetted, (C) partially wetted, and (D) fully wetted membrane.

phase's operating pressure is controlled to be similar to or greater than that of the gas phase to prevent the dispersion of gas bubbles in the liquid phase and avoid the phase-mixing phenomenon. Similarly, in liquid–liquid MC systems, two immiscible liquid phases are separated by the hydrophobic porous membrane filled with the gas phase. Since the mass transfer occurs just through the vapor phase in the hydrophobic membrane's pores, pores need to be hindered partially or entirely from the liquid phase and nonvolatile particles on the feed side. In nonwetting step [(a) in Figs. 9.4 and 9.5], the pores are entirely gas-filled during the process and cause the minimum resistance in diffusion throughout the pores of the membrane and allow the maximum amount for the mass transfer coefficient in comparison with the overall-wetting step (Zhang, Wang, Liang, & Tay, 2008; Mansourizadeh & Ismail, 2009; Himma & Wenten, 2017; Biniaz et al., 2019).

9.5 Novel approaches to membrane contactor systems

Being practical and efficient, MC systems have been studied comprehensively by numerous researchers over the past few years. Table 9.5 presents the latest information

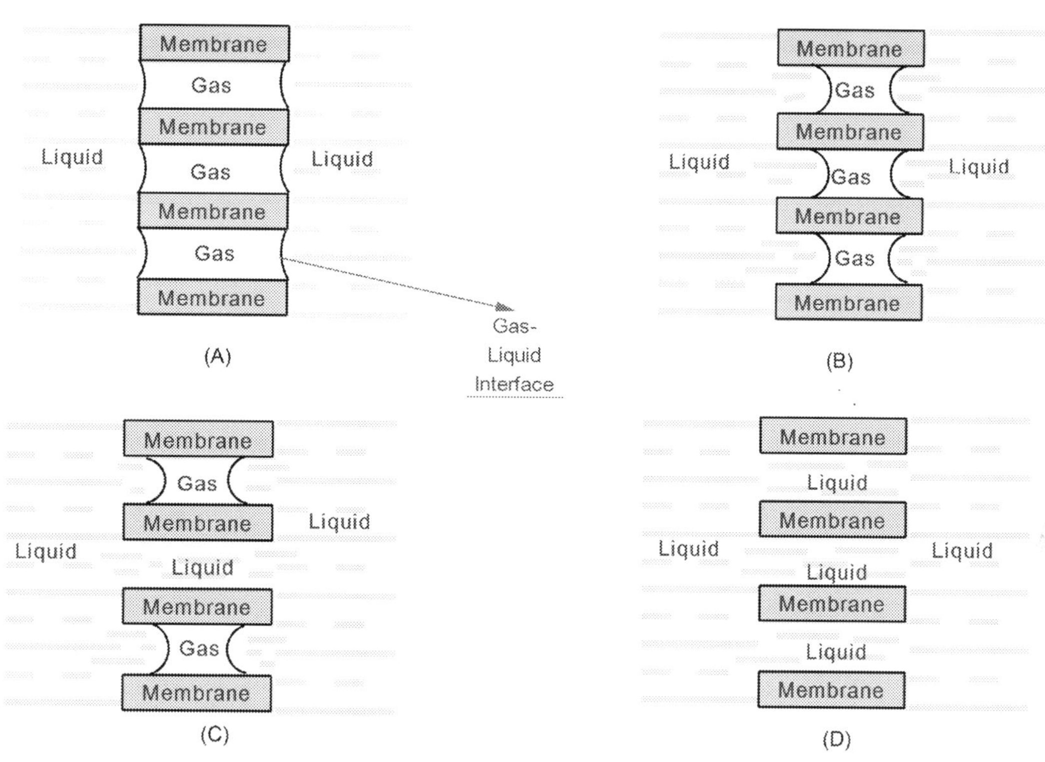

FIGURE 9.5 A schematic diagram of wettability phenomenon in liquid–liquid membrane contactor system. Steps: (A) nonwetted, (B) surface-wetted, (C) partially wetted, and (D) fully wetted membrane.

about the novel strategies conducted to enhance mass transfer, decrease the resistance, and consequently increase the system's efficiency since 2017.

9.6 Conclusions and future trends

The profound and practical significance of applying MC systems in different industrial sectors has rapidly increased because of their enormous benefits. This chapter presented a short review of the MCs and the mass transfer in the system. The mechanism of the mass transfer in the membrane pores is mainly diffusion in most conditions. The basic transfer equations through the liquid–liquid process and gas–liquid processes as two common types of MC processes were investigated by considering film and resistance-in-series theories. These equations should be modified for mass transfer through pores considering different resistance during the process. In an ideal process, the nonwetting condition greatly desires an MC to hinder the liquid phase's diffusion into the membrane pores. In these conditions, the pores are completely filled with the gas phase, and the resistance in the mass transfer through the membrane pores is negligible. Nonetheless, in the actual process, the liquid phase always diffuses partly or even entirely into membrane pores, known

TABLE 9.5 Novel approaches to promote mass transfer in membrane contactor systems since 2017.

Application	Highlights	Reference
Wastewater treatment	• A novel electrocatalytic MC was invented by coupling membrane contact ozonation with the electro-peroxone process for wastewater treatment. • A mechanism for mass transfer of O_3 and O_2 and the reaction resistance of the process were investigated to improve the performance of the process • The rate of removing nitrobenzene was considerably increased in the electrocatalytic membrane contact ozonation (ECMCO) process. • Compared with the conventional processes, the rate of mass transfer for O_2 was significantly improved in ECMCO. • Membrane thickness and porosity have a critical effect on the efficiency of ECMCO	Li et al. (2020)
Gas absorption	• A 3D computational fluid dynamics, that is, CFD code was used to calculate the shell-side fluid's resistance and compute the solute concentration, which transfers over a range of Reynolds numbers of 1 to 400 without adjustable parameters. • Mass transfer phenomenon was analyzed rigorously in the proposed HFMC • A formulation for the porous medium was utilized to upscale the results of the fiber scale to the contactor scale.	Pozzobon and Perré (2020)
Osmotic membrane distillation (OMD) and used in the apple juice industry	• Theoretical and experimental research was conducted using hydrophobic HFMC. • A model was designed for mass transport in OMD considering the resistance-in-series theory for the mass transfer of water vapor molecules through the hydrophobic porous HFMC and utilized in apple juice concentration. • The model of water vapor transfer was simulated in MATLAB and verified successfully with the experimental data. • The impacts of driving forces on mass transport and the membrane's porosity were investigated t the developed model. • The effects of operating parameters (e.g., the concentration of juice, concentration of stripping solution, and Reynolds number) on the flux of water vapor and overall mass transfer coefficient were studied. • The results indicated that increasing the diameter of membrane pores from 0.1 to 0.6 μm leads to a significant increase in mass transfer coefficient from 1.57×10^{-10} to 7.23×10^{-10} ms^{-1}. • Enhancing the activity coefficient from 0.1 to 0.5 significantly increases water vapor flux from 0.02 to 0.50 kg m^{-2} h^{-1}	Ahmad et al. (2020)

(Continued)

TABLE 9.5 (Continued)

Application	Highlights	Reference
The absorption of SO_2 into NaOH	• A multichannel ceramic membrane with a larger specific area and a higher throughput were investigated. • The proposed ceramic membrane is highly profitable in the industrialization of the SO_2 absorption process applying MC systems. • Different operating conditions for liquid and gas phases were examined to analyze the effectiveness of the mass transfer of SO_2 absorption in NaOH solution in an MC system having 19 hydrophilic membrane channels. • A 3D model was designed for mass transfer to analyze the particles' concentration distribution and optimize channel structure. For this purpose, some internal channels were blocked to decrease the resistance to mass transfer.	Kong, Gong et al. (2020)
Membrane distillation (MD), Membrane crystallization (MCr), Biopharmaceutical industry	• Two fluorinated polymers, namely Polyvinylidene fluoride-co-hexafluoropropylene blending with PVDF membrane, were fabricated using a two-stage phase separation method by interconnected channels decorating by polymer crystallites and utilizing nontoxic solvents. • Results confirmed that interface features of the membrane and high mass transfer rate could beneficially affect the system through the synergic mixture of immiscible compound materials. • The optimum blending amount of two materials was evaluated by chemical analysis, physical analysis, and mass-transfer property analysis of the proposed membrane. • The blending process's favorable outcome was proved for each of the features studied (toughness, wettability, crystallinity, flux, surface charge, surface roughness, and strength), showing a beneficial physical interaction among the particles without being affected by their thermodynamically stated immiscibility.	Meringolo, Poerio et al. (2019)
Gas separation and Wastewater treatment	• Novel strategies such as surface modifications of a hydrophobic polymer and ceramic membranes were investigated to enhance membranes' antiwetting characteristics in the MC system. • Results indicated that fabricating wetting-resistant hydrophobic membrane with novel and highly efficient components, such as HFs, dramatically increases MC systems' mass transfer rate and performance.	Goh, Naim, Rahbari-Sisakht, and Ismail (2019)

(Continued)

TABLE 9.5 (Continued)

Application	Highlights	Reference
CO_2 absorption	• Water-based nanofluids of carbon nanotubes (CNT) and SiO_2 were applied to enhance CO_2 absorption and the HF MC system's mass transfer rate. • A 2D mathematical model was designed. Molecular diffusion was considered in axial and radial flow directions in nonwetting conditions. • Model predictions were successfully confirmed and verified with experimental results. • CNT showed a much better absorption rate (up to 34%) than the SiO_2 nanoparticles due to their high adsorption capacity and hydrophobicity.	Rezakazemi, Darabi, Soroush, and Mesbah (2019)
Recovery of methane from anaerobic effluents	• A polypropylene MC was applied, and the effects of operating parameters and mode and fouling phenomenon on the long-term operation of the system were investigated utilizing N_2 or vacuum pressure as the sweep gas. • The results have been approved polypropylene MC system to be an effective and promising technology.	Henares, Ferrero, San-Valero, Martinez-Soria, and Izquierdo (2018)
Biogas upgrading	• Low mass-transport-resistance fluorinated TiO_2 and SiO_2 nanoparticles were fabricated in combination with PVDF components ($fTiO_2-SiO_2$)/ (PVDF) composite applied in HF membrane for biogas upgrading application in gas-liquid MC systems. • $fTiO_2-SiO_2$/PVDF complex HF membrane showed an extremely appropriate pore size of approximately 25 nm and a favorable contact angle of the water (approximately 124 degrees), which efficiently restricted wetting phenomena in the membrane. • The high CO_2 capture rates of $8.0-5.6 \times 10^{-3}$ mol $/m^2\ s^1$ were obtained because of the low mass transport resistance of the system.	Xu, Lin, Lee, Malde, and Wang (2018)
Absorption of CO_2 from CO_2/ N_2 mixture	• Silica nanoparticles with the high surface area were blended with HF membrane to improve its properties. • The vinyl membrane modified with silica nanoparticles significantly increased the hydrophobicity of membrane	Ghaee, Ghadimi et al. (2017)

(Continued)

TABLE 9.5 (Continued)

Application	Highlights	Reference
Biogas recovery from anaerobic membrane bioreactor	• An MC using an in-house manufactured HF membrane was applied as mass transfer devices for the recovery of CH_4. • A mathematical model was developed based on a resistance-in-series model to determine the overall coefficients of the mass transfer taking the simultaneous desorption of CO_2 and CH_4 into consideration. • The results indicated that increasing the velocity of liquid results in a significant reduction in the resistance of mass transport in the liquid as a consequent the rates of both CH_4 and CO_2 fluxes enhance remarkably. While raising gas velocity only leads to a considerable increase in the flux of CO_2. • The desorbed CO_2 enhanced the driving force of the mass transfer by decreasing CH_4 partial pressure in the gas phase and increasing the mass transfer coefficient in the gas phase to enhance CH_4 desorption.	Rongwong, Wongchitphimon, Goh, Wang, and Bae (2017)

as wetting phenomena. Although membrane technology satisfies the intensification of the process, particularly in water desalination and the petrochemical industry, efficient utilization of these types of technology at industrial applications still require further effort to address emerging and integrating innovative substances, economic analysis, and process control, novel design concepts, scaleup and practical evaluation of the fundamental operating parameters on actual pilot plants.

References

Ahmad, S., Marson, G. V., Zeb, W., Rehman, W. U., Younas, M., Farrukh, S., & Rezakazemi, M. (2020). Mass transfer modelling of hollow fiber membrane contactor for apple juice concentration using osmotic membrane distillation. *Separation and Purification Technology, 250*, 117209.

Biniaz, P., Makarem, M. A., & Rahimpour, M. R. (2020). Membrane reactors. In Maurizio Benaglia, & Alessandra Puglisi (Eds.), Catalyst Immobilization: Methods and Applications (pp. 307–324). Weinheim: Wiley-VCH. Available from https://doi.org/10.1002/9783527817290.ch9.

Biniaz, P., Roostaie, T., & Rahimpour, M. R. (2021). 3 - Biofuel purification and upgrading: Using novel integrated membrane technology. In M. R. Rahimpour, R. Kamali, M. Amin Makarem, & M. K. D. Manshadi (Eds.), *Advances in Bioenergy and Microfluidic Applications* (pp. 69–86). Elsevier. Available from https://doi.org/10.1016/B978-0-12-821601-9.00003-0.

Biniaz, P., Torabi Ardekani, N., Makarem, M. A., & Rahimpour, M. R. (2019). Water and wastewater treatment systems by novel integrated membrane distillation (MD). *ChemEngineering, 3*(1), 8.

Bird, R. B. (2002). Transport phenomena. *Applied Mechanics Reviews, 55*(1), R1–R4.

Curcio, E., & Drioli, E. (2005). Membrane distillation and related operations—a review. *Separation and Purification Reviews, 34*(1), 35–86.

Drioli, E., Criscuoli, A., & Curcio, E. (Eds.), (2011). *Membrane Contactors: Fundamentals, Applications and Potentialities*. Amsterdam: Elsevier.

Drioli, E., & Giorno, L. (Eds.), (2010). *Comprehensive Membrane Science and Engineering*. Amsterdam: Elsevier.

Drioli, E., & Giorno, L. (Eds.), (2018). *Encyclopedia of Membranes*. Berlin, Heidelberg: Springer.

Drioli, E., Di Profio, G., & Curcio, E. (2015). *Membrane-Assisted Crystallization Technology*. Singapore: World Scientific.

Faiz, R., & Al-Marzouqi, M. (2009). Mathematical modeling for the simultaneous absorption of CO_2 and H_2S using MEA in hollow fiber membrane contactors. *Journal of Membrane Science, 342*(1), 269–278. Available from https://doi.org/10.1016/j.memsci.2009.06.050.

Gaeta, S. (2009). Membrane contactors in industrial applications. In E. Drioli, & L. Giorno (Eds.), *Membrane Operations: Innovative Separations and Transformations* (pp. 499–512). Wiley-VCH Verlag GmbH & Co. KgaA. Available from https://doi.org/10.1002/9783527626779.ch22.

Ghaee, A., Ghadimi, A., Sadatnia, B., Ismail, A. F., Mansourpour, Z., & Khosravi, M. (2017). Synthesis and characterization of poly (vinylidene fluoride) membrane containing hydrophobic silica nanoparticles for CO_2 absorption from CO_2/N_2 using membrane contactor. *Chemical Engineering Research and Design, 120*, 47–57.

Goh, P., Naim, R., Rahbari-Sisakht, M., & Ismail, A. (2019). Modification of membrane hydrophobicity in membrane contactors for environmental remediation. *Separation and Purification Technology, 227*, 115721.

Gugliuzza, A., & Basile, A. (2013). Membrane contactors: Fundamentals, membrane materials and key operations. In A. Basile (Ed.), *Handbook of Membrane Reactors* (pp. 54–106). Woodhead Publishing. Available from https://doi.org/10.1533/9780857097347.1.54.

Happel, J. (1959). Viscous flow relative to arrays of cylinders. *AIChE Journal, 5*(2), 174–177.

Henares, M., Ferrero, P., San-Valero, P., Martinez-Soria, V., & Izquierdo, M. (2018). Performance of a polypropylene membrane contactor for the recovery of dissolved methane from anaerobic effluents: Mass transfer evaluation, long-term operation and cleaning strategies. *Journal of Membrane Science, 563*, 926–937.

Himma, N. F., & Gede Wenten, I. G. (2017). Superhydrophobic membrane contactor for acid gas removal. *Journal of Physics Conference Series, 877*(1), 012010. Available from https://doi.org/10.1088/1742-6596/877/1/012010.

Joscelyne, S. M., & Trägårdh, G. (2000). Membrane emulsification—a literature review. *Journal of Membrane Science, 169*(1), 107–117.

Kong, X., Gong, D., Ke, W., Qiu, M., Fu, K., Xu, P., ... Fan, Y. (2020). Investigation of the mass transfer characteristics of SO_2 absorption into NaOH in a multichannel ceramic membrane contactor. *Industrial & Engineering Chemistry Research, 59*(23), 11054–11062. Available from https://doi.org/10.1021/acs.iecr.0c01327.

Li, K., Zhang, Y., Xu, L., Liu, L., Wang, Z., Hou, D., ... Wang, J. (2020). Mass transfer and interfacial reaction mechanisms in a novel electrocatalytic membrane contactor for wastewater treatment by O_3. *Applied Catalysis B: Environmental, 264*, 118512.

Li, N. N., Fane, A. G., Ho, W. W., & Matsuura, T. (Eds.), (2011). *Advanced Membrane Technology and Applications*. John Wiley & Sons. Available from http://doi.org/10.1002/9780470276280.

Luis, P. (2018). Chapter 5 - Membrane contactors. In P. Luis (Ed.), *Fundamental Modelling of Membrane Systems* (pp. 153–208). Elsevier. Available from https://doi.org/10.1016/B978-0-12-813483-2.00005-8.

Luis, P., Garea, A., & Irabien, A. (2010). Modelling of a hollow fibre ceramic contactor for SO_2 absorption. *Separation and Purification Technology, 72*(2), 174–179.

Mansourizadeh, A., & Ismail, A. (2009). Hollow fiber gas–liquid membrane contactors for acid gas capture: A review. *Journal of Hazardous Materials, 171*(1–3), 38–53.

Meringolo, C., Poerio, T., Fontananova, E., Mastropietro, T. F., Nicoletta, F. P., De Filpo, G., ... Profio, G. D. (2019). Exploiting fluoropolymers immiscibility to tune surface properties and mass transfer in blend membranes for membrane contactor applications. *ACS Applied Polymer Materials, 1*(3), 326–334.

Nagy, E. (2019a). Chapter 11 - Membrane contactors. In E. Nagy (Ed.), *Basic Equations of Mass Transport Through a Membrane Layer* (2nd ed., pp. 337–345). Elsevier.

Nagy, E. (2019b). Chapter 18 - Membrane gas separation. In E. Nagy (Ed.), *Basic Equations of Mass Transport Through a Membrane Layer* (2nd ed., pp. 457–481). Elsevier.

Pozzobon, V., & Perré, P. (2020). Mass transfer in hollow fiber membrane contactor: Computational fluid dynamics determination of the shell side resistance. *Separation and Purification Technology, 241*, 116674.

Rezakazemi, M., Darabi, M., Soroush, E., & Mesbah, M. (2019). CO_2 absorption enhancement by water-based nanofluids of CNT and SiO_2 using hollow-fiber membrane contactor. *Separation and Purification Technology, 210*, 920–926.

Rongwong, W., Wongchitphimon, S., Goh, K., Wang, R., & Bae, T.-H. (2017). Transport properties of CO_2 and CH_4 in hollow fiber membrane contactor for the recovery of biogas from anaerobic membrane bioreactor effluent. *Journal of Membrane Science, 541*, 62–72.

Seader, J. D., Henley, E. J., & Keith Roper, D. (2011). *Separation Process Principles. Chemical and Biochemical Operations*. New Jersey: JohnWiley & Sons.

Seader, J. D., Henley, E. J., & Roper, D. K. (1998). *Separation Process Principles*. New York: Wiley,.

Seidel-Morgenstern, A. (2010). *Membrane Reactors: Distributing Reactants to Improve Selectivity and Yield*. John Wiley & Sons..

Xu, Y., Lin, Y., Lee, M., Malde, C., & Wang, R. (2018). Development of low mass-transfer-resistance fluorinated TiO_2-SiO_2/PVDF composite hollow fiber membrane used for biogas upgrading in gas-liquid membrane contactor. *Journal of Membrane Science, 552*, 253–264.

Zhang, H.-Y., Wang, R., Liang, D. T., & Tay, J. H. (2008). Theoretical and experimental studies of membrane wetting in the membrane gas–liquid contacting process for CO_2 absorption. *Journal of Membrane Science, 30*(1–2), 162–170.

CHAPTER

10

Transport phenomena in drug delivery membrane systems

Sara A.M. El-Sayed and Mostafa Mabrouk

Refractories, Ceramics and Building Materials Department, National Research Centre, Dokki-Giza, Egypt

Abbreviations

Å Angstrom
DDS drug delivery system
ED electrodialysis
GS gas separation
LM liquid membranes
MD membrane distillation
nm nanometer
PV pervaporation
μm micrometer

10.1 Introduction

Since the applications of the membranes are rapidly evolving, the current book chapter is devoted to clear issues related to the transport phenomena through drug delivery membrane systems.

10.1.1 Definition of a membrane

A membrane is the barrier that exists between two phases and is characterized by the permselectivity; phase 1 includes the feed and phase 2 includes the permeate (Fig. 10.1). This definition is based on a macroscopic point of view according to the separation process. However, this definition neither describes the structure of the membrane nor its function. The structure of the membrane could be either homogeneous or heterogeneous and its thickness could be also classified as either thin or thick. In a different context, it can

Current Trends and Future Developments on (Bio-) Membranes
DOI: https://doi.org/10.1016/B978-0-12-822257-7.00006-6

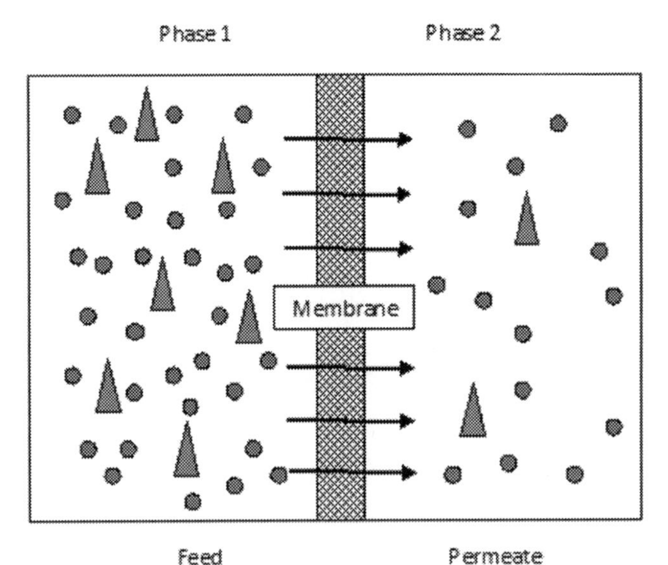

FIGURE 10.1 Schematic representation demonstrates the membrane selectivity. Source: *Adapted from Daramola, M.O., Aransiola, E.F., & Ojumu, T.V. (2012). Potential applications of zeolite membranes in reaction coupling separation processes.* Materials, *5(11), 2101–2136 (Daramola, Aransiola, & Ojumu, 2012).*

also be classified according to transportation process that takes place through membrane as either active or passive. Factors including concentration difference, temperature, or pressure difference could be considered to be the driving force that affect the transportation process (Mulder, 1991).

The concept of separation depends on the membrane's ability to allow one component to penetrate from the phase 1 to phase 2 in a more facile manner than the other components. However, it is important to note that membranes are not ideal barriers where not only the desired component is flowed through them, but also a little amount of other component or components may be flowed. Therefore the membrane efficiency is controlled by two parameters: the selectivity and the flow rate through the membrane. Flow rate is defined as the volume flowing through the membrane per unit area and time (Mulder, 1991).

The permeability of the membrane toward a certain component (selectivity) to flow is based on many parameters such as pore shape, size, and pore-size distribution of the membrane. Furthermore, the permeate volume (flux) is controlled by the membrane thickness (Radjabian & Abetz, 2020). With regard to the size of the retained solutes, the membrane process can be classified into microfiltration, ultrafiltration, nanofiltration, reverse osmosis, described in further detail later (Anis, Hashaikeh, & Hilal, 2019; Radjabian & Abetz, 2020). The concept of component transportation or rejection depends mainly on the membrane structure and the component features (Park, Kamcev, Robeson, Elimelech, & Freeman, 2017; Wu et al., 2012). For membrane, the membrane material properties such as hydrophilic/hydrophobic nature to the fluid, wettability, membrane thickness, interactions between membrane surface charged or uncharged components and the operating conditions (e.g., transmembrane pressure) govern the transportation process. On the other side, component physicochemical properties including size, charge, solubility and diffusivity, geometry, and surface functionality have great influence on the concepts of membrane transportation. The previously mentioned mechanism was applicable for the porous membrane; however, for the nonporous membranes (dense membranes), the concept of transportation is based on diffusion. The concept of diffusion mechanism is based on concentration gradient of permeates (Radjabian & Abetz, 2020).

10.1.2 Historical background

After World War I, Sartorius manufactured the first commercial membranes for practical applications (Zsigmondy & Bachmann, 1918). In the 1940s Kolff made the first membrane that was used for hemodialysis (Kolff et al., 1944). Loeb and Sourirajan (1962) developed an asymmetric membrane for industrial applications (Loeb & Sourirajan, 1962). These asymmetric membranes were composed of two layers; the top layer was very thin, with a thickness less than 0.5 μm, and it could determine the transport rate. This layer was supported with sublayer made of porous membrane with a thickness between 50 and 200 μm. At that time, it was worthy to highlight that the thickness of the actual barrier layer is inversely proportional to the permeation rate, so the permeation rate of the asymmetric membranes was higher than the symmetric (homogeneous) membranes of a comparable thickness (Mulder, 1991).

Since that time, two major types of membranes were generated for different applications that were also based on various transportation mechanisms. These generations of the membrane are summarized in the following scheme (Fig. 10.2). Membranes to can be used in the industrial technology, and they must be modified and developed through extensive investigation. On the basis of this scheme, the membranes classifications are discussed in detail in the following sections.

10.2 General classification of membranes

In accordance with the context, membranes could be classified as described above. In particular, they could be classified according to their nature or structure.

10.2.1 Membrane classification according to their nature

Membranes can be divided into biological or synthetic membranes, where their structure and functionality are completely different. Synthetic membranes can be classified according to their constituent material that includes both inorganic and organic or their composites (polymeric or liquid) (Fig. 10.3). Practically, a limited number of polymers can be used as a membrane material. The material of the membrane could be organic (all kinds of polymers) or inorganic (such as a ceramic, glass or a metal) (Abd El-Ghaffar & Tieama, 2017).

Biological membranes are the boundaries that exist in the human body and other living organisms. They exist in abundance and serve various functions for cell membranes, plasma, epithelial membranes, membranes of intracellular organelles, etc. Synthetic membranes can be manufactured from several materials such as metals (Dumee, He, & Lin, 2013), glass (Barascu, Kullmann, & Reinhardt, 2015), ceramics (Amin, Hassan, El-Sherbiny, & Abdallah, 2016), natural and synthetic polymers, or a composite of these materials with natural products such as collagen, albumin, and others (Piskin & Hoffman, 1986).

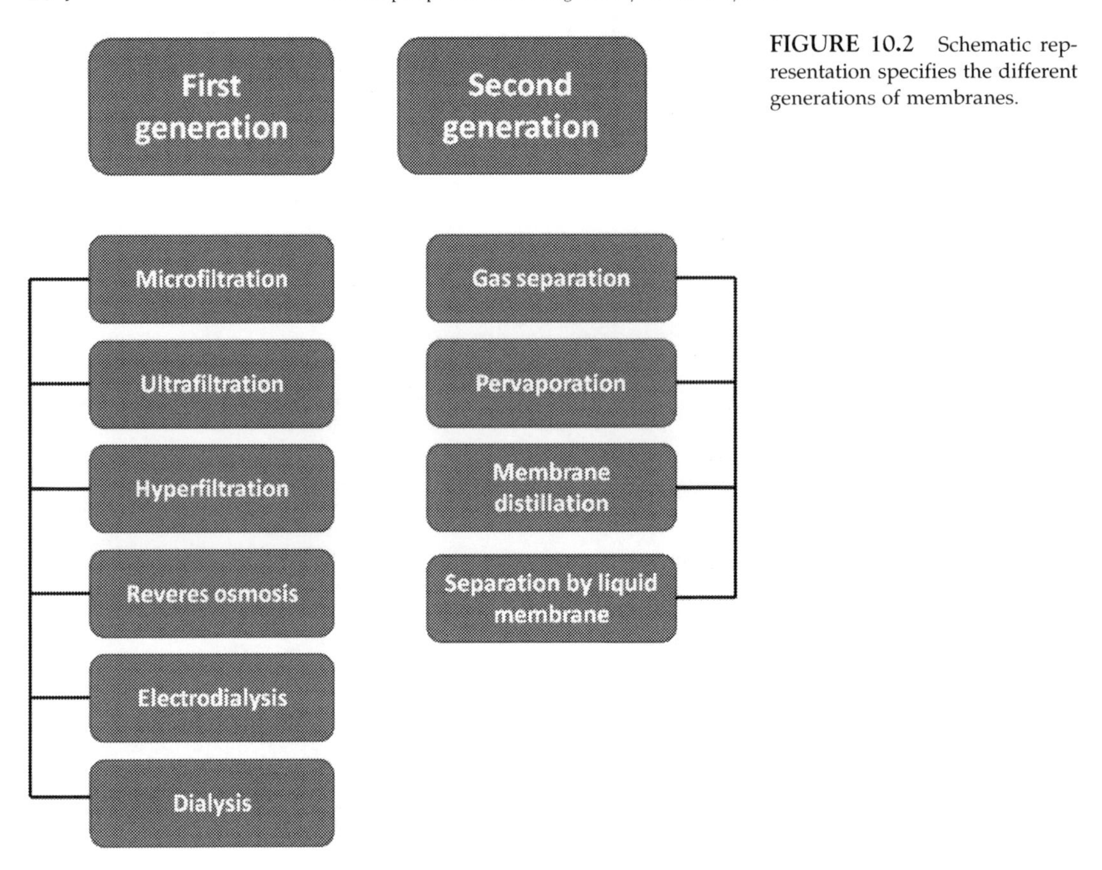

FIGURE 10.2 Schematic representation specifies the different generations of membranes.

10.2.2 Membrane classification according to their structure

Apart from their nature, membranes could be also classified according to their structure and morphology. Symmetric and asymmetric membranes are the main classes of membrane structures. It is worthy to highlight that depending on the membrane structure, the mechanism of the separation could be determined (Elson, Fried, & Dolbow, 2010).

Symmetric membranes have a thickness between 10 and 200 μm. The permeation rate decreases when the membrane thickness increases. Asymmetric membranes have two different layers with different thickness. The thickness of the top layer (very dense) ranges between 0.1 and 0.5 μm, coupled with a porous sublayer that is 50–150 μm thick. These membranes are characterized by their high selectivity and permeation rate (Wen, Xiao, & Sainath, 2016).

10.3 Transport phenomena in membranes

Transport phenomena in membranes arise from the membrane's ability to transport one component in a more facile manner than the other components because of the differences

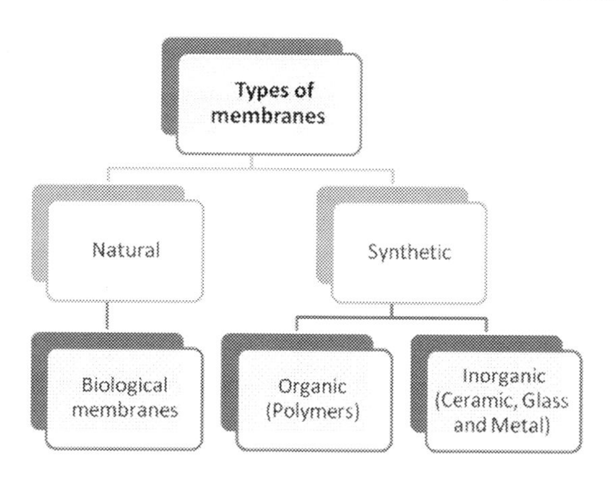

FIGURE 10.3 General classification describes membranes according to their nature. Source: *Adapted from Daramola, M.O., Aransiola, E.F., & Ojumu, T.V. (2012). Potential applications of zeolite membranes in reaction coupling separation processes.* Materials, *5(11), 2101–2136.*

in physical and/or chemical properties between the membrane and the permeating components. Driving force is the force that affects the components in the feed phase to transport through the membrane (Dudek, Borys, & Strzelewicz, 2020).

The application could be determined on the basis of the structure and the material of the membrane. The transportation process can include the particles from the microscopic scale till the molecules of an identical size or shape are achieved. The open membrane structure can be used when the size of the retained particles is more than 100 nm. This is because the hydrodynamic resistance of this membrane is low and it possesses small driving forces (low hydrostatic pressures), and these parameters are enough to obtain high flux (the volume flowing through the membrane per unit area and time). In order to adjust the quantity of diffused particles, molecules, or drugs through membranes, a high flux is required (Jeon, Yang, & Kim, 2012). There is a close relationship between the membrane thickness and its flux. A relatively low flux is expected when the membranes are thick, while a higher flux is expected when the membranes are thin because of their weak mechanical strength (Dumee et al., 2013). The dense membrane can be used when the macromolecules (the molecular weight in the range of 10^4–10^6) are the target to be separated because the hydrodynamic resistance increases as well as the pressure. Some membranes were prepared earlier from dense metal that showed huge implementation in the reactive separation as one of the important applications of membranes (Armor, 1989).

10.3.1 Transport mechanisms in synthetic membranes

The membrane separates the influent into two parts: the permeate and the concentrate. Permeate is the part that can be passed through the membrane, and the concentrate is the part that is rejected by the membrane. The transport mechanism is mainly determined according to the materials from which membranes are made. In addition, there are many factors that affect the transport rate, such as the concentration, the driving force, and the mobility of permeate molecules in the polymer matrix. The mobility depends on several factors, such as the size and shape of solutes, the temperature, chemical compositions, the chemical nature, microstructure and morphology of the membrane.

Driving force is necessary to drive the solutes and solvents through a membrane. This force can be represented in concentration, electrochemical gradient, electrical field and (partial, hydrostatic, or osmotic) pressure (Krishna, 1990).

The transportation process through the membranes can be classified into three kinds (Piskin & Hoffman, 1986). The first one is passive transportation in which transport through the membrane occurs by diffusion due to the chemical potential difference between the components on the opposite sides of the membrane. This difference in chemical potential is because of the differences in concentration, hydrostatic pressure, temperature, etc. The diffusion takes place in the direction of the gradient (from higher to lower). The second type is the facilitated transportation or carrier-mediated transportation in which transport is facilitated by using a carrier that is confined to the membrane. This carrier can interact with solutes on both sides of the membrane and can enhance the membrane selectivity. Finally, the third type is active transportation in which transport can occur against the gradient. This kind of transport is highly selective and requires energy such as adenosine triphosphate, which exists in the biological membrane. The last two types of transportations exist only in biomembranes.

10.4 Mechanism of particle transportation through membranes

There are many classifications of membranes depending on the type of the membrane. Membranes may be symmetric or asymmetric, homogenous or heterogeneous, solid or liquid, porous or nonporous. However, when it comes to the size of the particles that permeate the membrane, then the pore size of the membrane is the key parameter that can determine the effectiveness and efficiency of the membrane.

10.4.1 According to particle size

Porous membranes are classified according to the size of the component in the feed solution that is allowed to pass into four types from the largest to the smallest permeable particles (Fig. 10.4):

1. Microfiltration
2. Ultrafiltration
3. Nanofiltration
4. Reverse osmosis

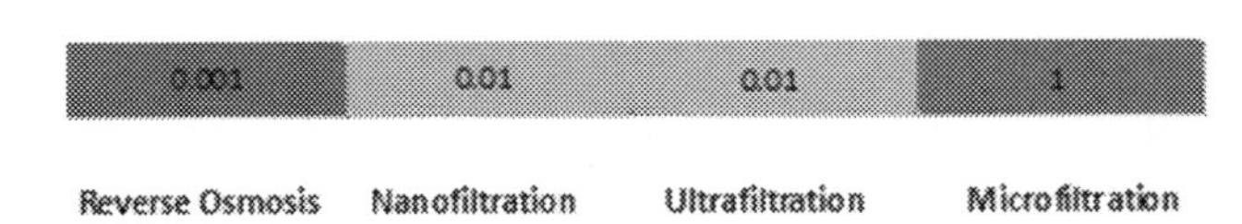

FIGURE 10.4 Classification of the porous membranes based on the permeable particles. Source: *Adapted from Porter, C.M. (1989). Handbook of Industrial Membrane Technology.*

10.4.1.1 Microfiltration

Microfiltration is the process of removing particles from fluids with particle size in the range of 0.025–10.0 μm by using a microporous membrane. The pore size cutoffs used for these types of separation are ranging between 0.05 and 1.0 μm (Porter, 1989). In biomedical applications, they have a common use as a part of drug delivering membranes; thus they have a crucial role in the delivery process as they dissolve when in contact with tear fluid (Orosz et al., 2004).

10.4.1.2 Ultrafiltration

Ultrafiltration is the process of separating the dissolved molecules and the tiny particles. This process is useful for the separation of the molecules of different sizes at least an order of magnitude. However, this process cannot be an effective method for separation of similar size molecules. To purify a material or collect a material, ultrafiltration membranes can be used. Ultrafiltration membranes are useful for the separation of materials with low molecular weight and proteins from buffer components for buffer exchange or desalting (Porter, 1989; Zydney, 1995).

10.4.1.3 Nanofiltration

Nanofiltration is a technology used for liquid separation. Nanofiltration can remove molecules in the range of 0.001 μm and reject molecules in the range of 1 nm (Porter, 1989; Zydney, 1995). Usually, this type of a process takes place through nanoporous membranes owing to the fact that they possess porous structure in the range of 1–100 nm (Langley & Hulliger, 1999).

10.4.1.4 Reverse osmosis

Reverse osmosis is a technique for separating the dissolved solids from solvents. The solvent can be driven through the reverse osmosis membrane toward lower concentration. The size of molecules that can be separated by this technique lies in the range of 1–10 Å. Reverse osmosis is used in several applications such as food, paper and tanning industries, chemical, textile, petrochemical, electrochemical, and in the treatment of tap water and wastewaters. This method is mainly used for water purification, for example, in desalination and wastewater treatment (Henley, Li, & Long, 1965; Porter, 1989; Ray, 1993). There is another type of membrane separation process known as dialysis. In this process, solutes are driven by concentration gradient unlike the reverse osmosis, in which the pressure is the driving force.

10.5 Methods of preparation of synthetic membranes

There are several techniques to manufacture a synthetic membrane whether from inorganic or organic material. Examples of these techniques include sintering, stretching, track-etching, template leaching, phase inversion, and coating (XueMei & Denis, 2019).

10.5.1 Sintering

Sintering is a technique that uses organic or inorganic materials for producing porous membranes. In this technique, the pressing powder is exposed to a high temperature depending on the material used (see Fig. 10.5A). During sintering, the interface between the contacting particles disappears. The porosity of the membrane formed depends on the particle size and particle-size distribution of the powder (pore former). The resulting membrane would possess a narrow pore-size distribution if the particle-size distribution of the former pore was also narrow (Guo, Berbano, & Guo, 2016). Generally, the size of the obtained pores could be greatly adjusted by utilizing this technique in wide range from hundreds of nanometers to tens of micrometers; it is thus considered extremely useful for preparing microfiltration membranes only.

10.5.2 Stretching

Stretching is a technique where polymeric material that is partially crystalline is used for making extruded film or foil and stretched perpendicular to the direction of the extrusion. The porosity of these membranes produced by this technique arises from the application of a mechanical stress that makes small ruptures. The obtained pore sizes of this technique lie between 0.1 and 3 μm. The porosity of the membranes created using this technique is much higher than that of the membranes obtained by sintering. The

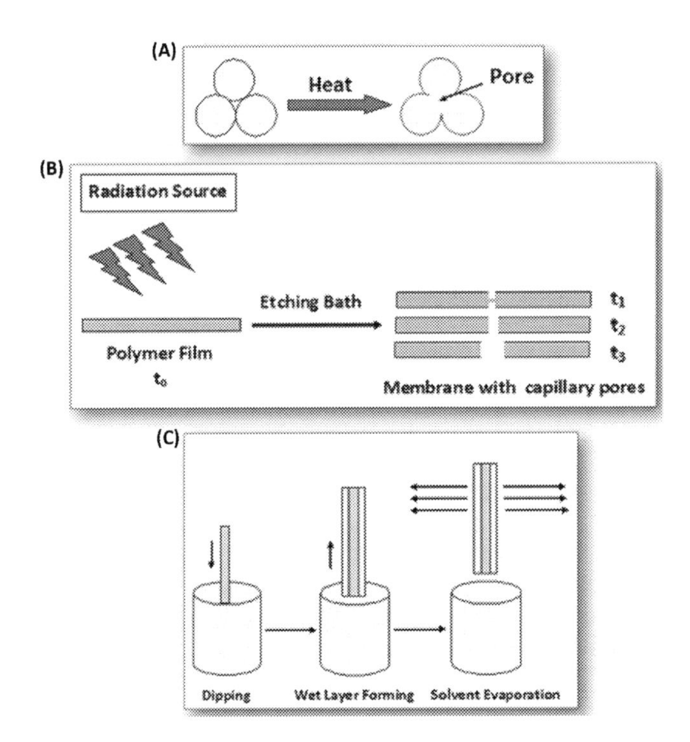

FIGURE 10.5 Demonstration of some preparation technique of membranes (A) sintering, (B) track-etching, and (C) coating.

biomedical field uses this technique usually in medication delivery, especially for soft tissues and topical applications (Kim, Jang, & Kwon, 1994).

10.5.3 Track-etching

Track-etching is a technique where the membrane pores formed are cylindrically parallel and have a uniform dimension (see Fig. 10.5B). High-energy particle radiation is applied to a film to induce apertures through the polymer matrix (often a polycarbonate). These membrane are highly flexible and find application in both industrial and scientific fields. In addition, this technique is a simple strategy that can yield reproducible pores with predetermined geometries; thus it is utilized quite frequently (Apel, Blonskaya, & Orelovitch, 2012).

10.5.4 Template leaching

Template leaching is a technique where a porous membrane can be obtained by dripping out one of the components from a film. Using this technique, porous glass membranes can be prepared (Kesting, 1985). The size of the fabricated pores in this technique largely depends on the template particle size, which is typically in the range of microns or submicrons. This explains the relatively bigger pore sizes obtained using this technique, thereby making it suitable for delivering macromolecules such as proteins and polypeptides.

10.5.5 Phase inversion

Phase inversion is a widely used commercial method for deriving membranes. This is a multilateral technique that allows different kinds of morphologies to be obtained. Asymmetric membranes are generally made using a phase inversion strategy in which a porous membrane is prepared from homogenous polymer solution. Various outer stimuli could be used to transform it into solid, for example, exposure to heat or to nonsoluble medium. The morphology of the subsequent layer structure that controls the rate of medication discharge is firmly impacted by the phase-inversion dynamics. In this manner, the capacity to plan productive medication conveyance frameworks dependent on asymmetric membrane coatings requires thorough investigation on the morphology of the membrane and the medication delivery (Altinkaya & Yenal, 2006).

10.5.6 Coating

Coating is a technique wherein a composite membrane can be produced. Dense membranes decrease the fluxes so the thickness must be decreased as much as possible, and this can be achieved by using coating technique (see Fig. 10.5C). This technique is highly suitable where the targets include sustained release and prolonged time of treatment. Successive membrane coatings were earlier reported for this reason, for example, nonporous layers (membranes) were used as outer layer for tramadol subcutaneous patches to increase its life time (Mabrouk, Beherei, & ElShebiney, 2018). Furthermore, nonporous layers of ciprofloxacin-loaded polyvinylpyrrolidone coatings were added to the surface of gelatin/alginate/brushite in order to make it antibacterial (Beherei, Shaltout, & Mobrouk, 2018).

10.6 Applications of membrane

There are several applications of membrane in industry including separation of chemical substances for a wide range as well as desalination and water treatment. In addition, it is very common in the biomedical industries such as for protein recovery, production of pharmaceutical tablets, and hemodialysis. Furthermore, there are some common applications in other different fields including detection and analysis of particulate contamination, ion-selective membrane electrode, specific gas probes, and controlled drug delivery system (DDS). This book chapter is concerned with controlled DDS using membranes.

10.6.1 Controlled drug delivery system through nanochannels

DDS attracted massive attention in the early 1970s; it is used for controlling the delivery of the drug. DDS offers perfectly managed delivery of the drugs and the dose and the rate of the delivered drugs are predefined in addition to its target (Mabrouk, Rajendran, & Soliman, 2019). The delivered drug aims to its target only (e.g., diseased tissues and tumors) without affecting other healthy organs; thus it minimizes the side effects of medications (Javad & Zohre, 2014). DDS is one of the most prominent uses of nanoporous membranes. The dose of the released drug is comparable to the pore size of the membrane (Piskin & Hoffman, 1986).

10.7 Transport phenomena in drug delivery membrane systems

Diameter of pores of the nanoporous membranes is extremely critical when they are used for medication delivery system as it defines the dose of a delivered medication. To keep up the portion of medication that is discharged to be in a restoring range for a drawn out time, a nanochannel with small pore diameters must be created. For instance, most protein medications, for example, insulin and hormones possess lower hydrodynamic diameters (nanometer size), which is fundamentally identified with their transportation mechanism (Piskin & Hoffman, 1986).

Typically, the most common used methods of drugs are tablets and intravenous injection. However, these classical methods have a number of disadvantages, such as the drugs quickly lose their activity, inferior solubility and biodistribution, and rapid dissolution of the medication in human body. In addition, they exhibit cytotoxicity and lead to destruction of tissue on extravasation and their side effects as well (Allen & Cullis, 2004). These restrictions can be handled by controlling the medication delivery to cells, tissue, and organs by using the appropriate DDS. The delivery systems can be gels (Gupta, Vermani, & Garg, 2002; Hoare & Kohane, 2008; Qiu & Park, 2001), polymeric micelles (Allen, Maysinger, & Eisenberg, 1999; Ganta, Devalapally, & Shahiwala, 2008; Kataoka, Harada, & Nagasaki, 2001), or implant devices that contain reservoirs

(Langer & Tirrell, 2004; Sershen & West, 2002). There are many advantages of the drug delivery membrane systems such as managing the releasing rate of drugs, increasing the time of drug releasing and stability, reduction in the dosage frequency for improving patient compliance, reduction in the side effects, and increasing the bioavailability of the drugs. According to the rate of the released quantities by applying stimuli, the medication delivery system can be classified into sustained and responsive.

There are three common problems when membranes come into contact with physiological fluids. These problems are biofouling, membrane degradation, and immune response caused by the membrane. Biofouling process occurs when the membrane is in contact with a biological environment where the cells, proteins, and other materials accumulate on a membrane surface (Lewis, 2000; Wisniewski & Reichert, 2000). Furthermore, the encapsulation of tissue occurs, which leads to fibroblast proliferation collagen synthesis and proliferation of blood vessels. These processes affect the transport of glucose molecules in biological environments due to formation vascular tissue capsule, which results in steric hindrance (Harrison, Turner, & Baltes, 1988).

Thus membrane surface modifications must be performed to solve the problem of biofouling. These modifications can be done by surface treatment such as by the use of coatings. For example, the membrane surface modification by polymer coating, which results in polar, electrically neutral, swellable, and flexible membranes which in turn results in interface formation between physiological environment and membrane surface (Mabrouk et al., 2019). Therefore it is important to select the perfect surface coating to control the problems created during the contact of membranes with the biological environment (Morra, 2000).

For the macromolecular and biopharmaceutical medications (peptides, plasmids, proteins, oligonucleotides, antibodies, and viruses), the nanoporous membranes can be used. The pore size of these membranes ranges from 10 to 300 nm. Once nanoporous membrane–loaded medications reach the target organs and cells, the diffusion mechanism provides a safe way to deliver the medications in this case. Furthermore, the nanosized channels permit the prolonged release of the medications by eliminating the burst release that takes place in the beginning. Therefore by organizing the porosity of the membranes, the release rate of the drugs can be controlled (Jeon et al., 2012). Owing to the importance of nanoporous membranes in overcoming the obstacles that confront the transportation pathway, it will be discussed in detail in the next section.

10.7.1 Fabrication methods of nanoporous membranes for drug delivery

There are two approaches for the fabrication of nanoporous membranes, including top-down and bottomup techniques. The former enhances the ability to control the uniform pore-size distribution and pore position as well. However, this technique takes long time for fabrication and costs very high. Thus the latter is used to obtain a high-density array of nanopores on a large scale (Jeon et al., 2012). There are several requirements that should be considered while fabricating nanoporous membranes in order to obtain a highly effective drug release (Adiga, Curtiss, & Elam, 2008; Adiga, Jin, & Curtiss, 2009; Tokarev & Minko, 2010). The first requirement is biocompatibility; the nanoporous membrane must be made from a biocompatible material and have the ability to maintain its shape till the end of the drug releasing, and then it must become biodegradable.

Furthermore, this material should be nontoxic, have the ability to be absorbed and extracted within the body so that additional surgery will not be required. In addition, the chemical and mechanical stability are required; it is very important to consider the mechanical strength of the material also in addition to its chemical dissolution as the drugs are stored through the pores of the membranes. Particularly, the failure of the mechanical or the chemical stability may expose the patient to a noncontrolled medication delivery that could result in serious side effects. Moreover, nanoporous membranes cut-off values could be highly uniform if the distribution of their pore sizes is highly uniform. In addition, depending on the size of the biomaterial (viruses, nucleotides, proteins, and synthetic medications), the pore sizes are designed. Finally, low tortuosity and short channel length are very critical for the medications to undergo rapid diffusion out of the membrane.

10.7.2 Long-term and sustained drug delivery

Sustained medication delivery system is the system that can control the rate of medication release to keep its concentration within the therapeutic range in the body. When the pores size is much larger than the target medication size, the movement of the medication molecules will be without any pore-size restriction. Otherwise, when the pore size decreases, the medication molecules movement is limited (Jeon et al., 2012).

10.7.3 Transport through semipermeable membranes

There is a remarkable benefit and noteworthy property of the nanoporous membranes in a DDS in addition to their use for sustained drug release. They can prevent the immunologic rejection during cell transplantation where they act as of transplanted cells. The insulin secretion for an encapsulated immunoisolation cage that contains two pieces of nanoporous membranes (Desai, Hansford, & Ferrari, 1999; Desai, West, & Cohen, 2004; Flamme, Mor, & Gong, 2005) with 18 nm channel size has been observed to have efficient insulin secretion even after two weeks. The insulin secretion for the nonencapsulated cells showed a dramatically decrease because of the attack by the antibody. This is because the antibodies could not pass through nanoporous membrane.

10.8 Conclusions and future trends

In the current book chapter, detailed information about membranes including definition, history, and classifications were discussed in its introduction. Next, the transportation phenomena and the mechanism of transportation through membranes were correlated with the permeate size. Furthermore, methods of membranes preparation and their applications, especially DDSs were also discussed. From the reviewed literature in this chapter, it was deduced that the drug transportation phenomena through membranes could be largely controlled using available advanced techniques for fabrication of nanoporous membranes. In addition, it is also worthy to highlight that some surface modifications for membranes could enhance their transportation efficiency by overcoming their limitations, especially in biological systems.

References

Abd El-Ghaffar, M. A., & Tieama, H. A. (2017). A review of membranes classifications, configurations, surface modifications, characteristics and its applications in water purification. *Chemical and Biomolecular Engineering, 2* (2), 57–82. Available from https://doi.org/10.11648/j.cbe.20170202.11.

Adiga, P., Curtiss, L. A., Elam, J. W., Pellin, M. J., Shih, C.-C., Shih, C.-M., et al. (2008). Nanoporous materials for biomedical devices. *The Journal of The Minerals, Metals & Materials Society (TMS), 60*, 26–32.

Adiga, S. P., Jin, C., Curtiss, L. A., Monteiro-Riviere, N. A., & Narayan, R. J. (2009). Nanoporous membranes for medical and biological applications. *Wiley Interdisciplinary Reviews. Nanomedicine and Nanobiotechnology, 1*(5), 568–581.

Allen, C., Maysinger, D., & Eisenberg, A. (1999). Nano-Engineering block copolymers aggregates for drug delivery. *Colloids and Surfaces B: Biointerfaces, 16*, 3–27.

Allen, T. M., & Cullis, P. R. (2004). Drug delivery systems: Entering the mainstream. *Science (New York, N.Y.), 303*, 1818–1822.

Altinkaya, S. A., & Yenal, H. (2006). In vitro drug release rates from asymmetric-membrane tablet coatings: Prediction of phase-inversion dynamics. *Biochemical Engineering Journal, 28*, 131–139.

Amin, S. K., Hassan, M., El-Sherbiny, S., & Abdallah, H. (2016). An overview of production and development of ceramic membranes. *International Journal of Applied Engineering Research, 11*, 7708–7721.

Anis, S. F., Hashaikeh, R., & Hilal, N. (2019). Microfiltration membrane processes: A review of research trends over the past decade. *Journal of Water Process Engineering, 32*, 100941.

Apel, P. Y., Blonskaya, I. V., Orelovitch, O. L., Sartowska, B. A., & Spohr, R. (2012). Asymmetric ion track nanopores for sensor technology. Reconstruction of pore profile from conductometric measurements. *Nanotechnology, 23*, 225503.

Armor, J. N. (1989). Catalysis with permselective inorganic membranes. *Applied Catalysis, 49*, 1–25.

Barascu, A., Kullmann, J., Reinhardt, B., Rainerb, T., Roggendorf, H., Syrowatkad, F., et al. (2015). Porous glass membranes with an aligned pore system via stretch forming in combination with thermally induced phase separation. *Glass Physics and Chemistry, 41*, 73–80. Available from https://doi.org/10.1134/S1087659615010058.

Beherei, H. H., Shaltout, A. A., Mobrouk, M., Abdelwahed, N. A. M., & Das, D. B. (2018). Influence of niobium pentoxide particulates on the properties of brushite/gelatin/alginate membranes. *Journal of Pharmaceutical Science, 107*(5), 1361–1371.

Daramola, M. O., Aransiola, E. F., & Ojumu, T. V. (2012). Potential applications of zeolite membranes in reaction coupling separation processes. *Materials, 5*(11), 2101–2136.

Desai, T. A., Hansford, D., & Ferrari, M. (1999). Characterization of nanoporous membranes for immunoisolation: Diffusion properties and tissue effects. *Journal of Membrane Science, 159*, 221–231.

Desai, T. A., West, T., Cohen, M., & Rampersaud, A. (2004). Nanoporous microsystems for islet cell replacement. *Advanced Drug Delivery Reviews, 56*(11), 1661–1673.

Dudek, G., Borys, P., Strzelewicz, A., & Krasowska, M. (2020). Characterization of the structure and transport properties of alginate/chitosan microparticle membranes utilized in the pervaporative dehydration of ethanol. *Polymers, 12*, 411.

Dumee, L. F., He, L., Lin, B., Ailloux, F.-M., Lemoine, J.-B., Velleman, L., et al. (2013). The fabrication and surface functionalization of porous metal frameworks—A review. *Journal of Materials Chemistry A, 1*, 15185–15206.

Elson, E. L., Fried, E., Dolbow, J. E., & Genin, G. M. (2010). Phase separation in biological membranes: Integration of theory and experiment. *Annual Review of Biophysics, 39*, 207–226. Available from https://doi.org/10.1146/annurev.biophys.093008.131238.

Flamme, K. E. L., Mor, G., Gong, D., Tempa, T. La., Fusaro, V. A., Grimes, C. A., et al. (2005). Nanoporous alumina capsules for cellular macroencapsulation: Transport and biocompatibility. *Diabetes Technology & Therapeutics, 7*, 684–694.

Ganta, S., Devalapally, H., Shahiwala, A., & Amiji, M. (2008). A review of stimuli-responsive nanocarriers for drug and gene delivery. *Journal of Controlled Release, 126*(3), 187–204.

Guo, J., Berbano, S. S., Guo, H., Baker, A. L., Lanagan, M. T., & Randall, C. A. (2016). Cold sintering process of composites: Bridging the processing temperature gap of ceramic and polymer materials. *Advanced Functional Materials, 26*, 7115–7121.

Gupta, P., Vermani, K., & Garg, S. (2002). Hydrogels: From controlled release to pH-responsive drug delivery. *Drug Discovery Today, 7*, 569–579.

Harrison, D. J., Turner, R. F. B., & Baltes, H. P. (1988). Characterization of perfluorosulfonic acid polymer coated enzyme electrodes and a miniaturized integrated potentiostat for glucose analysis in whole blood. *Analytical Chemistry, 60*, 2002–2007.

Henley, E. J., Li, N. N., & Long, R. B. (1965). Membrane separation processes. *Industrial & Engineering Chemistry Research, 57*(3), 18–29. Available from https://doi.org/10.1021/ie50663a004.

Hoare, T. R., & Kohane, D. S. (2008). Hydrogels in drug delivery: Progress and challenges. *Polymer, 49*, 1993–2007.

Javad, S., & Zohre, Z. (2014). Advanced drug delivery systems: Nanotechnology of health design a review. *Journal of Saudi Chemical Society, 18*, 85–99.

Jeon, G., Yang, S. Y., & Kim, J. K. (2012). Functional nanoporous membranes for drug delivery. *Journal of Materials Chemistry, 22*(30), 14814–14834.

Kataoka, K., Harada, A., & Nagasaki, Y. (2001). Block copolymer micelles for drug delivery: Design, characterization and biological significance. *Advanced Drug Delivery Reviews, 47*, 113–131.

Kesting, R. E. (1985). *Synthetic polymeric membranes*. New York: McGraw Hill.

Kim, J.-J., Jang, T.-S., Kwon, Y.-D., Kim, U. Y., & Kim, S. S. (1994). Structural study of microporous polypropylene hollow fiber membranes made by the melt-spinning and cold-stretching method. *Journal of Membrane Science, 93*(3), 209–215. Available from https://doi.org/10.1016/0376-7388(94)00070-0.

Kolff, W. J., Berk, H. T. H. J., Welle, N. M., van der Ley, A. J. W., van Dijk, M. E. C., & van Noordwijk, J. (1944). The artificial kidney: A dialyser with a great area. *Acta Medica Scandinavica, 117*(2), 121–134. Available from https://doi.org/10.1111/j.0954-6820.1944.tb03951.x.

Krishna, R. (1990). Multicomponent surface diffusion of adsorbed species: A description based on the generalized Maxwell-Stefan equations. *Chemical Engineering Science, 45*(7), 1779–1791.

Langer, R., & Tirrell, D. A. (2004). Designing materials for biology and medicine. *Nature, 428*, 487–492.

Langley, P. J., & Hulliger, J. (1999). Nanoporous and mesoporous organic structures: New openings for materials research. *Chemical Society Reviews, 28*, 279–291.

Lewis, A. L. (2000). Phosphorylcholine-based polymers and their use in the prevention of biofouling. *Colloids and Surfaces B, Biointerfaces, 18*, 261–275.

Loeb, S., & Sourirajan, S. (1962). Seawater demineralisation by means of an osmotic membrane. *Advances in Chemistry Series, 38*, 117–132. Available from https://doi.org/10.1021/ba-1963-0038.ch009.

Mabrouk, M., Beherei, H. H., ElShebiney, S., & Tanaka, M. (2018). Newly developed controlled release subcutaneous formulation for tramadol hydrochloride. *Saudi Pharmaceutical Journal, 26*, 585–592.

Mabrouk, M., Rajendran, R., Soliman, I. E., Ashour, M. M., Beherei, H. H., Tohamy, K. M., et al. (2019). Nanoparticle-and nanoporous-membrane-mediated delivery of therapeutics. *Pharmaceutics, 11*(6), 294.

Morra, M. (2000). On the molecular basis of fouling resistance. *Journal of Biomaterials Science, Polymer Edition, 11*(6), 547–569.

Mulder, J. (1991). *Basic principles of membrane technology*. Dordrecht: Springer. Available from https://doi.org/10.1007/978-94-017-0835-7.

Orosz, K. E., Gupta, S., Hassink, M., Abdel-Rahman, M., Moldovan, L., Davidorf, F. H., & Moldovan, N. I. (2004). Delivery of antiangiogenic and antioxidant drugs of ophthalmic interest through a nanoporous inorganic filter. *Molecular Vision, 10*, 555–565.

Park, H. B., Kamcev, J., Robeson, L. M., Elimelech, M., & Freeman, B. D. (2017). Maximizing the right stuff: The trade-off between membrane permeability and selectivity. *Science (New York, N.Y.), 356*(6343), eaab0530. Available from http://doi.org/10.1126/science.aab0530.

Piskin, E., & Hoffman, A. S. (Eds.), (1986). *Polymeric biomaterials*. (NATO ASI series, ser. E. Applied sciences; no. 106).

Porter, C. M. (1989). *Handbook of industrial membrane technology*.

Qiu, Y., & Park, K. (2001). Environment-sensitive hydrogels for drug delivery. *Advanced Drug Delivery Reviews, 53*, 321–339.

Radjabian, M., & Abetz, V. (2020). Advanced porous polymer membranes from self-assembling block copolymers. *Progress in Polymer Science, 102*, 101219.

Ray, M. S. (1993). Chemical engineering, volume 2: Particle technology and separation processes, 4th edn, by J. M. Coulson and J. F. Richardson. Pergamon Press, Oxford, UK. 1991. 968 pp. ISBN 0-08-037957-5. *Developments in Chemical Engineering and Mineral Processing*, 1(2-3), 172.

Sershen, S., & West, J. (2002). Implantable, polymeric systems for modulated drug delivery. *Advanced Drug Delivery Reviews, 54*(9), 1225–1235.

Tokarev, I., & Minko, S. (2010). Stimuli-responsive porous hydrogels at interfaces for molecular filtration, separation, controlled release, and gating in capsules and membranes. *Advanced Materials, 22*(31), 3446–3462.

Wen, L., Xiao, K., Sainath, A. V., Komur, M., Kong, X.-Y., Xie, G., et al. (2016). Engineered asymmetric composite membranes with rectifying properties. *Advanced Materials, 28*(4), 757–763. Available from https://doi.org/10.1002/adma.201504960.

Wisniewski, N., & Reichert, M. (2000). Methods for reducing biosensor membrane biofouling. *Colloids and Surfaces. B, Biointerfaces, 18*, 197–219.

Wu, D., Xu, F., Sun, B., Fu, R., He, H., & Matyjaszewski, K. (2012). Design and preparation of porous polymers. *Chemical Reviews, 112*, 3959–4015.

XueMei, T., & Denis, R. (2019). A review on porous polymeric membrane preparation. Part I: Production techniques with polysulfone and poly (vinylidene fluoride). *Polymers (Basel), 11*(7), 1160.

Zsigmondy, R., & Bachmann, W. (1918). Über Neue Filter. *Zeitschrift für anorganische und allgemeine Chemie, 103*(1), 119–128.

Zydney, A. L. (1995). Membrane handbook edited by W. S. Winston Ho, and Kamalesh K. Sirkar, Van Nostrand Reinhold, New York, 1992, 954 pp. $131.95. *AIChE Journal, 41*(10), 2343–2344. Available from https://doi.org/10.1002/aic.690411024.

CHAPTER

11

Transport phenomena in fixed and fluidized-bed inorganic membrane reactors

Alessio Caravella[1,2], *Katia Cassano*[1], *Stefano Bellini*[1], *Virgilio Stellato*[3] *and Giulia Azzato*[1]

[1]Department of Computer Engineering, Modelling, Electronics and Systems Engineering (DIMES), University of Calabria, Rende, Italy [2]Institute on Membrane Technology – National Research Council (ITM-CNR), University of Calabria, Rende, Italy [3]Department of Environmental Engineering (DIAM), University of Calabria, Rende, Italy

Abbreviations

List of symbols

Symbol	Description
A, B	Correlation coefficients in Eq. (11.6)
A_{wet}	Wet area [m^2]
a	Specific area [m^2/m^3]
a	Activity in Eq. (11.58)
a	Diameter ratio in Eq. (11.74)
$a_1, a_2, a_3, a_4, a_5, a_6, a_7$	Empirical parameters in Table 11.1 (Zheng et al., 1995)
C	Coefficient in Eq. (11.19)
C_ϵ	Closure coefficient in Eq. (11.49)
C_d	Drag coefficient
$C_{[H]}$	Concentration of atomic hydrogen [mol_H/m_{Pd}^3]
C_μ	Vector of surface loading [mol_i/kg]
$C_{\mu,i}$	Surface loading [mol_i/kg]
C_{pL}	Liquid-side specific heat [J/(mol K)]
D_c	Column diameter [m]
D_H	Atomic hydrogen diffusion coefficient [m^2/s]
D_e	Effective diffusion coefficient [m^2/s]
D_{ij}	Reynolds stress diffusion [m^2/s]

Current Trends and Future Developments on (Bio-) Membranes
DOI: https://doi.org/10.1016/B978-0-12-822257-7.00010-8

d_p	Particle diameter [m]
d_k	Vector of kinetic diameter of the species [m]
E	Activation energy [J/mol]
F_d	Drag force [N]
F_f	Friction force [N]
F_{gs}	Gas–Solid drag force
Fr_g	Froud number
f	Friction coefficient
g, g_0	Gravity constant [m/s^2]
$\Delta\overline{G}^0$	Gibbs free energy of hydrogen adsorption [J/mol]
h	Heat transfer coefficient [J/s m^2 K^1]
$\Delta\overline{H}^0$	Enthalpy of hydrogen adsorption [J/mol]
ID, OD	Inner and outer Diameters [m]
K	Correlation in Eq. (11.16)
K	Turbulent source coefficient in Eq. (11.53)
K	Equilibrium constant in Eq. (11.62) [$mol_{H2}\ m_{Pd}^{-3}\ mol_{H2}^{-1}\ m^3$]
K	Equilibrium constant in terms of activity
κ_{gs}	Gas–Solid drag coefficient [N s/m]
k	Turbulent fluctuation [m^2/s^2]
K_L, k_L	Liquid-side mass transfer coefficient [m/s]
k_s	Pseudo-thermal conductivity [J/s m K]
ℓ	Turbulence length scale [-]
L	Length [m]
M_i	Molar mass [mol/kg]
N_{Av}	Avogadro's number
N_i	Molar flux [mol/m^2 s]
n	Number of species
n_{Mem}	Number of membranes
Nu	Nusselt number
n_H	Concentration of the dissolved hydrogen atoms [mol_H/mol_{Pd}]
n_{S1}	Concentration of the interstitial sites [mol_{Pd}/mol_{Pd}]
$P_{k,c}$	Production term induced by shear [J/m^3 s]
Pr	Prandtl number
p	Pressure [Pa]
p_s	Particle pressure [Pa]
q_s	Pseudo-Fourier fluctuating kinetic energy flux [J/m^2 s^1]
R	Ideal gas constant [J/(mol K)]
Re	Reynolds number
Re_p	Particle Reynolds number
$\Delta\overline{S}^0$	Entropy of hydrogen adsorption [J/(mol K)]
S	Sieverts' solubility constant [mol_H/mol_Pd Pa-0.5]
S_{g0}	Zero-loading specific surface [m^2/kg]
S_{ij}	Strain-rate tensor
S_k, S_ϵ	Turbulence energy sources [J/m^3 s]
St	Stanton number
T	Temperature
t	Time [s]
t_{ij}	Stress tensor in Eq. (11.41)
u	Fluid velocity in Eq. (11.23) [m/s]
u_g	Gas velocity [m/s]
u_l	Liquid velocity [m/s]
u_r	Relative velocity [m/s]

u	Time-averaged velocity [m/s]
u', v', w'	Turbulent-fluctuation velocities [m/s]
V_f, V_s	Fluid and solid volumes [m^3]
$\overline{V}_H$	Partial molar volume of hydrogen [m^3/mol]
v	Velocity [m/s]
v_g, v_s	Gas and solid velocities [m/s]
$v_{r,s}$	Terminal velocity correlation [m/s]
x	Axial position [m]
z	Coordination number

Greek symbols

α	Distance between two consecutive sites in Eq. (11.75)
α	Phase fraction (for turbulent multiphase systems)
β	Fugacity coefficient
Γ	Adsorption thermodynamic factor
γ	Empirical exponent in Eq. (11.20)
γ	Dissipation of granular energy [J/kg]
γ	Activity coefficient in Eq. (11.58)
δ	Membrane thickness [m]
ε	Kinetic energy dissipation [J/kg]
ε	Voidage degree in Eq. (11.71)
ε_0	Zero-loading voidage degree in Eq. (11.72)
ε_g, ε_s	Void and Solid fractions
η	Small-scale length [m]
ϕ_{ij}	Pressure fluctuations in pressure–strain model [Pa]
ϕ_s	Geometrical parameter in Table 11.1 (Nguyen-Tien et al., 1985)
φ	Generic Variable in Eqs. (11.33)–(11.38)
μ	Viscosity [Pa s]
μ	Chemical potential in Eq. (11.58) and successive
μ^0	Standard chemical potential
μ_l	Liquid viscosity [Pa s]
μ_t	Turbulent viscosity [Pa s]
$\mu_{t,c}$	Eddy viscosity [Pa s]
ν	Kinematic viscosity [m^2/s]
ν	Poisson ratio in Eq. (11.59)
π	Permeance [mol/s m^2 Pa^n]
ρ	Density [kg/m^3]
σ	Covered surface fraction
σ_κ	Closure coefficient in Eq. (11.46)
σ_l	Surface tension of liquid [mN/m]
Θ	Fluctuating temperature [K]
θ_i	Surface coverage of a generic species
τ	Stress tensor [Pa]
τ	Turbulence time scale (in turbulence section) [s]
τ	Tortuosity
τ_0	Zero-loading tortuosity
$w(\varepsilon_g)$	Voidage degree correlation

Subscripts and superscripts

"$_{Av}$"	Average
"$_c$"	Continuous phase (in multiphase systems)
"$_{mf}$"	Minimum fluidization

"$_{GT}$" Gas translation
"$_{SD}$" Surface diffusion

Acronyms

BCC Body-centered cubic
DNS Direct numerical simulation
FCC Face-centered cubic
IC Inhibition coefficient
LES Large-eddy simulation
PRC Permeation reduction coefficient
RANS Reynolds-averaged Navier–Stokes

11.1 Introduction

This chapter reports an overview of momentum, heat and mass transport involved in fixed- and fluidized-bed membrane reactors. It is organized as follows. First, a brief overview is presented on the mass transfer mechanisms in the most important inorganic membranes used in membrane reactors, mainly metal and zeolite ones.

Afterward, the mechanisms of momentum, heat and mass transfer in fixed- and fluidized-beds are described, highlighting similarities and differences among configurations, flow regimes and particle types. Furthermore, as the most significant catalytic reactions occur in gas phase and important examples of membrane reactor prototypes are mostly used to selectively separate gaseous species through dense membranes as well as porous ones, this chapter focuses on gas-phase reactions only, although several examples of membrane reactors used to separate liquids can be found at a laboratory scale (photocatalytic systems, catalytic and inert membranes).

11.2 Overview of momentum transfer in catalytic reactors

In both fixed- and fluidized-bed reactors, momentum is transferred from the fluid phase to the particle surface, generating a drag force, which is created by the difference in velocity between phase interfaces. An in-depth knowledge of the way momentum is transferred between phases is defined in appropriate drag models, which represent the constitutive equations for momentum transfer.

11.2.1 Fixed-bed reactors

For fixed beds, the most commonly used friction model is represented by the Ergun equation (Eq. 11.1):

$$f_{\text{Ergun}} = \frac{150\varepsilon_s^2\mu_g}{\varepsilon_g d_p^2} + \frac{1.75\rho_g\varepsilon_s\left\|\vec{v}_g\right\|}{d_p} \tag{11.1}$$

which is a semiempirical equation arising from pairing the theoretical Carman–Kozeny model with empirical models valid at a very high velocity values (Bird, Stewart, &

Lightfoot, 1960). In the absence of catalytic particles, the friction can be calculated by the Fanning coefficient as follows (Bird et al., 1960):

$$f_{\text{Fanning}} = \begin{cases} \dfrac{16}{\text{Re}}, & \text{Re} < 2100 \\ \dfrac{0.0791}{\text{Re}^{0.25}}, & 2100 < \text{Re} < 10^5 \end{cases} \tag{11.2}$$

where the friction force is calculated by the Fanning friction factor defined on the basis of a characteristic kinetic energy and the total wet area, which is recalled to be the friction-providing exposed area (Bird et al., 1960).

In case of multimembrane tube-shell reactors (Fig. 11.1), friction force and wet area can be expressed as follows:

$$\begin{aligned} \left\| \vec{F}_f^{\text{Shell}} \right\| &\equiv f_{\text{Fanning}}^{\text{Shell}} \left(\frac{1}{2} \rho v^2 \right) A_{\text{wet}}^{\text{Shell}} \\ A_{\text{wet}}^{\text{Shell}} &= \pi L (n_{\text{Mem}} \text{OD}_{\text{Mem}} + \text{ID}_{\text{Shell}}) \end{aligned} \tag{11.3}$$

In multimembrane catalytic beds, the extent of friction provided by the shell and tubes surfaces is usually negligible with respect to the friction provided by the particles. However, the contribution of friction due to the membranes can become appreciable for a relatively high number of membranes, that is, in the (rare) case, where the exposed membrane area is comparable to the total area of the catalytic particles. In this case, the total friction force is the sum of the friction obtained by the Ergun equation and the friction obtained by the Fanning equation.

More recently, techniques of direct numerical simulation (DNS) were adopted to analyze the influence of end effects and confining walls in a packed bed on the mass transfer coefficient for Reynolds numbers <100 and a Schmidt number of around 2.5. The overall result was that the confining wall has a general significant influence for a bed-to-particle diameter ratio higher than around 11 (Bale et al., 2018).

To reduce the pressure drop per unit of length in packed beds, a novel packing arrangement was recently proposed by Guo, Sun, Zhang, Ding, and Liu (2017), who considered a structured packing that is highly ordered around the bed wall. As the resulting fluid dynamics was found to be quite different from – but more favorable than –the random one, the authors modified the Carman equation to fit the experimental data, suggesting that the proposed packing could be successfully used in a number of industrial applications.

Guo et al. analyzed the influence of a particular packing arrangement on both pressure drop and heat transfer coefficient. The particular packing considered consisted of guiding conduits highly accessible to fluid but inaccessible to particles placed within the packed bed.

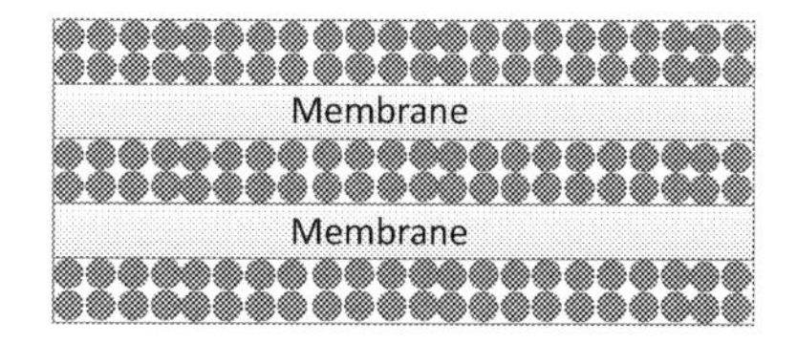

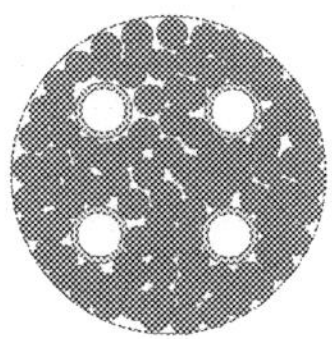

FIGURE 11.1 Sketch of a packed-bed multimembrane reactor.

They found that pressure drop reduced to around 16%–26% and the heat transfer coefficient dropped down by 4%–6%. An interesting aspect to consider is that this configuration is very similar to that in which we have a bundle of membranes within a packed bed. Therefore the results found in that study can be applied to estimate pressure drop and heat transfer coefficient in membrane reactors (Guo, Sun, Zhang, Ding, & Liu, 2019).

Guo, Sun, Li, and Yang (2014) investigated the 3D temperature distribution in packed beds composed of large catalytic particles for chemical looping combustion and low aspect ratio by conducting oxidation reactions. In particular, they found that the axial profile of temperature is driven by the internal diffusion in the presence of large particles. As well, they found serious channeling and nonuniform radial distribution of temperature.

Hofmann, Bufe, Brenner, and Turek (2016) showed that in milli-structured channels packed with particles of different shapes, the wall effect is relevant on pressure drop in laminar flow regime in the range $0.5 < Re < 12$ when the D/d ratio is included within 6–14.

Asakuma, Asada, Kanazawa, and Yamamoto (2016) analyzed the thermal conductivity of a packed bed by a pseudo-homogeneous approach and observed that heat transfer with thermal contact resistance is not dominant if the Biot number is near 100. In contrast, in the presence of large particles and pure-conductive heat transfer, several factors including microstructure, surface roughness and contact pressure are important.

Guo, Sun, Zhang, and Ding (2019) investigated the influence of confining wall on both pressure drop and heat transfer between particles and fluid for small values of D/d ratio and Re number <104 and analyzed several different literature correlations. They found that Carman's correlation and Wakao's one could lead to a better prediction among other correlations present in the open literature.

Li, Zou, Wang, Zhang, and Wang (2018) remarked that the bed coolability strictly depends on the friction between phases and developed a semiempirical model for predicting the extent of such a friction. In particular, they found that when a two-phase fluid passes through a small-particle bed (particle size of the order of 1.5–2 mm), the resistance owing to the friction increases with increasing flow rate according to the Reed model. In contrast, with particle size of the order of 4–8 mm, the resistance shows an up–down tendency typical of nonsteady stationary states, which can be difficult to predict.

Amiri, Ghoreishi-Madiseh, Hassani, and Sasmito (2019) investigated the mathematical behavior of packed beds considering several types of particles (uniform and nonuniform) with different particle-size distributions. They found that the currently available correlations are not able to predict correctly the flow behavior in the presence of a relatively large particle size. Therefore they developed a mathematical model pairing the Ergun and Forchheimer theories with a pore-scale approach, with the aim of evaluating correlations for both bed permeability and inertial resistance in the presence of very large particle size (of the order of 40–100 cm) and for porosity up to 0.7.

An interesting solution for an effective thermal management in highly exothermic catalytic reactors has been suggested by Fache, Marias, and Chaudret (2020), who analyzed the possible benefit in using magnetic induction to stabilize the thermal release from catalytic particles. In particular, they found that the induction-assisted catalytic reactors are less sensitive to changes in operating conditions. Furthermore, they observed that these type of reactors exhibit a superior performance in terms of yield before running away. This kind of technical solution could be successfully used in membrane reactors as well, where a poorly efficient thermal management can be detrimental for both membranes and catalyst.

11.2.2 Fluidized-bed reactors

The fluidization condition for a particle bed enhances the uniformity of composition and temperature distribution within the reaction environment. This particular fluid-dynamic regime allows minimizing gradients between fluid and solid phases, which holds both catalyst and membranes. If on the one hand the small gradients caused by fluidization tend to enhance the performance of reactor, on the other hand, the increased flow rate necessary to reach the fluidization conditions decreases the residence time, with a consequent potential decrease in conversion and yield.

Furthermore, there can be an erosion of materials due to the particles' circulation with respect to both catalyst and membrane. As for the former, the particle size can decrease by rupture during operation, increasing the contact catalytic area but, at the same time, increasing the pressure drop due to the reduced voidage degree.

A crucial topic in studying fluidized beds is to recognize the appropriate drag model that can describe the momentum transfer between the disperse phase (usually solid) and the gas phase. In fact, a precise drag model allows us to predict the macroscopic behavior of the particle bed in terms of hydrodynamic behavior and phenomena such as expansion, segregation, aggregation, and clustering. In particular, in the Euler–Euler approach, the drag force is expressed by means of the definition of the drag coefficient K_{gs}, by which the drag force is formally proportional to the difference of the velocities of gas and solid phases (Eq. 11.4):

$$\vec{F}_{gs} = K_{gs}(\vec{v}_g - \vec{v}_s) \quad (11.4)$$

Actually, the drag models in fluidized systems are generally much more complex than those used in the fixed ones; it is also because in several systems of interests, the three phases (solid, liquid and gas) are present at the same time. The most used drag models in the literature are briefly reported and described in the following subsections.

11.2.2.1 *Main drag models in the literature*

Syamlal-O'Brien model (1987)

$$K_{gs} = \frac{3}{4}\frac{\varepsilon_s \varepsilon_g \rho_g}{v_{r,s}^2 d_p} C_d\left(\frac{Re_p}{v_{r,s}}\right)|\vec{v}_s - \vec{v}_g| \quad (11.5)$$

$$v_{r,s} = 0.5\left(A - 0.06\mathrm{Re_p} + \sqrt{(0.06\mathrm{Re_p})^2 + 0.12\mathrm{Re_p}(2B - A) + A^2}\right) \quad (11.6)$$

$$A = \varepsilon_g^{4.14} \text{ and } \begin{matrix} B = 0.8 \quad \varepsilon_g^{1.28}, \varepsilon_g \le 0.85 \\ B = 0.8 \quad \varepsilon_g^{2.65} \quad \varepsilon_g > 0.85 \end{matrix}$$

where $v_{r,s}$ is the terminal velocity correlation for the solid phase.

Ding and Gidaspow model (1990)

$$K_{gs} = 150\frac{\varepsilon_s^2\mu_g}{\varepsilon_g^2 d_p^2} + 1.75\frac{\varepsilon_s\rho_g|\vec{v}_s - \vec{v}_g|}{\varepsilon_g d_p} \quad \text{for } \varepsilon_g \leq 0.8 \tag{11.7}$$

$$K_{gs} = \frac{3}{4}C_D\frac{\varepsilon_s\rho_g|\vec{v}_s - \vec{v}_g|}{d_p}\varepsilon_g^{-2.65} \quad \text{for } \varepsilon_g > 0.8 \tag{11.8}$$

where

$$C_D = \begin{cases} \frac{24}{\mathrm{Re}_p}\left[1 + 0.15(\mathrm{Re}_p)^{0.687}\right] \text{ for } \mathrm{Re}_p \leq 1000 \\ 0.44 \text{ for } \mathrm{Re}_p > 1000 \end{cases} \tag{11.9}$$

$$\mathrm{Re}_p = \frac{\varepsilon_g\rho_g d_p|\vec{v}_s - \vec{v}_g|}{\mu_g} \tag{11.10}$$

Energy-Minimization Multi-Scale (EMMS) model (Yang, Wang, Ge, & Li, 2003)

$$K_{gs} = 150\frac{\varepsilon_s^2\mu_g}{\varepsilon_g d_p^2} + 1.75\frac{\varepsilon_s\rho_g|\vec{v}_s - \vec{v}_g|}{d_p} \quad \text{for } \varepsilon_g < 0.75 \tag{11.11}$$

$$K_{gs} = \frac{3}{4}C_D\frac{\varepsilon_s\varepsilon_g\rho_g|\vec{v}_s - \vec{v}_g|}{d_p}\omega(\varepsilon_g), \quad \varepsilon_g \geq 0.74 \tag{11.12}$$

$$\omega(\varepsilon_g) = -0.5760 + \frac{0.0214}{4(\varepsilon_g - 0.7463)^2 + 0.0044}, \quad 0.74 < \varepsilon_g \leq 0.82 \tag{11.13}$$

$$\omega(\varepsilon_g) = -0.0101 + \frac{0.0038}{4(\varepsilon_g - 0.7789)^2 + 0.0040}, \quad 0.82 < \varepsilon_g \leq 0.97 \tag{11.14}$$

$$\omega(\varepsilon_g) = 1 + 32.8295(1 - \varepsilon_g)\varepsilon_g > 0.97 \tag{11.15}$$

Ayeni's model (Ayeni, Wu, Nandakumar, & Joshi, 2016)

$$\begin{gathered} K_{gs} = \frac{3\rho_g\varepsilon_g^2\varepsilon_s|\vec{v}_g - \vec{v}_s|}{d_p}C_D \\ C_D = \frac{6}{d_p}\left(\frac{3.6\varepsilon_s}{\varepsilon_g^4} + 1\right) + 0.11\left(\frac{40.91K^2\varepsilon_s^2}{\varepsilon_g^2} + 1\right) \\ K = \left[\frac{v_{mf}}{v_g} + \left(1 - \frac{v_{mf}}{v_g}\right)\left(1 - \frac{\varepsilon_s}{\varepsilon_{s,mf}}\right)\right] \end{gathered} \tag{11.16}$$

Beetstra-van der Hoef-Kuipers (BVK) model (2007)

$$K_{gs} = 10\frac{\varepsilon_g}{\varepsilon_s^2} + \varepsilon_s^2(1 + 1.5\sqrt{\varepsilon_g}) + \frac{0.413\mathrm{Re}_p}{24\varepsilon_s^2}\left(\frac{\frac{1}{\varepsilon_s} + 3\varepsilon_g\varepsilon_s + \frac{8.4}{(\mathrm{Re}_p)^{0.343}}}{1 + \frac{1000}{(\mathrm{Re}_p)^{0.5+2\varepsilon_g}}}\right) \tag{11.17}$$

Gibilaro, Di Felice, Waldram, and Foscolo model (1985)

$$K_{gs} = \left(\frac{17.3}{\mathrm{Re_p}} + 0.336\right)\frac{\rho_g\left|\vec{v}_s - \vec{v}_g\right|}{d_p}\varepsilon_s\varepsilon_g^{-1.8} \tag{11.18}$$

McKeen's and Pugsley model (2003)

$$K_{gs} = C\left(\frac{17.3}{\mathrm{Re_p}} + 0.336\right)\frac{\rho_g\left|\vec{v}_s - \vec{v}_g\right|}{d_p}\varepsilon_s\varepsilon_g^{-1.8} \tag{11.19}$$

It should be noticed that for $C = 1$, the McKeen model becomes identical to Gibilaro's et al. model (1985).

Di Felice's model (1994)

$$K_{gs} = \frac{3}{4}C_D\frac{\rho_g\varepsilon_s\left|\vec{v}_g - \vec{v}_s\right|}{d_p}\varepsilon_s^{-\gamma} \tag{11.20}$$

$$C_d = \left(0.63 + \frac{4.8}{\sqrt{\mathrm{Re_p}}}\right)^2 \tag{11.21}$$

$$\gamma = 3.7 - 0.65 \cdot e^{-\frac{\left[1.5-\log_{10}(\mathrm{Re_p})\right]^2}{2}} \tag{11.22}$$

11.2.3 Turbulence

This section reports an overview of turbulence and the most used model present in the open literature.

11.2.3.1 Definition and general concepts

A general classification of a flow field can be based on the different flow regimes: laminar, transient, and turbulent. In engineering applications, the turbulent regime is particularly important in addition to its characterization. The nature of turbulent regime implies a chaotic movement of molecules along with the formation of larger-scale eddy vortices, contrary to what happens in the laminar regime, which can be described as fluid layers that slide on one another. From a macroscopic point of view, when Reynolds number is less than 2100, the fluid flows in laminar regime, whereas for higher values, the regime is turbulent. Reynolds number is defined as shown in Eq. 11.23, which represents the ratio between inertial and viscous forces acting on the fluid:

$$\mathrm{Re} = \frac{UL}{\nu} \tag{11.23}$$

where U and L are fluid velocity and length scales of the flow, respectively. Inertial forces are proportional to U^2 and viscous forces to U. Turbulence is imagined as caused by an instability induced in laminar flow. To adequately describe the turbulent regime, a huge amount of information is generally required. In simpler situations, at least two parameters are needed: drag coefficient and transferred heat.

However, the mathematical complexity of turbulence modeling increases with increasing level of detail required in the flow description. A complete analytical description of turbulence is not feasible because of its chaotic nature. Moreover, even in the simplest description of turbulence, there are nonlinear terms in the Navier–Stokes equations that interact with viscous term for real fluids. The three-dimensional character of turbulence is underlined by the strong rotational nature of the local eddy vortices, whose description necessarily requires a statistical approach to be modeled (Wilcox, 1993).

Turbulence is a continuum phenomenon acting on different scale lengths. A characteristic parameter is the ratio between the smallest and the largest scale lengths, which decreases with increasing Reynolds number. In fact, the fluid dynamic structure of turbulent transport phenomena is characterized by several small-scale lengths. For example, part of a fluid can typically have turbulent length scales ranging from 10 to 100 μm. Eddies of larger scale are degraded into smaller-scales ones, which are maintained by the energy transport between scales.

Turbulent flows are distinguished by a variety of spatial scales ranging between the flux domain dimension and the scale of molecular diffusion. The small spatial-scale motions are responsible for the energy dissipation and also act at small time scales. Moreover, they are statistically independent of turbulent flows at large spatial and temporal scales but dependent on the energy provided by movements of large scale. The large-scale eddies are of the same order of the flow streamlines, whereas the smaller ones have size comparable with that of the viscous dissipation.

The ratio between the length of small- and large-scale eddies is a function of the fluid velocity and therefore of Reynolds number. Kolmogorov supposed that the motion at small scale is related to turbulent kinetic energy dissipation and turbulent viscosity. Energy of larger eddies is dissipated on smaller eddies down to viscous scales.

11.2.3.2 Turbulence models

All turbulent phenomena are characterized by energy transfer from large scales to scales of viscous dissipation. This transfer occurs without dissipation through nonlinear terms and pressure terms. In flows with high Reynolds numbers, there is an enormous dimensional difference between large and small scales and a dimension beyond which to apply a model can be chosen.

In computational fluid dynamics (CFD), there are three main techniques to model eddies and analyze turbulent flow:

1. DNS
2. Reynolds-averaged Navier–Stokes (RANS)
3. Large-eddy simulation (LES)

11.2.3.2.1 Direct numerical simulation model

In the DNS model, all eddy scales are solved. The Navier–Stokes equations are numerically solved without using any turbulent model. The computational costs are really high, which is the main limitation of DNS. The ratio between large- and small-scale lengths is:

$$\frac{L}{\eta} = R_e^{\frac{3}{4}} \tag{11.24}$$

This relationship indicates that for each length L, there must be $R_e^{\frac{3}{4}}$ nodes in the computational grid, and thus for a three-dimensional domain there are $R_e^{\frac{9}{4}}$ nodes. In addition, the simulation time must be of the order of the motion time in large scale, but the time step must be sufficiently small to capture small-scale dynamic. These estimates underline that DNS approach is unfeasible.

11.2.3.2.2 Reynolds-averaged Navier–Stokes model

RANS methods are based on averaged Navier–Stokes equations in which the Reynolds stress tensor is computed by means of a turbulent model. These models use a definition of turbulent viscosity with the hypothesis that it only depends on turbulent quantities and is independent of the mean flow field. Large scales are solved and small scales are modeled through appropriate turbulent models.

Large scales have not influential spatial gradients and have slow dynamics. This allows one to use a mesh with spacing that is independent of Reynolds number and a time step that is proportional to time scale of the entire geometry.

The LES approach is intermediate between RANS and DNS. All larger structures until inertial range are simulated directly and the smallest ones are modeled. As in the RANS technique, a "cut" is made between scales, but in LES this is independent of problem geometry and inertial range extension. The greatest assets make use of only one turbulent model because the motion scales in the inertial range have their dynamics independent of the geometry problem, obtaining a method of universal character. Obviously computational costs are higher than in RANS case. Currently, RANS models are the standard in industry application because they are able to simulate complex flow fields with acceptable computational costs and sufficiently accurate results.

When a turbulent flow is studied, space average, time average, and/or ensemble average may be used. Space averages are valid only for statistically homogeneous flows. Time averages are used for situations of statistically stationary turbulence with time scales larger than the fluctuations' time scale.

The third one is more complex than the others but is valid for inhomogeneous, nonstationary turbulence phenomena. In this case, the average is computed through an ensemble of turbulent flows with identical external conditions. Usually, time averages are selected because the mean quantities can be easily calculated as follows:

$$u(t) = U(t) + u(t)' \tag{11.25}$$

where $u(t)'$ is the fluctuating component. By definition, the time average of fluctuating component is equal to zero.

$$\frac{1}{\Delta t}\int_0^{\Delta t} u(t)' \mathrm{d}t = 0 \tag{11.26}$$

Although the average is null, the fluctuating term influences the flow field. The fluctuating component has a chaotic evolution around the mean value. As the values of all field variables (temperature, pressure, velocity, etc.) change chaotically in time, statistical methods are used. If a field variable is measured, in a fluid portion with turbulent motion, it must be measured for a given time T and repeated N times starting from the same initial conditions. Mean and fluctuating components can be expressed as follows:

$$\overline{u(t)} = \frac{1}{N}\sum_{i=1}^{N} u_i(t) \tag{11.27}$$

$$u'(\mathrm{t}) = u_i(t) - \overline{u(t)} \tag{11.28}$$

To quantify the fluctuation intensity, the standard deviation is evaluated:

$$\overline{u'(\mathrm{t})^2} = \frac{1}{N}\sum_{i=1}^{N} (u_i'(\mathrm{t}))^2 \tag{11.29}$$

Therefore in turbulent flow, each variable is composed of two terms, a mean quantity and a fluctuant quantity, this being known as Reynolds decomposition.

$$u_i(x,t) = U(x) + u_i'(x,t) \tag{11.30}$$

where:

$$U(x) = \frac{1}{\Delta t}\int_0^{\Delta t} u(t)\mathrm{dt} \neq 0 \tag{11.31}$$

$$\overline{u(t)'} = \frac{1}{\Delta t}\int_0^{\Delta t} u(t)'\mathrm{dt} = 0 \tag{11.32}$$

The time step Δt must be longer than the maximum period of the velocity fluctuations to make it possible to consider $\Delta t \rightarrow \infty$. Assuming that T_1 is the maximum period of the velocity fluctuations and T_2 is the slowest variation in the flow, the following relationship is valid $T_1 \ll \Delta t \ll T_2$. For two generic flow variables $u = \mathrm{U} + u'$ and $\varphi = \Phi + \varphi'$, the following rules are valid for time averaging:

$$U \cdot \varphi' = 0(or\ \ \Phi \cdot \mathrm{u}') \tag{11.33}$$

$$\overline{u' \cdot \varphi'} = 0 \tag{11.34}$$

$$\overline{u' \cdot \varphi'} \neq 0 \tag{11.35}$$

$$\overline{u \cdot \varphi} = U\Phi + \overline{u' \cdot \varphi'} \tag{11.36}$$

$$\vec{\nabla} \cdot \vec{\mathrm{u}} = \vec{\nabla} \cdot \vec{\mathrm{U}} \tag{11.37}$$

$$\vec{\nabla} \cdot (\vec{\mathrm{u}} \cdot \varphi) = \vec{\nabla} \cdot (\vec{\mathrm{U}}\Phi) + \vec{\nabla} \cdot (\vec{\mathrm{u}'} \cdot \varphi') \tag{11.38}$$

These rules are applied to the Navier–Stokes equations, thereby giving the equations named "*Reynolds-averaged equations.*" For incompressible fluids, we have:

$$\nabla \vec{U} = 0 \tag{11.39}$$

$$\rho\frac{\partial U_i}{\partial t} + \rho\vec{\nabla}(U_i \quad \vec{U}) = -\vec{\nabla}\mathrm{p} + \nabla t_{ij} + \nabla \tau_{ij} \tag{11.40}$$

$$t_{ij} = 2\nu S_{ij} = \mu \left[\frac{\partial u_i}{\partial x_j} + \frac{\partial u_j}{\partial x_i} \right] \tag{11.41}$$

$$\tau_{ij} = -\rho \overline{u_i^{'} \cdot u_j^{'}} \tag{11.42}$$

where S_{ij} is the strain-rate tensor and τ_{ij} is the Reynolds stress tensor. The former is a symmetric tensor, having six components that correspond to additional six unknown quantities. The time average of the nonlinear convective term generates the Reynolds stress tensor, which represents fluctuations velocity effect favoring the formation of eddies (Nygren, 2014). Therefore in a generic three-dimensional problem with turbulent flow, there are 10 unknown quantities: pressure, three velocity components, and the six components of Reynolds stress tensor. However, the available equations are mass conservation equation and three components of momentum conservation equation. Hence the problem is not closed mathematically. For system closure, the Reynolds stress tensor can be modeled in different ways, as briefly reported in the following subsections.

11.2.3.2.3 Standard $k-\varepsilon$ model

The $k-\varepsilon$ model is one of the possible two-equation models in RANS approach. This type of model computes turbulent kinetic energy k and turbulence length scale (or equivalent). For this reason, they are considered complete models. The first step for all the two-equation models is to use the Boussinesq approximation:

$$-\rho \overline{u_i^{'} \cdot u_j^{'}} = \mu_t \left(\frac{\partial U_{q,i}}{\partial x_j} + \frac{\partial U_{q,j}}{\partial x_i} \right) - \frac{2}{3} \rho k \delta_{ij} \tag{11.43}$$

With this approximation, Reynolds stresses are expressed as viscous stresses. In fact, in addition to the molecular viscosity, the Boussinesq approximation introduces the eddy viscosity μ_t. The definition of turbulent kinetic energy is derived from Prandtl postulate, for which a characteristic velocity scale for turbulence and kinetic energy of the turbulent fluctuation is computed on the basis of this velocity scale:

$$k = \frac{1}{2} \overline{u_i^{'} \cdot u_i^{'}} = \frac{1}{2} \left(\overline{u^{'2}} + \overline{v^{'2}} + \overline{w^{'2}} \right) \tag{11.44}$$

Considering a turbulence length scale ℓ and a turbulent kinetic energy k, a consequent definition of eddy viscosity can be obtained:

$$\mu_t = \text{constant} \cdot \rho k^{1/2} \ell \tag{11.45}$$

From this expression, it is possible to derive a transport equation for turbulent kinetic energy:

$$\rho \frac{\partial k}{\partial t} + \rho U_i \frac{\partial k}{\partial x_j} = \tau_{ij} \frac{\partial U_i}{\partial x_j} - \rho \varepsilon + \frac{\partial}{\partial x_j} \left[\mu + \frac{\mu_T}{\sigma_k} \frac{\partial k}{\partial x_j} \right] \tag{11.46}$$

where σ_k is a closure coefficient and ε is the kinetic energy dissipation for unit mass, defined as:

$$\varepsilon = -\frac{\partial k}{\partial t} \tag{11.47}$$

To close the problem, it is necessary to determine the dissipation ε or turbulence length scale ℓ for a complete description of turbulent flow. However, there exist several methods to calculate them. The k−ε model uses an exact equation for ε derived from the Navier−Stokes equations:

$$\rho\frac{\partial\varepsilon}{\partial t}+\rho U_i\frac{\partial\varepsilon}{\partial x_j}=-2\mu\left[\overline{u_{i,k}'u_{j,k}'}+\overline{u_{k,i}'u_{k,j}'}\right]\frac{\partial U_j}{\partial x_j}-2\mu\overline{u_k'u_{i,j}'}\frac{\partial^2 U_j}{\partial x_k x_j}-2\mu\nu\overline{u_{i,km}'u_{i,\mathrm{km}}'}$$
$$+\frac{\partial}{\partial x_j}\left[\mu\frac{\partial\varepsilon}{\partial x_j}-\overline{u_j'u_{i,m}'u_{i,m}'}-2\nu\overline{p_{i,m}'u_{j,m}'}\right] \tag{11.48}$$

This equation is quite complex to solve and is not feasible to solve with a satisfactory degree of accuracy. Hence on the basis of physical reasoning and dimensional analysis, another ε transport equation is obtained (Wilcox, 1993):

$$\rho\frac{\partial\varepsilon}{\partial t}+\rho U_j\frac{\partial\varepsilon}{\partial x_j}=C_{\epsilon1}\frac{\varepsilon}{k}\tau_{ij}\frac{\partial U_i}{\partial x_j}-C_{\epsilon2}\rho\frac{\varepsilon^2}{k}+\frac{\partial}{\partial x_j}\left[\left(\mu+\frac{\mu_T}{\sigma_\epsilon}\right)\frac{\partial\varepsilon}{\partial x_j}\right] \tag{11.49}$$

where the closure coefficients assume the following values: $C_{\epsilon1}=1.44$, $C_{\epsilon2}=1.92$, $\sigma_k=1.0$, and $\sigma_\epsilon=1.3$. In this approximation, various terms are parametrized as functions of large eddy scales. For this reason, the relationship between approximated equation and exact one does not deserve a particular consideration, and thus the former is fully adequate to model a turbulent flow (Wilcox, 1993).

For applying the standard k−ε model to a multiphase system, the equation needs some modifications. In fact, a turbulence model is applied to the continuous phase only. However, in a multiphase system, the equations of turbulent model have to be solved for all phases. The equations for turbulence kinetic energy k and turbulence energy dissipation rate ε are reported as follows:

$$\frac{\partial}{\partial t}\left((1-\alpha)\rho_c k_c\right)+\frac{\partial}{\partial x_j}\left((1-\alpha)\rho_c U_{i,c}k_c\right)=(1-\alpha)S_k-(1-\alpha)\left(P_{k,c}-\rho_c\varepsilon_c\right)$$
$$+\frac{\partial}{\partial x_j}\left[(1-\alpha)\left(\mu_c+\frac{\mu_{T,c}}{\sigma_k}\right)\frac{\partial k_c}{\partial x_j}\right] \tag{11.50}$$

$$\frac{\partial}{\partial t}\left((1-\alpha)\rho_c\varepsilon_c\right)+\frac{\partial}{\partial x_j}\left((1-\alpha)\rho_c U_{i,c}\varepsilon_c\right)=(1-\alpha)S_\epsilon-(1-\alpha)\frac{\varepsilon_c}{k_c}\left(C_{\epsilon1}P_{k,c}-C_{\epsilon2}\rho_c\varepsilon_c\right)$$
$$+\frac{\partial}{\partial x_j}\left[(1-\alpha)\left(\mu_c+\frac{\mu_{T,c}}{\sigma_\epsilon}\right)\frac{\partial\varepsilon_c}{\partial x_j}\right] \tag{11.51}$$

where $P_{k,c}$ is the production term induced by shear, S_k and S_ϵ are the source terms due to bubble-induced turbulence. For each phase, eddy viscosity is estimated as:

$$\mu_{t,c}=C_\mu\rho_c\frac{{k_c}^2}{\varepsilon_c} \tag{11.52}$$

Since the energy lost through the drag force of the bubbles is converted to turbulent kinetic energy, the turbulent kinetic energy source can be expressed as:

$$S_k=\mathrm{K}F_d\mathrm{U}_r \tag{11.53}$$

where K is a coefficient to tune the turbulent source, F_d is the drag force, and U_r is the relative velocity. A different expression is developed for S_ε considering the time scale of bubble-induced turbulence τ:

$$S_\varepsilon = \frac{C_\varepsilon}{\tau} S_k \tag{11.54}$$

In multiphase flows, the time scale is related to the turbulent eddy lifetime before passing to a smaller scale as well as to velocity and length scale of bubbles. There are a number of diversified approaches to model this term, which correspond to different expressions of dissipation energy source term. One of the most common expression is:

$$S_\varepsilon = C_\varepsilon \frac{k^{0.5}}{d_B} S_k \tag{11.55}$$

For liquid phases, a multiphase formulation for Reynolds stress tensor is additionally used in addition to the $k-\varepsilon$ model.

$$\frac{\partial}{\partial t}\left((1-\alpha)\rho_c R_{ij}\right) + \frac{\partial}{\partial x_j}\left((1-\alpha)\rho_c U_{i,c} R_{ij}\right) = (1-\alpha)S_{ij} + (1-\alpha)\left(P_{ij} + {}_{ij} - \varepsilon_{ij}\right) + \frac{\partial}{\partial x_j}\left[(1-\alpha)D_{ij}\right] \tag{11.56}$$

$$R_{ij} = \frac{\tau_{ij}^{Re}}{\rho_c} \tag{11.57}$$

where D_{ij} is the Reynolds stress diffusion and p_{ij} is the pressure–strain model for pressure fluctuations (Colombo & Fairweather, 2015).

11.3 Overview on gas transport in membrane reactors

In the following subsections, several aspects related to the mass transfer inside and outside membranes are presented and discussed.

11.3.1 Mass transfer among phases in fluidized beds

The literature approach to mass transfer among phases in fluidized beds can be distinguished into (1) gas–liquid and (2) wall-to-bed mass transfer.

As for the former, several experimental works were carried out to evaluate the mass transfer coefficient (Asfour & Nhaesi, 1990; Dhanuka & Stepanek, 1980; Kim & Kim, 1990; 1990; Alvarez-Cuenca, Nerenberg, & Asfour, 1984; Nguyen-Tien, Patwari, Shumpe, & Deckwer, 1985; Patwari, Nguyen-Tien, Schumpe, & Deckwer, 1986; Chang, Kang, & Kim, 1986; Schumpe, Deckwer, & Nigam, 1989; Kim, Park, & Kim, 1990; Kang, Fan, Min, & Kim, 1991; Miyahara, Lee, & Takahasi, 1993; Zheng, Chen, Feng, & Hoffmann, 1995), whose correlations are summarized in Table 11.1.

TABLE 11.1 Main literature correlations for the mass transfer coefficient

Nguyen-Tien et al. (1985)	$\phi_s = \varepsilon_s/(1-\varepsilon_g) k_L a = 0.39\left(1 - \frac{\phi_s}{0.58}\right) U_g^{0.67}$ $0.018\,\mathrm{m/s} \le U_g \le 0.16\,\mathrm{m/s}; 0.05\,\mathrm{mm} \le d_p \le 1\,\mathrm{mm}$
Patwari et al. (1986)	$k_L a = 1.68 \cdot 10^{-2} U_g^{0.36} \mu_l^{-1.30} D_e^{0.5}$
Schumpe et al. (1989)	$k_L a / D_e^{0.5} = 2988 U_g^{0.44} U_l^{0.42} \mu_l^{-0.34} U_t^{0.71}$ $0.017 \le U_g(\mathrm{m/s}) \le 0.118; 0.03 \le U_l(\mathrm{m/s}) \le 0.16; 0.001 \le \mu_l(\mathrm{Pa\ s}) \le 0.119;$ $0.08 \le U_t(\mathrm{m/s}) \le 0.6$
Kim and Kim (1990)	$\frac{k_L a d_p^2}{D_e} = 0.0015\left(\frac{P_v d_p^4}{v_l^3}\right)^{0.67}\left(\frac{U_l^2 \rho_{av} d_p}{W_{sLm}}\right)^{0.1}\left[1 + 0.036\left(\frac{V_f}{V_s}\right)^{1.18} - 1.149 \cdot 10^{-3}\left(\frac{V_f}{V_s}\right)^{2.09}\right]$
Kim et al. (1990)	$k_L a = 0.73 U_g^{0.87} U_l^{0.45} d_p^{0.71}\left[1 + 0.036\left(\frac{V_f}{V_s}\right)^{1.11} - 1.348 \cdot 10^{-3}\left(\frac{V_f}{V_s}\right)^{2.09}\right]$ $0.02 \le U_g \le 0.2\,\mathrm{m/s}; 0.02 \le U_l \le 0.1\,\mathrm{m/s}; 1.0 \le d_p \le 6.0\,\mathrm{mm}$ $0.0 \le V_f/V_s \le 0.2$
Kang et al. (1991)	$k_L a = 0.256 U_g^{0.56} U_l^{0.41} \mu_p^{-0.52} d_p^{0.47} (1 + R_b)^{1.68}$
Lee, Kim, & Kim (1993)	*Bubble-disintegrating regime:* $k_L a = 2.36 \cdot 10^{-5}\ U_g^{0.686} U_l^{0.469} d_p^{0.788} \sigma_l^{-1.532} \mu_l^{-0.548}$ $\frac{k_L a d_p^2}{D_e} = 4.51 \cdot 10^{-5}\left(\frac{v_l}{D_e}\right)^{0.5}\left(\frac{E_D d_p^4}{v_l^3}\right)^{0.507}\left(\frac{U_l^2 \rho_l d_p}{\sigma_l}\right)^{0.457}$ Bubble-coalescing or slug-flow regime $k_L a = 1.10 \cdot 10^{-6}\ U_g^{0.940} U_l^{0.381} d_p^{0.790} \sigma_l^{-2.273} \mu_l^{-0.671}$ $\frac{k_L a d_p^2}{D_e} = 4.19 \cdot 10^{-5}\left(\frac{v_l}{D_e}\right)^{0.5}\left(\frac{E_D d_p^4}{v_l^3}\right)^{0.483}\left(\frac{U_l^2 \rho_l d_p}{\sigma_l}\right)^{0.436}$ $0.01 \le U_g(\mathrm{m/s}) \le 0.20; 0.01 \le U_l(\mathrm{m/s}) \le 0.12; 1.0 \le d_p(\mathrm{mm}) \le 8.0$ $42.6 \le \sigma_l(\mathrm{mN/m}) \le 72.4; 0.001 \le \mu_l(\mathrm{Pas}) \le 0.119$
Zheng et al. (1995)	$k_L a = a_1 U_g^{a_2} U_p^{a_3} (1 - \bar{\varepsilon}_s)^{a_s} \cdot \left(1 + a_6\sqrt{h} + a_7 h\right)$

11.3.2 External mass transfer between membrane and fluid bulk in the absence and presence of inhibitors

Owing to the membrane permselectivity, some species in mixture are able to pass through preferentially with respect to the other ones. For this reason, the less-permeating species tend to accumulate near the membrane surface, creating in this way a resistance to the transport of the more permeating ones. This phenomenon is usually referred to as *"concentration polarization,"* and it can seriously decrease the permeating flux with respect to the pure-species condition. This is particularly crucial for metal membranes, which have a virtually infinite permeability toward hydrogen with respect to all the other gas species (Caravella and Sun, 2016; Caravella et al., 2016; Caravella, Barbieri, & Drioli, 2009).

For this type of membrane, the situation becomes more complex in the presence of inhibitor species, which competitively adsorb on the metal surface reducing the possibility for hydrogen to dissociatively adsorb.

To model the hydrogen permeation in the presence of inhibition only, Wang, Flanagan, and Shanahan (2004) developed the following permeation law, which was obtained by correcting the permeance according to the Langmuir adsorption model:

$$N_{H_2} = \left[\frac{1}{1+c \cdot \exp\left(\frac{\Delta E_{CO}}{RT}\right)}\right]^2 \overline{P}^0_{H_2} \exp\left(-\frac{E_a}{RT}\right) \frac{\sqrt{P_{H_2,Ret}} - \sqrt{P_{H_2,Per}}}{\delta} \tag{11.58}$$

Likewise, Barbieri, Scura, Lentini, De Luca, and Drioli (2008) presented a novel permeation equation accounting for the CO adsorption on Pd-alloy surface:

$$J_{H_2} = \left\{\left[1 - \alpha(T)\frac{K_{CO}P_{CO}}{1 + K_{CO}P_{CO}}\right]\pi_0 \exp\left(-\frac{E_a}{RT}\right)\right\}\left[\sqrt{P^{Feed}_{H_2}} - \sqrt{P^{Perm}_{H_2}}\right] \tag{11.59}$$

which was successively used in Caravella, Scura, Barbieri, and Drioli (2010) to describe the coadsorption of CO and H_2. In that work, the definition of the concentration polarization coefficient was extended to include the presence of inhibitors, defining as well the inhibition coefficient (IC, Eq. 11.60–11.62) and the permeation reduction coefficient (PRC, Eqs. 11.63–11.64), which can include the combined influence of both polarization and inhibition.

$$\pi^{Inhibited}_{Membrane} = \left(1 - \alpha\frac{K_{CO}\,P_{CO}}{1 + K_{CO}\,P_{CO}}\right)\pi^{Clean}_{Membrane} \tag{11.60}$$

$$IC = \alpha\frac{K_{CO}\,P_{CO}}{1 + K_{CO}\,P_{CO}} \tag{11.61}$$

$$\pi^{Inhibited}_{Membrane} = (1 - IC)\,\pi^{Clean}_{Membrane} \tag{11.62}$$

$$PRC = 1 - \frac{\pi^{Inhibited}_{Bulk}(\text{Elementary Steps})}{\pi^{Clean}_{Membrane}(\text{Elementary Steps})} \tag{11.63}$$

$$PRC = 1 - (1 - CPC)(1 - IC) \tag{11.64}$$

Similar to Caravella et al. (2010), Catalano, Giacinti Baschetti, and Sarti (2010) developed a permeation model accounting for the presence of adsorbing CO based on a Langmuir kinetic approach (Eqs. 11.65–11.67).

$$r_{CO}^{ads} = k_{CO}^{ads} p_{CO}^{int} n_S (1 - \theta_{CO} - \theta_H) \tag{11.65}$$

$$r_{CO}^{des} = k_{CO}^{des} n_S \theta_{CO} \tag{11.66}$$

$$r_H^{ads} = 2 S_0 k_H^{ads} p_{H_2} n_S^2 \left(\frac{1}{1 + \lambda_{CO} P_{CO}^{int}} \right)^2 (1 - \theta_H)^2 \tag{11.67}$$

At the same time, Mejdell, Chen, Peters, Bredesen, and Venvik (2010) modeled the influence of adsorbed CO on hydrogen permeation using the Sieverts–Langmuir approach within a microkinetic model, obtaining the following expression for the surface coverage of CO:

$$\theta_{CO} = \frac{K_{CO} P_{CO}}{1 + K_{CO} P_{CO} + \sqrt{K_{H_2} P_{H_2}}} \tag{11.68}$$

Later, Abir and Sheintuch (2014) and Patrascu and Sheintuch (2015) developed a first-principle multicomponent permeation law comprising a multiadsorption model based on the Langmuir approach.

$$\Theta = \frac{1 + \sqrt{K_H P_{H_2}}}{1 + \sqrt{K_H P_{H_2}} + K_{H_2O} P_{H_2O} + \sqrt[3]{K_{CO} P_{CO}} + \sqrt{K_{CO_2} P_{CO_2}} + K_{CH_4} P_{CH_4}} \tag{11.69}$$

As well, Boon, Pieterse, van Berkel, van Delft, and van Sint Annaland (2015) used a similar theoretical methodology based on the multicomponent Sips adsorption model (Do, 1998), obtaining the following expression for the surface coverage of the generic adsorbed species:

$$\theta_i = \frac{(b_i p_i)^{1/n_i}}{1 + \frac{\theta_H}{1 - \theta_{Tot}} + \sum_{i \neq H_2} (b_i p_i)^{1/n_i}}, \quad i \neq H_2 \tag{11.70}$$

11.3.3 Peculiar aspects on the effect of hydrogen dissolved in metal membranes

The most used metal membranes in membrane reactors are the Pd-alloy ones thanks to their peculiar performance in separating hydrogen from other gases. Furthermore, there are several other nonPd alloys that have been currently studied for hydrogen separation, among which vanadium alloys represent an important type of membranes showing interesting performance in terms of permeability and solubility. It must be emphasized that even the nominal nonPd alloys actually have an ultra-thin Pd top layer to promote the hydrogen dissociation. The details of these types of membranes are reported in the following subsections.

11.3.3.1 *Effect of hydrogen interstitial site occupancy*

Hydrogen absorbed in metals occupies interstitial sites, as shown by neutron diffraction and other evidences (Alefeld & Volkl, 1978). The face-centered cubic (FCC) lattice shows one octahedral (O) interstitial site per metal atom and two tetrahedral (T) interstitial site per metal atom, as well as the hexagonal close-packed (HCP) lattice. In contrast, the body-centered cubic (BCC) lattice, characterizing the structures of vanadium, niobium, and tantalum, has an atom on the cubic cell corners and a single one at the center. With such a configuration, a total of three octahedral sites and six tetrahedral ones per metal atom are found.

In FCC lattices, the octahedral sites are larger than the tetrahedral ones, whereas the opposite is observed in the BCC lattice. Therefore in BCC lattices belonging to the Groups IV and V, the occupation of tetrahedral sites by the H atoms is favored over the occupation of octahedral ones, whereas the opposite occurs for H in Pd (Dolan, 2010). As a matter of fact, Cser, Török, Krexner, Prem, and Sharkov (2004) observed that the H_2 molecule dissociates in H-atoms on palladium surface, which are preferentially absorbed into the octahedral sites of the FCC lattice.

In the same way, a study on a 50%–50% atomic ratio vanadium–deuterium system conducted by neutron diffraction revealed that c.90% of the number of dissolved deuterium atoms occupy the BCC-lattice tetrahedral sites, whereas the remainder 10% occupy the octahedral positions (Oriani, 1994). Furthermore, using a density functional theory analysis, Lu, Gou, Bai, Zhang, and Chen (2017) found that the H_2 solution energy in the T-sites of pure vanadium is −0.332 eV against the value of −0.149 eV for the O-sites, demonstrating that hydrogen is preferentially absorbed in the T-sites of BCC metals.

The introduction of one hydrogen atom inside the lattice increases the volume of the system, but only by one-fifth of the atomic volume.

Depending on the particular operating conditions, hydrogen absorbs preferentially in certain types of sites. The deep understanding for the reasons for such a changing behavior allows understanding the complex mechanism of hydrogen diffusion in metals (Fukai, 1984).

As the interstitial dissolution sites represent a finite population, the dissolved hydrogen occupation can be described by the equilibrium thermodynamic. In this way, one can express the activity of the dissolved hydrogen in a more general manner, passing through the chemical potential.

$$\mu = \mu^0 + \mathrm{RTln}(a) = \mu^0 + \mathrm{RTln}\left(\gamma \frac{\theta_1}{1-\theta_1}\right) \tag{11.71}$$

where $\theta_1 = n_H/n_{S_1}$, n_H is the dissolved hydrogen concentration, and n_{S_1} is the concentration of the class of interstitial sites (tetrahedral, octahedral, etc.) that produces the lowest Gibbs free energy, which for palladium are the octahedral ones. Therefore it implies that for θ_1 tending to 1, the chemical potential μ tends to infinity. In other words, as the occupancy degree of the available sites increases, it is progressively more difficult to occupy the remaining ones, which have a higher chemical potential, that is, one needs to overcome greater activation energies for the filling process. In practice, the source of hydrogen at a certain point is not able to occupy the lowest-energy class of interstitial sites anymore, and thus the next-higher energy class ones start to be gradually filled (Oriani, 1994).

Indeed, McLennan, Gray, and Dobson (2008) confirmed that the tetrahedral sites in palladium–deuterium systems are partially occupied when the deuterides PdD_x are formed above the critical point, in spite of the former theory, which expected only one class of sites – for a Pd–X system, the octahedral ones – to be filled. Moreover, they observed that the maximum occupancy of tetrahedral sites at a D/Pd atomic ratio of 0.6, at which one-third of all D atoms are placed in tetrahedral sites.

Additionally, the H–Pd interactions induce energy difference among sites of the same class. As an instance, it is possible that H atoms occupy tetrahedral sites placed on alternate planes, forming in this way a symmetric sublattice, as that occurring in Ta–H systems (Oriani, 1994).

When interstitial sites are occupied, a certain stress is induced in the lattice, which usually expands upon dissolution of H atoms. The so-called *ball-in-hole* model can satisfactorily describe the dissolution phenomenon. In particular, this model assumes the dissolved atoms as rigid *balls*, whereas the dissolution sites are considered to be elastic and spherical *holes*. Since at the beginning, the hole is smaller than the ball, the latter produces a force toward the cavity walls forcing the metal matrix to increase its volume. If ΔV_1 is the volume of the misfitting ball, the volume of the cavity ΔV_2 is given by:

$$\Delta V_2 = \frac{3(1-\nu)}{1+\nu}\Delta V_1 \tag{11.72}$$

where ν is the lattice Poisson ratio. For a typical value of atomic ratio $n_H/n_{Pd} = 0.3$, ΔV_2 is larger than ΔV_1 by 50%. Fukai (1984) found that such an augmented volume is rather incompressible: at around 130 GPa, the hydride volume decreases by about 74% of its original value, while pure vanadium decreases by almost 69%; thus the two values are very close.

Indeed, putting hydrogen by simply applying a pressure hinders the lattice expansion, while sending hydrogen by subjecting the metal to cathodic charging does not cause decrease of the expanding effect, because of the absence of pressure on the external metal surface which would act as a compressing force.

11.3.3.2 Interactions of hydrogen with lattice imperfections

The dissolved hydrogen can interact with possible interstitial defects, which can be divided into three types: zero-, one- and two-dimensional ones. The first ones (point defects) are local defects where an atom is missing (vacancy) or is trapped into an interstitial void (self-interstitials).

Indeed, as temperature increases, the atoms' position changes in the lattice and becomes more frequent and random. It can even occur that the vacancy population is higher than the equilibrium value, like after quenching from high temperature.

A peculiar observation was made by Kirchheim (1986), who prepared two palladium samples: a defect-free one and a sample plastically stretched (up to 50%) by cold drawing/rolling to increase the defect extent. Specifically, he measured the hydrogen partial molar volume $\overline{V}_H$ in both samples.

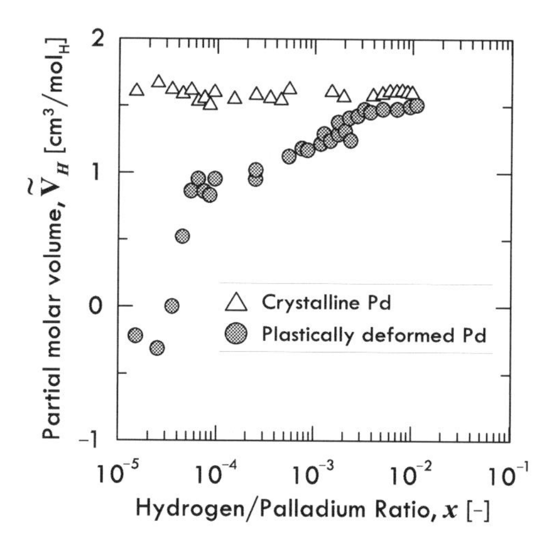

FIGURE 11.2 Partial molar volume of hydrogen as a function of H/Pd atomic ratio. *Source: Original data from Kirchheim R., 1986. Interaction of hydrogen with external stress fields,* Acta Metallurgica, *34: 37–42.).*

The results are summarized in Fig. 11.2 (Kirchheim, 1986), where it can be noticed that below a certain value of hydrogen concentration (40 ppm), $\overline{V}_H$ is even negative, which means that the lattice shrinks instead of expanding. Kirchheim explained this apparently strange observation by considering the chemi-physical interactions involved in the filling phenomenon of vacancies by hydrogen. Specifically, if the H atom is sufficiently close to the nearest Pd atom, the Pd–H attractive forces are dominant, and thus the lattice shrinks. From the value of 40 ppm onward, the lattice starts growing positively, as the effect of site expansion becomes dominant over the former.

Mutschele and Kirchheim (1987) observed a Gaussian distribution for the Pd–H interaction energies, calculating an expected value E_1 of 9.2 kJ/mol$_H$, whereas the Pd–H interaction energy within the grains E_0 is equal to 3.9 kJ/mol$_H$. The difference between E_1 and E_0 (5.3 kJ/mol$_H$) is the segregation energy, which is the energy necessary for the H atom to pass from the metal site to the defect.

Another type of defects is represented by microvoids, arising when an external stress is applied to a ductile material. If the stress is applied continuously, microcracks can be formed by void coalescence.

Furthermore, we can distinguish the interactions between H atoms and lattice defects into chemical and elastic. The chemical defects are defects involving relatively large atomic displacements, whilst the elastic ones are defects involving small displacements sufficiently far from defects to be able to apply the linear elasticity theory (Oriani, 1994).

11.3.3.3 Hydrogen permeability and solubility in metals

Permeability is the ability of a certain material to let a species pass through and depends on other two intrinsic properties, that is, diffusivity and solubility. It is the best

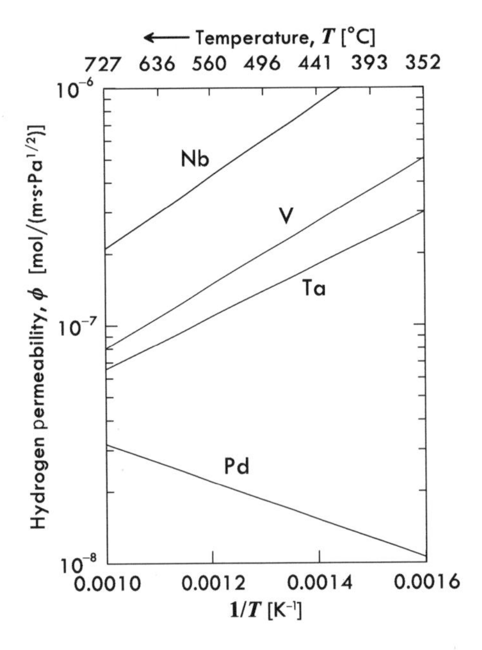

FIGURE 11.3 Permeabilities of selected metals as a function of inverse temperature. *Source: Original data from Buxbaum, R. E., Kinney, A. B. (1996). Hydrogen transport through tubular membranes of palladium-coated tantalum and niobium. Industrial & Engineering Chemistry Research, 35, 530–537.*

metric to compare the intrinsic mass transport properties of different materials when the transport is limited by the bulk diffusion.

Fig. 11.3 (Buxbaum & Kinney, 1996) shows the permeability as a function of temperature for several metals. In particular, we can observe that the early transition metals (Nb, V, Ta) with a BCC structure show the highest permeability. Although these materials could easily overcome the performances of palladium in terms of the sole permeability, in practice, they are subjected to serious embrittlement, which is a chemi-physical process by which the introduced hydrogen in the metallic solid generates microcracks leading to mechanical failure of the metal and represents a crucial limit preventing from obtaining sufficiently high hydrogen flux (Dolan, 2010).

As for the permeability function with temperature, the BCC metals show a decreasing permeability with increasing temperature, whereas the opposite is observed for palladium. This effect can be explained considering the decreasing solubility trend with increasing temperature.

At moderate pressures, the concentration of hydrogen dissolved in solid metals $C_{[H]}$ is well described by the empirical relation well known as Sieverts' law:

$$C_{[H]} = S \cdot \sqrt{p_{H_2}} \tag{11.73}$$

where S (temperature-dependent Sieverts' constant) represents a measure of the solubility of hydrogen into the solid metal. Oriani (1994) pointed out how the square root correlation came out from the nature of the problem, which involves for each H_2 molecule, the dissociation into two H atoms, as proved by Sieverts:

$$H_2(g) \rightleftharpoons 2[H] \tag{11.74}$$

The equilibrium constant for this reaction can be written as follows:

$$K = \frac{C_{[H]}^2}{C_{H_2}} \quad (11.75)$$

Through the ideal gas equation, one can write:

$$K' = K \cdot RT = \frac{C_{[H]}^2}{p_{H_2}} \quad (11.76)$$

Considering the thermodynamic of the problem, it is more accurate to write the previous equation in terms of activity and fugacity, described as:

$$a = \gamma C \quad (11.77)$$

$$f = \beta p \quad (11.78)$$

where γ and β are the related coefficient which could be approximated to unity for low values of concentrations and pressure, respectively. Hence

$$\mathcal{K} = \frac{a_{[H]}^2}{f_{H_2}} \quad (11.79)$$

And finally:

$$C_{[H]} = \left(\frac{1}{\gamma}\sqrt{\frac{\mathcal{K}}{\beta}}\right) \cdot \sqrt{p_{H_2}} = \mathbf{S} \cdot \sqrt{p_{H_2}} \quad (11.80)$$

So that S arises from this last equation. The $\mathbf{K}$ term can be expressed as:

$$\ln\mathcal{K} = -\frac{\Delta\overline{G}^0}{RT} \quad (11.81)$$

where $\Delta\overline{G}^0$ is the change of molar Gibbs free energy between standard states for the dissociation reaction. Passing through the natural logarithms, we obtain:

$$\ln\mathbf{S} = -\frac{\Delta\overline{G}^0}{RT} - \ln\left(\gamma\beta^{1/2}\right) = -\frac{\Delta\overline{H}^0}{RT} + \frac{\Delta\overline{S}^0}{R} - \ln\left(\gamma\beta^{1/2}\right) \quad (11.82)$$

where $\Delta\overline{H}^0$ and $\Delta\overline{S}^0$ are the molar variation between standard states of enthalpy and entropy, respectively. The former term classifies two categories of metals, conventionally called *endothermic* and *exothermic occluders* of hydrogen (Oriani, 1994). The substantial difference between these two is that metals for which $\Delta\overline{H}^0 < 0$ (exothermic), show a solubility decreasing with increasing temperature, and those for which $\Delta\overline{H}^0 > 0$ (endothermic), display the opposite behavior.

In the case of metals in which only a simple solution of hydrogen occurs, the solubility increases as the temperature increases. This is actually the case of endothermic occluders, for example, Mn, Fe, Co, and Ni, which only form a metal–hydrogen solid solution in which the hydrogen is located at random interstices (Dolan, 2010), the extent of the solid solution varying according to the hydrogen pressure and the temperature. However, the transition metals such as Pd, Ta, V, and Nb are exothermic occluders, which means that the solubility decreases with increasing temperature. This is related to their tendency to form ordered hydrides at high hydrogen concentrations, whose formation process is exothermic (Cotterill, 1961).

Permeability can be seen as the product of two terms: the diffusivity and the solubility. Diffusion coefficient of hydrogen through metals can be determined using a range of methods. The diffusivities of palladium, niobium, vanadium, and tantalum are illustrated in Fig. 11.4 (Volkl and Alefeld, 1978). Analyzing this graph, it is visible how, from a certain temperature on, diffusivities of Nb, Pd, and V are comparable, even if vanadium displays greater values.

What makes the difference is the hydrogen solubility in such metals. Fig. 11.5 shows the solubility isotherms for V, Ta, a V–Ni alloy (85:15) at 400°C and for palladium at 340°C. As we can observe, V absorbs much more hydrogen than palladium. For a hydrogen partial pressure of 7 bar, the H-to-V ratio at 400°C is around 0.6, which means a number of 6 H atoms per 10 V atoms (38÷62). In contrast, in the same conditions, the H-to-Pd ratio at 340°C is around 0.02 (2÷98).

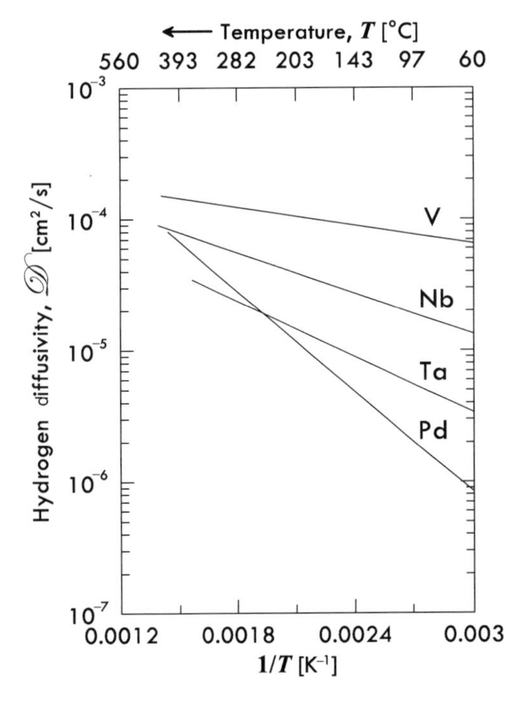

FIGURE 11.4 Diffusion coefficients of selected metals as a function of inverse temperature. *Source: Original data from Volkl, J., Alefeld, G. (1978). Diffusion of hydrogen in metals. In G. Alefeld, J. Volkl (Eds.), Hydrogen in metals. I: Basic properties (pp. 321–348). Berlin, Heidelberg: Springer Berlin Heidelberg.*

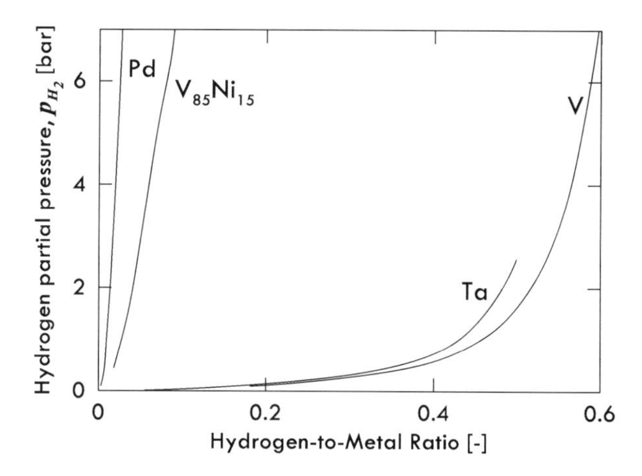

FIGURE 11.5 Solubility isotherms for selected metals and alloys. *Source: Original data from Dolan, M.D. (2010). Non-Pd BCC alloy membranes for industrial hydrogen separation. Journal of Membrane Science, 362, 12–28.*

Therefore the dominant contribution to permeability is solubility for BCC metals, whereas diffusivity controls permeability in the FCC ones (Pd in particular) (Dolan, 2010).

An interesting aspect to notice is that the solubility of vanadium to hydrogen significantly reduces by adding a small percentage of nickel to the V lattice. Specifically, in the same conditions, the hydrogen-to-metal ratio reduces from 0.6 to 0.09, reducing in turn the hydrogen solubility and thus the tendency to embrittlement (Dolan, 2010).

11.3.4 Aspects on mass transport of gases in microporous ceramic membranes

Microporous ceramic membranes can be used to separate gas species in membrane reactors and in conditions of relatively high temperatures, at which organic membranes are seriously degraded. The selective transport of gases in these types of membranes is based on several different transport mechanisms occurring in series–parallel to each other: surface diffusion, Knudsen diffusion, and gas-translation diffusion. The most complete model for surface diffusion was developed by Krishna (1990), who adopted a multicomponent approach based on the Maxwell–Stefan one to describe the transport of adsorbed species over the surface (Krishna and Wesselingh, 1997; Krishna, 1990; Kaptejin, Moulijn, & Krishna, 2000; Bakker, van den Broeke, Kapteijn, & Moulijn, 1997):

$$-\rho\theta_i\frac{\nabla\mu_i}{\mathrm{RT}} = -\rho\sum_{j=1}^{n}\Gamma_{\mathrm{ij}}\nabla\theta_j = \sum_{j=1}^{n}\frac{C_{\mu,j}N_i^{\mathrm{SD}} - C_{\mu,i}N_j^{\mathrm{SD}}}{C_{\mu\mathrm{s},i}C_{\mu\mathrm{s},j}D_{\mathrm{ij}}^{\mathrm{SD}}} + \frac{N_i^{\mathrm{SD}}}{C_{\mu\mathrm{s},i}D_i^{\mathrm{SD}}} \tag{11.83}$$

As for the Knudsen diffusion, its original form, which describes the transport of a single species along the pores driven by elastic collisions with the pore walls, was modified by Caravella, Zito, Brunetti, Drioli, and Barbieri (2016) to include the effect of pore blocking caused by the reduced porosity and the augmented tortuosity due to the presence of the other adsorbed species in mixture [Eqs (11.71)–(11.74), Fig. 11.6].

$$D_{i,\mathrm{eff}}^{\mathrm{Kn}} = \frac{\varepsilon(\underline{C}_{\mu},\underline{d}_k)}{\tau(\underline{C}_{\mu},\underline{d}_k)}\frac{d_p(\underline{C}_{\mu},\underline{d}_k)}{3}\sqrt{\frac{8\mathrm{RT}}{\pi M_i}} \tag{11.84}$$

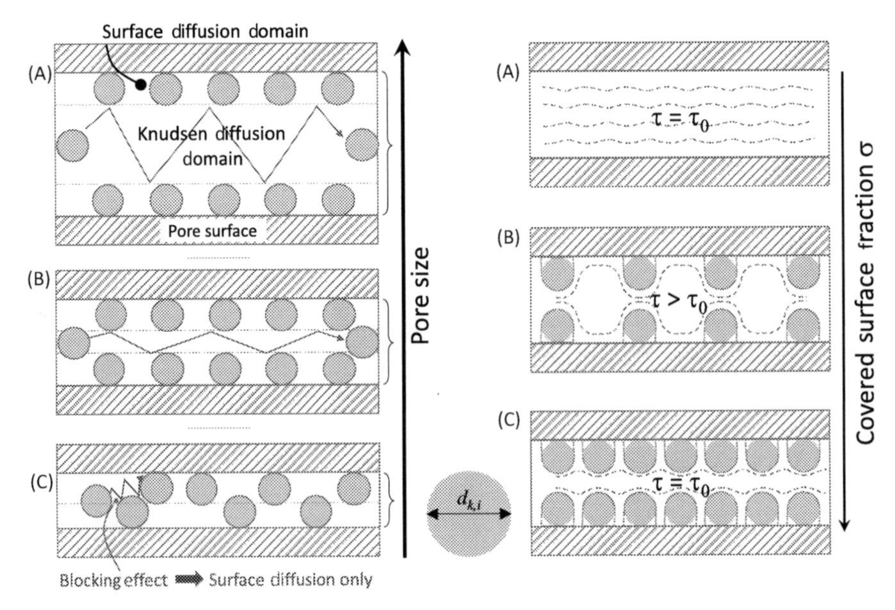

FIGURE 11.6 Sketches of some peculiar aspects of the mass transport within microporous membranes. *Source: Adapted and reprinted from Caravella, A., Zito, P. F., Brunetti, A., Drioli, E., Barbieri, G. (2016). A novel modelling approach to surface and Knudsen multicomponent diffusion through NaY zeolite membranes. Microporous and Mesoporous Materials, 235, 87–99, with permission of Elsevier.*

$$\varepsilon(\underline{C}_{\mu}, \underline{d}_k) = \varepsilon_0 - \rho(1-\varepsilon_0)N_{Av}\sum_{i=1}^{n} d_{k,i}^3 C_{\mu,i} \tag{11.85}$$

$$d_p(\underline{C}_{\mu}, \underline{d}_k) = \frac{\frac{4\varepsilon_0}{\rho(1-\varepsilon_0)} - 4N_{\mathrm{Av}}\sum_{i=1}^{n} d_{k,i}^3 C_{\mu,i}}{S_{g0} + N_{\mathrm{Av}}\left(\frac{\pi}{4}+2\right)\sum_{i=1}^{n} d_{k,i}^2 C_{\mu,i}} \tag{11.86}$$

$$\sigma(\underline{C}_{\mu}, \underline{d}_k) = \frac{\pi}{4S_{g0}} N_{\mathrm{Av}}\sum_{i=1}^{n} d_{k,i}^2 C_{\mu,i}$$

$$a = \frac{1}{\mathrm{n}d_p}\sum_{i=1}^{n} d_{k,i} \tag{11.87}$$

$$\tau(\underline{C}_{\mu}, \underline{d}_k) = \tau_0 + \frac{a\sigma(1-\sigma)}{\varepsilon^a}$$

Regarding the gas-translation diffusion, the molecules diffusing in the porous structure keep their own species identity overcoming the energy barrier to move from one site to another one. This implies an activated Knudsen-type diffusion coefficient:

$$D_i^{\mathrm{GT}} = \frac{\varepsilon(\underline{C}_{\mu}, \underline{d}_k)}{\tau(\underline{C}_{\mu}, \underline{d}_k)}\frac{\alpha}{z}\sqrt{\frac{8\mathrm{RT}}{\pi M_i}}\exp\left(-\frac{\tilde{E}_i^{\mathrm{GT}}}{\mathrm{RT}}\right) \tag{11.88}$$

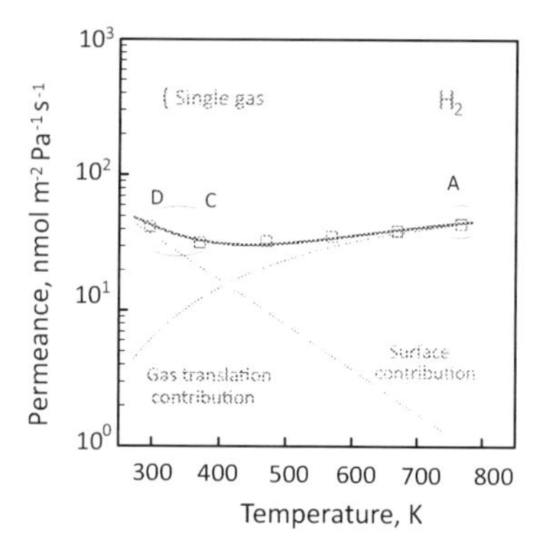

FIGURE 11.7 Single-gas permeation flux of hydrogen as a function of temperature. *Source: Reprinted from Zito, P.F., Caravella, A., Brunetti, A., Drioli, E., Barbieri, G. (2018). Discrimination among gas translation, surface and Knudsen diffusion in permeation through zeolite membranes. Journal of Membrane Science, 564, 166–173, with permission of Elsevier.*

$$J_i^{GT} = \frac{\varepsilon(\underline{C}_\mu, \underline{d}_k)}{\tau(\underline{C}_\mu, \underline{d}_k)} \frac{\alpha}{z} \sqrt{\frac{8RT}{\pi M_i}} \exp\left(-\frac{\tilde{E}_i^{GT}}{RT}\right) \frac{\Delta C_i}{\delta} \tag{11.89}$$

where the parameter z is the coordination number, which is the mean number of adjacent cage site, α is the distance between two consecutive sites and E^{GT} is the activation energy of the gas-translation diffusion mechanism. The combination of the just-recalled diffusion mechanisms leads to a mechanism that can describe the complex behavior of permeance with temperature (see the example of hydrogen in Fig. 11.7) (Zito, Caravella, Brunetti, Drioli, & Barbieri, 2018).

The existence of the gas-translation diffusion depends on the particular ratio of particle/kinetic diameter as well as on the species/wall interaction energy. In particular, such a mechanism can occur for each species at unitary-order diameters ratio and at relatively high temperature, where the species/wall interactions characterizing the surface diffusion become sufficiently weak for the gas translation to become dominant. In this case, the apparent activation energy of diffusion shifts from the adsorption site-to-site activation energy to the cage-to-cage one typical of gas translation.

11.4 Heat transfer among phases

In reactors involving three phases, the solid particles become fluidized by the action of two flowing fluid phases. If we have a solid–liquid–gas system, gas forms bubbles, which are segregated from the particle-containing continuous phase. In case of solid–liquid–liquid system, the immiscible liquid is the segregated phase, whereas the other contains the solid particles.

The study of three-phase systems is important in reactions between gas and liquid phases that need a certain specific catalyst to be promoted. Such kind of systems are applied in biochemical and petrochemical processes such as the hydrogenation and hydrodesulfurization of waste oil, Fisher–Tropsch reactions, fermentation processes, which involve, for example, cells or other biomaterials deposited on pellets, water as growth environment and oxygen (in aerobic growth processes), treatment of wastewater and liquefaction of coal (Fan, 1989; Kim, Kang, & Kwon, 1986).

Out of the solid–liquid–gas systems and the solid–liquid–liquid one, the former is certainly more difficult to study because of the more complex resultant flow (Tarmy and Coulaloglou, 1992). This aspect is particularly important for heat transport within the single phases (see Table 11.2 for typical constitutive equations) and heat transfer among phases, for which some used correlations of literature are reported in Table 11.3 (Verma, Deen, Padding, & Kuipers, 2013).

The specific fluid dynamic conditions of fluidization tend to make the temperature profile inside reactor uniform in nonisothermal processes, thanks to the deep contact among phases. A crucial aspect that has been studied by a number of worldwide research group focuses on the analysis of fluid dynamics influence on the heat transfer coefficient. In particular, several groups have been paying their attention on the wall-to-bed heat transfer coefficient (Chiu and Ziegler, 1983; Chiu and Ziegler, 1985; Ostergaard, 1964; Viaswanathan, Kakar, & Murti, 1965; Kato, Kago, Uchida, & Morooka, 1980; Kato, Uchida, Kago, & Morooka, 1981; Kato, Taura, Kago, & Morooka, 1984; Muroyama, Fukuma, & Yasunishi, 1984; Muroyama, Fukuma, & Yasunishi, 1986), whereas others have focused on measuring the heater-to-bed heat transfer coefficient (Armstrong, Baker, & Bergougnou, 1976; Baker, Armstrong, & Bergougnou, 1978; Kang, Suh, & Kim, 1983; Kang, Suh, & Kim, 1985; Kim, Lee, & Kim, 1988). To combine the former with the latter, Muroyama et al. (1986) and Kim et al. (1988) proposed a two-resistance model able to theoretically predict the overall heat transfer coefficient, for which a number of semiempirical correlations were proposed in the literature (Chiu and Ziegler, 1985; Kim and Laurent, 1988; Suh and Deckwer, 1989; Suh, Jin, & Kim, 1985). In most cases, the characteristic temperature difference that the heat transfer coefficient is based on is the difference between the bulk of the fluidized bed and the heater surface. In contrast, some groups measured the real radial profile temperature (Chiu and Ziegler, 1983; Kang and Kim, 1988; Muroyama et al., 1986). The main literature correlations for the estimation of the heat transfer coefficient are encapsulated in Table 11.3.

TABLE 11.2 Constitutive equations for the two-fluid model (Nieuwland, van Sint Annaland, Kuipers, Van, & Swaaij, 1996).

Particle pressure: $p_s = [1 + 2(1 + e_n)\varepsilon_s g_0]\varepsilon_s \rho_s \Theta$
Bulk viscosity: $\lambda_s = \frac{4}{3}\varepsilon_s \rho_s d_p g_0 (1 + e_n)\sqrt{\frac{\Theta}{\pi}}$
Shear viscosity: $\mu_s = 1.016\frac{5\pi}{96}\rho_s d_p \sqrt{\frac{\Theta}{\pi}}\frac{\left(1 + \frac{8}{5}\left(\frac{(1 + e_n)}{2}\varepsilon_s g_0\right)\right)\left(1 + \frac{8}{5}\varepsilon_s g_0\right)}{\varepsilon_s g_0} + \frac{4}{3}\varepsilon_s \rho_s d_p g_0 (1 + e_n)\sqrt{\frac{\Theta}{\pi}}$
Dissipation of granular energy due to inelastic particle–particle collision: $\gamma = 3\left(1 - e_n^2\right)\varepsilon_s^2 \rho_s g_0 \Theta\left[\frac{4}{d_p}\sqrt{\frac{\Theta}{\pi}} - (\nabla \cdot \bar{u}_s)\right]$
Pseudo-Fourier fluctuating kinetic energy flux: $\bar{q}_s = -k_s \nabla \Theta$
Pseudo-thermal conductivity: $k_s = 1.02513\frac{75\pi}{384}\rho_s d_p \sqrt{\frac{\Theta}{\pi}}\frac{\left(1 + \frac{6}{5}(1 + e_n)\varepsilon_s g_0\right)\left(1 + \frac{12}{5}\varepsilon_s g_0\right)}{\varepsilon_s g_0} + 2\varepsilon_s \rho_s d_p g_0 (1 + e_n)\sqrt{\frac{\Theta}{\pi}}$

TABLE 11.3 Main heat transfer coefficient correlations in the literature

Reference	Correlations
Baker et al. (1978)	$h = 1977U_l^{0.070}U_g^{0.059}d_p^{0.106}$ (mm/s; mm)
Kato et al. (1981)	$\mathrm{Nu}' = 0.044(\mathrm{Re}'\mathrm{Pr})^{0.78} + 2.0(\mathrm{Fr}_g)^{0.17}$ $\mathrm{Nu}' = \frac{hd_p\varepsilon_l}{K_L(1-\varepsilon_l)}, Re' = \frac{\rho_l d_p U_l \varepsilon_l}{\mu_l(1-\varepsilon_l)}, Pr = \frac{\mu_l C_{Pl}}{K_L}, \mathrm{Fr}_g = \frac{U_g^2}{gd_p}$
Chiu and Ziegler (1983)	$\mathrm{St}_{m3} = \mathrm{St}_{m2} = 0.1234\mathrm{Re}_{m2}^{-0.305}Pr^{-2/3}$
Kang et al. (1983)	$h = 2290U_g^{0.10}U_g^{0.05}\mu_l^{-0.18}d_p^{0.04}$
Muroyama et al. (1984)	$\frac{h_3}{h_2} = 1 + 0.0413\left(\frac{U_g}{U_l}\right)^{0.30}\left[\frac{\frac{d_p}{D_c}(\rho_s-\rho_l)}{\rho_l}\right]^{0.61}$
Chiu and Ziegler (1985)	$\mathrm{Nu} = 0.762\mathrm{Re}_m^{0.646}Pr^{0.638}U_R^{0.266}\phi_s^{-1}\left(1 - \frac{\varepsilon_{l2}}{\varepsilon_{l3}}\right)$
Kang et al. (1985)	$\mathrm{Nu} = 0.036Re_l^{0.81}Pr^{0.65}$
Saberian-Broudjcnni, Wild, and Charpentier (1985)	$h = 2.13\cdot 10^{-4}\mathrm{U_l^{0.338}U_g^{0.540}}\rho_l^{1.235}\mu_l^{0.162}\mathrm{K_L^{-0.04}C_{Pl}^{1.04}d_p^{-0.930}}(\rho_s-\rho_l)^{-0.357}$
Suh et al. (1985)	$h = 0.0647\left[K_L\rho_l C_{Pl}\left\{\frac{[(U_l+U_g)(\varepsilon_l\rho_l+\varepsilon_g\rho_g+\varepsilon_s\rho_s)-U_l\rho_l]g}{\varepsilon_l\mu_l}\right\}^{0.5}\right]^{0.5}$
Kim et al. (1986)	$h = 0.0722\left[K_{sl}\rho_{sl}C_{Psl}\left\{\frac{[(U_{sl}+U_g)(\varepsilon_{sl}\rho_{sl}+\varepsilon_g\rho_g+\varepsilon_s\rho_s)-U_{sl}\rho_{sl}]g}{\varepsilon_{sl}\mu_{sl}}\right\}^{0.5}\right]^{0.5}$
Muroyama et al. (1986)	$j_H' = 0.137Re_{l,g}' - 0.271$ $j_H' = \frac{h_w}{\rho_l C_{Pl}U_l}\varepsilon_l Pr^{2/3}, Re_{l,g}' = Re_l(\varepsilon_g+\varepsilon_l)/[\varepsilon_l(1-\varepsilon_l-\varepsilon_g)]$
Hatate et al. (1987)	$\mathrm{Nu}Pr_l^{-1/3}(\mu_b/\mu_w)^{-0.14}/Re_l^{0.2}Re_g^{0.055} = 10.5\exp\{-3.18\cdot 10^{-4}(\mathrm{C_s}-22)^2\}$
Magiliotou, Chert, and Fan (1988)	$h = 0.1\left[K_L\rho_l C_{Pl}\left(\frac{P_v}{v_l}\right)^{0.5}\right]^{0.5} + 0.285\left[K_L\rho_l C_{Pl}\left\{\varepsilon_s^{1/3}(U_l-U_{mf})/(\varphi_s d_{eq})\right\}\right]^{0.5}$
Kim et al. (1990)	$h = 0.0685\cdot\cdot\left(\mathrm{K_L}\rho_l\mathrm{C}_{Pl}\left\{\frac{\{(\mathrm{U_l-U_g})\{[1-\left(\frac{V_f}{V_s}\right)/\left(1+\frac{V_f}{V_s}\right)]\varepsilon_s\rho_s+\varepsilon_f\rho_f+\varepsilon_l\rho_l+\varepsilon_g\rho_g\}-\mathrm{U_l}\rho_l\}\mathrm{g}}{\varepsilon_l\mu_l}\right\}^{1/2}\right)^{1/2}$ $0.02 \le U_g \le 0.14\mathrm{m/s}; 0.02 \le U_l \le 0.09\mathrm{m/s}; 1.0 \le d_p \le 6.0\mathrm{mm};$ $0.0 \le V_f/V_s \le 0.20; 1500 \le \rho_f \le 1800\mathrm{kg/m^3}$
Zaidi, Deckwer, Mrani, and Benchekchou (1990)	$\mathrm{Nu} = 0.042Re_l^{0.720}Pr_l^{0.860}Fr_g^{0.067}$ $0.81 \le U_g \le 14.4\mathrm{cm/s}; 1.27 \le U_l \le 9.0\mathrm{cm/s}; 3.7 \le \mu_l \le 300\mathrm{mPa};$ $d_p = 3$ and 5 mm $h = 1800U_g^{0.11}\mu_l^{-0.14}U_l^{1.03(0.65-\varepsilon)}d_p^{0.58(\varepsilon-0.68)}$ $0.01 \le U_g \le 0.14\mathrm{m/s}; 0.013 \le U_l \le 0.09\mathrm{m/s}$
Kumar, Kusakabe, and Fan (1993)	$\mathrm{Nu}' = 5.56\cdot 10^{-2}(\mathrm{Re}'\mathrm{Pr})^{0.709}\left[\frac{\rho_s-\rho_l}{\rho_l}\right]^{-0.156}$ where $\mathrm{Nu}' = h_{av,2}d_p\varepsilon_l/K_L(1-\varepsilon_l), Re'Pr = d_pU_l\rho_l C_{Pl}/K_L(1-\varepsilon_l)$ $59.96 \le Re'Pr \le 15472.7(\mathrm{for}\,Pr = 5.83); 1.04 \le \rho_s \le 2.50\mathrm{g/cm^3}$

11.5 Conclusion and future trends

The overview provided in this chapter evidenced the complex and intimate relationships among the transport phenomena involved in different types of membrane reactors. In particular, the fluid dynamics of the fluidized-bed ones is much more complex than the fixed-bed ones, and this causes several additional design problems related to the presence of the membranes. In fact, the flowing solid particles can cause serious damage to membranes because of the surface erosion. To avoid this detrimental phenomenon, a new generation of membranes provided with sufficiently resistant protective layers should be developed. This is particularly important for supported metal membranes, whose selective layer cannot bear the continuous rubbing with the catalyst particles.

References

Abir, H., & Sheintuch, M. (2014). Modeling H_2 transport through a Pd or Pd/Ag membrane, and its inhibition by co-adsorbates, from first principles. *Journal of Membrane Science, 466*, 58–69.

Alefeld, G., & Volkl, J. (Eds.), (1978). *Hydrogen in metals I: Basic Properties*. Berlin: Springer-Verlag.

Alvarez-Cuenca, M., Nerenberg, M. A., & Asfour, A. A. (1984). Mass transfer effects near the distributor of three-phase fluidized beds. *Industrial & Engineering Chemistry Fundamentals, 23*(4), 381–386.

Amiri, L., Ghoreishi-Madiseh, S. A., Hassani, F. P., & Sasmito, A. P. (2019). Estimating pressure drop and Ergun/Forchheimer parameters of flow through packed bed of spheres with large particle diameters. *Powder Technology, 356*, 310–324.

Armstrong, E. R., Baker, C. G. J., & Bergougnou, M. A. (1976). Heat transfer and hydrodynamics studies on three-phase fluidized beds. In D. U. Keairns (Ed.), *Fluidization technology* (pp. 453–457). New York: McGrawHill.

Asakuma, Y., Asada, M., Kanazawa, Y., & Yamamoto, T. (2016). Thermal analysis with contact resistance of packed bed by a homogenization method. *Powder Technology, 291*, 46–51.

Asfour, A. A., & Nhaesi, A. H. (1990). An improved model for mass transfer in three-phase fluidized beds. *Chemical Engineering Science, 45*, 2895–2900.

Ayeni, O. O., Wu, C. L., Nandakumar, K., & Joshi, J. B. (2016). Development and validation of a new drag law using mechanical energy balance approach for DEM–CFD simulation of gas–solid fluidized bed. *Chemical Engineering Journal, 302*, 395–405.

Baker, C. G. J., Armstrong, E. R., & Bergougnou, M. A. (1978). Heat transfer in three-phase fluidized beds. *Powder Technology, 21*, 195–204.

Bakker, W. J. W., van den Broeke, L. J. P., Kapteijn, F., & Moulijn, J. A. (1997). Temperature dependence of one-component permeation through a silicalite-1 membrane. *AIChE Journal. American Institute of Chemical Engineers, 43*, 2203–2214.

Bale, S., Tiwari, S., Sathe, M., Berrouk, A. S., Nandakumar, K., & Joshi, J. (2018). Direct numerical simulation study of end effects and D/d ratio on mass transfer in packed beds. *International Journal of Heat and Mass Transfer, 127*, 234–244.

Barbieri, G., Scura, F., Lentini, F., De Luca, G., & Drioli, E. (2008). A novel model equation for the permeation of hydrogen in mixture with carbon monoxide through Pd-Ag membranes. *Separation and Purification Technology, 61*, 217–224.

Beetstra, R., van der Hoef, M. A., & Kuipers, J. A. M. (2007). Numerical study of segregation using a new drag force correlation for polydisperse systems derived from lattice-Boltzmann simulations. *Chemical Engineering Science, 62*, 246–255.

Bird, R. B., Stewart, W. E., & Lightfoot, E. N. (1960). *Transport phenomena*. New York.: Wiley.

Boon, J., Pieterse, J. A. Z., van Berkel, F. P. F., van Delft, Y. C., & van Sint Annaland, M. (2015). Hydrogen permeation through palladium membranes and inhibition by carbon monoxide, carbon dioxide, and steam. *Journal of Membrane Science, 496*, 344–358.

Buxbaum, R. E., & Kinney, A. B. (1996). Hydrogen transport through tubular membranes of palladium-coated tantalum and niobium. *Industrial & Engineering Chemistry Research*, *35*, 530–537.

Caravella, A., Barbieri, G., & Drioli, E. (2009). Concentration polarization analysis in self-supported Pd-based membranes. *Separation and Purification Technology*, *66*, 613–624.

Caravella, A., Melone, L., Sun, Y., Brunetti, A., Drioli, E., & Barbieri, G. (2016). Concentration polarization distribution along pd-based membrane reactors: A modelling approach applied to water-gas shift. *International Journal of Hydrogen Energy*, *41*, 2660–2670.

Caravella, A., Scura, F., Barbieri, G., & Drioli, E. (2010). Inhibition by CO and polarization in Pd-based membranes: A novel permeation reduction coefficient. *Journal of Physical Chemistry B*, *114*, 12264–12276.

Caravella, A., & Sun, Y. (2016). Correct evaluation of the effective concentration polarization influence in membrane-assisted devices. Case study: H_2 production by water gas shift in Pd-membrane reactors. *International Journal of Hydrogen Energy*, *41*, 11653–11659.

Caravella, A., Zito, P. F., Brunetti, A., Drioli, E., & Barbieri, G. (2016). A novel modelling approach to surface and Knudsen multicomponent diffusion through NaY zeolite membranes. *Microporous and Mesoporous Materials*, *235*, 87–99.

Catalano, J., Giacinti Baschetti, M., & Sarti, G. C. (2010). Hydrogen permeation in palladium-based membranes in the presence of carbon monoxide. *Journal of Membrane Science*, *362*, 221–233.

Chang, S. K., Kang, Y., & Kim, S. D. (1986). Mass transfer in two- and three-phase fluidized beds. *Journal of Chemical Engineering of Japan*, *19*, 524–530.

Chiu, T. M., & Ziegler, E. N. (1983). Heal transfer in three phase fluidized beds. *AIChE Journal. American Institute of Chemical Engineers*, *29*, 677–685.

Chiu, T. M., & Ziegler, E. N. (1985). Liquid holdup and heat transfer coefficient in liquid solid and three-phase fluidized beds. *AIChE Journal. American Institute of Chemical Engineers*, *31*, 1504–1509.

Colombo, M., & Fairweather, M. (2015). Multiphase turbulence in bubbly flows: RANS simulations,. *International Journal of Multiphase Flow*, *77*, 222–243.

Cotterill, P. (1961). The hydrogen embrittlement of metals. *Progress in Materials Science*, *9*, 205–301.

Cser, L., Török, G., Krexner, G., Prem, M., & Sharkov, I. (2004). Neutron holographic study of palladium hydride. *Applied Physics Letters*, *85*, 1149–1151.

Dhanuka, V. R., & Stepanek, J. B. (1980). Simultaneous measurement of interfacial area and mass transfer coefficient in three-phase fluidized beds. *AIChE Journal. American Institute of Chemical Engineers*, *26*, 1029–1038.

Di Felice, R. (1994). The voidage function for fluid-particle interaction systems. *International Journal of Multiphase Flow.*, *20*, 153–159.

Ding, J., & Gidaspow, D. (1990). A bubbling fluidization model using kinetic theory of granular flow. *AIChE Journal. American Institute of Chemical Engineers*, *36*, 523–538.

Do, D. D. (1998). *Adsorption analysis: Equilibria and kinetics*. London: Imperial College Press, ISBN 1-86094-137-3.

Dolan, M. D. (2010). Non-Pd BCC alloy membranes for industrial hydrogen separation. *Journal of Membrane Science*, *362*, 12–28.

Fache, A., Marias, F., & Chaudret, B. (2020). Catalytic reactors for highly exothermic reactions: Steady-state stability enhancement by magnetic induction. *Chemical Engineering Journal*, *390*, 124531.

Fan, L. S. (1989). *Gas-Liquid-Solid fluidization engineering*. Stoneham, MA.: Butterworths.

Fukai, Y. (1984). Site preference of interstitial hydrogen in metals. *Journal of the Less Common Metals*, *101*, 1–16.

Gibilaro, L. G., Di Felice, R., Waldram, S. P., & Foscolo, P. U. (1985). Generalized friction factor and drag coefficient correlations for fluid-particle interactions. *Chemical Engineering Science*, *40*, 1817–1823.

Guo, X., Sun, Y., Li, R., & Yang, F. (2014). Experimental investigations on temperature variation and inhomogeneity in a packed bed CLC reactor of large particles and low aspect ratio. *Chemical Engineering Science*, *107*, 266–276.

Guo, Z., Sun, Z., Zhang, N., & Ding, M. (2019). Influence of confining wall on pressure drop and particle-to-fluid heat transfer in packed beds with small D/d ratios under high Reynolds number. *Chemical Engineering Science*, *209*, 115200.

Guo, Z., Sun, Z., Zhang, N., Ding, M., & Liu, J. (2017). Pressure drop in slender packed beds with novel packing arrangement. *Powder Technology*, *321*, 286–292.

Guo, Z., Sun, Z., Zhang, N., Ding, M., & Liu, J. (2019). Influence of flow guiding conduit on pressure drop and convective heat transfer in packed beds. *International Journal of Heat and Mass Transfer*, *134*, 489–502.

Hatate, Y., Tajari, S., Fujita, T., Fukumoto, T., Ikari, A., & Hano, T. (1987). Heal transfer coefficient in three-phase vertical upflows of gas-liquid -fine solid particle particles system. *Journal of Chemical Engineering of Japan, 20*, 568–574.

Hofmann, S., Bufe, A., Brenner, G., & Turek, T. (2016). Pressure drop study on packings of differently shaped particles in milli-structured channels. *Chemical Engineering Science, 155*, 376–385.

Kang, Y., Fan, L. T., Min, B. T., & Kim, S. D. (1991). Promotion of oxygen transfer in three-phase fluidized-bed bioreactors by floating bubble breakers. *Biotechnology and Bioengineering, 7*, 580–586.

Kang, Y., & Kim, S. D. (1988). Solid flow transition in liquid and three-phase fluidized beds. *Particulate Science and Technology, 6*, 133–144.

Kang Y., Suh I. S., & Kim S. D., (1983). *Heat transfer characteristics of a three-phase fluidized bed*. In Procs. PACItEC83, 2: 1–6.

Kang, Y., Suh, L. S., & Kim, S. D. (1985). Heat transfer characteristics of three phase fluidized beds. *Chemical Engineering Communications, 34*(1-6), 1–13.

Kaptejin, F., Moulijn, J. A., & Krishna, R. (2000). The generalized Maxwell-Stefan model for diffusion in zeolites: Sorbate molecules with different saturation loadings. *Chemical Engineering Science, 55*, 2923–2930.

Kato, Y., Kago, T., Uchida, K., & Morooka, S. (1980). Wall-bed heat transfer characteristics of three-phase packed and fluidized bed. *Kagaku Kogaku Ronbunshu, 6*, 579–584.

Kato, Y., Taura, Y., Kago, T., & Morooka, S. (1984). Heat transfer coefficient between bed and inserted vertical tube wall in three-phase fluidized bed. *Kagaku Kogaku Ronbunshu, 4*, 427–431.

Kato, Y., Uchida, K., Kago, T., & Morooka, S. (1981). Liquid holdup and heat transfer coefficient between bed and wall in liquid-solid and gas-liquid-solid fluidized beds. *Powder Technology, 28*, 173–179.

Kim, J. O., & Kim, S. D. (1990). Bubble breakage phenomena, phase holdups and mass transfer in three-phase fluidized beds with floating bubble breakers. *Chemical Engineering and Processing: Process Intensification, 28* (2), 101–111.

Kim, J. O., & Kim, S. D. (1990). Gas liquid mass transfer in a three-phase fluidized bed with floating bubble breakers. *The Canadian Journal of Chemical Engineering, 68*, 368–375.

Kim, J. O., Park, D. H., & Kim, S. D. (1990). Heat transfer and wake characteristics in three-phase fluidized beds with floating bubble breakers. *Chemical Engineering and Processing: Process Intensification, 28*, 113–119.

Kim, S., & Laurent, A. (1988). L'etat des connaissances sur le transfert de chaleur dans les lits fluidises triphasiques. *Entropie, 143–144*, 5–22, 1991. The state of knowledge on heat transfer in three-phase fluidized beds. Int. Chem. Eng., 31: 284-302.

Kim, S. D., Kang, Y., & Kwon, H. K. (1986). Heat transfer characteristics in two- and three-phase slurry fluidized beds. *AIChE Journal. American Institute of Chemical Engineers, 32*, 1397–1400.

Kim, S. D., Lee, Y. J., & Kim, J. O. (1988). Heat transfer and hydrodynamics studies on two- and three-phase fluidized beds of floating bubble breakers. *Experimental Thermal and Fluid Science, 1*, 237–242.

Kirchheim, R. (1986). Interaction of hydrogen with external stress fields. *Acta Metallurgica, 34*, 37–42.

Krishna, R. (1990). Multicomponent surface diffusion of adsorbed species. A description based on the generalized Maxwell-Stefan diffusion equations. *Chemical Engineering Science, 45*, 1779–1791.

Krishna, R., & Wesselingh, J. A. (1997). The Maxwell-Stefan approach to mass transfer. *Chemical Engineering Science, 52*, 861–911.

Kumar, S., Kusakabe, K., & Fan, L. S. (1993). Heat transfer in three-phase fluidized beds containing low-density particles. *Chemical Engineering Science, 48*, 2407–2418.

Lee, D. H., Kim, J. O., & Kim, S. D. (1993). Mass transfer characteristics and phase holdup in three phase fluidized beds. *Chemical Engineering Communications, 119*, 179–196.

Li, L., Zou, X., Wang, H., Zhang, S., & Wang, K. (2018). Investigations on two-phase flow resistances and its model modifications in a packed bed. *International Journal of Multiphase Flow, 101*, 24–34.

Lu, Y., Gou, M., Bai, R., Zhang, Y., & Chen, Z. (2017). First-principles study of hydrogen behavior in vanadium-based binary alloy membranes for hydrogen separation. *International Journal of Hydrogen Energy, 42*, 22925–22932.

Magiliotou, M., Chert, Y. M., & Fan, L. S. (1988). Bed-immersed object heat transfer in a three-phase fluidized bed. *AIChE Journal. American Institute of Chemical Engineers, 34*, 1043–1047.

McKeen, T., & Pugsley, T. (2003). Simulation and experimental validation of a freely bubbling bed of FCC catalyst. *Powder Technology, 129*, 139–152.

McLennan, K. G., Gray, E. M. A., & Dobson, J. F. (2008). Deuterium occupation of tetrahedral sites in palladium. *Physical Review B – Covering Condensed Matter and Materials Physics, 78*, 1–9.

Mejdell, A. L., Chen, D., Peters, T. A., Bredesen, R., & Venvik, H. J. (2010). The effect of heat treatment in air on CO inhibition of a ~3 μm Pd-Ag (23 wt.%) membrane. *Journal of Membrane Science, 350*(1-2), 371–377.

Miyahara, T., Lee, M. S., & Takahasi, T. (1993). Mass transfer characteristics of a three-phase fluidized bed containing low-density and/or small particles. *International Chemical Engineering, 33*, 680–686.

Muroyama, K., Fukuma, M., & Yasunishi, A. (1984). Wall-to-bed heat transfer coefficient in gas-liquid-solid fluidized beds. *The Canadian Journal of Chemical Engineering, 62*(2), 199–208.

Muroyama, K., Fukuma, M., & Yasunishi, A. (1986). Wall-to-bed heat transfer in liquid-solid and gas liquid solid fluidization beds part II: Gas liquid-solid fluidized beds. *The Canadian Journal of Chemical Engineering, 64*, 409–418.

Mutschele, T., & Kirchheim, R. (1987). Segregation and diffusion of hydrogen in grain boundaries of palladium. *Scripta Metallurgica, 21*(2), 135–140.

Nguyen-Tien, K., Patwari, A. N., Shumpe, A., & Deckwer, W. D. (1985). Gas-liquid mass transfer in fluidized particle beds. *AIChE Journal. American Institute of Chemical Engineers, 31*, 194–201.

Nieuwland, J. J., van Sint Annaland, M., Kuipers, J. A. M., Van., & Swaaij, W. P. M. (1996). Hydrodynamic modeling of gas/particle flows in riser reactors. *AIChE Journal. American Institute of Chemical Engineers, 42*, 1569–1582.

Nygren A. C. L. (2014). *Simulation of bubbly flow in a flat bubble column - Evaluation of interface and turbulence closure models.* M.Sc. Thesis. Lund University, Faculty of Engineering, Department of Chemistry and Chemical Engineering URL: https://lup.lub.lu.se/student-papers/search/publication/4730957.

Oriani, R. A. (1994). The physical and metallurgical aspects of hydrogen in metals. *Fusion Technology, 26*, 235–266.

Ostergaard K. (1964). *Fluidization*. Soc. Chem. Ind., London, p. 58.

Patrascu, M., & Sheintuch, M. (2015). On-site pure hydrogen production by methane steam reforming in high flux membrane reactor: Experimental validation, model predictions and membrane inhibition. *Chemical Engineering Journal, 262*, 862–874.

Patwari, A. N., Nguyen-Tien, K., Schumpe, A., & Deckwer, W. D. (1986). Three-phase fluidized beds with viscous liquid: Hydrodynamics and mass transfer. *Chemical Engineering Communications, 40*, 49–65.

Saberian-Broudjcnni, G., Wild, M. N., & Charpentier, C. (1985). Contribution to the study of wall heat transfer in reactors with a gas liquid-solid fluidized bed at low liquid velocities. *The Canadian Journal of Chemical Engineering, 63*, 553–564.

Schumpe, A., Deckwer, W. D., & Nigam, K. D. P. (1989). Gas-liquid mass transfer in three-phase fluidized beds with viscous pseudoplastic liquids. *The Canadian Journal of Chemical Engineering, 67*, 873–877.

Suh, I. S., & Deckwer, W. D. (1989). Unified correlation of heat transfer coefficient in three-phase fluidized beds. *Chemical Engineering Science, 44*, 1455–1458.

Suh, I. S., Jin, G. T., & Kim, S. D. (1985). Heat transfer coefficients in three phase fluidized beds. *International Journal of Multiphase Flow, 11*, 255–259.

Syamlal, M., & O'Brien, T. (1987). *The derivation of a drag coefficient formula from velocity-voidage correlations*. Tech. Note, United States Dep. Energy, Off. Foss. Energy, NETL, Morgantown W.V.

Tarmy, B. I., & Coulaloglou, C. A. (1992). Industrial view of gas/liquid/solid reactor development. *Chemical Engineering Science, 47*, 3231–3246.

Verma, V., Deen, N. G., Padding, J. T., & Kuipers, J. A. M. (2013). Two-fluid modeling of three-dimensional cylindrical gas-solid fluidized beds using the kinetic theory of granular flow. *Chemical Engineering Science, 102*, 227–245.

Viaswanathan, S., Kakar, A. S., & Murti, P. S. (1965). Effect of dispersing bubbles into liquid fluidized beds on heat transfer and hold-up at constant bed expansion. *Chemical Engineering Science, 20*, 903–910.

Volkl, J., & Alefeld, G. (1978). Diffusion of hydrogen in metals. In G. Alefeld, & J. Volkl (Eds.), *Hydrogen in metals. I: Basic properties* (pp. 321–348). Berlin, Heidelberg: Springer Berlin Heidelberg.

Wang, D., Flanagan, T. B., & Shanahan, L. K. (2004). Permeation of hydrogen through pre-oxidized Pd membranes in the presence and absence of CO. *Journal of Alloys and Compounds, 372*, 158–164.

Wilcox, D. C. (1993). *Turbulence modelling for CFD*. La Canada CA: DCW Industries, Inc..

Yang, N., Wang, W., Ge, W., & Li, J. (2003). CFD simulation of concurrent-up gas–solid flow in circulating fluidized beds with structure-dependent drag coefficient. *Chemical Engineering Journal, 96*, 71–80.

Zaidi, A., Deckwer, W. D., Mrani, A., & Benchekchou, B. (1990). Hydrodynamics and heat transfer in three-phase fluidized beds with highly viscous pseudoplastic solutions. *Chemical Engineering Science, 45*, 2235–2238.

Zheng, C., Chen, Z., Feng, Y., & Hoffmann, H. (1995). Mass transfer in different flow regimes of three-phase fluidized beds. *Chemical Engineering Science, 50*, 1571–1578.

Zito, P. F., Caravella, A., Brunetti, A., Drioli, E., & Barbieri, G. (2018). Discrimination among gas translation, surface and Knudsen diffusion in permeation through zeolite membranes. *Journal of Membrane Science, 564*, 166–173.

CHAPTER

12

Mass transport through capillary, biocatalytic membrane reactor

Endre Nagy and Imre Hegedüs

University of Pannonia, Research Institute of Biomolecular and Chemical Engineering, Laboratory of Chemical and Biochemical Processes, Veszprem, Hungary

Notation

C concentration in the membrane layer, kg/m^3, mol/m^3
C_δ^o outlet reactant concentration, kg/m^3, mol/m^3
D diffusion coefficient, m^2/s-
J mass transfer rate, kg/m^{2}s, mol/m^{2}s
J_o mass transfer rate without biochemical reaction, kg/m^{2}s, mol/m^{2}s
K_M Michaelis-Menten constant, kg/m^3, mol/m^3
N number of sublayers, ($N = 1000$), -
r radial coordinate, m
r_o lumen radius, m
r_m $= r_o + \delta$, m
R dimensionless local coordinate, ($R = r/r_o$), -
v_{max} maximum reaction rate, kg/m^{2}s, mol/m^{2}s

Greek letters

δ membrane thickness, m
ϑ reaction modulus, -

Superscript

$*$ interface
o bulk

Subscript

i ith sublayer
o physical

Current Trends and Future Developments on (Bio-) Membranes
DOI: https://doi.org/10.1016/B978-0-12-822257-7.00014-5

12.1 Introduction

The catalytic membrane reactor as a promising novel technology is widely recommended in the chemical and the biochemical industry (Charcosset, 2006; Giorno & Drioli, 2000; Giorno, Mazzei, & Drioli, 2009; Marcano & Tsotsis, 2002; Rios, Belleville, & Paolucci-Jeanjean, 2007). A number of reactions have been investigated by means of chemical/biochemical catalytic membrane reactors, such as dehydrogenation of alkanes to alkenes, partial oxidation reactions using inorganic or organic peroxides, as well as partial hydrogenations, hydration, etc. As catalytic membrane reactors for these reactions, intrinsically catalytic membranes can also be used (e.g., zeolite or metallic membranes) or membranes that have been made catalytic by dispersion or impregnation of catalytically active particles such as metallic complexes, metallic clusters or activated carbon, biocatalytic enzymes, zeolite particles, etc. throughout dense polymeric- or inorganic membrane layers (Marcano & Tsotsis, 2002). Several studies discussed the mass transport through membrane reactors, mostly with inorganic reactions (Bronetti, Capannelli, & Comite, 2002; Brunetti, Caravella, Barbiella, & Drioli, 2007; Seidel-Morgenstein, 2010). In the majority of the previous experiments, the reactants can be separated from each other by the catalytic/biocatalytic membrane layer/reactor. In another promising realization of this catalytic process is when the membrane itself is catalyst and thus the reaction takes place during the reactant transport/product(s) through the catalytic/biocatalytic membrane reactor (Cabral & Tramper, 1994; Giorno et al., 2009; Westermann & Melin, 2012). Main point of this study is to discuss the reactant/product transport through biocatalytic membrane reactor, where the transport is accompanied by biochemical reaction.

12.1.1 Biocatalytic membrane reactors

Biocatalytic reactors with enzymes are often combined with membrane technologies. The main advantage of membrane technologies is that the simultaneous biochemical reaction and separation of the product component(s) from the reaction mixture by membrane layer can shift the reaction equilibrium and with it, it can increase the reaction efficiency. Two main groups of biocatalytic membrane reaction processes are distinguished, namely membrane bioreactors, where the reaction process and the separation are realized separately as well as biocatalytic membrane reactors, where the bioreaction and the product separation take place in a biocatalytic membrane layer/reactor (Mazzei, Piacentini, Gebreyohannes, & Giorno, 2017).

1. *Membrane bioreactors*: bioreactor devices can work with free or immobilized enzymes, while separation of product component(s) is realized separately by membrane separation equipment. The separation of the product from the reaction mixture can be eliminated, for example, the product inhibition and therefore can increase the efficiency of the reaction process, for example, 95% of microcrystalline cellulose, was pretreated in ionic liquid (1-butyl-3-methylimidazolium chloride), and then it was hydrolyzed by cellulase enzymes, while glucose was separated from the remaining cellulose-enzyme mixture by ultrafiltration membrane (Lozano, Bernal, Jara, & Belleville, 2014).

2. *Biocatalytic membrane reactors* contain catalytic enzymes, bounded in the gel layer on the membrane surface or into the internal surface of the porous support layer of asymmetric membranes, thus during the bioreaction process, where product diffuses through the catalytic membrane layer or transport can additionally take place by forced flow, applying hydraulic pressure difference. For example, numerous patents and papers, have been submitted on enzymatic hydrolysis of cellulose, applying biocatalytic membrane reactors for it (Irfan, Ghazanfar, Ur Rehman, & Asma, 2019; Jing, Guo, Xia, Liu, & Wang, 2019; Mazzei et al., 2017; Zhou et al., 2019).

Wide scale of industrial technologies uses biocatalytic membrane reactors, for example, pharmaceutical and food industry as, for example, biogas and bioethanol production (Asif et al., 2019). Environmental applications of biocatalytic membrane reactor are also important. Removal of pollutant molecules with low concentrations is also possible, for example, breaking down low concentration pharmaceutical pollutant even at ppm concentration ranges (Nagy, 2017).

Membranes used for enzyme immobilization could be both symmetric and asymmetric ones (Giorno & Drioli, 2000). The shape of the membrane could be flat sheet, spiral, tubular or combination of them (e.g., plate-and-frame or tube-or-shell modules) (Giorno & Drioli, 2000). However, flow bioreactors, for example, micro or mesoreactors frequently contain membrane elements with immobilized enzymes (Tamborini, Fernandes, Paradisi, & Molinari, 2018). Enzyme immobilization can improve lifetime, stability and reusability of enzymes, but it usually highly reduces its catalytic activity, for example, mutant lipase bounded directly into epoxy resin retains 43.6% of its original catalytic activity, after ten cycles. Immobilization can also limit mass transport of substrate molecules through the membrane and these processes are usually expensive and time-consuming (Chapman, Ismail, & Dinu, 2018).

Immobilization could realize as (1) attachment of the enzyme to the membrane internal surface; (2) entrapment the enzymes into the membrane pores or into polymer matrix and (3) enzyme gelation on the outer surface of the membrane as a protein layer (Fig. 12.1) (Cen, Liu, Xue, & Zheng, 2019; Nagy, 2017; Thakur & Thakur, 2018).

1. *Attachment of biocatalyst* to the internal/external surface of membrane is also named as fouling-like immobilization techniques [see Fig. 1.1 from (A) to (D)]. Fouling is a great general problem of membranes layers because it can seriously reduce the flow across the membrane, but the phenomena can also be used for enzyme immobilization (Cen et al., 2019). Attachment of biocatalyst to membrane can be accomplished by:
   a. *Physical interactions* (adsorption of enzyme molecules into the support layer of the membrane) (Jesionowski, Zdarta, & Krajewska, 2014). Linkages between adsorbed enzyme molecules and the inner porous parts of membrane are stabilized by weak interactions (e.g., van der Waals forces, H-bonds). These immobilization techniques are easy and cheap, for example, Chen, Ma, Zhu, Chen, and Huang (2019) immobilized lipase in RGM-PSF polysulfone hollow-fiber membrane (with perfect radial gradient pores) by cross-flowing of enzyme solution through the membrane. The whole amount lipase was immobilized and its activity value was as high as that of free enzymes. In most cases, however, the enzyme immobilization by adsorption could block the active sites of the enzymes and therefore their catalytic activity could be highly reduced.

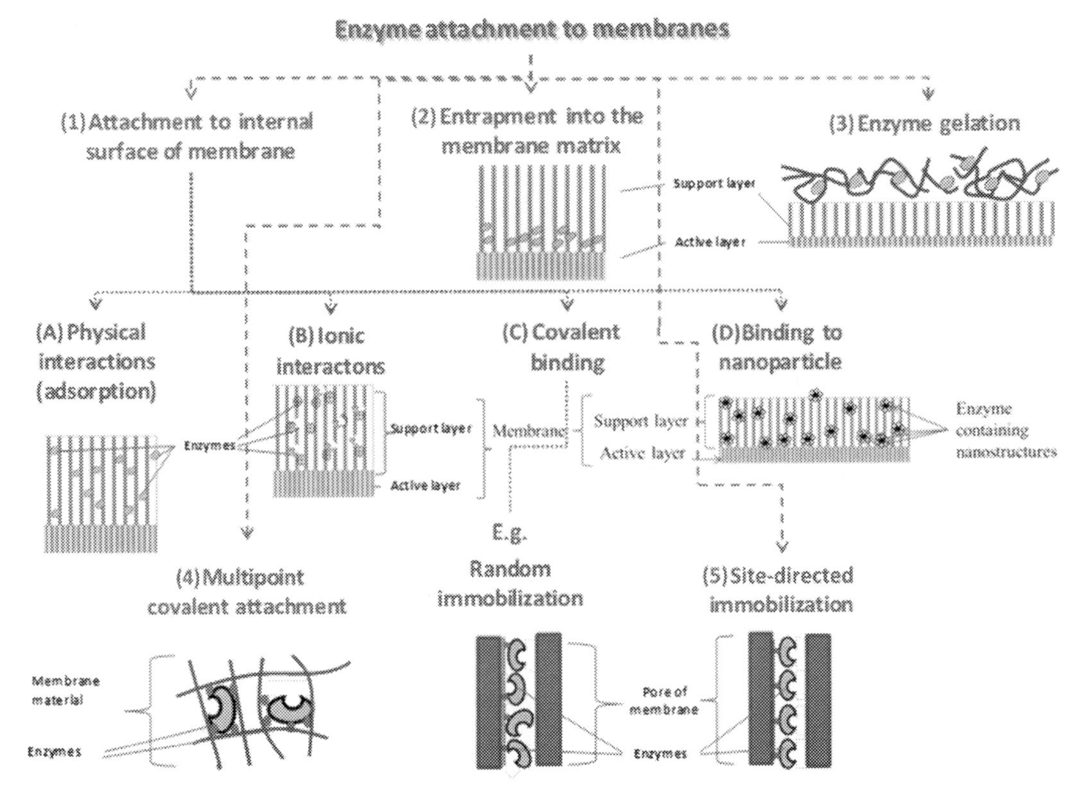

FIGURE 12.1 Types of attachment of enzyme molecules to membranes for synthesis of enzyme containing biocatalytic membrane reactors: (1) attachment of enzyme molecules to the internal surface of the membranes could be carried out by (a) physical interactions (by adsorption); (b) by electrostatic ("ionic") interactions; (c) by covalent bond(s), for example, random immobilization; or (d) through binding of enzyme molecules to nanoparticles. (2) However, entrapment of enzyme molecules into the membrane matrix; (3) enzyme gelation; (4) multipoint attachment of enzyme to membrane matrix and (5) site-directed immobilization are also frequently used methods for creation biocatalytic membrane reactor with enzyme.

b. Immobilization with *ionic interactions* is other simple method (Mazzei, Giorno, Piacentini, Mazzuca, & Drioli, 2009); they immobilized β-glucosidase in asymmetric capillary membrane (polysufone with cutoff of 30 kDa) by cross flow of enzyme solution and it was found that immobilized and free enzymes have more or less the same kinetics.

c. Attachment of enzymes to the membrane *by covalent binding* is strong enough to protect it from re-moving (leaching) and helps enzymes to retain their native conformation (Garcia-Galindo et al., 2019; Giorno & Drioli, 2000; Osbon & Kumar, 2016; Sigurdardóttir et al., 2018).

Covalent binding of enzymes onto or into the membrane could be realized *by cross-linking* of enzymes to the membrane with cross-linker agent. In this case membrane is previously covalently modified usually with amino, carboxyl, or aldehyde groups (Cen et al., 2019). For example, Vitola et al. (2019) immobilized thermostable phosphortriesterase mutant enzymes in polymeric membrane by glutaraldehyde cross-linker in order to hydrolyze organophosphate pesticides in water and immobilized enzyme had been retain its stability for 1 year.

Nanoparticle-mediated covalent immobilization of enzymes to membrane is also advantageous method. For example, Gebreyohannes et al. (2018) prepared cellulase enzyme containing biocatalytic membrane reactor, where cellulolytic enzymes were bounded to magnetic nanoparticles and this catalytic nanoparticles were immobilized on the external membrane surface. Then this catalytic layer can easily be removed by means of magnetic field after the desired reaction.

2. *Entrapment of enzymes* in membrane can also be realized by different manner (Fig. 1.2).
   a. The simplest method is entrapment of enzyme molecules *within the pores* in the sponge layer of asymmetric membrane, where the pore size of active layer of membrane is lower than the size of enzyme molecules (Chakraborty et al., 2016; Giorno et al., 2009). For example, Lopez and Matson (1997) entrapped lipase in asymmetric hollow fiber membrane (PAN) to generate medical drug for several years (75 tons/year).
   b. Entrapment also possible in polymeric surfaces *(fiber entrapment)*. For example, Adhikari, Schraft, and Chen (2017) entrapped alcohol dehydrogenase enzymes with polycationic [polymer poly(2-(dimethylamino) ethyl methacrylate] in single-wall carbon nanotube and reduced graphene oxide thin film.
   c. Promising enzyme entrapment methods are *enzyme encapsulation* into metal-organic frameworks (Fig. 1.3); see for example, (Drout, Robison, & Farha, 2019) or enzyme containing monoliths (Dizge, Demirdogen, & Ocakoglu, 2019) or enzyme-activated beads (Dizge et al., 2019). Glucoamylase enzyme was immobilized in Ca-alginate bed as a support layer and activity has detected after 36 days and 30 cycle (Eldin Mohy, Seuror, Nasr, & Tieama, 2011).
3. *Gel formation* (enzyme gelation) is sol-gel reaction, in order to fix the enzyme in the inner side of the membrane (Yoo, Feng, Kim, & Yagonia, 2017) (Fig. 1.3). For example, tetraethyl orthosilicate was reacted with water and hydrolyzed. The resulting spatial gel entrapped 88% of lipase enzymes and immobilized enzyme increases its lifetime 2.7 times (Yagonia, Park, & Yoo, 2014).

Besides, some other interesting methods exist to improve activity and stability of enzymes attached to membranes.

4. For example, *multipoint covalent attachment* of enzymes to membrane avoids the unfolding of enzyme molecule caused by heat, pH-changes or organic solvents and therefore increase its lifetime (Bilal, Asgher, Cheng, Yan, & Iqbal, 2019; Reiser, Muheim, Hardeggea, Frank, & Fiechter, 1994) (Fig. 1.4). Carrier materials for multipoint attachment could be nanoparticles, graphene, mesoporous and ceramic materials or electrospun polymer-based fibers (Bilal et al., 2019). For example, Sóti et al. (2016) immobilized lipases in polyvinyl fiber membrane (produced by electrospinning) with multipoint attachment and the immobilized lipase has retained more than 80% of its original activity after ten cycles.
5. *Site-directed immobilization* (site-specific or oriented immobilization) of enzymes could also increase the efficiency of membrane processes, because significant portion of randomly immobilized enzyme can be inactive because their active centers are hidden and are not

accessed to reach substrates, therefore the enzymatic activity of randomly immobilized enzymes might be highly reduced (Butterfield, Bhattacharyya, Daunert, & Bachas, 2001) (Fig. 1.5). Contrary to it, site-specific immobilized catalytic enzyme bioreactors, which based on specific molecular recognition, leave the active center of the enzymes free and therefore the bound between the enzyme and the membrane does not reduce seriously the enzymatic activity (Garcia-Galindo et al., 2019; Mazzei et al., 2017; Sigurdardóttir et al., 2018).

6. *Biomimetic membranes* usually enhance highly the lifetime of immobilized enzymes. Biomimetic membranes mean generally artificial membranes, which can mimic characteristic behavior of biological membranes. Biological membranes usually have great selectivity, recognize well specific molecules as signals for information transfer and regulate finely the velocity of transport through the membrane (Piacentini, Mazzei, Drioli, & Giorno, 2017). Biological (and biomimetic) membranes have local molecular environment, which help enzyme molecules to retain their activity for longer period of time (Shen, Saboe, Sines, Erbakan, & Kumar, 2014). Biocatalytic membrane bioreactors containing polymeric membrane with immobilized enzymes are also a good example for biomimetic membranes (Piacentini et al., 2017; Shen et al., 2014). For example, β-glucosidase enzyme immobilized to polysulfone capillary membrane (cutoff 30 kDa) has better stability values than that of the same free enzyme, though its kinetic parameter values are the same as that of the free enzymes (Mazzei et al., 2009).
7. Another method, which can avoid the activity loss of enzyme activity, is *enzyme immobilization* when *enzyme molecules are cross-linked with another one* without linking of enzyme molecules to any carrier or support materials and therefore biocatalytic activity values of stabilized enzymes could not decrease as seriously as activity and stability values of enzymes linked to carrier molecules. This method is also frequently used for enzyme immobilization technique and it is easy and effective alternate of carrier-bonded enzyme immobilization method (Sheldon, 2007).
   a. This cross-linkage could be carried out as direct linkage between each enzyme molecules, where every enzyme molecules are cross-linked with another one (*Cross Linked Enzymes*, CLE). Unfortunately CLE technology usually results in very low activity values and low mechanical stability of cross-linked enzymes (Chakraborty et al., 2016; Sheldon, 2007).
   b. Contrary to CLE methods, small enzyme crystals (with diameter less than 1 μm) could be stabilized by direct covalent linkage between one another (*Cross-Linked Enzyme Crystals*, CLECs). This stabilization form of enzymes results in good catalytic activity and stability values (Sheldon, 2007). Similarly, enzyme aggregates stabilized by weak physical interactions (with diameter less than 1 μm) could also be stabilized by direct covalent linkage between aggregated enzyme particles (*Cross-Linked Enzyme Aggregates*, CLEAs) and CLEAs have also good activity and stability values (Sheldon, 2007). Enzyme crystals and enzyme aggregates are formed by addition of salts or organic solvents or multicharged polymers to enzyme solution and thus physically bound aggregated enzymes are created. Cross linkage between small enzyme crystals or aggregate particles could also be easily synthesized by addition of cross-linking agents (e.g., glutaraldehyde) to enzyme crystals or aggregates (Asif et al., 2019). For example, nitrilase enzyme was immobilized as CLEA and used in ultrafiltration membrane bioreactor and more than 90% conversion rate was detected even after 52 hours operating time (Malandra et al., 2009).

One of the main promising areas of the application of enzyme containing membrane bioreactors are multiphase enzymatic reactions where water-insoluble substrates are hydrolyzed (Coutte et al., 2016). For example, localization of immobilized β-D-glucosidase enzyme in biphasic polysulfone capillary membrane was well-studied (Mazzuca, Giorno, Spadafora, Mazzei, & Drioli, 2006), and the reaction conditions were optimized (Giorno & Drioli, 2000; Mazzei, Drioli, & Giorno, 2012; Nagy, Dudás, Mazzei, Drioli, & Giorno, 2015). Optimal biocatalytic process summarizes factors added from process optimization, optimized enzyme immobilization into/onto membranes and optimized enzyme engineering (e.g., genetically modified enzymes) (Chapman et al., 2018).

12.1.2 Enzyme immobilization

Enzyme molecules, as biocatalysts represent specific reaction routes during a biocatalytic process; they usually do not form by-products and they need moderated physical conditions (room temperature, usually water as reaction medium, do not need expensive organic solvents, etc.); they are often cheaper than the classical catalysts used for the main chemical reactions. According to their environmentally friendly features, enzymes are ideal biocatalysts for green chemistry, as well. Green chemistry tries to exchange the toxic and environmentally pollutant organic reagents and metal catalysts into not hazardous ones as microbes or enzymes (Zimmerman, Anastas, Erythropel, & Leitner, 2020). Besides, due to their increasing importance and central role, enzymes have a great demand in industry (mainly in pharmaceutical, food processing, detergent, biofuel and paper/pulp sector). Enzyme global market presumably will over 6 billion $ in 2021 (Chapman et al., 2018).

However enzymes have several disadvantages, first of all, their short life-time and great sensitivity against little changes in conditions of their environment (changes in pH and temperature values or concentration values of ions in enzyme solution). In the other hand, enzymes are sometimes expensive.

Some technologies exist to improve the efficiency of enzymatic industrial processes, for example, reaction medium engineering, enzyme engineering, substrate engineering, etc. (Sheldon & Woodley, 2018). These technologies can usually help for enzyme to extend its lifetime for one or more process cycles. Therefore the price of enzymes could extensively be reduced and the whole process could be more economic.

1. *Reaction media engineering* changes the physical characteristics of reaction medium (e.g., with salt concentration of water) and it can help for easy separation of enzymes and products at the end of the catalytic process (Fig. 12.2A). Enzyme molecules are solved usually in water therefore their removal from the reactant sludge for their reuse is usually difficult after the biocatalytic process, for applying it in a second cycle. Therefore numerous attempts exist to exchange water solvent to other reaction medium, which can also dissolve enzyme molecules but it cannot dissolve the product (s) and therefore enzymes and products are separated. Some nonaqueous reaction medium exists (e.g., organic solvents or ionic liquids), where enzymes have acceptable or good activity and products are easily separable.
2. *Enzyme engineering* intends to improve enzyme properties (heat stability, reaction stability, biocatalytic activity, or selectivity) making it more effective ones than that of

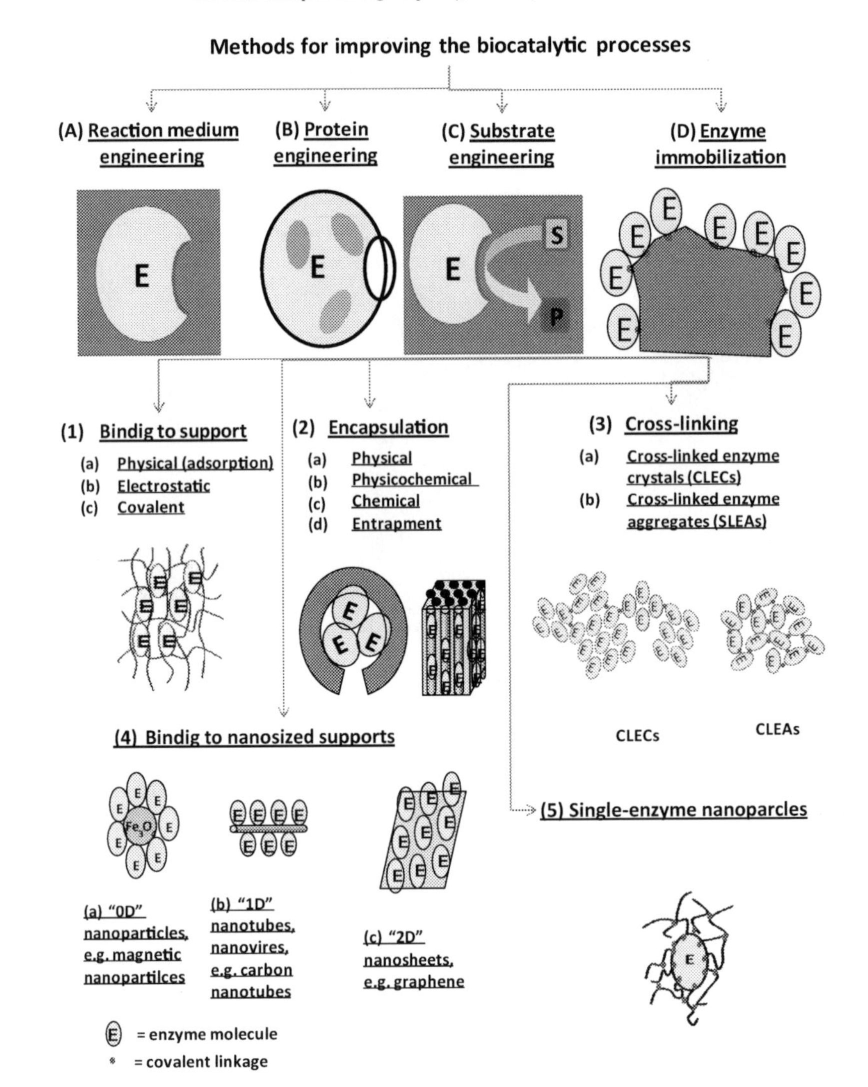

FIGURE 12.2 Methods for improve efficiency of biocatalytic industrial processes (A) Reaction medium engineering changes reaction medium for effective separation of products; (B) Protein engineering increases effectivity of enzyme molecule (enhances its stability or activity); (C) substrate engineering optimize biocatalytic reactions or could find new reaction routes; (D) enzyme immobilization attaches enzyme molecules to a support material. This attachment is usually carried out as (1) Binding physical or chemical manner to the carrier; (2) Encapsulation of enzyme molecules by physical and/or chemical attachment or by entrapment into a matrix; (3) cross-linking of enzyme crystals (CLECs) or enzyme aggregates (CLEAs) with each other by covalent linkage without inner support material; (4) attachment of enzyme molecules to nanosized support (e.g., nanoparticles, nanorods, or nanosheets) open new perspectives; (5) single enzyme nanoparticles summarize advantages of encapsulation and immobilization.

the native enzymes. The surface of enzyme are modified during its engineering, when one more amino acids are changed in the enzyme structure by genetic engineering or direct evolution that transform the enzyme more massive or effective than native ones (Fig. 12.2B). These changes of enzyme activity could then catalyze new reaction routes, for example, synthesis of pharmaceuticals.

3. *Substrate engineering* works on optimization of biocatalytic reaction (find the optimal pH, temperature, etc.) or sometimes on exploration of new biocatalytic reaction routes that means easier synthesis steps to produce the same chemicals as that the traditional synthesis methods (Fig. 12.2C). For example, lipases usually catalyze hydrolysis of triglycerides, but lipases can catalyze hydrolysis of different esters or synthesis of esters (reverse reaction), as well.
4. Both separation of enzymes from water soluble products and their reusability as biocatalysts in one or more next reaction cycles, at the end of a given reaction process can easily do, when enzymes are *immobilized to a carrier*, which is easily separable from the reaction mixture, after the biocatalytic process (Fig. 12.2D). Other principal advantageous feature of immobilized enzymes is, that they could not aggregate in their immobilized form under those conditions (e.g., pH, temperature, ionic strange, etc.), where the native ones (not immobilized enzymes) can easily be aggregated. Therefore the enzyme immobilization could increase the life-time of enzymes and can also decrease their vulnerability in case of temperature and pH-changes (increase the stability of enzymes) (Chakraborty et al., 2016). This procedure is realized usually by binding of enzyme molecules to previously fabricated carrier support. Enzyme binding to support is usually carried out (see also previous sublayer) (1) by physical adsorption, (2) by chemical linkage, (3) by entrapment of them in a previously synthesized matrix (e.g., polymer matrix), or (4) by binding enzymes to each other as cross-linked enzyme crystals (CLECs) or cross-linked enzyme aggregates (CLEAs) is also a possible way for their immobilization and easier separation.

The micro-environment of enzyme-molecules during immobilization is changed and therefore characteristic features of immobilized enzymes (its biochemical activity, selectivity, pH-stability and heat stability, etc.) can also be changed. Moreover, in some cases the pH-optima and temperature-optima of immobilized enzymes could also be improved. Accordingly, the prevention of unfolding pH-stability and temperature stability of immobilized enzymes are usually higher even under extreme conditions (at acidic or basic pH-values or at temperature values higher by 10°C–20°C than its optimal values) (Yoo et al., 2017).

Covalent bonds between immobilized enzyme molecules and carriers change the ternary structure of enzyme molecules to more rigid form, than that of the native one but the covalent binding can also prevent enzyme molecules from unfolding (denaturation), therefore the stability of immobilized enzymes (the lifetime of its active period) is usually increased, though their activity values are usually decreased. Therefore biocatalytic activity of enzymes in their immobilized form is usually lover than activity of the same enzymes in their free form (without immobilization): usually less than 50%, but in most cases less than 10%. Sometimes, however, biocatalytic activity of immobilized enzymes can be higher than that in its native form (e.g., lipase enzymes) (Yoo et al., 2017).

1. *The binding of immobilized enzyme* molecules to carrier supports can have different types as:

a. *Non covalent* ones, which can be formed by:

i. *physical interaction* between enzymes and carrier surface (adsorption), for example, very promising methods are the immobilization of enzymes in metal-organic frameworks (Liang, Wu, Xiong, Zong, & Lou, 2020). For example, Cao et al. (2016) immobilized lipase enzyme (originated from *Bacillus subtilis*) into hierarchically organized porous metal-organic material by adsorption and it could retain more than 90% of its initial catalytic activity and the reaction process worked by about 99% of its original conversion after 10 cycles.

ii. *Electrostatic ("ionic") interaction* between enzyme molecules and support material, for example, Awad, Ghanem, Abdel Wahab, and Wahba (2020) modified a surface of κ-carrageenan gel by highly polycationic hyperbranched polyamidoamid (PAMAM) polymer and then they attached protease enzyme to this gel by ionic interaction for immobilization. Thermal stability of immobilized protease enzyme (half life time of its activity at higher temperatures) was increased by more than three times, at 70°C, comparing it to its original value, at optimal temperature (40°C).

b. *Enzyme immobilization by covalent bond(s)* could be carried out in numerous cases, for example, Mahmoodi, Saffar-Dastgerdi, and Hayati (2020) immobilized laccase enzyme in carbon nanotube-nanozeolite composite by covalent cross-linkage with glutaraldehyde and this enzyme nanocomposite has retained 84% of its activity at 80°C, after 5 cycles and 69% after 10 cycles, related it to its activity, at optimal temperature (45°C) (Fig. 2.1).

2. *Encapsulation* of enzyme molecules into a closed or partially separated room (Fig. 12.2/2) can avoid the enzymes from interaction with hydrophobic interphases (e.g., air bubbles or oil drops); it could be realized by

a. *Physical interactions* between enzyme molecules and matrices. It could be realized as centrifugal extrusion, electrostatic encapsulation, fluidized bed, freeze drying, thermal gelation, solvent evaporation, spray drying or vacuum encapsulation; for example, Olímpio, Mendes, Trevisan, and Garcia (2020) encapsulated pancreatin enzymes (hydrolytic amylase, lipase and protease) in hydrogel by electrostatic interactions in order to preserve their initial activity (between 50%–90%) after extreme acidic pH in stomach;

b. *Physico-chemical interactions*, namely coacervation, desolvation, layer-by-layer deposition, sol-gel transformations or supercritical fluidization; for example, Love et al. (2020) encapsulated enzyme coacervates in unilamellar vesicles; these separate liquid/liquid phases as artificial cells; this separation was pH-sensitive and dormant enzyme functions could also activate by this vesicles;

c. *Chemical interactions* between enzyme molecules and encapsulating matrices; that is, emulsification, ionic cross-linking, polycondensation, polymerization; for example, Priyanka and Rastogi (2020) encapsulated α-amylase in emulsion liquid membrane as enzyme catalytic membrane bioreactor for decomposition of starch to maltose and at about 70% of its initial activity remained, after 5 cycles;

d. *Microencapsulation* is also a well-known process for enzyme encapsulation and it creates enzyme containing micrometer size vesicles; for example Garcia et al. (2019) stabilized phytase enzyme by microencapsulation for medical usage and after digest system simulations, it retained 90% of initial activity of microencapsulated enzyme;

e. *Entrapment* is a physically closed boundary *of enzymes* in a very small place; it could realized by gels or micrometer/nanometer sized channels in porous materials (mesoporous or nanoporous materials):
   i. good example for *entrapment into gels* is work of Falcone, Shao, Rashid, and Kraatz (2019), where enzymes were entrapped—horseradish peroxidase and α-amylase—in hydrogels and its activity was reduced down to about 60%–70% of its initial one, after 6 cycles;
   ii. another example for enzyme *entrapment in porous matrial* is presented by Shao, Chen, Ying, and Zhang (2020) who immobilized carbonic anhydrase enzyme in mesoporous molecular sieves and the enzyme remained active 2–3 times longer period than the native one.

3. *CLEAs and CLECs* (see previous subchapter) (Fig. 12.2/3).
4. Numerous further methods exist to enhance efficiency of immobilized enzymes, for example, enzyme immobilization or encapsulation in/on *nanomaterials* could realize relatively high enzyme concentration because of the high surface/volume ratio of nanomaterials (Fig. 12.2/4) (Liu & Dong, 2020). Nanosized support materials are
   a. "zero dimensional" *nanoparticles* (e.g., magnetic nanoparticles),
   b. one dimensional *nanotubes* or *nanopores* (e.g., carbon nanotubes),
   c. two-dimensional *nanosheets* (e.g., graphene sheets).
5. *Multipoint covalent attachment* of enzyme molecules to a pre-existing support could effectively increase the enzyme stability. For example, Fernández-Lorente et al. (2015) immobilized enzymes by glyoxil groups, which reacted with surface amino groups (mainly with side chain primary amino groups of lysine amino acids) at multipoint on the surface of enzyme molecule. This stabilization retains 50%–90% of original activity of different free (not immobilized) enzymes (β-galactosidase, penicillin G-acylase, trypsin, etc.) and its stability (half-life-time) increased with at least one order of magnitude.
6. *Oriented immobilization* (see previous subchapter).
7. *Single enzyme nanoparticles* a special group of enzyme immobilization (Fig. 12.2/5). This method applies a few nanometer thick polymer layer around every single enzyme molecules or around little amount of enzyme molecules. This layer is bounded covalently to the surface of enzyme molecule, which is porous enough not to hinder the more or less free transport of substrate molecules to the active center of the enzymes and leaves the product to diffuse freely away from the enzyme's active site. This synthesis method composes advantages of multipoint covalent attachment, encapsulation and the flexible (biomimetic) polymeric support and addition of these stabilizing factors supports strongly enzyme molecule. For example, immobilized β-D-xylosidase enzyme as single enzyme nanoparticle using acrylamide-bisacrylamide random copolymer and the immobilized enzyme had about 25 times longer half life time than that of free enzymes at its optimal temperature (50°C) and it held about 40 times longer half life time than that of free enzymes, at 80°C.

12.2 On mass transport through a biocatalytic membrane layer

From a chemical engineering point of view, it is important to predict the mass transfer rate of the reactant, entering the membrane layer from the upstream phase, and also to predict the downstream mass transfer rate on the permeate side of the catalytic membrane as a function of the physico-chemical parameters. If this transfer (permeation) rate is known as a function of the reaction rate constant, it can be substituted into the boundary conditions of the differential mass balance equations for the upstream and/or the downstream phases. This chapter extends previous investigations by including the effect of the cylindrical mass transport accompanied by Michaelis-Menten kinetics. The main purpose of this chapter is to present and to show an approach solution, applying biocatalyst as nanoparticles or simple enzyme(s) as well as to discuss the effect of the physical and chemical parameters of the biocatalytic, capillary membrane layer on the mass transfer rates, focusing on the effect of the variable cylindrical space. The approach solution method presented in this chapter can also be applied to higher order chemical reactions (Nagy, 2019). Nagy's book (2019) discusses the effect of chemical reaction rate, the effect of the convective flow, the effect of variable diffusion coefficient, convective velocities, etc., for flat-sheet membrane and applying the above describe solution methodology (Nagy, 2019). There was not published paper on the diffusive mass transport through cylindrical membrane, to our knowledge. That is why we proposed as topics of this chapter for this type of mass transport.

12.3 Theoretical part

The capillary, hollow fiber membrane is a most often applied membrane module in the separation industry, due to its very advantageous surface/volume ratio. The very thin lumen radius and membrane thickness can provide very favorable mass transport conditions and separation conditions. The asymmetric inorganic/polymeric membrane layer excellently applicable for immobilizing biocatalytic nanoparticles, enzymes, into the porous membrane layer, thus it can be an effective biocatalytic membrane reactor. This proves the importance of the mass transport rate through a cylindrical membrane layer. The schematic illustration of a capillary, with the important notations, is given in Fig. 12.3. The component transport can takes place either from the porous or the selective sides of a biocatalytic cylindrical membrane depending on the location of immobilized biocatalyst. The inlet of the substrate solution should generally be occurring on the porous side of the membrane, in order to avoid enzyme washing away from the porous layer. It is assuming here that the component(s), poduct(s), reactant(s) has simple diffusive transport, accompanied by biochemical reaction(s), from the lumen side of the membrane into the shell side. The bioreaction, assuming Michaelis-Menten kinetics with or without product or substrate resistance, has nonlinear kinetics. Its analytical solution is not known, thus authors have developed a quasianalytical approach solution for it.

The differential mass balance equation in steady-state condition, assuming a single component reaction with diffusive transport and variable mass transport parameters [$D(C,r)$, $v_{max}(r)$] is as (Crank, 1975; Nagy, 2019):

$$D(C,r)\left(\frac{d^2C}{dr^2}+\frac{1}{r}\frac{dC}{dr}\right)-\frac{v_{max}(r)}{K_M+C}C=0 \tag{12.1}$$

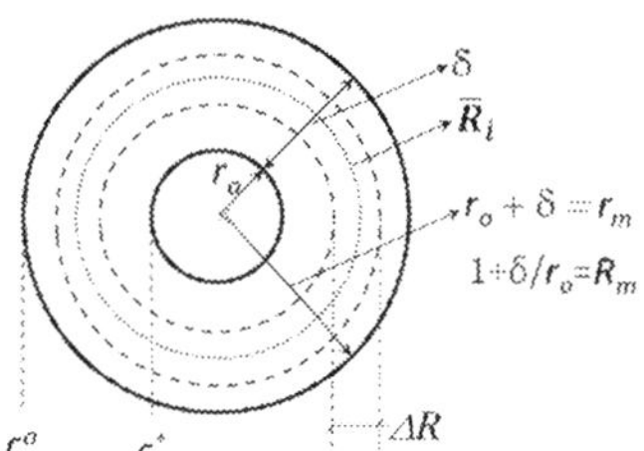

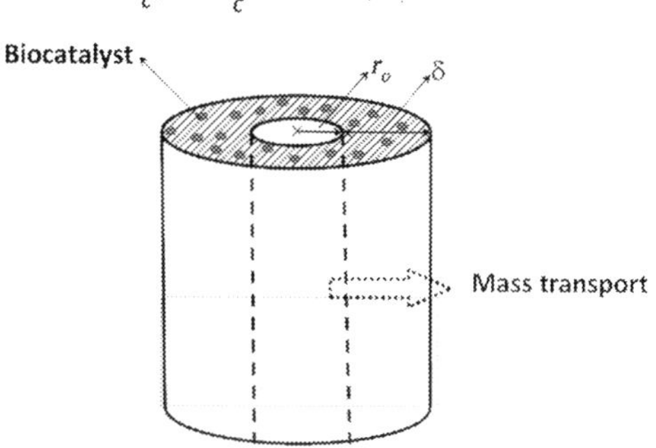

FIGURE 12.3 Schematic illustration of the physical model of the diffusive component transport through cylindrical membrane layer.

Rewriting Eq. (12.1) in general dimensionless form, one can get as ($R = r/ro_o$):

$$\frac{d^2C}{dR^2} + \frac{1}{R}\frac{dC}{dR} - \frac{v_{\max}(R)}{K_M + C}\frac{r_o^2}{D(C,R)}C = 0 \tag{12.2}$$

The solution of the above differential equation by constant parameters can analytically be solved by Bessel function (Nagy, 2019; O'Neil, 1987). We have looked for such an approach solution, which can be used in case of variable mass transfer coefficient (e.g., variable diffusion coefficient, due to, for example, anisotropic membrane, and/or variable reaction rate constant). Changes of these coefficients are then involved in the value of the bioreaction term. For the approach solution the following simplification was applied: the membrane layer was divided into N (during our simulation its value was chosen to be 1000), very thin sublayers, with constant values of the variable diffusion coefficient, reaction rate constant, $\overline{R}_i$ average radius [its algebraic mean value is considered in every sublayer, ($\overline{R}_i = 1 + (i - 0.5)\Delta R$; $\Delta R = \delta/(Nr_o)$; r_o is the lumen radius of the cylindrical membrane, m)]. The obtained differential equation, with constant parameters, for the ith sublayer ($i = 1,\ldots,N$) is as:

$$\frac{d^2C}{dR^2} + \frac{1}{\overline{R}_i}\frac{dC}{dR_i} - \Psi_i^2 C = 0 \tag{12.3}$$

where

$$\Psi_i = \sqrt{\frac{r_o^2}{D\left(C_{i-1}, \overline{R}_i\right)}\frac{v_{\max}\left(\overline{R}_i\right)}{K_M + C_{i-1}\left(C_{i-1}, \overline{R}_i\right)}} \tag{12.4}$$

Hereinafter the concentration and radius dependency of the mass transport and the kinetic parameters are neglected for the sake of the convenience. In case of bioreactor the concentration dependence is not really important, while the local coordinate dependency occurs in case of anisotropic membrane, only. Thus it can be got as:

$$\vartheta_i = \sqrt{\frac{r_o^2}{D}\frac{v_{max}}{K_M + C_{i-1}(\overline{R}_i)}} \tag{12.5}$$

Thus the differential balance equation to be solved is as:

$$\frac{d^2C}{dR^2} + \frac{1}{\overline{R}_i}\frac{dC}{dR} - \vartheta_i^2 C = 0. \tag{12.3a}$$

The methodology of solution was applied by numerous Nagy's works with chemical/biochemical reactions, without (Nagy, 2019, chapter 12.6) and with convective velocity (Nagy, 2009a,b, 2011, 2019, pp. 227–284; Nagy & Kulcsár, 2009; Nagy, Lepossa, & Prettl, 2012) for flat-sheet membrane layer. This is the first case of its application for cylindrical membrane without convective velocity through the membrane. The solution with Eq. (12.4) is the same as with source term by Eq. (12.5). For elimination of the second term in Eq. (12.3a) let the reader introduce the following new variable:

$$C = \tilde{C}\, exp\left(-\frac{1}{2\overline{R}_i}R\right). \tag{12.6}$$

Replacing the C value given in Eq. (12.6) into Eq. (12.3a), one can get the following simplified differential equation to be solved:

$$\frac{d^2\tilde{C}}{dR^2} - \Theta_i^2\tilde{C} = 0 \tag{12.7}$$

where

$$\Theta_i = \sqrt{\frac{1}{4\overline{R}_i^2} + \vartheta_i^2} \tag{12.7a}$$

Solution of Eq. (12.7) can be easily obtained by well-known mathematical expression as it follows:

$$\tilde{C} = T_i e^{\theta_i R} + S_i e^{-\theta_i R} \tag{12.8}$$

Taking into account Eq. (12.6), one can get the concentration distribution of the reactant as:

$$C = T_i e^{\tilde{\lambda}_i R} + S_i e^{\lambda_i R} \quad R_i \le R \le R_{i+1} \tag{12.9}$$

with

$$\tilde{\lambda}_i = \Theta_i - \frac{1}{2\overline{R}_i} \quad \lambda_i = -\Theta_i - \frac{1}{2\overline{R}_i} \tag{12.10}$$

T_i and P_i parameters of Eq. (12.10) can be determined by means of the internal, for the ith sublayer (with $1 \le i \le N$-1) and two external boundary conditions, namely at $R = 1$ and $R = 1 + \delta/r_o$, where δ is the cylindrical membrane thickness (see Fig. 12.3). The boundary conditions at the internal interfaces of the sublayers ($1 \leq i \leq N\text{-}1$; $R_i = 1 + i\Delta R$; $\Delta R = \delta/[Nr_o]$) can be obtained from the following two equalities, considering the interface concentrations between the sublayers and the mass transfer rates. Thus they can be written for the concentrations and the mass transfer rates, $R = R_i$, as:

For the internal concentrations:

$$T_i e^{\tilde{\lambda}_i R_i} + S_i e^{\lambda_i R_i} = T_{i+1} e^{\tilde{\lambda}_{i+1} R_i} + S_{i+1} e^{\lambda_{i+1} R_i} (i = 1 - N - 1) \tag{12.11}$$

For the internal mass transfer rates:

$$T_i \tilde{\lambda}_i e^{\tilde{\lambda}_i R_i} + S_i \lambda_i e^{\lambda_i R_i} = \left(T_{i+1} \tilde{\lambda}_{i+1} e^{\tilde{\lambda}_{i+1} R_i} + S_{i+1} \lambda_{i+1} e^{\lambda_{i+1} R_i} \right) \frac{\overline{R}_{i+1}}{\overline{R}_i} (i = 1 - N - 1) \tag{12.12}$$

where C_i is the reactant concentration of the ith sublayer in the catalytic membrane reactor at $R = R_i$. Value of $\overline{R}_{i+1}/\overline{R}_i$ expresses that the two sides of the interface between the sublayers has different values of the average radius, $\overline{R}$. Accordingly, the specific mass transfer rates on the two sides of an interface are not equal to each other, depending on the sudden change in the interface. It is worth noting that the boundary condition at the outlet surface (on the permeate side of the catalytic membrane) depends on the operation condition, namely how the permeated component(s) is removed from the outlet membrane surface or there is or not sweeping phase on it (Nagy, 2019). In this study a sweeping phase is applied on the permeate side of the membrane layer, which remove the permeate phase from the membrane surface inducing diffusive flux on the outlet membrane surface. Accordingly, the diffusive mass transport exists on the outlet membrane surface, thus the inlet and the outlet boundary conditions are as:

$$\text{at} \quad R = 1 \quad C^o = T_1 e^{\tilde{\lambda}_1} + S_1 e^{\lambda_1} \tag{12.13}$$

and

$$\text{at} \quad R = R_m \equiv 1 + \delta/r_o \quad C_{m,p} = T_N e^{(\tilde{\lambda}_N R_m)} + S_N e^{(\lambda_N R_m)} \tag{12.14}$$

Applying the internal [Eqs. (12.11), (12.12)] and the external [Eqs. (12.13), (12.14)] boundary conditions one can get three $N \times N$ dimensional determinants (see e.g., Nagy, 2019). These determinants are used for determination of the T_i and S_i ($i = 1\text{-}N$) values, applying the known determinant's laws. The solution of this special problem (not shown here in details) is as follows (note $C_{m,p} = \Phi C_p$):

$$T_1 = \left(\prod_1^N \left[\frac{E_i^T}{E_i^O} \right] C^o - \frac{1}{\prod_{i=1}^{N-1} e^{-\Delta R/(2\overline{R}_i)} E_i^O} C_{m,p} \right) \frac{1}{2\cosh(\Theta_1 \Delta R) e^{(-\Delta R/2\overline{R}_1)} e^{\lambda_1} \prod_1^N E_i^O} \tag{12.15}$$

$$S_1 = \left(\prod_1^N \left[\frac{E_i^S}{E_i^O} \right] C^o - \frac{1}{\prod_{i=1}^{N-1} e^{-\Delta R/(2\overline{R}_i)} E_i^O} C_{m,p} \right) \frac{1}{2\cosh(\Theta_1 \Delta R) \left[e^{-\Delta R/2\overline{R}_1} \right] e^{\tilde{\lambda}_1} \prod_1^N E_i^O} \tag{12.16}$$

Values of different variables are listed in Appendix 12.A and Appendix 12.B. Knowing the values of T_1 and S_1 the values of the other T_i and S_i parameters, for predicting the concentration distribution, can be obtained by means of the internal surface conditions [Eqs. (12.11) and (12.12); $i = 2, N\text{-}1$]. Thus, for example, the value of T_2 and S_2 are calculated in the knowledge of their previous parameter values. To obtain it, one gets two algebraic equations from the boundary conditions with two unknown parameters, which can then be expressed and predicted easily.

12.4 Evaluation of the predicted results

The main purpose of this section is to show how the cylindrical space affects the mass transport process, the concentration distribution, the inlet mass transfer rate at different kinetic parameters, taking into account the effect of the lumen radius, convective flow and the membrane thickness. The developed mathematical solution makes possible to predict the different cases (with or without inhibitions) of Michaelis-Menten kinetics. Here we will apply the Michaelis-Menten kinetics without inhibition for the sake of more clear illustration of the role of the cylindrical biocatalytic membrane layer. The first-order and the zero-order reactions, as limiting cases and also the general Michaelis-Menten kinetics are shown a few cases for illustration of the effect of values of lumen radius, r_o, and the membrane thickness, δ, on the mass transport process.

The approach solution (continuous lines) and the exact one (dotted lines) of the physical mass transport can be compared to each other in special cases. A brief comparison of the physical mass transport is illustrated in the Appendix 12.A (Fig. 12.A1), in them the concentration distribution obtained by the two methods, at $r_o = \delta = 300$ μm and also $r_o = \delta = 500$ μm is shown. As can be seen there is certain deviation between the curves. These data show that the deviation between the models can strongly decrease with the increase of the lumen radius. The inlet, physical mass transfer rates are for the both prediction methods at 500 and 300 μm membrane's data as $J^o = 0.84 \times 10^{-5}$ mol/m^{2}s and $J^o = 1.06 \times 10^{-5}$ mol/m^{2}s as well as $J^o = 0.94 \times 10^{-5}$ mol/m^{2}s and $J^o = 1.26 \times 10^{-5}$ mol/m^{2}s, considering both the exact and approach solution, respectively. The relative deviations for the two lumen radius are 12%, 19%, due to the decreasing spherical effect with increasing lumen radius. The deviation gradually decreases down to zero with the increase of the lumen radius.

The biochemical reaction, depending strongly on the value of reaction modulus, decreases the spherical effect on the concentration distribution, and it strongly increases the mass transfer rate, at given values of lumen radius and biocatalytic membrane thickness. At low values of the lumen radius and that of the membrane thickness, the change of the space volume and accordingly the change of the internal surface area is rather strong at the chosen value of ΔR, which can significantly affect concentration change. This can then see later at Figs. 12.4A and B, as well as in Fig. 12.5, at which the physical, specific mass transfer rate is plotted as a function of lumen radius.

12.4.1 The effect of the lumen radius on the mass transport

The effect of the change in the space volume, and thus on the specific internal surface, strongly depends on the δ/r_o values, and the individual values of both the lumen radius

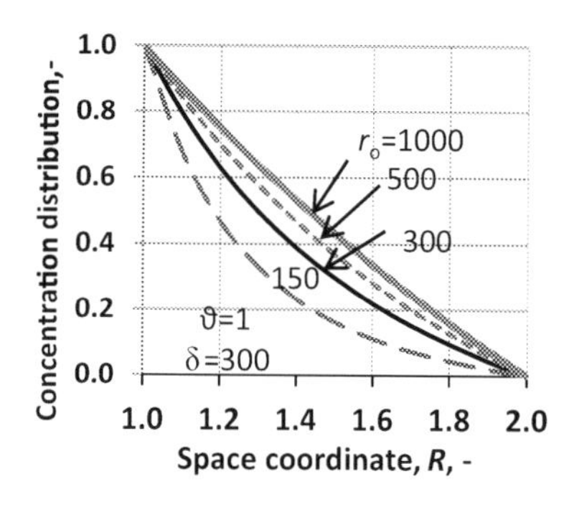

(A)

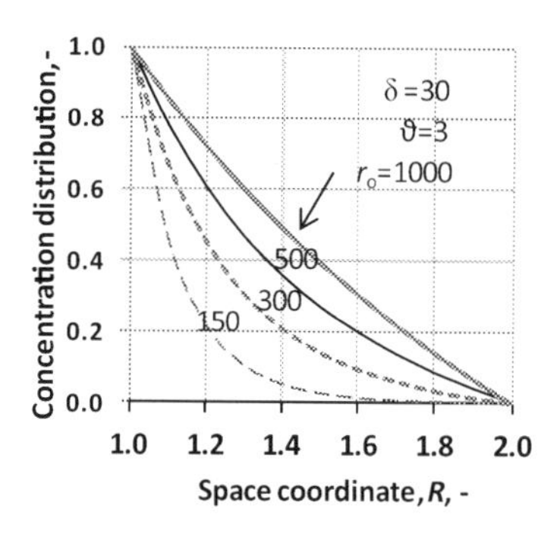

(B)

FIGURE 12.4 (A) Concentration distribution as a function of cylindrical space coordinate, at different values of lumen radius, at $\vartheta = 1$($D = 1.9 \times 10^{-9}$ m^2/s; $\delta = 300$ μm). (B) Concentration distribution as a function of cylindrical space coordinate, at different values of lumen radius, at $\vartheta = 3$($D = 1.9 \times 10^{-9}$ m^2/s; $\delta = 300$ μm).

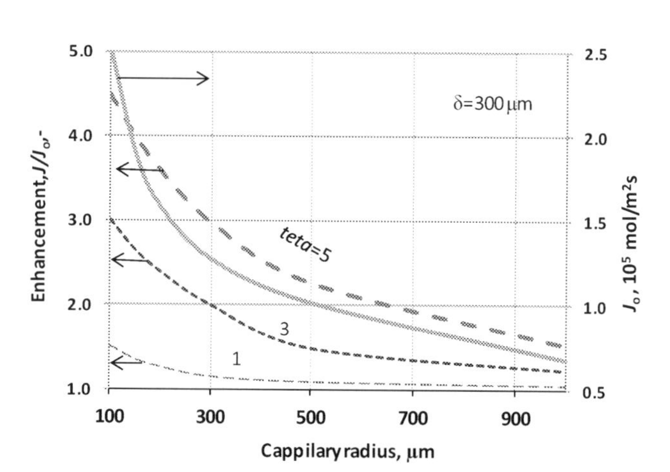

FIGURE 12.5 Change of enhancement as well as the physical mass transfer rate as a function of the lumen radius, at different values of reaction modulus, and at $\delta = 300$ μm ($D = 1.9 \times 10^{-9}$ m^2/s)

and the biocatalytic membrane thickness. During the approach solution, the thickness of the individual sublayers was changed, because the number of the sublayers was chosen to be constant ($N = 1000$, and thus $\Delta R = \delta/(Nr_o)$, and $\overline{R}_i = i\Delta R - \Delta R/2$). According to this last expression for the average value of radius, $\overline{R}_i$, the internal surface of the sublayers was chosen to be constant across the whole range of sublayer, and thus the average values of the internal surface for every single sublayer was hold to be constant between R_i and R_{i+1} interval, it changed at every inlet point of the R_ith sublayer (value of the average radius of the ith sublayer, $\overline{R}_i$, was constant between R_i and R_{i+1} with $i = 1$ and N). Accordingly, the two neighboring sublayers have different internal mass transfer surfaces, which then also has changed the specific, inlet and the outlet mass transfer rate, at internal surface of the sublayers, at $R = R_i$ [i = 1-(N-1)] [see Eq. (12.12) to it].

The effect of the lumen radius on the concentration distribution is illustrated by two figures, at two different values of the reaction modulus, namely $\vartheta = 1$ and 3 as well as at $\delta = 300$ μm and four different values of the lumen radius (Figs. 12.4A and B). Reaction modulus with $\vartheta = 1$ means practically an intermediate reaction modulus, while at $\vartheta = 3$, the value of the reaction modulus falls already into the fast reaction rate regime.

Accordingly, as it can be expected, the effect of the biochemical reaction rate is significantly different and strongly depends on the values of the lumen radius. Lower values curves have increasing inlet concentration gradient, and thus increasing specific, inlet mass transfer rate. Note, if one wants to compare the outlet and the inlet mass transfer rates, between two neighboring sublayers, for example, between the R_i and R_{i+1} at place of R_i, it should also be taken into account the difference in the internal surface, namely it is $2\pi\overline{R}_i$ in the whole range of the R_i sublayer, and that is $2\pi\overline{R}_{i+1}$ throughout in the R_{i+1} sublayer, at for example, even at its inlet ($R = R_{i+1}$) and outlet side i.e. at $R = R_{i+1}$. These figures clearly shows the important effect of the of space curvature, this can then lower the effect of the biochemical reaction rate as it will be shown later, for example, at Fig. 12.5.

12.4.2 The effect of the lumen radius and the membrane thickness

Variation of the lumen radius can change significantly the value of the physical mass transfer coefficient. For example, the physical mass transfer rate strongly changes as a function of r_o, the effect of the capillary radius gradually decreases with the increase of the lumen radius. At for example, $r_o > 1000\ \mu m$ the spherical effect can practically be neglected. Thus the effect of the reaction modulus is not hindered/accelerated by the capillary effect.

Fig. 12.5 illustrates the effect of the biochemical reaction on the mass transfer rate enhancement as a function of the lumen radius, at three different values of reaction modulus between $\vartheta = 1$ and $\vartheta = 5$. The continuous curves illustrate the physical, inlet mass transfer rates, at value of $R = 1$ as a function of the lumen radius, in Fig. 12.5. This latter curve gives important information on the variation of J^o physical, specific mass transfer rate as a function of the lumen radius. As it can be seen, the physical specific, inlet mass transfer rate quickly decreases as a function of the increasing lumen radius. Note that this figure shows increased value of the inlet mass transfer rate, with taking into account the increase of the real mass transfer surface. The effect of the chemical reaction is also shown, as a function of the inlet, specific mass transfer rates, as a function of the lumen radius, at different values of the reaction modulus, at $\delta = 300\ \mu m$. This figure clearly illustrates the decrease of the specific mass transfer rates. As one can see, the specific, inlet mass transfer rate, enhancement, significantly decreases as a function of the lumen radius, illustrating clearly the lowering effect of the curvature. This should mean that the effect of the biochemical reaction rate is much higher, at high values of the lumen radius, because the curvature effect gradually decreases with the increasing value of r_o. Value of the enhancement decreases strongly, as a function of the lumen radius, at given values of the reaction modulus. Note, obviously, the absolute values of the mass transfer rates gradually increase with the reaction modulus.

How the reaction modulus affects the value of enhancement illustrated in Fig. 12.6, at constant value of the reaction modulus, at $r_o = 300\ \mu m$ and at different values of the membrane reactor's thickness. As it is obvious, the larger value of δ means larger space for the reaction, lesser curvature effect, which means increasing effect of the biochemical reaction rate. Accordingly, the value of enhancement essentially increases with the increase of the

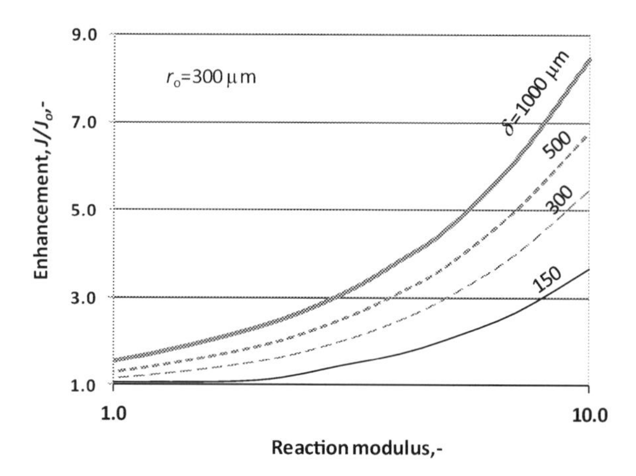

FIGURE 12.6 Enhancement as a function of the reaction modulus, ϑ, at different values of the biocatalytic membrane thickness ($D = 1.9 \times 10^{-9}$ m^2/s)

membrane thickness. This tendency well illustrates by this figure, at given value of reaction modulus. As can be seen the enhancement increasing with intensifying tendency with the reaction modulus. On the other hand, the rate of enhancement is much less at lower values of membrane thickness, due to the larger effect of the curvature. In this fast reaction rate regime, the reactant concentration is practically equal close to zero in the most portion of the membrane layer; the increase in the transport rate is the consequence of the larger concentration gradient close to the membrane inlet surface.

12.4.3 Some results with Michaelis-Menten kinetics

In the previous subchapters, the discussed data were obtained as the limiting case of this kinetics, namely the first-order reaction. Accordingly, the effect of the reactant concentration is defined by Eq. (12.1), in the denominator of the kinetic expression, by constant kinetic parameters, namely v_{max}, K_M. Question arises how the concentration in the denominator can affect the value of enhancement, illustrating by two different figures. The literature publishes rarely reaction data, which compare data in the whole reaction rate regime that involves the effect of the limiting cases as well as that of the general Michaelis-Menten kinetics. It is well known that the concentration independent case, zero-order reaction, can have much higher effect on the mass transport than the other two cases. The effect of the denominator in the Michaelis-Menten kinetics is illustrated by the Fig. 12.7, at constant values of the kinetic parameters ($v_{max} = 2$ mol/m^2s, $K_M = 1$ mol/m^3), but at different reactant concentration in the denominator of the MM-kinetics. In the case of first-order reaction the concentration in the denominator was chosen to be zero, while in case of the zero-order reaction, value of K_M was chosen to be zero. Accordingly, the value of $\vartheta = 2$, for first-order and close to zero-order reactions, while its value was changed between 1and 2, in case of general-order reaction, due to the change of the zero outlet concentration. Comparing the concentration distribution between the first-order (as limiting case of the Michaelis-Menten kinetics, $K_M >> C$) and general-order, Michaelis-Menten kinetics, the deviation between them is rather moderate. While the effect of the zero-order reaction is much higher due to the fact, that the reaction rate is constant and independent of the concentration values, in this case. This shows that the usage of zero-order reaction during enzyme catalyzed reaction can only be recommended, when the K_M value really negligible in the denominator of the kinetic expression.

How the reaction modulus can have the value of the enhancement, as a function of the lumen radius is illustrated by Fig. 12.8, at $\delta = 300\ \mu m$. During the calculation of the enhancement the values of v_{max}/K_M ratio were always chosen to be equal to 2. Accordingly, both values of v_{max} and K_M were changed multiplying the ratio by a given values between 1 and 10 (namely 1, 2, 3, 5), accordingly, for example, $v_{max}/K_M = 10/5 = 2$. Accordingly, the role of the concentration in the denominator gradually decreases, due to the fact that values of K_M became larger and larger comparing it to that of denominator's concentration, which changes between 0 and 1. The relatively strong effect on the

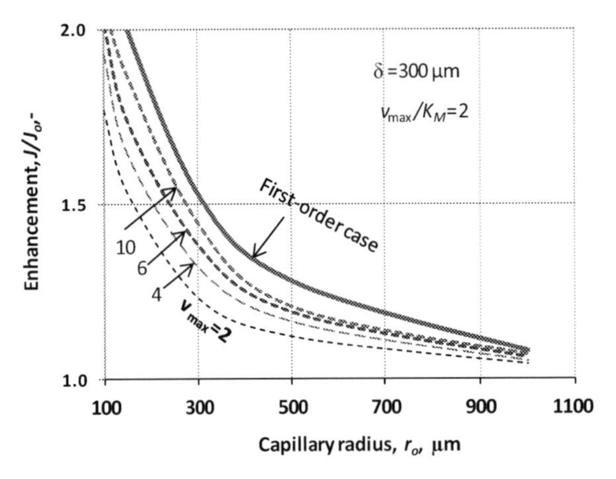

FIGURE 12.7 Concentration distribution as a function of space coordinate obtained by the general Michaelis-Menten kinetics (broken line; $v_{max} = 2\ mol/m^2s$; $K_M = 1$), and its limiting cases, namely at first-order (C values in the denominator was chosen to be equal to 0 in the Michaelis-Menten kinetic equation, i.e. in expression of ϑ) and zero-order biochemical reaction (continuous line; $K_M = 0$) ($D = 1.9 \times 10^{-9}\ m^2/s$; $\delta = 300\ \mu m$; $r_o = 300\ \mu m$).

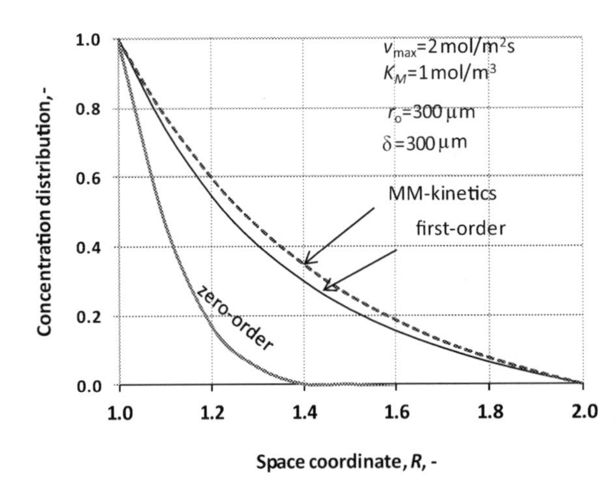

FIGURE 12.8 Enhancement as a function of the lumen radius at different values of ratio of v_{max}/K_M as well as at first-order limiting case, $v_{max}/K_M = 2/1$; ($D = 1.9 \times 10^{-9}\ m^2/s$; $\delta = 300\ \mu m$).

enhancement is obtained with the increase of the K_M values, at $v_{max}/K_M = 2$. This is true in the whole investigated lumen radius range. The curves approach gradually to limiting case of first-order reaction. On the other hand, further decrease of the K_M value below unity means that the effect of K_M will gradually be negligible ($K_M << C_i$), thus values of ϑ reaction modulus will gradually be determined by value of v_{max} (it is obvious that further decrease of v_{max} values, holding the $v_{max}/K_M = 2$, the enhancement will tends to unity in the whole lumen radius range; not shown here). These results confirm that the usage of the limiting cases might be recommended to be thinking over carefully.

12.5 Concluding remarks

This study focuses on the diffusive mass transport process, accompanied by biochemical reaction, through biocatalytic cylindrical membrane layer. An important aspect during the mass transport is how the increasing space, assuming the transport from the lumen side of the membrane, the membrane thickness and the biochemical reaction itself can affect the mass transport, and accordingly the inlet mass transfer rate. For prediction of data an approach solution was shown and applied to our simulations. Thus for prediction of the concentration distribution and the values of enhancement, explicit, closed mathematical expression was developed as a function of the space coordinate, and that of the lumen radius and the reaction modulus, in cases of general Michaelis-Menten kinetics. As it is shown, values of the lumen, that is, the capillary radius, the membrane thickness can have significant effect on the inlet mass transfer rate, on the concentration distribution, inside of the membrane bioreactor, as well as on the reaction efficiency. Accordingly, the correct choice of values of these two parameters requests careful prediction of the effect of these parameters, depending on the biochemical reaction rates.

Acknowledgment

The Hungarian National Development Agency, grants OTKA 116727 and NKFIH-1158–6/2019, is gratefully acknowledged for the financial support.

References

Adhikari, B. R., Schraft, H., & Chen, A. (2017). A high-performance enzyme entrapment platform facilitated by a cationic polymer for the efficient electrochemical sensing of ethanol. *Analyst, 142*, 2595–2602.

Asif, M. B., Hai, F. I., Jegatheesan, V., Price, W. E., Nghiem, L. D., & Yamamoto, K. (2019). *Applications of membrane bioreactors in biotechnology processes. Current trends and future developments on (bio-) membranes*. Elsevier Inc.

Awad, G. E. A., Ghanem, A. F., Abdel Wahab, W. A., & Wahba, M. I. (2020). Functionalized κ-carrageenan/ hyperbranched poly(amidoamine)for protease immobilization: Thermodynamics and stability studies. *International Journal of Biological Macromolecules, 148*, 1140–1155.

Bilal, M., Asgher, M., Cheng, H., Yan, Y., & Iqbal, H. M. N. (2019). Multi-point enzyme immobilization, surface chemistry, and novel platforms: A paradigm shift in biocatalyst design. *Critical Reviews in Biotechnology, 39*, 202–219.

Bronetti, A., Capannelli, G., & Comite, A. (2002). Catalytic membrane reactors for the oxide hydrogenation of propane: Experimental and modeling study. *Journal of Membrane Science, 197*, 75–88.

Brunetti, A., Caravella, A., Barbiella, G., & Drioli, E. (2007). Simulation study of water shift reaction in a membrane reactor. *Journal of Membrane Science, 306*, 329–340.

Butterfield, D. A., Bhattacharyya, D., Daunert, S., & Bachas, L. (2001). Catalytic biofunctional membranes containing site-specifically immobilized enzyme arrays: A review. *Journal of Membrane Science, 181*, 29–37.

Cabral, J. M. S., & Tramper, J. (1994). In J. M. S. Cabral, D. Best, I. Boross, & J. Tramper (Eds.), *Applied biocatalysis* (pp. 330–370). Switzerland: Harwood Academic Publisher.

Cao, Y., Wu, Z., Wang, T., Xiao, Y., Huo, Q., & Liu, Y. (2016). Immobilization of: Bacillus subtilis lipase on a Cu-BTC based hierarchically porous metal-organic framework material: A biocatalyst for esterification. *Dalton Transactions, 45*, 6998–7003.

Cen, Y. K., Liu, Y. X., Xue, Y. P., & Zheng, Y. G. (2019). Immobilization of enzymes in/on membranes and their applications. *Advanced Synthesis and Catalysis, 361*, 5500–5515.

Chakraborty, S., Rusli, H., Nath, A., Sikder, J., Bhattacharjee, C., Curcio, S., & Drioli, E. (2016). Immobilized biocatalytic process development and potential application in membrane separation: A review. *Critical Reviews in Biotechnology, 36*, 43–48.

Chapman, J., Ismail, A. E., & Dinu, C. Z. (2018). Industrial applications of enzymes: Recent advances, techniques, and outlooks. *Catalysts, 8*, 20–29.

Charcosset, C. (2006). Membrane processes in biotechnology: An overview. *Biotechnology Advances, 24*, 482–492.

Chen, P. C., Ma, Z., Zhu, X. Y., Chen, D. J., & Huang, X. J. (2019). Fabrication and optimization of a lipase immobilized enzymatic membrane bioreactor based on polysulfone gradient-pore hollow fiber membrane. *Catalysts, 9*.

Coutte, F., Lecouturier, D., Firdaous, L., Kapel, R., Bazinet, L., Cabassud, C., & Dhulster, P. (2016). *Recent trends in membrane bioreactors. Current developments in biotechnology and bioengineering: Bioprocesses, bioreactors and controls*. Elsevier B.V..

Crank, J. (1975). *The mathematics of diffusion* (pp. 69–88). Oxford: Clarendon Press.

Dizge, N., Demirdogen, R. E., & Ocakoglu, K. (2019). Developments and applications in enzyme activated membrane reactors: A review. *European Journal of Engineering and Natural Sciences, 3*, 26–41.

Drout, R. J., Robison, L., & Farha, O. K. (2019). Catalytic applications of enzymes encapsulated in metal–organic frameworks. *Coordination Chemistry Reviews, 381*, 151–160.

Eldin Mohy, M. S., Seuror, E. I., Nasr, M. A., & Tieama, H. A. (2011). Affinity covalent immobilization of glucoamylase onto ρ-benzoquinone- activated alginate beads: II. Enzyme immobilization and characterization. *Applied Biochemistry and Biotechnology, 164*, 45–57.

Falcone, N., Shao, T., Rashid, R., & Kraatz, H. B. (2019). Enzyme entrapment in amphiphilic myristyl-phenylalanine hydrogels. *Molecules (Basel, Switzerland), 24*.

Fernández-Lorente, G., Lopez-Gallego, F., Bolivar, J., Rocha-Martin, J., Moreno-Perez, S., & Guisan, J. (2015). Immobilization of proteins on highly activated glyoxyl supports: Dramatic increase of the enzyme stability via multipoint immobilization on pre-existing carriers. *Current Organic Chemistry, 19*, 1719–1731.

Garcia, J. D. G. G., Iliná, A., Ventura, J., Michelena, G., Nava, E., Espinoza González, C., & Martínez, J. L. (2019). Microencapsulation of *Aspergillus niger* phytases produced in triticale by solid fermentation and microencapsulates characterization. *African Journal of Biotechnology, 18*, 564–584.

Garcia-Galindo, I., Gómez-García, R., Palácios-Ponce, S., Ventura, J., Boone, D., Ruiz, H. A., … Aguilar-González, C. N. (2019). *New features and properties of microbial cellulases required for bioconversion of agro-industrial wastes. Enzymes in food biotechnology*. Elsevier Inc.

Gebreyohannes, A. Y., Dharmjeet, M., Swusten, T., Mertens, M., Verspreet, J., Verbiest, T., … Vankelecom, I. F. J. (2018). Simultaneous glucose production from cellulose and fouling reduction using a magnetic responsive membrane reactor with superparamagnetic nanoparticles carrying cellulolytic enzymes. *Bioresource Technology, 263*, 532–540.

Giorno, L., & Drioli, E. (2000). Biocatalytic membrane reactors: Applications and perspectives. *Trends in Biotechnology, 18*, 339–349.

Giorno, L., Mazzei, R., & Drioli, E. (2009). *Biochemical Membrane Reactors in Industrial Processes. Membrane operations: Innovative separations and transformations* (pp. 397–409). Wiley.

Irfan, M., Ghazanfar, M., Ur Rehman, A., & Asma, S. (2019). Approaches to enhance industrial production of fungal cellulases. In M. Srivastava, N. Srivastava, P. W. Ramteke, & P. K. Mishra (Eds.), *Approaches to enhance industrial production of fungal cellulases* (pp. 37–52). Cham: Springer.

Jesionowski, T., Zdarta, J., & Krajewska, B. (2014). Enzyme immobilization by adsorption: A review. *Adsorption, 20*, 801–821.

Jing, Y., Guo, Y., Xia, Q., Liu, X., & Wang, Y. (2019). Catalytic production of value-added chemicals and liquid fuels from lignocellulosic biomass. *Chem, 5*, 1–27.

Liang, S., Wu, X. L., Xiong, J., Zong, M. H., & Lou, W. Y. (2020). Metal-organic frameworks as novel matrices for efficient enzyme immobilization: An update review. *Coordination Chemistry Reviews, 406*, 213149.

Liu, D.-M., & Dong, C. (2020). Recent advances in nano-carrier immobilized enzymes and their applications. *Process Biochemistry, 92*, 464–475.

Lopez, J. L., & Matson, S. L. (1997). A multiphase/extractive enzyme membrane reactor for production of diltiazem chiral intermediate. *Journal of Membrane Science, 125*, 189–211.

Love, C., Steinkühler, J., Gonzales, D. T., Yandrapalli, N., Robinson, T., Dimova, R., & Tang, T. Y. D. (2020). Reversible pH responsive coacervate formation in lipid vesicles activates dormant enzymatic reactions. *Angewandte Chemie, 59*, 5950–5957.

Lozano, P., Bernal, B., Jara, A. G., & Belleville, M. P. (2014). Enzymatic membrane reactor for full saccharification of ionic liquid-pretreated microcrystalline cellulose. *Bioresource Technology, 151*, 159–165.

Mahmoodi, N. M., Saffar-Dastgerdi, M. H., & Hayati, B. (2020). Environmentally friendly novel covalently immobilized enzyme bionanocomposite: From synthesis to the destruction of pollutant. *Composites Part B: Engineering, 184*, 107666.

Malandra, A., Cantarella, M., Kaplan, O., Vejvoda, V., Uhnáková, B., Štěpánková, B., … Martínková, L. (2009). Continuous hydrolysis of 4-cyanopyridine by nitrilases from Fusarium solani O1 and *Aspergillus niger* K10. *Applied Microbiology and Biotechnology, 85*, 277–284.

Marcano, J. G. S., & Tsotsis, T. T. (2002). *Catalytic membranes and membrane reactors*. Veinheim: Wiley-VCH.

Mazzei, R., Drioli, E., & Giorno, L. (2012). Enzyme membrane reactor with heterogenized β-glucosidase to obtain phytotherapic compound: Optimization study. *Journal of Membrane Science, 390–391*, 121–129.

Mazzei, R., Giorno, L., Piacentini, E., Mazzuca, S., & Drioli, E. (2009). Kinetic study of a biocatalytic membrane reactor containing immobilized β-glucosidase for the hydrolysis of oleuropein. *Journal of Membrane Science, 339*, 215–223.

Mazzei, R., Piacentini, E., Gebreyohannes, A., & Giorno, L. (2017). Membrane bioreactors in food, pharmaceutical and biofuel applications: State of the art, progresses and perspectives. *Current Organic Chemistry, 21*.

Mazzuca, S., Giorno, L., Spadafora, A., Mazzei, R., & Drioli, E. (2006). Immunolocalization of β-glucosidase immobilized within polysulphone capillary membrane and evaluation of its activity in situ. *Journal of Membrane Science, 285*, 152–158.

Nagy, E. (2009a). Mathematical modeling of biochemical membrane reactors. In E. Drioli, & L. Giorno (Eds.), *Membrane operations, innovative separation and transformations* (pp. 309–334). Weinheim: Wiley-VCH Verlag.

Nagy, E. (2009b). Basic equations of mass transfer through biocatalytic membrane layer. *Asia-Pacific Journal of Chemical Engineering, 21*, 270–278.

Nagy, E. (2011). Mass transfer through catalytic membrane layer. In A. El-Amin (Ed.), *Mass transfer in multiphase systems and its applications* (pp. 677–716). Rijeka: INTECH OPEN access Publisher.

Nagy, E. (2017). Survey on biocatalytic membrane reactor and membrane aerated biofilm reactor. *Current Organic Chemistry, 21*, 1713–1724.

Nagy, E. (2019). *Basic equation of mass transfer through a membrane layer*. Amsterdam: Elsevier.

Nagy, E., Dudás, J., Mazzei, R., Drioli, E., & Giorno, L. (2015). Description of the diffusive-convective mass transport in a hollow fiber biphasic biocatalytic membrane reactor. *Journal of Membrane Science, 482*, 144–157.

Nagy, E., & Kulcsar, E. (2009). Mass transport through biocatalytic membrane layer. *Desalination, 245*, 422–436.

Nagy, E., Lepossa, A., & Prettl, Z. S. (2012). Mass transfer through bio-catalytic membrane reactor. *Industrial & Engineering Chemistry Research, 5*(1), 1635–1646.

O'Neil, P. V. (1987). *Advanced engineering mathematics*. Belmont, CA: Wadsworth Inc.

Olímpio, F. M. P., Mendes, A. A., Trevisan, M. G., & Garcia, J. S. (2020). Preparation and delayed release study on pancreatin encapsulated into alginate, carrageenan and pectin hydrogels. *Journal of the Brazilian Chemical Society, 31*, 320–330.

Osbon, Y., Kumar, M., (2016). *Biocatalysis and strategies for enzyme improvement*. Intech i, 13.

Perry, J. (1968). Vegyészmérnökök Kézikönyve *(Chemical engineers' handbook)*. Budapest: Műszaki Kiadó.

Piacentini, E., Mazzei, R., Drioli, E., & Giorno, L. (2017). *1.1 From biological membranes to artificial biomimetic membranes and systems, . Comprehensive membrane science and engineering* (1, pp. 1–16). Elsevier.

Priyanka, B. S., & Rastogi, N. K. (2020). Encapsulation of β-amylase in water-oil-water enzyme emulsion liquid membrane (EELM) bioreactor for enzymatic conversion of starch to maltose. *Preparative Biochemistry & Biotechnology, 50*, 172–180.

Reiser, J., Muheim, A., Hardeggea, M., Frank, G., & Fiechter, A. (1994). Aryl-alcohol Dehydrogenase from the white-rot fungus phanerochaete chrysosporium. *The Journal of Biological Chemistry, 269*, 28152–28159.

Rios, G. M., Belleville, M.-P., & Paolucci-Jeanjean, D. (2007). Membrane engineering in biotechnology: Quo vamus? *Trends in Biotechnology, 25*, 242–246.

Seidel-Morgenstein, A. (2010). *Membrane reactors*. Weiheim: Wiley-WCH.

Shao, P., Chen, H., Ying, Q., & Zhang, S. (2020). Structure–Activity relationship of carbonic anhydrase enzyme immobilized on various silica-based mesoporous molecular sieves for CO_2 absorption into a potassium carbonate solution. *Energy & Fuels, 34*(2), 2089–2096.

Sheldon, R. A. (2007). Enzyme immobilization: The quest for optimum performance. *Advanced Synthesis and Catalysis, 349*, 1289–1307.

Sheldon, R. A., & Woodley, J. M. (2018). Role of biocatalysis in sustainable chemistry. *Chemical Reviews, 118*, 801–838.

Shen, Y. X., Saboe, P. O., Sines, I. T., Erbakan, M., & Kumar, M. (2014). Biomimetic membranes: A review. *Journal of Membrane Science, 454*, 359–381.

Sigurdardóttir, S. B., Lehmann, J., Ovtar, S., Grivel, J. C., Negra, M. D., Kaiser, A., & Pinelo, M. (2018). Enzyme immobilization on inorganic surfaces for membrane reactor applications: Mass transfer challenges, enzyme leakage and reuse of materials. *Advanced Synthesis and Catalysis, 360*, 2578–2607.

Sóti, P. L., Weiser, D., Vigh, T., Nagy, Z. K., Poppe, L., & Marosi, G. (2016). Electrospun polylactic acid and polyvinyl alcohol fibers as efficient and stable nanomaterials for immobilization of lipases. *Bioprocess and Biosystems Engineering, 39*, 449–459.

Tamborini, L., Fernandes, P., Paradisi, F., & Molinari, F. (2018). Flow bioreactors as complementary tools for biocatalytic process intensification. *Trends in Biotechnology, 36*, 73–88.

Thakur, V. K., & Thakur, M. K. (2018). *Polymer gels. Science and fundamentals*. Singapore: Springer.

Vitola, G., Mazzei, R., Poerio, T., Porzio, E., Manco, G., Perrotta, I., ... Giorno, L. (2019). Biocatalytic membrane reactor development for organophosphates degradation. *Journal of Hazardous Materials, 365*, 789–795.

Westermann, T., & Melin, T. (2012). Flow-through catalytic membrane reactors- Principles and applications. *Chemical Engineering and Processing, 48*, 17–28.

Yagonia, C. F. J., Park, K., & Yoo, Y. J. (2014). Immobilization of Candida antarctica lipase B on the surface of modified sol-gel matrix. *Journal of Sol-Gel Science and Technology, 69*, 564–570.

Yoo, Y. J., Feng, Y., Kim, Y. H., & Yagonia, C. F. J. (2017). *Fundamentals of enzyme engineering* (pp. 1–209). Springer.

Zhou, M., Ju, X., Li, L., Yan, L., Xu, X., & Chen, J. (2019). Immobilization of cellulase in the non-natural ionic liquid environments to enhance cellulase activity and functional stability. *Applied Microbiology and Biotechnology, 103*, 2483–2492.

Zimmerman, J. B., Anastas, P. T., Erythropel, H. C., & Leitner, W. (2020). Designing for a green chemistry future. *Science (New York, N.Y.), 367*, 397–400.

Further reading

Guisan, J. M., López-Gallego, F., Bolivar, J. M., Rocha-Martín, J., & Fernandez-Lorente, G. (2020). The science of enzyme immobilization. *Methods in Molecular Biology, 2100*, 1–26.

Nagy, E. (2008). Mass transport with varying diffusion- and solubility through a catalytic membrane layer. *Chemical Engineering Research and Design, 86*, 723–730.

Nagy, E. (2010). Convective and diffusive mass transport through anisotropic, cylindrical membrane layer. *Chemical Engineering and Processing, 49*, 716–721, Spec. Issue "Intensified Transport Complex Geometries.

Appendix 12.A

Physical mass transport through cylindrical membrane

The mass transport through cylindrical space, without chemical/biochemical reaction can be given by analytical solution. The differential mass balance equation for this diffusive transport, without chemical reaction, is as:

$$\frac{d^2C}{dr^2}+\frac{1}{r}\frac{dC}{dr}=0 \tag{12.A1}$$

The general solution of the above differential equation is (Nagy, 2019):

$$C = T\,ln\,r + S \tag{12.A2}$$

The T and S parameters can be predicted by suitable boundary conditions; accordingly ($R = r/r_o$):

$$\text{at} \quad R = 1 \quad \text{then} \quad C = C^o \tag{12.A3}$$

and

$$\text{at} \quad R = 1+\delta/r_o = R_m \quad \text{then} \quad C = C^o_\delta \tag{12.A4}$$

Thus one can get for values of parameters T and S as:

$$T=\frac{C^o - C^o_\delta}{ln\,r_o - ln\,r_m}; \quad S=\frac{C^o_\delta\,ln\,r_o - C^o\,ln\,r_m}{ln\,r_o - ln\,r_m}\equiv\frac{C^o\,ln\,R_m}{ln\,R_m} \tag{12.A5}$$

Accordingly, the diffusive, specific, inlet mass transfer rate will be as

$$J^o = -D\frac{dC}{dR}\bigg|_{R=1} = \frac{D}{r_o}\frac{\left(C^o - C^o_\delta\right)}{ln\,R_m}. \tag{12.A6}$$

Applying the above expressions, both the concentration distribution and the inlet mass transfer rate can be predicted at different values of lumen radius and membrane thickness. The concentration distribution calculated by the analytical solution [Eq. (12. A5)], as well as by the approach solution presented, is illustrated in Fig. 12.A1. As can be seen there is a certain deviations between the model results, especially at low lumen radius (it reaches 19% at $r_o = 300$ μm), which gradually decreases with the increase of the lumen radius.

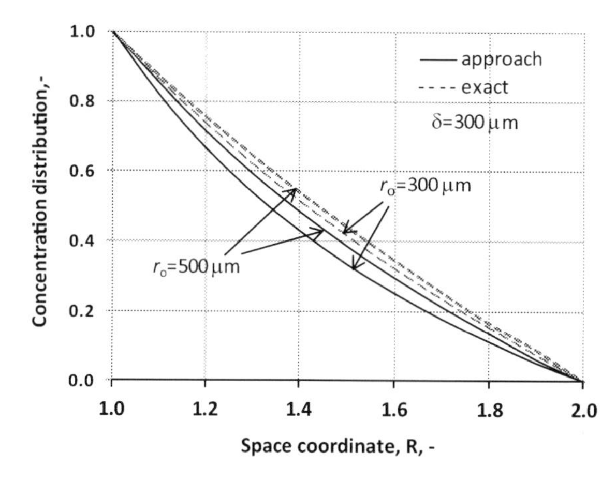

FIGURE 12.A1 Concentration distribution through the membrane layers obtained by the exact and the approach solutions.

Appendix 12.B

The approach solution methodology of the mass transport process is discussed in previous studies (Nagy, 2019) by its different aspects. The essence of the methodology is to obtain approaching differential mass balance equation with constant parameters, written it for every single, very thin sublayer, which then can already be solved analytically. But, in such a way, we get $2 \times N$ algebraic equations, using the suitable boundary conditions for them, which are to be solved. These algebraic systems then can be solved by known mathematical means (Perry, 1968).

The simplified differential equation obtained, assuming the Michaelis-Menten kinetics, is as:

$$\frac{d^2C}{dR^2} + \frac{1}{\overline{R}_i}\frac{dC}{dR_i} - \vartheta_i^2 C = 0 \tag{12.B1}$$

with

$$\vartheta_i = \sqrt{\frac{r_o^2}{D}\frac{v_{\max}}{K_M + C_{i-1}(\overline{R}_i)}}. \tag{12.B2}$$

Basic expressions of solution are given in the main text. The details of the solution of algebraic equation system are also briefly shown in Nagy (2019). The T and S parameters obtained for the first sublayer as follows:

$$T_1 = \left(\prod_1^N \left[\frac{E_i^T}{E_i^O}\right] C^o - \frac{1}{\prod\limits_{i=1}^{N-1} e^{-\Delta R/(2\overline{R}_i)} E_i^O} C_{m,p} \right) \frac{1}{2\cosh(\Theta_1 \Delta R) e^{(-\Delta R/2\overline{R}_1)} e^{\lambda_1} \prod_1^N E_i^O} \tag{12.B3}$$

$$S_1 = \left(\prod_1^N \left[\frac{E_i^S}{E_i^O}\right] C^o - \frac{1}{\prod\limits_{i=1}^{N-1} e^{-\Delta R/(2\overline{R}_i)} E_i^O} C_{m,p} \right) \frac{1}{2\cosh(\Theta_1 \Delta R) \left[e^{-\Delta R/2\overline{R}_1}\right] e^{\overline{\lambda}_1} \prod_1^N E_i^O} \tag{12.B4}$$

All other parameters can then be simply calculated from layer to layer upwards from $N = 2$ up to N, knowing always the T_i and S_i values of the previous layer. Thus both the concentration distribution, the mass transfer rates can be predicted by given values of the membrane mass transport parameters and the reaction kinetics parameters.

The obtained values of different variables, for calculation of values of T_1 and S_1 parameters by expressions (12.B3) and (12.B4), are listed below:

$$E_i^j = \left\langle -\frac{\tan h(\Theta_i \Delta R)}{\Theta_i}\frac{\overline{R}_i}{\overline{R}_{i-1}}\left(\frac{F_{i-1}^j}{E_{i-1}^j} - \frac{1}{2\overline{R}_{i-1}}\right) - 1 \right\rangle; \quad j = T, S, O; \quad i = 2, \ldots, N \tag{12.B5}$$

$$F_i^j = \left\langle -\frac{A_i}{\Theta_i}\frac{\overline{R}_i}{\overline{R}_{i-1}}\left(\frac{F_{i-1}^j}{E_{i-1}^j} - \frac{1}{2\overline{R}_{i-1}}\right) - B_i\right\rangle; \quad j = T, S, O; \quad i = 2, \ldots, (N-1) \tag{12.B6}$$

$$A_i = \frac{1}{2\overline{R}_i}\tanh(\Theta_i\Delta R) - \Theta_i; \quad B_i = \Theta_i \tanh(\Theta_i\Delta R) - \frac{1}{2\overline{R}_i} \quad i = 2, \ldots, N \tag{12.B7}$$

The initial values of E_i^j and F_i^j, namely Ω_1^j and ψ_1^j ($j = T,S,O$) are as:

$$E_1^T = e^{\lambda_1\Delta R} \quad E_1^S = e^{\tilde{\lambda}_1\Delta R} \quad E_1^O = -\tanh(\Theta_1\Delta R) \tag{12.B8}$$

$$F_1^T = \lambda_1 e^{\lambda_1\Delta R} \quad F_1^S = \tilde{\lambda}_1 e^{\tilde{\lambda}_1\Delta R} \quad F_1^O = \frac{1}{2\overline{R}_1}\tanh(\Theta_1\Delta R) - \Theta_1 \tag{12.B9}$$

The value of A_i, ($i = 1,..,N$) and B_i (with $i = 2,\ldots,N\text{-}1$) as well as E_i^j (with $j = T,S,O$ and $i = 1,\ldots,N\text{-}1$) and F_i^j (with $j = T,S,O$ and $i = 1,\ldots,N$) are listed above. It is important to note, that calculation of the E_i^j and F_i^j values for $i = 1$ to $N\text{-}1$ (or $i = 1,\ldots,N$ in case of E_N^j) needs very accurate calculation process. Every, calculated variable should be given or calculated, even, for example, also the value of ΔR [$= \delta/(r_o N)$] either, with very accurately. Each steps of the calculation was carried out by homemade computer program written in *Quick basic computer language* with accuracy of 14 decimals. This is the maximal accuracy by this program. To get the concentration distribution, the values of T_i and/or S_i, for $i = 2,\ldots,N$, should be determined (the N value was here chosen to be 1000; note that in reality it is enough to predict either the value T_1 or S_1 because if you know one of these two parameters, the other one can be obtained from the boundary condition given for the interface, namely at $R = 1$: $T_1 e^{\tilde{\lambda}_1} + S_1 e^{\lambda_1} = 1$).

CHAPTER

13

Transport phenomena in photocatalytic membrane reactors

Enrica Fontananova and Valentina Grosso
Institute on Membrane Technology of the National Research Council (ITM-CNR), Rende (CS), Italy

Abbreviations

AOP	Advanced oxidation process
CB	Conduction band
e^-	Electron
Eg	Energy band gap
e_{CB}^-	Photogenerated electron in conduction band
PCMRs	Photocatalytic membrane reactors
MF	Microfiltration
NF	Nanofiltration
PES	Polyethersulfone
PSF	Polysulfone
PVDF	Polyvinylidenefluoride
PVDF-HFP	PVDF-co-hexafluoropropylene
PVDF-TrFE	Polyvinylidenefluoride–trifluoroethylene
PTFE	Polytetrafluoroethylene
POM	Polyoxometalate
RR	Recovery ratio
UF	Ultrafiltration
h^+	Photogenerated hole
h_{VB}^+	Photogenerated hole in valence band
VB	Valence band

List of symbols

Cf	Solute concentration in the feed
Cp	Solute concentration in the permeate
Ci	Concentration
C_m	Concentration at the membrane interface

DOI: https://doi.org/10.1016/B978-0-12-822257-7.00008-X

C_b Concentration in the bulk of the solution
$D_{i,m}$ Diffusion coefficient
F Faraday constant,
h Planck's constant
J Flux
J_∞ Limiting value of the flux
k Mass transfer coefficient
K_p Constant depending on the pore shape and tortuosity
$K_{i,c}$ Hindrance factor
L Phenomenological coefficient
P Pressure
Q_p Permeate flow rate
Q_f Feed flow rate
R Universal gas constant
R_g Resistance of the mass transfer
R_m Membrane resistance
r Pore radius
S Specific surface area
T Temperature,
V_i Partial volume
V Solvent velocity
χ Axial coordinate
γ_i Activity coefficient
z_i Valence

Greek symbols

δ Membrane thickness
ε Porosity
μ Electrochemical potential
η Viscosity
ψ Chemical potential

13.1 Introduction

Photocatalysis is today recognized as an environment-friendly, sustainable and energy-saving technology for environmental remediation and fine chemical synthesis (Guo, Qi, & Liu, 2017; Maldotti, Molinari, & Amadelli, 2002; Molinari, Argurio, Bellardita, & Palmisano, 2017). There are various aspects that make photocatalytic membrane reactors (PCMRs) a green approach for catalytic chemical conversion: (1) the opportunity of combining PCMRs with other technologies in an integrated process; (2) the prospect of operating at room temperature and pressure; (3) the use of molecular oxygen; (4) the use of green and safe photocatalyst (e.g., TiO_2); (5) the possibility to avoid the formation of recalcitrant byproducts by the complete destruction of contaminants in advanced oxidation processes (AOPs); (6) the prospect of using renewable solar energy; (7) the possibility of tailoring the residence time of the substrates in the reactor and, as a consequence, reaction selectivity, by working on membrane structure and driving force applied (König, 2017; Molinari et al., 2019; Palmisano, Yurdakal, Augugliaro, Loddo, & Palmisano, 2007).

In several photocatalytic processes, a photocatalyst is suspended in a liquid medium. The photocatalyst-recovering step from the solution, at the end of the operation, is often the main limiting step for application of photoreactors at a large scale (Mozia, Morawski, Molinari, Palmisano, & Loddo, 2013). The recovery of the catalytic particles, necessary in order to prevent their wash out and consequent decrease of their amount in the reactor system, can be carried out by a membrane in a PCMR.

This membrane permits continuous operation because the recovery of the photocatalyst and the reaction take place in the same time. This allows to get a system that is more efficient with respect to other separation technologies in terms of material recovery, easiness of scaleup, energy costs, reduction of the environmental impact, and selective removal of the components (Molinari, Caruso, & Poerio, 2009; Molinari, Caruso, Argurio, & Poerio, 2008).

The most investigated configurations of PCMRs combine a pressure-driven membrane operation, such as nanofiltration (NF), ultrafiltration (UF), and microfiltration (MF), with a photocatalytic process (Mascolo et al., 2007; Molinari, Borgese, Drioli, Palmisano, & Schiavello, 2002; Tang & Chen, 2004).

In a PCMR, the membrane has the dual role of maintaining the photocatalyst (in suspension or immobilized) and substrates/products in the reaction environment controlling their residence time. From this point of view, it is the fundamental role of transport properties of the membranes on the process efficiency. In the case of PCMR applied in AOP of organic pollutants, the residence time must be adequate to guarantee complete degradation of all the organic substances present in solution. When the PCMR is used in a synthetic pathway, the membrane might work as a barrier for the selective separation of the product, thereby reducing its secondary reactions that lead to undesired byproducts (Mozia, Argurio, & Molinari, 2018).

13.2 Fundamental aspects of photocatalytic membrane reactors

In this section, after the presentation of the fundamental aspects of a photocatalytic process, the main functions of the membrane in the PCMR and the foremost techniques used for photocatalyst immobilization in/on membrane will be discussed.

13.2.1 Main aspects of a photocatalytic process

Photocatalytic reactions can be directed either toward the formation of new substances or decomposition of undesirable material (König, 2017). As a consequence, photocatalytic processes can be roughly distinguished in two classes: photocatalytic synthesis and photocatalytic degradation.

A variety of photocatalysts have been investigated in literature including metals, metal oxides, carbon-based nanostructures, quantum dots, metal–organic frameworks. Among these, the metal oxide semiconductor photocatalysts are the most widely applied in photocatalysis.

Photocatalysts are activated by exposure to light with appropriate energy. The electronic structure of a semiconductor photocatalyst is in fact characterized by a valence

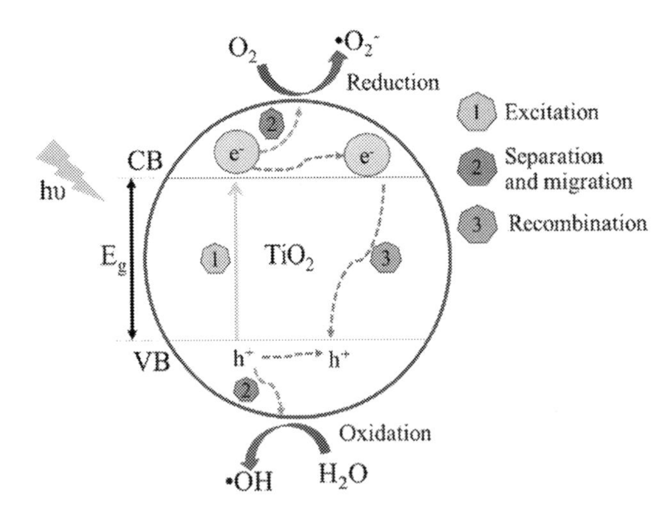

FIGURE 13.1 Mechanism of semiconductor (TiO_2) activation under light irradiation He et al. (2021). Source: *Reproduced with permission from He J., A. Kumar, M. Khan, I.M.C. Lo, (2021) Critical review of photocatalytic disinfection of bacteria: From noble metals- and carbon nanomaterials-TiO2 composites to challenges of water characteristics and strategic solutions. Science of the Total Environment, 758, 143–953.*

band (VB) and a conduction band (CB), which are separated by an energy band gap (Eg). When a semiconductor particle is excited by irradiation with photons of energy ($h\surd$) equal or higher than its Eg, valence electrons (e^-) are promoted from VB to CB ($e_{CB}{}^-$), thereby leaving a positive hole ($h_{VB}{}^+$) in the VB (Fig. 13.1) (He, Kumar, Khan, & Lo, 2021; Herrmann, 2005; Molinari, Argurio, & Palmisano, 2015).

Among the metal oxides photocatalysts, titanium dioxide (TiO_2) is the most used thanks its nontoxic nature, high chemical stability, and elevated availability in the market. The main limitations of TiO_2 are the elevated electron–hole recombination ratio, which reduces the performance of the photocatalytic system and a large bandgap (3.2 eV) that infers a low adsorption capacity for visible light (Corredor, Rivero, Rangel, & Ortiz, 2019; Hairom, Mohammad, & Kadhum, 2014; Intarasuwan, Amornpitoksuk, Suwanboon, & Graidist, 2017; Lee & Park, 2013). To avoid these limitations, some strategies, such as doping with a cocatalyst, have been adopted in literature. Efficient composite photocatalysts have also been synthesized by combining the semiconductor photocatalyst with other materials, including carbonaceous materials such as carbon nanotubes and graphene (Leary & Westwood, 2011).

The generally accepted mechanism of the photocatalytic oxidation of organic compounds in water using TiO_2 under UV–Vis light activation is a multistep electron transfer mechanism in which radical species are involved (Analitica et al., 1993; Doll & Frimmel, 2005; Konstantinou & Albanis, 2004):

$$TiO_2 + h\upsilon \rightarrow TiO_2\left(e_{CB}{}^- + h_{VB}{}^+\right) \quad (13.1)$$

$$TiO_2\ \left(h_{VB}{}^+\right) + H_2O \rightarrow TiO_2 + H^+ + OH^\bullet \quad (13.2)$$

$$TiO_2\ \left(h_{VB}{}^+\right) + OH^- \rightarrow TiO_2 + OH^\bullet \quad (13.3)$$

$$TiO_2\ \left(e_{CB}{}^-\right) + O_2 \rightarrow TiO_2 + O_2^{\bullet -} \quad (13.4)$$

$$O_2^{\bullet -} + H^+ \rightarrow HO_2^{\bullet} \tag{13.5}$$

$$HO_2^{\bullet} + HO_2^{\bullet} \rightarrow H_2O_2 + O_2 \tag{13.6}$$

$$TiO_2\ (e_{CB}^{-}) + H_2O_2 \rightarrow OH^{\bullet} + OH^{-} \tag{13.7}$$

$$H_2O_2 + O_2^{\bullet} \rightarrow OH^{\bullet} + OH^{-} + O_2 \tag{13.8}$$

$$H_2O_2 + h\upsilon \rightarrow 2OH^{\bullet} \tag{13.9}$$

$$\text{Organic compound} + OH^{\bullet} \rightarrow \text{degradation products} \tag{13.10}$$

$$\text{Organic compound} + TiO_2\left(h_{VB}^{+}\right) \rightarrow \text{oxidation products} \tag{13.11}$$

$$\text{Organic compound} + TiO_2(e_{CB}^{-}) \rightarrow \text{reduction products} \tag{13.12}$$

Furthermore, various transport phenomena are involved in a solid/liquid photocatalysis process (Nasrollahi, Ghalamchi, Vatanpour, & Khataee, 2021):

1. Diffusion of reactants from the bulk liquid through a boundary layer to the solution–catalyst interface.
2. Inter- and/or intra-particle diffusion of reactants to the active surface sites of the catalyst.
3. Adsorption of at least one of the reactants.
4. Reactions in the adsorbed phase.
5. Desorption of the product(s).
6. Removal of the products from the interface region to the bulk solution.

13.2.2 Reactor configuration and membrane function

The PCMR can be divided generally into two main groups: (1) reactors with catalyst suspended in a solution (often indicated as slurry PCMR) and (2) reactors with catalyst supported in/on the membrane (Fig. 13.2). In the case (1), the membrane only has separation functions and the light source can be above or inside the feed tank. The efficiency of the photocatalytic process is usually much higher than in case of the fixed-bed reactors owing to the higher exposure surface. Nevertheless, the catalyst particles have to be separated from the treated solution after the catalytic process. In case (2), the membrane also has catalytic function (catalytically active membrane) and the light source can be above or inside the membrane unit or alternatively, above or inside an additional vessel placed between the feed tank and the membrane unit (Mozia, 2010; Mozia, Tomaszewska, & Morawski, 2007). The main disadvantage of this configuration with respect to the first one is the reduction of photocatalytically active surface accessible for the reacting species.

PCMR can be further classified as (1) integrative-type PCMR and (2) split-type PCMR (Molinari, Lavorato, & Argurio, 2015; Sarasidis, Plakas, Patsios, & Karabelas, 2014; Zheng, Shen, Shi, Cheng, & Yuan, 2017). In the former, the photocatalytic reaction and membrane separation processes are carried out in a single apparatus (Fig. 13.2). In split-type PCMRs

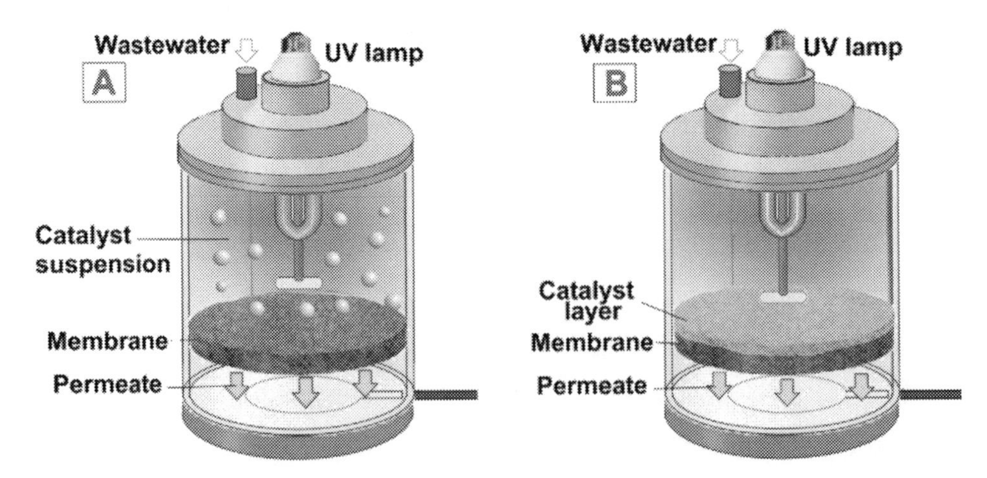

FIGURE 13.2 Example of integrative-type photocatalytic membrane reactors with: (A) photocatalytic membrane reactor with catalyst suspended, (B) photocatalytic membrane reactor with catalyst supported Nasrollahi et al. (2021). Source: *Reproduced with permission from Nasrollahi N., L. Ghalamchi, V. Vatanpour, A. Khataee, (2021) Photocatalytic-membrane technology: A critical review for membrane fouling mitigation, Journal of Industrial and Engineering Chemistry, 93, 101–116.*

the photocatalytic reaction and membrane separation take place into two separate apparatuses (Fig. 13.3).

PCMRs can operate in dead-end and cross flow modes. When the PCMR operates in dead-end mode and catalyst particles are dispersed in solution, they are retained by the membrane and collect on its surface. The formed cake layer reduces the membrane permeability and the photocatalytic performance. On the contrary, in the cross-flow mode, the feed flows tangentially to the membrane reducing membrane fouling and cake-layer formation.

Depending on the specific configuration of the PCMR and the type of reaction, the membrane can have different functions. Separation and recovery of photocatalyst are the principal roles when a photocatalyst is in suspension. Furthermore, during the degradation of organic pollutants in AOPs, the membrane should be able to retain the compounds and products/byproducts in the reactor until their complete mineralization. Finally, in the case in which photocatalysis is applied in a fine synthesis, the role of the membrane is often the separation of the product(s) from the reaction environment (Molinari et al., 2010).

When the membrane is catalytically active, because of being made of material having catalytic activity, or because a catalyst is immobilized in/on the membrane, the membrane can incorporate the dual functions of catalysis and separation (Mozia et al., 2007). Furthermore, a specific advantage of this type of catalytic membrane is that membrane fouling could be reduced and permeate flux increased, because organic compounds which are responsible for the formation of gel layer and filtration cake can be also decomposed (Fig. 13.4).

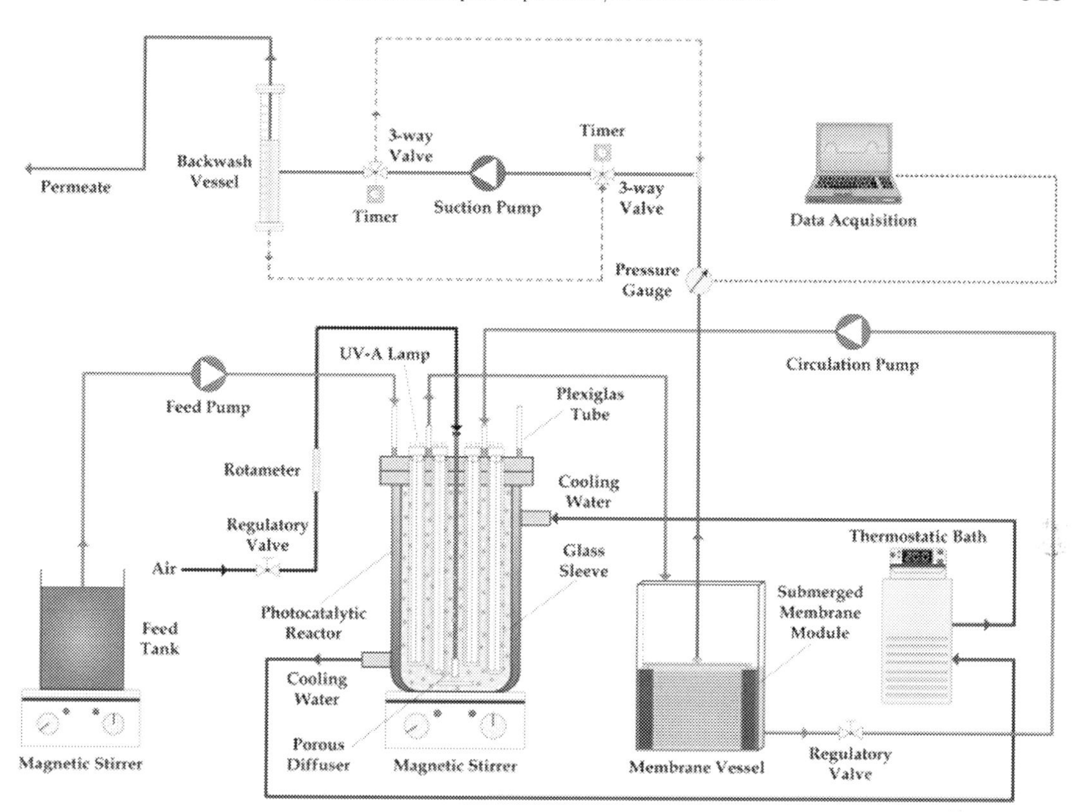

FIGURE 13.3 Schematic diagram of split-type photocatalytic membrane reactor Sarasidis et al. (2014). Source: *Reproduced with permission from Sarasidis V.C., K.V. Plakas, S.I. Patsios, A.J. Karabelas, (2014) Investigation of diclofenac degradation in a continuous photo-catalytic membrane reactor. Influence of operating parameters, Chemical Engineering Journal, 239, 299–311.*

13.2.3 Photocatalytic membranes

The choice of the membrane to be used in a PCMR is a key aspect in the design of these integrated systems combing photocatalysis with membrane separation. Both polymeric and inorganic membranes have been applied in PCMRs. In particular, the latter ones are used owing to their high chemical and thermal stability properties that make them attractive for this application. However, because of their lower price and an easier manufacturing, polymeric membranes still dominate the market of PCMR.

The polymeric membrane material must be stable to degradation induced by light irradiation and by the radical species formed in solution. Among the different polymers used in PCMR, partially fluorinated polymers, such as polyvinylidenefluoride (PVDF), polyvinylidenefluoride–trifluoroethylene PVDF-TrFE, polytetrafluoroethylene (PTFE), and PVDF-co-hexafluoropropylene (PVDF-HFP), represent a viable option in terms of chemical stability. On the contrary, aromatic polymers, such as polysulfone (PSF) and

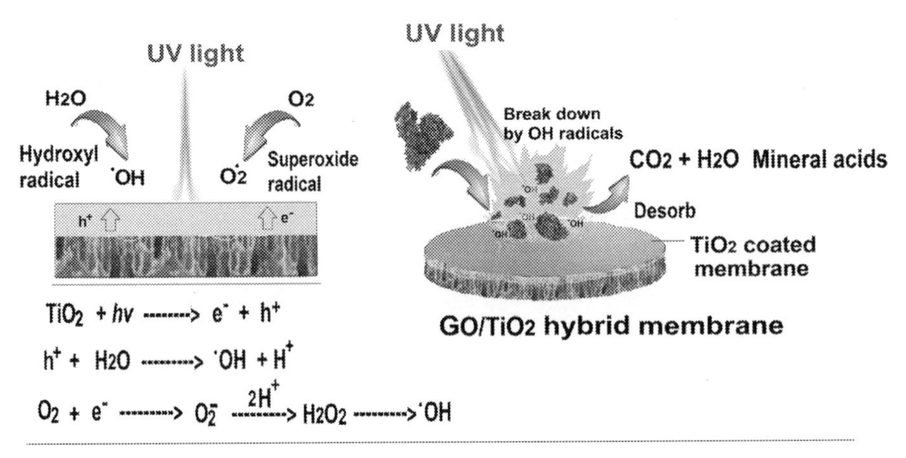

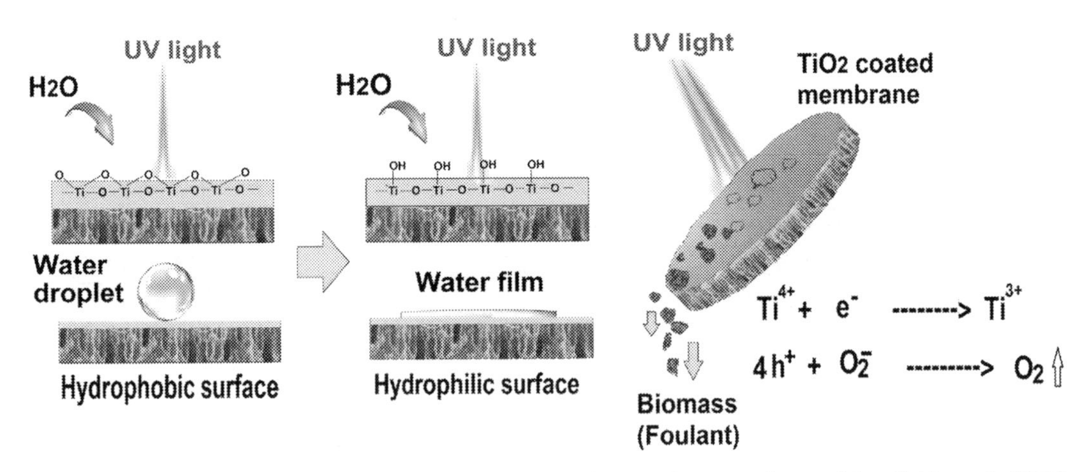

FIGURE 13.4 Antifouling mechanism of TiO_2-coated photocatalytic membranes Nasrollahi et al. (2021). Source: *Reproduced with permission from Nasrollahi N., L. Ghalamchi, V. Vatanpour, A. Khataee, (2021) Photocatalytic-membrane technology: A critical review for membrane fouling mitigation, Journal of Industrial and Engineering Chemistry, 93, 101–116.*

polyethersulfone (PES), often show a limited stability owing to the attack by radical species under UV irradiation (Bonchio, Carraro, Scorrano, Fontananova, & Drioli, 2003; Drioli & Fontananova, 2010; Romay, Diban, Rivero, Urtiaga, & Ortiz, 2020).

As previously mentioned, a photocatalyst can be suspended in the solution or immobilized in/on membrane. In the first case, the presence of photocatalyst in the feed solution determines high efficiency of the process because the active surface area is high and permits an effective contact between the photocatalyst and the organics species (Zheng et al., 2017). The major disadvantages in this configuration type are the mass transfer resistance increase caused by deposition of the photocatalyst nanoparticles on the membrane surface and the light scattering due to the particles suspended in solution (Molinari et al., 2015). The immobilization of a photocatalyst in/on a membrane circumvents these problems and allows a facile recovery and reuse of the photocatalyst. Furthermore, the microstructured and functionalized environment offered by the membrane can positively influence the catalyst performance.

Phototactically active membranes can be classified into three main groups as a function of the catalyst location: (1) photocatalyst coated/grafted on the membrane surface (Fontananova et al., 2006); (2) photocatalyst blended with a polymer matrix (mixed matrix membrane) (Bonchio et al., 2003) and (3) free-standing photocatalytic membranes in which the membrane is manufactured with a photocatalytic material (Zheng et al., 2017).

Concerning the first two types of photocatalytic membranes, their design requires to consider additional aspects with respect to the preparation of a simple separative membrane (Drioli & Fontananova, 2010):

1. Transparency of the polymeric matrix in the region of absorbance of the photocatalyst.
2. Catalyst stability during the immobilization procedure.
3. Influence of the catalyst particles on the membrane properties (mechanical, transport, etc.).
4. Effect of the catalyst loading, dimensions, and distribution on the process performance
5. Leaching of the catalyst in liquid phase.

However, specific advantages in terms of selectivity and efficiency can be achieved with these membranes that justify these additional efforts. An interesting example is the heterogenization of the decatungstate ($W_{10}O_{32}^{-4}$), a photocatalytic polyoxometalate (POM, polyanionic metal oxide cluster of early transition metals) applied in oxidation catalysis for fine chemistry and wastewater treatments (Drioli & Fontananova, 2010). A membrane-induced structure reactivity trend was observed and exploited in the batch heterogeneous photooxidation of water-soluble alcohols using polymeric catalytic membranes prepared embedding decatungstate within porous membranes made of PVDF and polydimethylsiloxane (Bonchio et al., 2003). The efficiency of the catalytic process was influenced and tailored by the selective sorption and diffusion of the substrate in membrane as a function of the membrane material and structure (Bonchio et al., 2003; Drioli & Fontananova, 2010). PVDF porous asymmetric membranes functionalized with decatungstate were also successfully applied in the aerobic photooxidation of phenol in water and used as model of organic pollutant (Drioli et al., 2008). The phenol degradation rate depended on the catalyst loading in membrane and the transmembrane pressure applied in the cross-flow UF operation. This last factor indeed influences the catalyst/substrate contact time. The catalyst heterogenized in membrane exhibited a higher catalytic activity in comparison with the homogeneous catalyst. This result can be ascribed to the selective absorption of the organic substrate from the aqueous phase on the hydrophobic PVDF membrane that increases the effective substrate concentration around the catalytic sites, allowing an intensive contact in the flow-through catalytic membrane reactor (Drioli et al., 2008).

Polymeric photocatalytic membranes operating in water with O_2 were prepared by anchoring Keggin-type phosphotungstic acid ($H_3PW_{12}O_{40}$) on the surface of plasma-activated polymeric membranes made of PVDF (Fontananova et al., 2006). These photocatalytic membranes were successfully applied in the aerobic degradation reaction of phenol in a flow-through PCMR (Fontananova et al., 2006).

PCMR technology also has promising applications in the field of CO_2 reuse for the production of valuable products (Brunetti & Fontananova, 2019). TiO_2 (anatase:rutile crystalline phase 60%:40%) was immobilized in Nafion membranes by sonication of the catalyst nanoparticles into a hydroalcoholic solution of the polymer, solution casting, and solvent evaporation method (Pomilla et al., 2018). The hydrophilic domains of the Nafion polymer favours membrane hydration that, combined with the elevated CO_2 permeance of the perfluorinated polymer matrix, guarantees the accessibility of reactants to the catalyst nanoparticles dispersed in the membrane. The photocatalytic TiO_2–Nafion membranes were applied in the CO_2 photoreduction with water to obtain methanol under UV–Vis irradiation in a continuous membrane photoreactor. A good methanol flow rate/TiO_2 weight of 45 mmol $(g_{catalyst}\ h)^{-1}$ was obtained operating at 2 bar of feed pressure and 45°C (Pomilla et al., 2018). The reduction of CO_2 was also carried out by a photocatalytic Nafion membrane containing exfoliated carbon nitride (C_3N_4) as metal-free photocatalyst (Brunetti et al., 2019; Pomilla et al., 2018). The effect of the H_2O/CO_2 feed molar ratio and contact time (defined as the ratio of the catalyst weight dispersed in the membrane to CO_2 feed flow rate, considered to be the limiting reagent) on the species production, reaction selectivity, and converted carbon, was investigated (Fig. 13.5) (Pomilla et al., 2018).

The highest value of total converted carbon/catalyst weight ratio was 47.6 μmol $(g_{catalyst}\ h)^{-1}$ with methanol and ethanol as main products. This value was obtained at H_2O/CO_2 feed molar ratio of 5–2 seconds as contact time. The performance of the PCMR was about 10 times higher in terms of carbon conversion than the batch system. This result was due to the fine dispersion of catalyst in the polymer matrix which, upon operating in continuous mode, allowed the removal in continuous of the products from the reaction sites, thereby increasing the conversion and reducing undesired secondary reactions (Pomilla et al., 2018). Instead, higher contact times caused a partial oxidation of the alcohols, favouring HCHO production.

13.3 Mass transport mechanisms in the main pressure-driven membrane operations involved in photocatalytic membrane reactor

The two main parameters describing the membrane's separation performance are the flux and the selectivity, which are, respectively, the rate at which a given component is transported through the membrane and the ability to separate a given component from others.

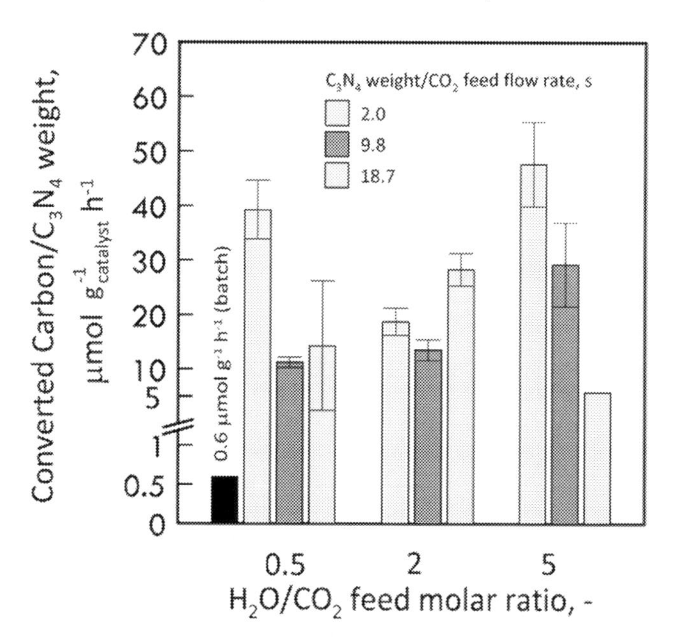

FIGURE 13.5 Total converted carbon as a function of H_2O/CO_2 feed molar ratio and contact time Pomilla et al. (2018). Source: *Reprinted with permission from Pomilla F.R., A. Brunetti, G. Marcì, E.I. Garcıéa-López, E. Fontananova, L. Palmisano, G. Barbieri, (2018) CO2 to liquid fuels: Photocatalytic conversion in a continuous membrane reactor, ACS Sustainable Chemistry & Engineering, 6(7), 8743–8753.*

The transport of different species through a membrane is a nonequilibrium process, and the separation of the different species is due to differences in their transport rate.

The driving forces in membrane processes are several and include gradient in temperature, concentration, pressure, and electrical potential (Mulder, 1991; Drioli, Curcio, Fontananova). The electrochemical potential μ (which including the chemical potential) is a thermodynamic function that includes all these parameters. For a single component i transported, the flux J_i can be described by a semiempirical equation:

$$J_i = -L \cdot \frac{d\mu_i}{dx} \tag{13.13}$$

where $\frac{d\mu_i}{dx}$ is the electrochemical potential gradient and L is a phenomenological coefficient.

Driving force and flux are interconnected in multicomponent system (Kedem & Katchalsky, 1961) and near to the equilibrium, a linear equation similar to the previous one can be used:

$$J_i = -L_{ij} \frac{\delta\mu_i}{\delta_\chi} \tag{13.14}$$

Where L_{ij} is a phenomenological coefficient.

A number of PCMRs combine a photocatalytic process with a pressure-driven membrane operation such as NF, UF, and MF. The size of the particles or molecules separated increase from NF to MF and consequently, the pore sizes and pressure applied diminish (Table 13.1).

NF membranes are used to retain in the reactor low–molecular weight molecules, substrates, or intermediates of the reaction, as well as to efficiently retain the catalyst in the reactor (Molinari, Pirillo, Falco, Loddo, & Palmisano, 2004).

TABLE 13.1 Typical pore size and driving force applied in pressure driven membrane processes.

Membrane processes	Pore size (nm)	Pressure applied (bar)
Nanofiltration	<2	10–25
Ultrafiltration	1–100	1–10
Microfiltration	50–10,000	<2

The selective transport in NF membranes is the result of several mechanisms: solution–diffusion process, sieving effects, and electrostatic interactions between electrically charged species and the surface charge of the membrane.

The permeate flux in NF can be expressed using the Nernst–Plank equation (Mulder, 1991; Drioli, Curcio, Fontananova):

$$J_i = -c_i D_{i,m} \frac{d(ln\gamma_i)}{dx} - \frac{c_i D_{i,m}}{RT} V_i \frac{dP}{dx} - D_{i,m} \frac{dc_i}{dx} - \frac{z_i c_i D_{i,m}}{RT} F \frac{d\Psi}{dx} + K_{i,c} c_i V \tag{13.15}$$

where J_i is the flux, c_i is the concentration, γ_i is the activity coefficient, $D_{i,m}$ is the diffusion coefficient, z_i is the valence, V_i is the partial volume, and $K_{i,c}$ is the hindrance factor for *i-th* component through the membrane, x is the axial coordinate, F is the Faraday constant, P is the pressure, R is the universal gas constant, T is the temperature, ψ is the chemical potential, and V is the solvent velocity.

The parameter that expresses the separation ability of the membrane is the rejection defined as:

$$\text{Rejection} = \frac{C_f - C_p}{C_f} = 1 - \frac{C_p}{C_f} \tag{13.16}$$

where C_p is the concentrations of the *i-th* component in the permeate and C_f is the concentration in the feed, respectively (Mulder, 1991).

The recovery ratio (RR) is defined as:

$$\text{RR} = \frac{Q_p}{Q_f} \tag{13.17}$$

where Q_p indicates the permeate flow rate and Q_f is the feed flow rate. To have a high efficiency of the membrane process the recovery ratio has to be as high as possible. However, the osmotic pressure limit reduces the maximum value of the concentration in the retentate and prevents the possibility to achieve zero liquid discharge (i.e., complete recovery of the solvent) in pressure-driven membrane operation.

Another membrane operation frequently used in PCMRs is the UF process applied for a selective removal of the reaction products from feed solution and/or the photocatalyst particles retention in the reactor. The transport mechanism in UF mainly exploits differences in size and shape of components with respect to the membrane pores. UF membranes are typically porous with an asymmetric structure (Fontananova et al., 2006; Drioli et al., 2008).

Assuming that the pores are cylindrical, the convective flow through the membrane can be expressed by the Hagen–Poiseuille equation (Mulder, 1991):

$$J_w = \frac{\varepsilon r^2 \Delta P}{8 \eta \delta} \tag{13.18}$$

where ε is the porosity, δ is the membrane thickness, ΔP is the difference of pressure, r is the pore radius, η is the viscosity.

On the contrary, assuming that the pores are interstices between close-packed spheres similar to sintered structures, the flow can be expressed using the Carman–Kozeny equation (Mulder, 1991):

$$J_w = \frac{\varepsilon^3}{K_p \eta S^2 (1-\varepsilon)^2} \frac{\Delta P}{\delta} \tag{13.19}$$

where K_p is a constant depending on the pore shape and tortuosity, and S is the specific surface area.

PCMR can also be combined with MF membranes. MF is a low-pressure membrane operation that can separate colloidal and suspended micrometer-size particles in the solvated size range of 0.02–10 μm. In this case, the retention of the substrates and/or the photocatalyst particles occurs by sieving mechanism.

The volumetric flux through MF membranes is proportional to the pressure and can be described both by Hagen–Poiseuille and Carman–Kozeny (Eqs. 13.18 and 13.19).

One of the main limitations of liquid-phase pressure-driven membrane operations, including several PCMRs, is the fouling. Membrane fouling decreases permeate flux and productivity, increases energy consumption, and reduces membrane lifetime (Pidou et al., 2009). Feed composition, reactors design, temperature, operation conditions, light intensity, membrane type, and pH of the solution can potentially affect the fouling.

In a pressure-driven membrane operation, retained substances can accumulate in proximity of the membrane surface; thus a concentration gradient between the membrane interface and the bulk of the solution leads to a diffusive flow back of the solutes. Steady-state concentration profile is established when the convective transport of solutes to the membrane surface is counterbalanced by a diffusive flux of the retained substances back into the bulk solution. Using the film model, if the solute is completely rejected by the membrane, the relationship between the concentration profile and the flux can be expressed as follows (Mulder, 1991; Drioli, Curcio, Fontananova):

$$\frac{c_m}{c_b} = \exp\left(\frac{J}{k}\right) \tag{13.20}$$

Where c_m is the concentration at the membrane interface, c_b is the and the concentration in the bulk of the solution, J is the volumetric transmembrane flux, and k is the mass-transfer coefficient.

In the PCMRs configuration in which the photocatalyst is suspended, the photocatalyst particles can also contribute to the formation of a cake layer precipitating on the membrane surface. This phenomenon determines the formation of a dense cake layer that further increases the resistance of the mass transfer (R_g) in addition to that of the membrane (R_m)

In accordance with the solution concentration in the bulk and the mass transfer coefficient, the flux can increase with pressure up to a limiting value (J_∞). This value is given by the following equation (Maldotti et al., 2002):

$$J_\infty = \frac{\Delta P}{\mu(R_m + R_g)} \tag{13.21}$$

On the other hand, several authors have reported the possibility to mitigate membrane fouling by the self-cleaning effect of membranes functionalized with photocatalysts (Nasrollahi et al., 2021).

13.4 Conclusion and future trends

The synergistic combination of a catalytic conversion with a membrane-based separation in a PCMR offers several advantages in comparison to traditional photoreactors: an easier recovery and reuse of the catalyst, simpler downstream and upstream processing, reduced formation of byproducts, compact process equipment, the possibility of tailoring the contact between reactants and catalyst working on the applied driving force and membrane structure with consequent benefits in terms of conversion and selectivity.

The main limitations of this innovative technology are related to the durability of the membrane and membrane fouling, which reduce the process efficiency. These key issues can be addressed by the development of selective and durable (photocatalytic) membranes produced at acceptable costs, with the ability to operate with high efficiency (elevated fluxes and selectivity) over long times, combined with a smart membrane reactor design based on an application-based multidisciplinary approach.

The development of new photocatalysts with tailored properties, including elevated affinity for the membrane material in the case of heterogenous catalysts immobilized in/on the membranes, represents a further step for the large-scale application of PCMRs. Moreover, it is essential to improve the environmental sustainability of these integrated processes favoring catalysts that can use solar light and avoid the use of toxic solvents, not only as reaction solvents but also in the membrane preparation protocols.

References

Analitica, C., Torino, U., Borgarello, E., Tinucci, L., Ricerche, E. N. I., & Milanese, S. D. (1993). Photocatalytic activity and selectivity of titania colloids and particles prepared by the sol-gel technique: Photooxidation of phenol and atrazine. *Langmuir: The ACS Journal of Surfaces and Colloids, 9*, 2995–3001.

Bonchio, M., Carraro, M., Scorrano, G., Fontananova, E., & Drioli, E. (2003). Heterogeneous photooxidation of alcohols in water by photocatalytic membranes incorporating decatungstate. *Advanced Synthesis & Catalysis, 345*, 1119–1126.

Brunetti, A., & Fontananova, E. (2019). CO_2 conversion by membrane reactors. *Journal of Nanoscience and Nanotechnology, 19*, 3124–3134.

Brunetti, A., Pomilla, F. R., Marcì, G., Garcia-Lopez, E. I., Fontananova, E., Palmisano, L., & Barbieri, G. (2019). CO_2 reduction by C_3N_4-TiO_2 Nafion photocatalytic membrane reactor as a promising environmental pathway to solar fuels. *Applied Catalysis B: Environmental, 255*, 1–8.

Corredor, J., Rivero, M. J., Rangel, C. M., & Ortiz, I. (2019). Comprehensive review and future perspectives on the photocatalytic hydrogen production. *Journal of Chemical Technology & Biotechnology, 94*, 3049–3063.

Doll, T. E., & Frimmel, F. H. (2005). Cross-flow microfiltration with periodical back-washing for photocatalytic degradation of pharmaceutical and diagnostic residues – evaluation of the long-term stability of the photocatalytic activity of TiO_2. *Water Research, 39*, 847–854.

Drioli, E., Curcio, E., & Fontananova, E. (2010). Mass transfer operation – membrane separations. In John Bridgwater, Martin Molzahn, & Ryszard Pohorecki (Eds.), *Chemical Engineering*. EOLSS Publications.

Drioli, E., & Fontananova, E. (2010). Catalytic membranes embedding selective catalysts: Preparation and applications. In P. Barbaro, & F. Liguori (Eds.), *Heterogenized Homogeneous Catalysts for Fine Chemicals Production. Catalysis by Metal Complexes* (vol. 33, pp. 203–229). Dordrecht: Springer. Available from https://doi.org/10.1007/978-90-481-3696-4_6.

Drioli, E., Fontananova, E., Bonchio, M., Carraro, M., Gardan, M., & Scorrano, G. (2008). Catalytic membranes and membrane reactors: An integrated approach to catalytic process with a high efficiency and a low environmental impact. *Chinese Journal of Catalysis, 29*(11), 1152–1158.

Fontananova, E., Donato, L., Drioli, E., Lopez, L. C., Favia, P., & D'Agostino, R. (2006). Heterogenization of polyoxometalates on the surface of plasma-modified polymeric membranes. *Chemistry of Materials: A Publication of the American Chemical Society, 18*(6), 1561–1568.

Guo, Y., Qi, P. S., & Liu, Y. Z. (2017). A review on advanced treatment of pharmaceutical wastewater. *IOP Conference Series: Earth and Environmental Science, 63*, 012025.

Hairom, N. H. H., Mohammad, A. W., & Kadhum, A. A. H. (2014). Effect of various zinc oxide nanoparticles in membrane photocatalytic reactor for Congo red dye treatment. *Separation and Purification Technology, 137*, 74–81.

He, J., Kumar, A., Khan, M., & Lo, I. M. C. (2021). Critical review of photocatalytic disinfection of bacteria: From noble metals- and carbon nanomaterials-TiO_2 composites to challenges of water characteristics and strategic solutions. *Science of the Total Environment, 758*, 143–953.

Herrmann, J. M. (2005). Heterogeneous photocatalysis: State of the art and present applications In honor of Pr. R. L. Burwell Jr. (1912–2003), Former Head of Ipatieff Laboratories, Northwestern University, Evanston (Ill). *Topics in Catalysis, 34*, 49–65. Available from https://doi.org/10.1007/s11244-005-3788-2.

Intarasuwan, K., Amornpitoksuk, P., Suwanboon, S., & Graidist, P. (2017). Photocatalytic dye degradation by ZnO nanoparticles prepared from $X_2C_2O_4$ (X = H, Na and NH_4) and the cytotoxicity of the treated dye solutions. *Separation and Purification Technology, 177*, 304–312.

Kedem, O., & Katchalsky, A. (1961). A physical interpretation of the phenomenological coefficients of membrane permeability. *The Journal of General Physiology, 45*(1), 143–179.

König, B. (2017). Photocatalysis in organic synthesis – past, present, and future. *European Journal of Organic Chemistry, 2017*(15), 1979–1981. Available from https://doi.org/10.1002/ejoc.201700420.

Konstantinou, I. K., & Albanis, T. A. (2004). TiO_2-assisted photocatalytic degradation of azo dyes in aqueous solution: Kinetic and mechanistic investigations: A review. *Applied Catalysis B: Environmental, 49*(1), 1–14.

Leary, R., & Westwood, A. (2011). Carbonaceous nanomaterials for the enhancement of TiO_2 photocatalysis. *Carbon, 49*(3), 741–772.

Lee, S., & Park, S. (2013). TiO_2 photocatalyst for water treatment applications. *Journal of Industrial and Engineering Chemistry, 19*(6), 1761–1769.

Maldotti, A., Molinari, A., & Amadelli, R. (2002). Photocatalysis with organized systems for the oxofunctionalization of hydrocarbons by O_2. *Chemical Reviews, 102*(10), 3811–3836.

Mascolo, G., Comparelli, R., Curri, M. L., Lovecchio, G., Lopez, A., & Agostiano, A. (2007). Photocatalytic degradation of methyl red by TiO_2: Comparison of the efficiency of immobilized nanoparticles vs conventional suspended catalyst. *Journal of Hazardous Materials, 142*(1-2), 130–137.

Molinari, R., Argurio, P., Bellardita, M., & Palmisano, L. (2017). *Photocatalytic processes in membrane reactors. Comprehensive Membrane Science and Engineering* (2nd ed., pp. 165–193). Elsevier. Available from https://doi.org/10.1016/B978-0-12-409547-2.12220-6.

Molinari, R., Caruso, A., Argurio, P., & Poerio, T. (2008). Degradation of the drugs Gemfibrozil and Tamoxifen in pressurized and de-pressurized membrane photoreactors using suspended polycrystalline TiO_2 as catalyst. *Journal of Membrane Science, 319*(1–2), 54–63.

Molinari, R., Caruso, A., & Poerio, T. (2009). Direct benzene conversion to phenol in a hybrid photocatalytic membrane reactor. *Catalysis Today, 144*(1–2), 81–86.

Molinari, R., Lavorato, C., Argurio, P., Szymański, K., Darowna, D., & Mozia, S. (2019). Overview of photocatalytic membrane reactors in organic synthesis, energy storage and environmental applications. *Catalysts*, *9*(3), 239. Available from https://doi.org/10.3390/catal9030239.

Molinari, R., Lavorato, C., & Argurio, P. (2015). Photocatalytic reduction of acetophenone in membrane reactors under UV and visible light using TiO_2 and P/TiO_2 catalysts. *Chemical Engineering Journal*, *274*, 307–316. Available from https://doi.org/10.1016/j.cej.2015.03.120.

Molinari, R., Pirillo, F., Falco, M., Loddo, V., & Palmisano, L. (2004). Photocatalytic degradation of dyes by using a membrane reactor. *Chemical Engineering and Processing: Process Intensification*, *43*(9), 1103–1114.

Molinari, R., Borgese, M., Drioli, E., Palmisano, L., & Schiavello, M. (2002). Hybrid processes coupling photocatalysis and membranes for degradation of organic pollutants in water. *Catalysis Today*, *75*, 77–85.

Molinari, R., Argurio, P., & Palmisano, L. (2015). 7 - Photocatalytic membrane reactors for water treatment. In A. Basile, A. Cassano, & N. K. Rastogi (Eds.), *Advances in Membrane Technologies for Water Treatment* (pp. 205–238). Oxford: Woodhead Publishing. Available from https://doi.org/10.1016/B978-1-78242-121-4.00007-1.

Mozia, S., Morawski, A. W., Molinari, R., Palmisano, L., & Loddo, V. (2013). 6 - Photocatalytic membrane reactors: Fundamentals, membrane materials and operational issues. In Angelo Basile (Ed.), *Handbook of Membrane Reactors, Woodhead Publishing Series in Energy* (vol. 2, pp. 236–295). Oxford: Woodhead Publishing. Available from https://doi.org/10.1533/9780857097347.1.236.

Mozia, S., Tomaszewska, M., & Morawski, A. W. (2007). Photocatalytic membrane reactor (PMR) coupling photocatalysis and membrane distillation—Effectiveness of removal of three azo dyes from water. *Catalysis Today*, *129*(1–2), 3–8.

Mozia, S., Argurio, P., & Molinari, R. (2018). Chapter - 4 PMRs utilizing pressure-driven membrane techniques. In A. Basile, S. Mozia, & R. Molinari (Eds.), *Current Trends and Future Developments on (Bio-) Membranes* (pp. 97–127). Elsevier. Available from https://doi.org/10.1016/B978-0-12-813549-5.00004-9.

Mozia, S. (2010). Photocatalytic membrane reactors (PMRs) in water and wastewater treatment. A review. *Separation and Purification Technology*, *73*(2), 71–91.

Mulder, J. (1991). *Basic Principles of Membrane Technology*. Dordrecht: Springer. Available from https://doi.org/10.1007/978-94-017-0835-7.

Nasrollahi, N., Ghalamchi, L., Vatanpour, V., & Khataee, A. (2021). Photocatalytic-membrane technology: A critical review for membrane fouling mitigation. *Journal of Industrial and Engineering Chemistry*, *93*, 101–116.

Palmisano, G., Yurdakal, S., Augugliaro, V., Loddo, V., & Palmisano, L. (2007). Photocatalytic selective oxidation of 4-methoxybenzyl alcohol to aldehyde in aqueous suspension of home-prepared titanium dioxide catalyst. *Advanced Synthesis & Catalysis*, *349*(6), 964–970. Available from https://doi.org/10.1002/adsc.200600435.

Pidou, M., Parsons, S. A., Raymond, G., Jeffrey, P., Stephenson, T., & Jefferson, B. (2009). Fouling control of a membrane coupled photocatalytic process treating greywater. *Water Research*, *43*(16), 3932–3939.

Pomilla, F. R., Brunetti, A., Marcì, G., Garcıéa-López, E. I., Fontananova, E., Palmisano, L., & Barbieri, G. (2018). CO_2 to liquid fuels: Photocatalytic conversion in a continuous membrane reactor. *ACS Sustainable Chemistry & Engineering*, *6*(7), 8743–8753.

Romay, M., Diban, N., Rivero, M. J., Urtiaga, A., & Ortiz, I. (2020). Critical issues and guidelines to improve the performance of photocatalytic polymeric membranes. *Catalyst*, *10*(5), 570–605.

Sarasidis, V. C., Plakas, K. V., Patsios, S. I., & Karabelas, A. J. (2014). Investigation of diclofenac degradation in a continuous photo-catalytic membrane reactor. Influence of operating parameters. *Chemical Engineering Journal*, *239*, 299–311.

Tang, C., & Chen, V. (2004). The photocatalytic degradation of reactive black 5 using TiO_2/UV in an annular photoreactor. *Water Research*, *38*(11), 2775–2781.

Zheng, X., Shen, Z. P., Shi, L., Cheng, R., & Yuan, D. H. (2017). Photocatalytic membrane reactors (PMRs) in water treatment: Configurations and influencing factors. *Catalysts*, *7*, 224. Available from https://doi.org/10.3390/catal7080224.

CHAPTER

14

Transport phenomena in polymeric membrane reactors

Brent A. Bishop, Oishi Sanyal and Fernando V. Lima

Department of Chemical and Biomedical Engineering, West Virginia University, Morgantown, WV, United States

Abbreviations

CMS	Carbon molecular sieve
GPU	Gas permeation unit
PBI	Polybenzimidazole
WGS	Water–gas shift
WGS-MR	Water–gas shift membrane reactor

Nomenclature

C_{CO_2}	Carbon capture fraction
D_h	Hydraulic diameter
d_i	Inner tube diameter
d_o	Outer tube diameter
D_p	Catalyst particle diameter
f	Darcy friction factor
$F_{i,s}$	Component molar flow rate of component *i* in the shell
$F_{i,t}$	Component molar flow rate of component *i* in the tube
$\overline{H}_{perm}$	Total enthalpy lost from the tube (gained by the shell) due to permeation
H_s	Shell-side enthalpy
H_t	Tube-side enthalpy
J_i	Molar flux of component i
$P_{i,o}$	Permeability of component i
P_{mem}	Pressure in the membrane material
P_s	Shell-side pressure
P_t	Tube-side pressure
Q_i	Permeance of component i
R_{H_2}	Hydrogen recovery fraction
$R_{i,j}$	Molar reaction rate of component *i* in reaction *j*

Current Trends and Future Developments on (Bio-) Membranes
DOI: https://doi.org/10.1016/B978-0-12-822257-7.00004-2

T_s	Shell-side temperature
T_t	Tube-side temperature
U	Overall heat transfer coefficient
v_s	Superficial velocity
z	Axial dimension
$\alpha_{H_2/i}$	Hydrogen selectivity through membrane relative to component i
δ_{mem}	Membrane thickness
ϵ	Void fraction of catalyst packing
μ	Fluid viscosity
ρ	Fluid density
χ_{CO}	CO conversion

14.1 Introduction

In recent years, the concept of *process intensification* by means of combining two or more unit operations has gained considerable attention owing to the prospects of enhanced energy efficiency, modularity, and reduced CO_2 footprint (Basile, 2013; Bishop & Lima, 2020, 2021; Lin, 2018; Radcliffe, Singh, Berchtold, & Lima, 2016). By combining reaction and separation in a single unit operation and enhancing the conversions of equilibrium-limited reactions, membrane reactors effectively serve as intensified units. These units have been primarily developed for H_2 generation from hydrocarbons (Itoh & Haraya, 2000). Developing suitable membrane materials and processes is therefore important for this application, especially since they must be operated at high temperatures. This book chapter focuses on developing a general computational model for a tubular membrane reactor, using a case study of polybenzimidazole (PBI) membranes applied to water–gas shift (WGS) reactions. This model is based on nonisothermal countercurrent membrane reactors and provides a solid platform for membrane development with optimum transport properties. The central theme of membrane design for WGS reactions usually involves achieving high H_2/CO_2 separation performance (Li, Singh, Dudeck, Berchtold, & Benicewicz, 2014) under "dry" and humid conditions (Lei et al., 2021; Omidvar et al., 2019).

While polymer membranes typically undergo decomposition at high temperatures, pure PBI and PBI-derived membranes exhibit attractive separation performance up to ~350°C (Singh, Ganpat, Dudeck, & Berchtold, 2020). The rigid molecular structure of PBI offers significant size-sieving capabilities to these membranes, helping to discriminate between these similarly sized molecules in terms of kinetic diameters ($H_2 \rightarrow 2.89$ Å and $CO_2 \rightarrow 3.30$ Å). Experimental data reported by Berchtold, Singh, Young, & Dudeck (2012) have been used as the baseline values for the development and validation of the proposed model. The model aims to guide the design of membranes with the goal of maximizing H_2 generation along with CO_2 capture by optimizing membrane transport properties (permeability and selectivity) and flow configurations.

The model was built using the AVEVA Process Simulation platform. A control volume was selected and a submodel was developed in the equation-oriented environment that utilizes the mass and energy transport equations discussed below. This platform allows for a very complex model to be simulated that includes countercurrent, nonisothermal, and bidirectional permeation to be simultaneously solved.

The remainder of this chapter is organized as follows: First, a derivation of the mass and energy transport of polymeric membranes is given. Simulations are then performed to identify the current benefits and shortcomings of pure-PBI membranes. Lastly, this chapter provides conclusions and outline some directions for future work.

14.2 Transport phenomena for the general membrane reactor case

Before attempting to model the transport in polymeric membrane reactors, it is the best to begin with the equations for any general membrane system. Initially, the general membrane system studied consists of a double pipe shell and tube configuration with the "pipe" in this case being surrounded by a permeable material. This arrangement is shown in Fig. 14.1.

14.2.1 Modeling flow in the membrane reactor

The process side, which is a packed bed, is modeled using the Ergun equation, that is, Eq. (14.1), while the sweep gas side uses the Colbrooke equation, that is, Eq. (14.2) for modeling the pressure drop.

Ergun equation for tube:

$$\frac{dp_t}{dz} = \frac{150\mu}{D_p{}^2}\frac{(1-\epsilon)^2}{\epsilon^3}v_s + \frac{1.75\rho}{D_p}\frac{(1-\epsilon)}{\epsilon^3}v_s|v_s| \tag{14.1}$$

Where μ is the fluid's viscosity, D_p is the catalyst particle diameter, ε is the void fraction of the catalyst, v_s is the superficial velocity of the fluid, and ρ is the fluid density.

Colebrook equation for shell:

$$\frac{dp_s}{dz} = \frac{64f\mu v_s}{2D_h{}^2} \tag{14.2}$$

Where D_h is the hydraulic diameter of the shell side and f is the Darcy friction factor and can be solved with the implicit formula:

$$\frac{1}{\sqrt{f}} = -2\log\left(\frac{\epsilon}{3.7D_h} + \frac{2.51}{Re\sqrt{f}}\right) \tag{14.3}$$

These equations can be solved at any point along the shell or the tube sides of the membrane reactor to get the pressure drop and subsequently, the volumetric flow rates.

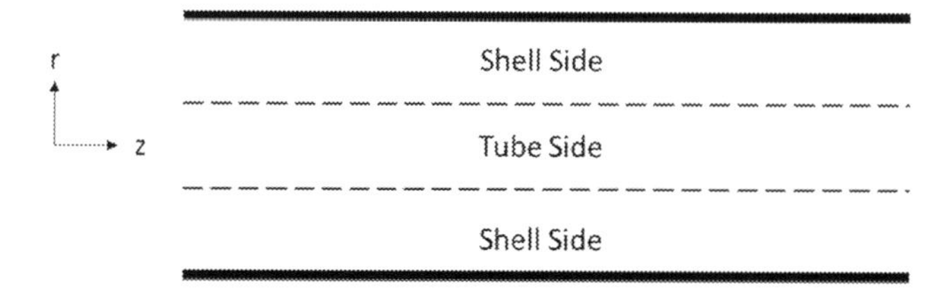

FIGURE 14.1 Shell and tube configuration for the membrane reactor.

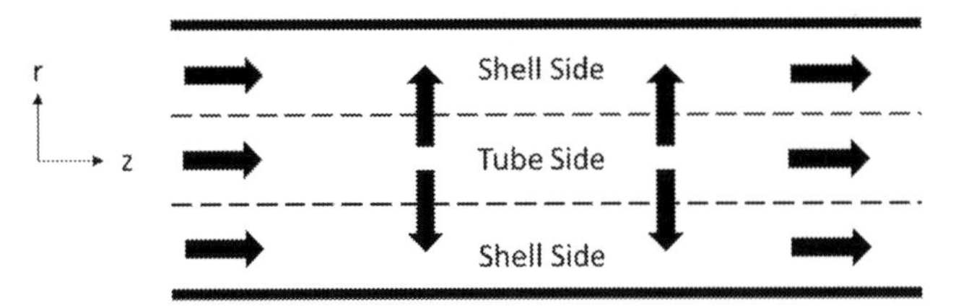

FIGURE 14.2 Arrows depicting the assumed directions of flow and permeation for the mole and energy balance derivations.

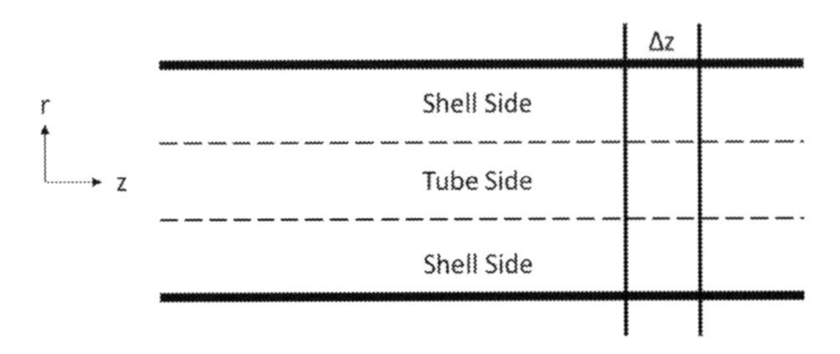

FIGURE 14.3 Depiction of the control volume for the membrane reactor.

Assuming that the flow rates of the tube and shell sides are known, derivations of more complex transport phenomena are possible. Membrane reactors are especially challenging to model as nonisothermal and countercurrent systems because of the process nonlinearities that arise under these conditions. To derive the mole and energy balances for the membrane reactor, the following set of conventions are adopted:

1. Flow on the tube and shell sides is assumed to be in the positive axial direction.
2. The reaction rate of a component, R_i, is assumed to be the rate of consumption of that component due to reaction.
3. The direction of permeation is assumed to be in the positive radial direction (i.e., from the tube into the shell).

A graphical representation of these conventions is shown in Fig. 14.2.

To simplify the derivations, a thin slice of the membrane reactor is chosen as the control volume, as depicted in Fig. 14.3.

14.2.2 The differential component mole balance

The control volume contains three regions: the tube side, the membrane, and the shell side, and each of these regions must satisfy a mole balance. Starting with the tube side, there are four contributions to changes in the accumulation of a component i inside the tube: (1) the flow of a component at z; (2) the flow of a component at $z + \Delta z$; (3) the consumption of that component within the volume of the tube due to reactions; and (4) the permeation of that component through the surface of the

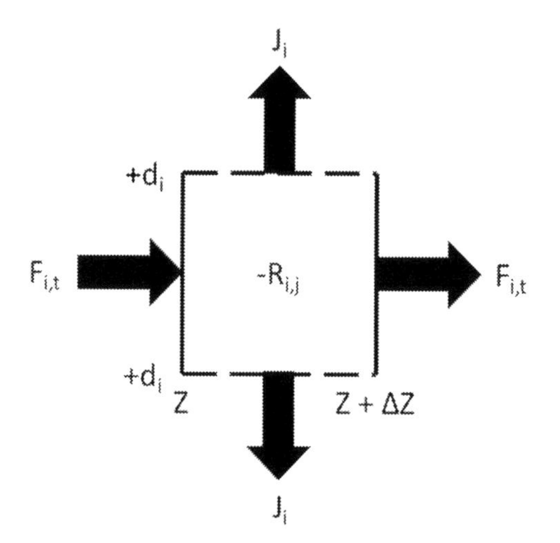

FIGURE 14.4 Tube side component mole balance of the selected control volume.

membrane. A graphical representation for the mole balance of the tube side is provided in Fig. 14.4.

Using Fig. 14.4, the steady-state mole balance of component i on the tube side is found to be:

$$F_{i,t}\big|_z - F_{i,t}\big|_{z+\Delta z} - \sum_{j=1}^{N} \frac{\pi}{4} d_i^2 \Delta z * R_{i,j} - \pi d_i \Delta z * J_i = 0 \tag{14.4}$$

where d_i is the inner diameter of the tube, $R_{i,j}$ is the reaction rate of component i in reaction j, and J_i is the mole flux of component i into the membrane. Dividing by Δz and taking the limit as Δz goes to zero, the following differential mole balance is derived:

$$\frac{dF_{i,t}}{dz} = -\frac{\pi}{4} d_t^2 \sum_{j=1}^{N} R_{i,j} - \pi d_i * J_i \tag{14.5}$$

Similarly, the shell side of the membrane reactor works the same way as the tube side but differs in two ways. The shell side is used here for capturing a component permeating out of the tube side so the sign of the permeation term is flipped and there are no reactions. This results in the following steady-state mole balance for the shell side of the membrane reactor:

$$F_{i,s}\big|_z - F_{i,s}\big|_{z+\Delta z} + \pi(d_i + 2\delta_{\text{mem}})\Delta z * J_i = 0, \tag{14.6}$$

where δ_{mem} is the membrane thickness. Once again, dividing by Δz and taking the limit as Δz goes to zero, the following differential mole balance for the shell side is derived:

$$(\pm)\frac{dF_{i,s}}{dz} = \pi(d_i + 2\delta_{\text{mem}}) * J_i \tag{14.7}$$

In this case, the differential term could either be positive or negative depending on cocurrent or countercurrent operation, respectively.

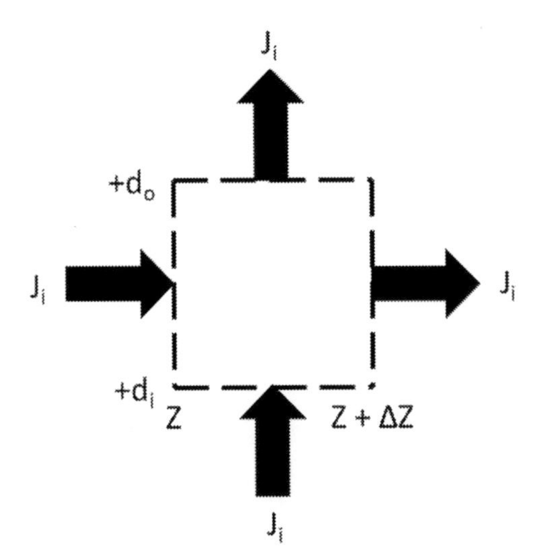

FIGURE 14.5 Graphical depiction of the steady-state component mole balance for the membrane.

Finally, the mole balance for the membrane itself should be considered. Material can enter the membrane in the control volume from four directions: from the tube, the shell, the membrane at z, and the membrane at $z + \Delta z$ as depicted in Fig. 14.5.

Therefore the steady-state mole balance for the membrane is:

$$\left(J_i\big|_z - J_i\big|_{z+\Delta z}\right) * \frac{\pi}{4}\left(d_o^{\,2} - d_i^{\,2}\right) + \left(J_i\big|_{d_i} * d_i - J_i\big|_{d_o} * d_o\right)\pi\Delta z = 0 \tag{14.8}$$

In this equation, the terms have been grouped on the basis of direction of permeation either axially along the membrane or radially through the membrane. However, a scaling argument could be made that the permeation in the axial direction is negligibly small relative to the radial permeation. Most permeation models for membranes assume a form described in the following equation:

$$J_i = \frac{Q_i}{\delta}\left(P_2^{\,n} - P_1^{\,n}\right) \tag{14.9}$$

For $n = 1$ (like for polymer membranes that work through Fickian diffusion), the molar flux of a material through the membrane is directly proportional to the pressure drop across the distance of permeation. To ignore the axial contribution of flux in the membrane, the following statement would need to hold:

$$\frac{P_t - P_s}{\delta_{\text{mem}}} \gg \frac{dP_{\text{mem}}}{dz} \tag{14.10}$$

Considering the PBI polymer membrane addressed in this work as an example, the above equation simplifies to the following:

$$\frac{42[\text{atm}] - 21[\text{atm}]}{1 * 10^{-7}[m]} = 2.1 * 10^{8}\,\frac{\text{atm}}{m} \gg \frac{dP_{\text{mem}}}{dz}$$

In this example, the pressure drop across the membrane is several orders of magnitude larger than the pressure drop along the length of the membrane, and therefore the flux of material in the axial direction can be ignored. Although this was specifically set for the PBI membrane in focus, generally membranes are sufficiently thin such that this condition is always satisfied. This means that Eq. (14.5) simply reduces down to show that the flux from the tube into the membrane is equal to the flux from the membrane into the shell, and therefore it is equal to the flux from the tube to the shell. Assuming the inner diameter and outer diameter to be approximately equal owing to the thin membrane, the final differential component mole balances for the general membrane reactor system are:

Tube side differential component mole balance (at steady state)

$$\frac{dF_{i,t}}{dz} = -\frac{\pi}{4}{d_t}^2\sum_{j=1}^{N} R_{i,j} - \pi d_t * J_i \tag{14.5}$$

Shell-side differential component mole balance (at steady state)

$$(\pm)\frac{dF_{i,s}}{dz} = \pi d_t * J_i \tag{14.11}$$

14.2.3 The differential energy balance

There are many important contributions that make up the energy balance for the membrane reactor. Starting with the tube side again, there are four contributions: (1) the enthalpy flowing into the tube side of the control volume at z; (2) the enthalpy flowing out of the tube side of the control volume at $z + \Delta z$; (3) the enthalpy flowing in or out of the tube side of the control volume through the membrane; and (4) the heat transferred because of a temperature difference between the tube and shell side.

The third contribution (denoted as H_{perm} in Fig. 14.6) accounts for the Joule–Thomson effect, where a gas can change temperature because of an isenthalpic change in pressure. It can be

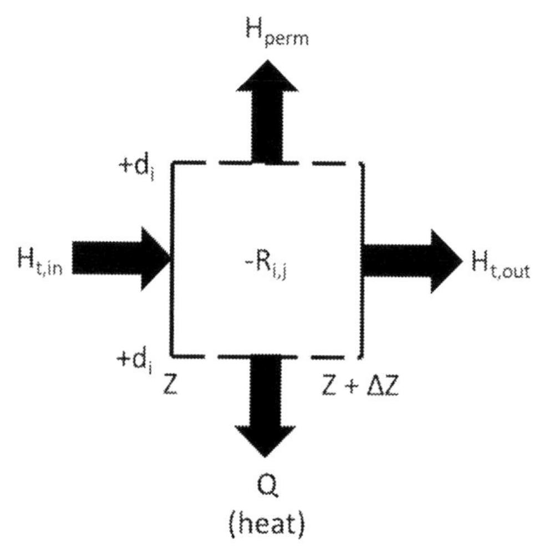

FIGURE 14.6 Graphical depiction of the steady-state energy balance of the tube side.

calculated by determining the difference in the enthalpy of the component permeating through the membrane at the pressure it started at and where it ended at multiplied by the flux. The enthalpy lost from the tube side because of permeation can be solved with the following integral:

$$\overline{H}_{\text{perm}} = \sum_i J_i \pi d_t \Delta z \int_{P_t}^{P_s} \left(\frac{dH}{dP}\right)_{T_i} dP \tag{14.12}$$

where $\overline{H}_{\text{perm}}$ is the total enthalpy exchanged because of permeation through the membrane and T_i is the temperature of the side of the reactor the component i originated. Also, the equation for the energy gained by the shell side through this process is equal and opposite to this integral.

Because the shell side must receive the heat and enthalpy lost by the tube side, it has an equivalent energy balance but with opposite signs for the permeation and heat exchange terms:

Tube-side energy balance

$$F_t H_t|_z - F_t H_t|_{z+\Delta z} - \sum_i J_i \pi d_t \Delta z \int_{P_t}^{P_s} \left(\frac{dH}{dP}\right)_{T_i} dP - U\pi d_t \Delta z (T_t - T_s) = 0 \tag{14.13}$$

Shell-side energy balance

$$F_s H_s|_z - F_s H_s|_{z+\Delta z} + \sum_i J_i \pi d_t \Delta z \int_{P_t}^{P_s} \left(\frac{dH}{dP}\right)_{T_i} dP + U\pi d_t \Delta z (T_t - T_s) = 0 \tag{14.14}$$

where F_t and F_s are the total molar flow rate of the tube and shell sides, respectively, and H_t and H_s are the total molar enthalpies of the tube and shell sides, respectively, U is the overall convective heat transfer coefficient, and T_t and T_s are the temperatures of the tube and shell sides, respectively. As before, the equations can be divided by Δz and the limit taken as Δz goes to zero to derive the differential energy balances for the general membrane system.

Tube-side differential energy balance (at steady state)

$$\frac{d(F_t H_t)}{dz} + \sum_i J_i \pi d_t \Delta z \int_{P_t}^{P_s} \left(\frac{dH}{dP}\right)_{T_i} dP + U\pi d_t \Delta z (T_t - T_s) = 0 \tag{14.15}$$

Shell-side differential energy balance (at steady state)

$$(\pm)\frac{d(F_s H_s)}{dz} - \sum_i J_i \pi d_t \Delta z \int_{P_t}^{P_s} \left(\frac{dH}{dP}\right)_{T_i} dP - U\pi d_t \Delta z (T_t - T_s) = 0 \tag{14.16}$$

14.3 Case study: polymer-based, water–gas shift membrane reactor

By utilizing the differential mole and energy balances for the generalized membrane reactor, any membrane reactor system can be modeled by defining the set of R_i associated with the catalyst and the J_i associated with the desired membrane material. In the case of polymer membranes, a Fickian model for diffusion is an example of a potential model for J_i and takes on the form:

Fickian diffusion model

$$J_i = \frac{Q_i}{\delta_{\text{mem}}} (P_{t,i} - P_{s,i}) \tag{14.17}$$

where Q_i is the permeance of the component i through the membrane. In this case, the flux for any component in the membrane reactor is dependent on its partial pressure driving force across the membrane and its permeance through that material. To explore this further, a simulation study was performed of a PBI membrane reactor packed with a copper-based catalyst for low-temperature WGS kinetics (Radcliffe, Singh, Berchtold, & Lima, 2016). PBI was selected for its high thermal resistance, making it a good polymer candidate to be paired with an exothermic reaction such as WGS. The setup for this simulation is shown in Fig. 14.7.

In the ideal case, the syngas entering the reactor would have a 1:1 ratio of CO to H_2O with all the H_2 produced permeating out of the reactor and into the steam sweep gas. This would leave a pure CO_2 product in the retentate and none in the permeate, and all the hydrogen would be captured in the steam sweep gas and easily separated by condensing the steam. Although some expensive metallic membrane materials such as palladium can essentially achieve this using ion-transport means (as opposed to Fickian diffusion), this is not the case for polymeric membranes. For an example model of a PBI membrane, the following component permeances are observed and shown in Table 14.1.

A brief scan over the values shows a major challenge confronted by polymer membrane reactors. The sweep gas, steam, is three times more permeable through the membrane than hydrogen. This means that the "ideal" case mentioned previously is unachievable. Wherever there is PBI membrane present in the reactor to remove

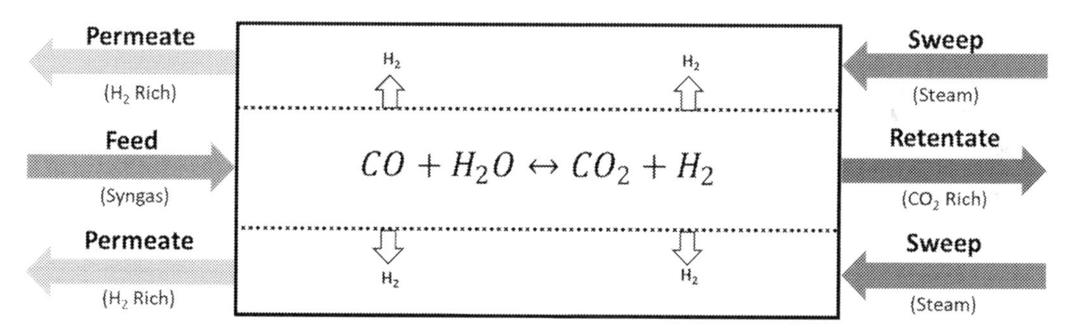

FIGURE 14.7 Schematic of a countercurrent, water–gas shift membrane reactor with polybenzimidazole membrane.

TABLE 14.1 Component permeance ($Q_{i,o}$), permeability ($P_{i,o}$), and component selectivity ($\alpha_{H_2/i}$) values observed at 300°C in literature (Radcliffe, Singh, Berchtold, & Lima, 2016) reported in gas permeation units (GPU) and barrer, respectively.

Component (i)	$Q_{i,o}$(GPU)	$P_{i,o}$(barrer)	$\alpha_{H_2/i}$
H_2	250.0	25.00	–
CO_2	8.9	0.89	28
H_2O	750.0	75.00	0.33
CO	2.5	0.25	100
N_2	2.5	0.25	100

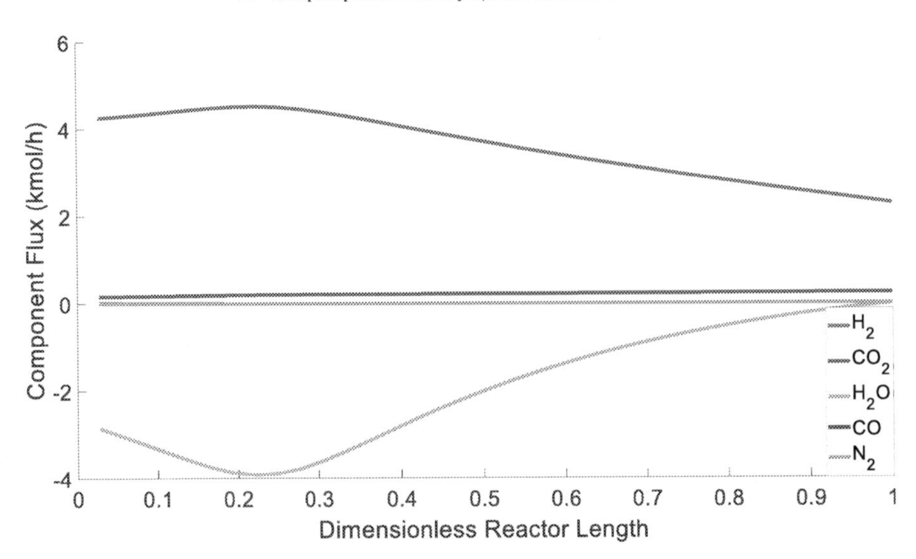

FIGURE 14.8 Flux profiles (from tube to shell) of each component in the water–gas shift membrane reactor using polybenzimidazole polymer membrane. In this case, the syngas and sweep gas mass flow rates are set equal to 3600 kg/h (Case 2).

hydrogen, there is a stronger driving force attempting to inject steam into tube side, and therefore the desired pure CO_2 product is not achieved. Also, there is a driving force present to remove CO_2 from the tube side and enter the permeate. This is an especially undesired scenario because the carbon is not completely captured, and the easy recovery of the hydrogen from the permeate is now very difficult with CO_2 present. Fig. 14.8 shows the flux profiles for each component and this tradeoff between hydrogen recovery from the process and steam injection into the process.

Although steam injection into the tube side of the membrane reactor (represented by negative values for the steam profile) may at first feel like only a negative phenomenon, that could also have a positive effect. In the same way that removing hydrogen pushes the equilibrium of WGS to the right, so does adding more steam to the process. This becomes more obvious when the flow rate of the sweep gas is lowered. These results are shown in Fig. 14.9.

In this case in Fig. 14.9, there are two different modes of operation that take place. In the first 62% of the reactor, all components are permeating from the tube into the shell side, and in the final 38% of the reactor, the steam injection issue mentioned previously is observed. Comparing these two modes, the highest observed permeation of hydrogen takes place where steam injection also occurs and becomes stagnant when the direction of permeation for water is in a direction that is "more desired." However, injecting steam at the end of the reactor would mostly serve the role of adding impurity to the tube side rather than being a potential reactant. Therefore flipping this behavior is investigated by injecting steam at the beginning of the reactor and removing hydrogen throughout the reactor, as shown in Fig. 14.10, by using a high-sweep gas to process gas ratio.

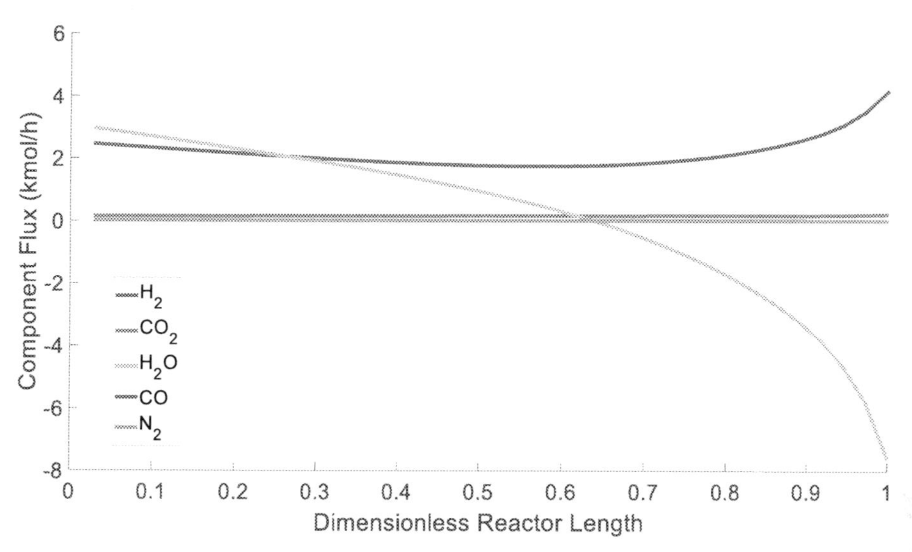

FIGURE 14.9 Component flux profiles when the sweep gas mass flow rate is reduced to 800 kg/h (Case 1).

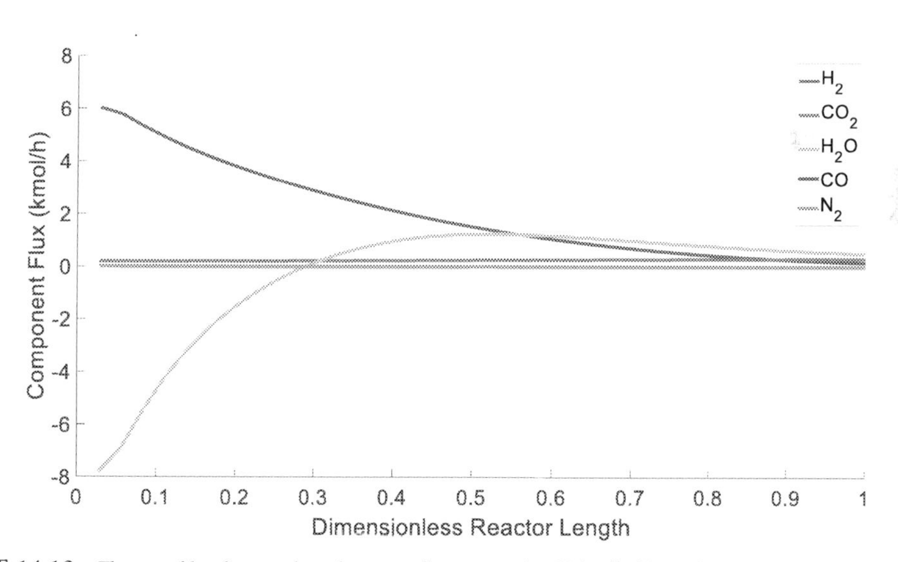

FIGURE 14.10 Flux profiles for a reduced syngas flow rate of 1600 kg/h (Case 3).

This indeed leads to improved membrane reactor performance in terms of hydrogen recovery and carbon monoxide conversion but with a decrease in carbon captured. The performance objectives for this system can be defined by Eqs. (14.18) and (14.19):

$$R_{H_2} = \frac{H_2 \text{ in permeate}}{(H_2 + CO) \text{ in feed}} = \frac{F_{H_2,p}}{F_{H_2,f} + F_{CO,f}} \tag{14.18}$$

$$C_{CO_2} = \frac{\text{carbon in retentate}}{\text{carbon in feed}} = \frac{F_{CO,r} + F_{CO_2,r}}{F_{CO,f} + F_{CO_2,f}} \tag{14.19}$$

in which R_{H_2} and C_{CO_2} are the hydrogen recovery and carbon capture fractions, respectively and the subscripts f, p, and r denote the feed, permeate, and retentate streams, respectively. These results are summarized in Table 14.2.

Although steam injection at the beginning of the reactor leads to improved performance in producing and recovering hydrogen in the polymer-based WGS-MR, it also leads to increases in impurities of steam in the retentate and CO_2 in the permeate. These dilemmas pose a few questions for directions in polymer membrane research:

1. How much more does the permeance of CO_2 need to be decreased to have both high carbon capture and hydrogen recovery?
2. How much does the H_2O permeance need to be decreased to minimize the amount of water lost into the reactor?

To answer the first question, a study is conducted to determine the permeance of CO_2 required such that 90% carbon capture is achievable while also meeting other desired benchmarks. These other benchmarks include the carbon monoxide conversion to be above 98%, the combined purity of CO_2 and H_2O in the retentate to remain above 95%, and the concentration of hydrogen in the retentate to remain below 4%. The current desired benchmarks for the WGS-MR system are summarized in Table 14.3.

TABLE 14.2 A comparison of three key performance standards for the water–gas shift membrane reactor: carbon capture, hydrogen recovery, and carbon monoxide conversion, respectively.

Case	Sweep gas(kg/h)	Process gas(kg/h)	C_{CO_2}(%)	R_{H_2}(%)	χ_{CO}(%)
1	800	3600	93.9	47.8	79.5
2	3600	3600	93.1	77.9	99.7
3	3600	1600	81.8	98.9	99.9

TABLE 14.3 Summary of desired benchmarks for the water–gas shift membrane reactor system (Bishop & Lima, 2020, 2021; Lima, Daoutidis, Tsapatsis, & Marano, 2012; Radcliffe, Singh, Berchtold, & Lima, 2016).

Benchmark	Desired value (%)
Carbon capture	$\geq$90
Hydrogen recovery	$\geq$95
CO conversion	$\geq$98
$CO_2 + H_2O$ retentate purity	$\geq$95
H_2 retentate purity	$\leq$4

The study was conducted by utilizing a sequential quadratic programming optimizer available in the AVEVA Process Simulation platform. The selectivity of H_2 to CO_2 was varied by changing the permeance of CO_2 and holding the permeance of H_2 constant. With a similar range of H_2 permeance, a membrane with higher H_2/CO_2 selectivity could, in principle, lead to increased carbon capture. With that hypothesis in mind, the selectivity was gradually increased by five units, and the optimizer would run to find a design for the membrane reactor that maximized the carbon capture, and then the value was recorded. The results of this study are shown in Fig. 14.11.

Fig. 14.11 shows that a selectivity of 45 is required to achieve all of the benchmarks for the WGS-MR. This means according to the model, a goal for polymer membranes would be to reduce the permeance of CO_2 by a factor of 2, which is already achievable with current technology through the use of "modified" PBI membranes (Berchtold, Singh, Young, & Dudeck, 2012).

To answer the second question of how much the permeance of water must be decreased to reduce the amount of sweep gas lost to the reactor, another study is proposed. For this study, the selectivity of hydrogen to carbon dioxide from the previous study is fixed at 50. This is because at a value of 45, some of the benchmarks are not attainable when the water's permeance begins to decrease. The selectivity of hydrogen to water is then varied by changing the permeance of water while fixing the permeance of hydrogen, and the percentage of sweep gas lost to the reactor is recorded and shown in Fig. 14.12.

Interestingly, the amount of steam that leaks into the reactor increases slightly as the water permeance is initially decreased. For the PBI membrane considered in this study, the value of the selectivity of hydrogen to water is of about 0.33. Fig. 14.12 suggests that the selectivity would have to be increased to about 1.1 before any benefits could begin to be noticed. This means the water permeance would have to be reduced by a factor of 3 or greater before any benefits can be realized, requiring the membrane to be more selective to hydrogen than the steam sweep gas (i.e., a selectivity greater than 1).

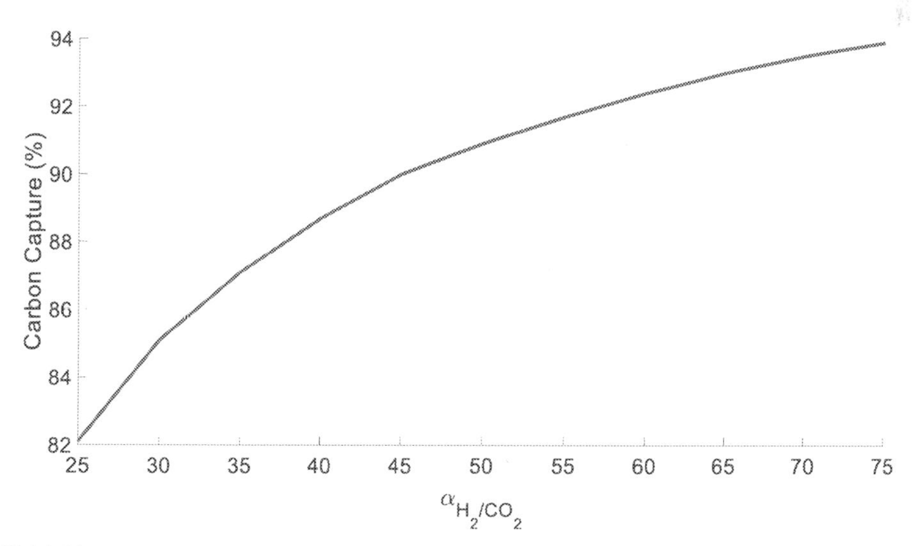

FIGURE 14.11 Maximum possible carbon capture given selectivity of hydrogen to carbon dioxide.

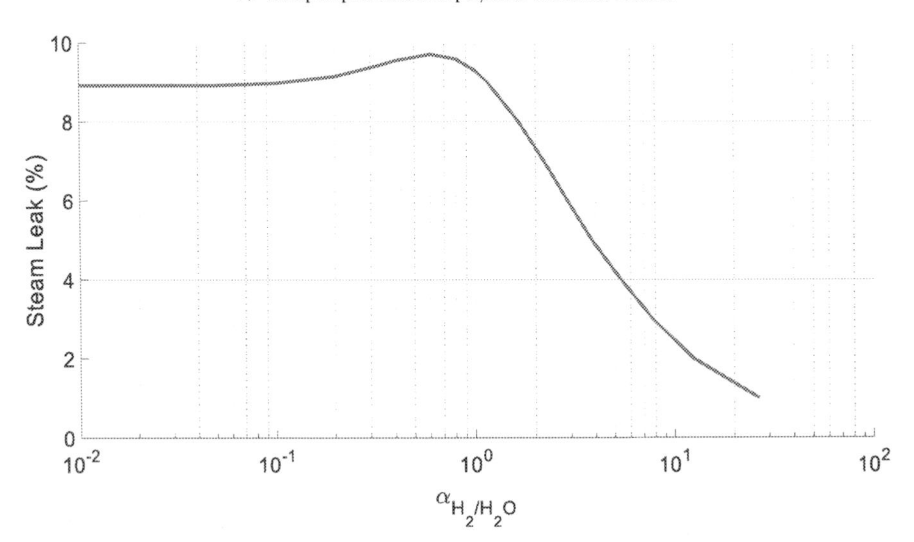

FIGURE 14.12 Percentage of sweep gas that leaks into the tube of the reactor as a function of hydrogen to water selectivity.

14.4 Conclusions and future trends

Although some membrane materials like the modified PBI membranes may be able to achieve the requisite CO_2 selectivity to meet the carbon capture specification, this model has identified that significant work is still required to reduce the membrane selectivity to water. This is because steam is used as the sweep gas and can leak into the tube side of the unit. This leak can bring benefits as water is a reactant in the WGS reaction; however, this means that additional steam must be sent to the unit as makeup, likely leading to increased capital costs.

Clearly, membranes with higher H_2/CO_2 and H_2/H_2O selectivities would lead to enhancing carbon capture and minimizing water permeation into the reactor. Recent work on PBI hollow fiber membranes has shown significant improvements in terms of reduced skin thickness and therefore increasing H_2 permeances (i.e., 400 GPU at $T = 350°C$) but with modest (~24) H_2/CO_2 selectivities (Singh, Ganpat, Dudeck, & Berchtold, 2020). Improvements in selectivities could be achieved by controlled modifications to the pristine PBI membranes – for example, via addition of polyprotic acids (H_3PO_4, H_2SO_4, etc.) (Zhu et al., 2018). Such modifications are typically associated with high H_2/CO_2 selectivities (~140 for highest doping levels) but low permeabilities. A combination of the above two approaches could, in principle, lead to membranes with desired transport properties, as recommended by the developed transport model.

An alternative approach involves the creation of rigid carbon molecular sieve (CMS) membranes via high-temperature (>500°C) pyrolysis of polymers under essentially inert atmosphere. In fact, under optimum pyrolysis conditions (i.e., at 900°C), PBI-derived CMS membranes show ~two times higher H_2 permeabilities and ~six times higher H_2/CO_2 selectivities as compared to pure PBI membranes (Omidvar et al., 2019). Like polymer membranes, CMS membranes have the advantage of being processed into scalable hollow fiber formats, which allow packing into high-performance membrane modules. This allows for easy processability as compared to other inorganic materials, which otherwise show high intrinsic separation performance (Lin, 2018). Moreover, carbon-based membranes are also stable at high testing temperatures, and are therefore appropriate materials for WGS membrane reactors (Itoh & Haraya, 2000). Performance improvements beyond PBI-derived CMS membranes could be achieved by using other polymer precursors like polyimides (Sanyal et al., 2018) and polymers of intrinsic micro porosity (Swaidan, Ma, Litwiller, & Pinnau, 2013). Most recently, cellulose-derived CMS hollow fiber membranes have been shown to provide high H_2/CO_2 selectivities ~ 84 at $130°C$ along with H_2 permeances ~150 GPU (Lei et al., 2021). Beyond intrinsic transport performance, cost considerations must also be included in the overall material selection. Such analyses are beyond the scope of this chapter, but a thorough review of various membrane reactor materials along with their respective costs has been provided by Lin (2018).

Going forward, the proposed modeling approach will be used to develop models for other advanced materials to further optimize the designs and in turn, guide future membrane material development. Future efforts will also involve thorough techno-economic analysis of various scenarios to guide a holistic development in this field.

References

Basile, A. (2013). *Handbook of Membrane Reactors: Fundamental Materials Science, Design and Optimisation*. Elsevier.

Berchtold, K. A., Singh, R. P., Young, J. S., & Dudeck, K. W. (2012). Polybenzimidazole composite membranes for high temperature synthesis gas separations. *Journal of Membrane Science, 415-416*, 265–270.

Bishop, B. A., & Lima, F. V. (2020). Modeling, simulation, and operability analysis of a nonisothermal, countercurrent, polymer membrane reactor. *Processes, 8*(1), 78.

Bishop, B. A, & Lima, F. V. (2021). Novel module-based membrane reactor design approach for improved operability performance. *Membranes, 11*(2), 157.

Itoh, N., & Haraya, K. (2000). A carbon membrane reactor. *Catalysis Today, 56*(1–3), 103–111.

Lei, L., Pan, F., Lindbråthen, A., Zhang, X., Hillestad, M., Nie, Y., ... Guiver, M. D. (2021). Carbon hollow fiber membranes for a molecular sieve with precise-cutoff ultramicropores for superior hydrogen separation. *Nature Communications, 12*(1), 1–9.

Li, X., Singh, R. P., Dudeck, K. W., Berchtold, K. A., & Benicewicz, B. C. (2014). Influence of polybenzimidazole main chain structure on H_2/CO_2 separation at elevated temperatures. *Journal of Membrane Science, 461*, 59–68.

Lima, F. V., Daoutidis, P., Tsapatsis, M., & Marano, J. J. (2012). Modeling and optimization of membrane reactors for carbon capture in integrated gasification combined cycle units. *Industrial & Engineering Chemistry Research, 51*(15), 5480–5489.

Lin, Y. (2018). Inorganic membranes for process intensification: Challenges and perspective. *Industrial & Engineering Chemistry Research, 58*(15), 5787–5796.

Omidvar, M., Nguyen, H., Huang, L., Doherty, C. M., Hill, A. J., Stafford, C. M., ... Lin, H. (2019). Unexpectedly strong size-sieving ability in carbonized polybenzimidazole for membrane H_2/CO_2 separation. *ACS Applied Materials & Interfaces, 11*(50), 47365–47372.

Radcliffe, A. J., Singh, R. P., Berchtold, K. A., & Lima, F. V. (2016). Modeling and optimization of high-performance polymer membrane reactor systems for water–gas shift reaction applications. *Processes, 4*(2), 8.

Sanyal, O., Zhang, C., Wenz, G. B., Fu, S., Bhuwania, N., Xu, L., ... Koros, W. J. (2018). Next generation membranes—Using tailored carbon. *Carbon, 127*, 688–698.

Singh, R. P., Ganpat, J. D., Dudeck, K. W., & Berchtold, K. A. (2020). Macrovoid-free high performance polybenzimidazole hollow fiber membranes for elevated temperature H_2/CO_2 separations. *International Journal of Hydrogen Energy, 45*(51), 27331–27345.

Swaidan, R., Ma, X., Litwiller, E., & Pinnau, I. (2013). High pressure pure-and mixed-gas separation of CO_2/CH_4 by thermally-rearranged and carbon molecular sieve membranes derived from a polyimide of intrinsic microporosity. *Journal of Membrane Science, 447*, 387–394.
Zhu, L., Swihart, M. T., & Lin, H. (2018). Unprecedented size-sieving ability in polybenzimidazole doped with polyprotic acids for membrane H_2/CO_2 separation. *Energy & Environmental Science, 11*(1), 94–100.

CHAPTER

15

Transport phenomena in polymer electrolyte membrane fuel cells

Irene Gatto, Alessandra Carbone and Enza Passalacqua
CNR ITAE, Messina, Italy

Abbreviations

AEM	anion-exchange membranes
AEMFC	anion-exchange membrane FC
AFC	alkaline fuel cells
Am-sPAEKS	sulfonated poly(arylene ether ketone sulfone) containing amino groups
DABCO	1,4-diazabicyclo[2.2. 2]octane
DC	direct current
DH-PEFC	direct hydrogen PEFC
DM-PEFC	direct methanol PEFC
DVB	divinylbenzene
EIS	electrochemical impedance spectroscopy
ETFE	poly(ethylene-alt-tetrafluoroethylene)
EW	equivalent weight
FC	fuel cell
FEP	poly(tetrafluoroethylene-*co*-hexafluoropropylene)
IEC	ionic exchange capacity
IEM	ionic exchange membranes
ILs	ionic liquids
LSC	long side chain
MCFC	molten carbonate fuel cells
PA	phosphoric acid
PAA	poly(acrylic acid)
PAEs	polyarylenethers
PAFC	phosphoric acid fuel cells
PBI	polybenzimidazole
PEEK	polyetheretherketone
PEEKK	Polyetheretherketoneketone

Current Trends and Future Developments on (Bio-) Membranes
DOI: https://doi.org/10.1016/B978-0-12-822257-7.00007-8

PEFC	polymer electrolyte fuel cells
PEM	proton exchange membranes
PEMFC	proton exchange membrane fuel cell
PEO	poly(ethylene oxide)
PES	polyether sulfone
PFSA	perfluorosulfonic acid
PILs	polymerized ionic liquids
PIs	polyimides
PPQ	polyphenylquinoxalines
PPs	polyphenylenes
PPZs	polyphosphazenes
PSF	poly(bisphenol-A sulfone)
PSt	polystyrene
PVA	poly(vinyl alcohol)
PVDF	polyvinylidene fluoride
PVP	poly(vinylpyrrolidone)
QA	quaternary ammonium
RH	relative humidity
SCNT	single-walled carbon nanotubes
SD	sulfonation degree
SOFC	solid oxide fuel cells
sPEEK	sulfonated polyetheretherketone
sPEK	sulfonated polyetherketone
SPI	sulfonated polyimides
sPPO	poly(substituted-phenylene oxide)
SSC	short side chain
TEA	trimethylamine
TMA	trimethylamine
W_{up}	water uptake

Nomenclature

$a\omega$	water activity
D'	intradiffusion coefficient of water
D_λ	diffusion coefficient of water in the membrane
E_a	activation energy
λ	number of absorbed water molecules for sulfonic group
m_{dry}	mass of dry membrane
m_{wet}	mass of wet membrane
n_{H2O}	number of water molecules inside the membrane
n_{HSO3^-}	number of sulfonic acid sites
pKa	acid dissociation constant
R	ideal gas constant
S	swelling
s	expansion factor of the membrane
σ	conductivity of membrane
T	the temperature and
T_g	glass transition temperature
V_l	volume of liquid water
V_v	volume of membrane
W	sample width

15.1 Introduction

Fuel cell (FC) is a device that converts the chemical energy of fuel directly into electricity by electrochemical reactions, without combustion. Different classifications can be done for FCs on the basis of the electrolyte used, the type of ions migrating through the electrolyte, the type of the reactants (e.g., primary fuels and oxidants), the operative conditions of temperature and pressure, the primary fuels feed in FC system (direct and indirect). The most common classification is based on the nature of the electrolyte and includes the following types: (1) alkaline fuel cells (AFCs) with an alkaline solution electrolyte (such as sodium NaOH or potassium hydroxide KOH), (2) phosphoric acid fuel cells (PAFCs) with an acidic solution electrolyte (with a phosphoric acid solution electrolyte), (3) PEFCs in which the electrolyte consists of a polymeric membrane, (4) molten carbonate fuel cells (MCFC) with molten carbonate salt electrolyte, and (5) solid oxide fuel cells (SOFC) with ceramic ion conducting electrolyte in solid oxide form. Table 15.1 summarizes the main types of FCs.

Among the abovementioned FC types, the most interesting and recently studied are PEFCs.

In PEFC, a solid ionic exchange membrane (IEM), which has the ability of ions transferring, is used as a polymeric electrolyte.

IEMs are typically composed of hydrophobic substrates with immobilized positively or negatively charged functional groups, and exchangeable counter-ions. Depending on the type of counter-ions, IEMs are broadly classified into proton exchange membrane FC (PEMFC) and anion-exchange membrane FC (AEMFC). In both configurations, when the functional groups attached onto the polymer backbone are sufficiently hydrated, a mobility of cations or anions occurs from one group to an adjacent one leading to ion conductivity through the membrane. The most commonly functional groups contained in PEMs are: sulfonic acid, phosphoric acid, and carboxylic acid groups (Xu, Wu, & Wu, 2008), while to

TABLE 15.1 Scheme of different fuel-cell typologies.

Fuel cell	Electrolyte	Operating temperature (°C)	Electrochemical reactions
AFC	Aqueous solution of KOH soaked in a matrix	90–100	Anode: $H_2 + 2OH^- \rightarrow 2H_2O + 2e^-$ Cathode: $\frac{1}{2} O_2 + H_2O + 2e^- \rightarrow 2OH^-$ Cell: $H_2 + \frac{1}{2}O_2 \rightarrow H_2O$
PAFC	Phosphoric acid soaked in a matrix	175–200	Anode: $H_2 \rightarrow 2\,H^+ + 2e_-$ Cathode: $\frac{1}{2} O_2 + 2\,H^+ + 2e^- \rightarrow H_2O$ Cell: $H_2 + \frac{1}{2}O_2 \rightarrow H_2O$
PEFC	Solid organic polymer	30–80	Anode: $H_2 \rightarrow 2\,H^+ + 2e^-$ Cathode: $\frac{1}{2} O_2 + 2\,H^+ + 2e^- \rightarrow H_2O$ Cell: $H_2 + \frac{1}{2}O_2 \rightarrow H_2O$
MCFC	Solution of lithium, sodium, and/or potassium carbonates soaked in a matrix	600–1000	Anode: $H_2 + CO_3^{2-} \rightarrow H_2O + CO_2 + 2e^-$ Cathode: $\frac{1}{2} O_2 + CO_2 + 2e^- \rightarrow CO_3^{2-}$ Cell: $H_2 + \frac{1}{2}O_2 + CO_2 \rightarrow H_2O + CO$
SOFC	Solid zirconium oxide with a small amount of yttria	600–1000	Anode: $H_2 + O_{2-} \rightarrow 2H_2O + 2e_-$ Cathode: $\frac{1}{2} O_2 + 2e^- \rightarrow O^{2-}$ Cell: $H_2 + \frac{1}{2}O_2 \rightarrow H_2O$

obtain AEMs, quaternary ammonium cations, imidazole, and guanidinium cations are generally anchored onto the polymer backbones (Ran et al., 2015).

In an FC, the IEM is placed between the electrodes and selectively conducts protons from anode to cathode (for PEMFC); Fig. 15.1A or hydroxide ions from cathode to anode (for AEMFC); Fig. 15.1B, hindering the transport of electrons and blocking permeation of fuels (Ran et al., 2017).

The PEMFCs have higher efficiency compared to AEMFC. Moreover, PEMs exhibit high chemical and thermal stability and high proton conductivity, even if they are strongly conditioned from FC operating conditions. Unfortunately, strong acidic conditions, mild cell temperature (<80°C), and a strong dependence on water of the proton conduction of PEMFC operation require the use of Platinum-based electrocatalysts for fuel oxidation, which significantly increases costs, pure H_2 as fuel, and a complicated humidifying system to avoid the membrane dry-out affecting the long-term stability of PEMs (Hickner, Ghassemi, Kim, Einsla, & McGrath, 2004). To overcome these challenges, anhydrous PEMs for high-temperature FC were proposed, but heat management owing to the increase in temperature is a critical factor (Li, He, Jensen, & Bjerrum, 2003). On the other hand, AEMs for AEMFC are emerging because of the possibility of using cheap catalysts and different fuels (Hickner, Herring, & Coughlin, 2013). In any case, the low hydroxide conductivity and low stability in alkaline environment are serious issues regarding the performance of AEMs (Kreuer, 2013). In fact, anions have low mobility with respect to protons, thereby reducing the anion conductivity of membrane.

In the following section, a description of the transport phenomena that occur in PEMs will be elucidated. Moreover, the main structural and chemical, physical and

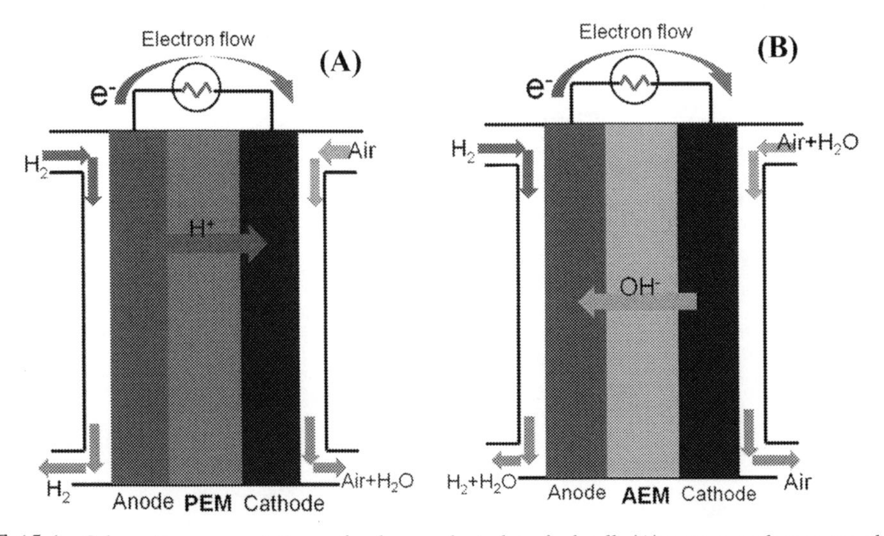

FIGURE 15.1 Schematic representations of polymer electrolyte fuel cell: (A) proton exchange membrane fuel cell and (B) anion-exchange membrane fuel cell. Source: *Reprinted with permission from Ran, J., Wu, L., He, Y., Yang, Z., Wang, Y., Jiang, C., ... Xu, T. (2017). Ion exchange membranes: New developments and applications, Journal of Membrane Science, 522 (2017) 267–291.*

electrochemical characteristics of the main proton and anion-exchange polymers that are currently used in PEFCs, will be reported.

15.2 Transport phenomena

The membrane is the most important component used in PEFCs. The main transport phenomena that occur within it are the transport of water and ions and the two phenomena are strictly related. Maintaining an optimal level of water is a critical problem in the membrane because sufficient water is needed to maintain a high conductivity. In fact, the ions' transport depends on the amount of water contained within the membrane, and the performance of the FC is directly controlled by those phenomena. The proton and hydroxide ion are the fastest diffusing ions in liquid water, but at room temperature, hydroxide mobility is only 57% of the proton mobility. The activation energy, for hydroxide mobility, is only 0.4 kcal/mol larger than that for proton mobility (Agmon, 2000). Owing to their similarities, hydroxide transport mechanism and transport of protons are considered analogous to each other; therefore in the following paragraphs, the water transport and the proton conduction will be discussed in detail.

15.2.1 Water transport

There are four different types of water transport across the PEM. The first is the back diffusion, where the water transport is the result of the flow driven by the concentration gradient from the cathode side to the anode side. The second is the electroosmotic drag; in this case, the water passes from the anode side to the cathode side through the membrane because of proton transport (Dadda, Abboudi, Zarrit, & Ghezal, 2014; Zawodzinski et al., 1993). The third type of water transport is the hydraulic permeation. This effect is caused by the pressure gradient of the gas or capillary between the anode and the cathode side due to the transport of water through the membrane. The last mode to transport water is the thermo-osmosis flux, which is governed by the temperature gradient.

The back diffusion and electroosmotic drag are the principal phenomena that occur in PEFC (Meng, 2006). These two phenomena are strongly correlated and accurately determine the overall water flow in the membrane. Water transportation in the membrane by electroosmosis and back diffusion has been widely studied (Kim & Mench 2009; Lee, Han, & Hwang, 2008). The effects of the thermo-osmosis and hydraulic permeation are generally negligible compared with the effects of back diffusion and electroosmotic drag (Dai et al., 2009; Zaffou, Kunz, & Fenton, 2006). These mechanisms determine the water content in the membrane, which is essential to allow suitable proton conductivity. Otherwise, the membrane will be subjected to a dry-out and the performance will be strongly lowered to unacceptable levels.

Since the membrane protonic conductivity—directly related to the performance—depends on the level of hydration, it is important to understand the water transport mechanism in the cell and how it affects the cell's performance (Zamel & Li 2008). Generally, the membranes are constituted of hydrophobic backbone with highly hydrophilic regions, which can absorb large amount of water, leading to a dilute acid environment where the H^+ ions are able to move.

Although the hydrated regions are rather separated, movement of the proton through the long structure of the supporting molecule is still possible (Barbir, 2005).

In a PEMFC, water can exist in three phases: solid, gas, and liquid, which are not contemporarily present in every domain of the cell. PEM is a solid that does not contain voids and can chemically incorporate water as hydronium ions (H_3O^+). It has been found experimentally that Nafion, the most popular PEM based on perfluorosulfonic acid (PFSA), is able to absorb water even in conditions of oversaturation, until reaching the flooding limit, as described in the well-known Schroeder's paradox (Hu, Li, Xu, Huang, & Ouyang, 2016); for this reason, it is complicated to estimate the water content in the membrane when water is (or probably is) present in the liquid phase inside the membrane.

All the transport properties of the polymeric membranes depend on the hydration levels. In particular, the protonic conductivity is strongly affected by the amount of water absorbed by the membrane; thus it is very important to measure or calculate the degree of hydration or water content. There are different parameters related to the water content in the membrane. The first one is λ, that is, the number of absorbed water molecules for sulfonic group:

$$\lambda = nH_2O/nHSO_3^- \tag{15.1}$$

where nH_2O is the number of water molecules inside the membrane, and $nHSO_3^-$ is the number of sulfonic acid sites accessible to water molecules in the membrane.

The parameter that expresses the water present in the liquid phase is the swelling S:

$$S = \frac{V_1}{V_v} \lim_{x \to \infty} \tag{15.2}$$

V_l is the volume of liquid water and V_v is the volume of membrane.

Another crucial parameter is the diffusion coefficient of water in the membrane D_λ.

$$D_\lambda = D'/(1+s\lambda)^2 d \ \ln a\omega/d\lambda \tag{15.3}$$

where D' is the intradiffusion coefficient of water in the Nafion membrane, s is the expansion factor of the membrane due to the water sorption, and $a\omega$ the water activity assuming a thermodynamic equilibrium with the water inside the membrane.

Other two very important parameters for PEMs are the water uptake (W_{up}) and ionic exchange capacity (IEC). W_{up} represents the total water content within the membrane. It is calculated by using the following equation:

$$W_{up} = \left(\frac{\infty(m_{wet} - m_{dry})}{m_{dry}}\right) \times 100 \tag{15.4}$$

where m_{wet} and m_{dry} represent the mass of membrane in wet and dry conditions, respectively.

IEC is the measure of relative concentration of acid groups within the PEM and is related to the membrane's ability to exchange H_3O^+ ions located in the aqueous environment through the $-SO_3^-$ sulphonyl groups anchored in their structure. It is defined as the ratio between the equivalent of exchangeable ions and the dry weight of the membrane and can be calculated by the inverse of equivalent weight (EW) of the polymers.

$$\mathrm{IEC} = \frac{1}{\mathrm{EW}} = \mathrm{mole}\frac{\mathrm{H}^+}{m_{\mathrm{dry}}} \tag{15.5}$$

where mole H^+ represents number of moles of acid groups, and m_{dry} represent the weight of dry polymer sample in grams.

All these parameters together give us information about the amount and the typology of water present into the PEM. In particular, the λ value provides information about the typology of water bound to the $-SO_3^-$ group. In fact, Kreuer (1997) reported that for the PFSA membranes, when $\lambda \leq 3$, the water molecules are the primary hydration sphere; when $3 \geq \lambda \leq 14$, the water is loosely bound; for its value over 14, the water is in the second phase. For polyaromatic polymers, these values are different. It is possible to have the primary hydration phase for $\lambda \leq 5$; the loosely bound water is for λ included from 5 to 10; for its value over 10, we have water as the second phase.

This means that the first absorbed water molecules cause the sulfonic groups dissociation, resulting in the formation of hydronium ions.

Owing to the separation between hydrophobic and hydrophilic phase, at low water content, it is reasonable to consider the water absorbed by the membrane as mainly located in the sulfonic groups. As a consequence, the continuous water phase cannot be formed and the conductivity values are extremely low.

This trend changes when loosely bound water is present; in fact, the counter-ion cluster dimensions increase, leading to an excess of mobile protons able to move over the entire cluster. When the water content further increases and λ values reach the water as a second phase, a coalescence of the ion clusters occurs, leading to the formation of a continuous phase with properties similar to those of bulk water.

These mechanisms are experimentally confirmed by water absorption isotherms, which highlight how the water mobility and self-diffusion values approach the bulk water as well as the protonic charge carriers.

The free-water phase is shielded from the sulfonic groups by the strongly bound water molecules of the primary hydration shell. Fig. 15.2(A) and (B) report the water absorption isotherm for Nafion and sulfonated polyaromatic membranes, in which are highlighted the different kinds of water.

The different behavior of Nafion and polyaromatics such as sulfonated polyetheretherketones (sPEEKs) can be explained considering the different microstructures and the pKa of the acidic functional groups. In fact, the low phase separation of sPEEK than Nafion corresponds to narrower, less connected hydrophilic channels and to larger separations between less acidic sulfonic acid functional groups. At high water content, this entails a reduction of electroosmotic drag and water permeation, while a high proton conductivity is maintained Kreuer (2001).

To quantify and understand the nature of water contained in the PEM, a number of techniques have been used such as small-angle X-ray scattering spectroscopy (Kreuer, 2001), atomic force microscopy (O'Dea, Economou, & Buratto, 2013), nuclear magnetic resonance spectroscopy (Galitskaya et al., 2020; Roy et al., 2017; Zhao, Majsztrik, & Benziger, 2011), dielectric relaxation

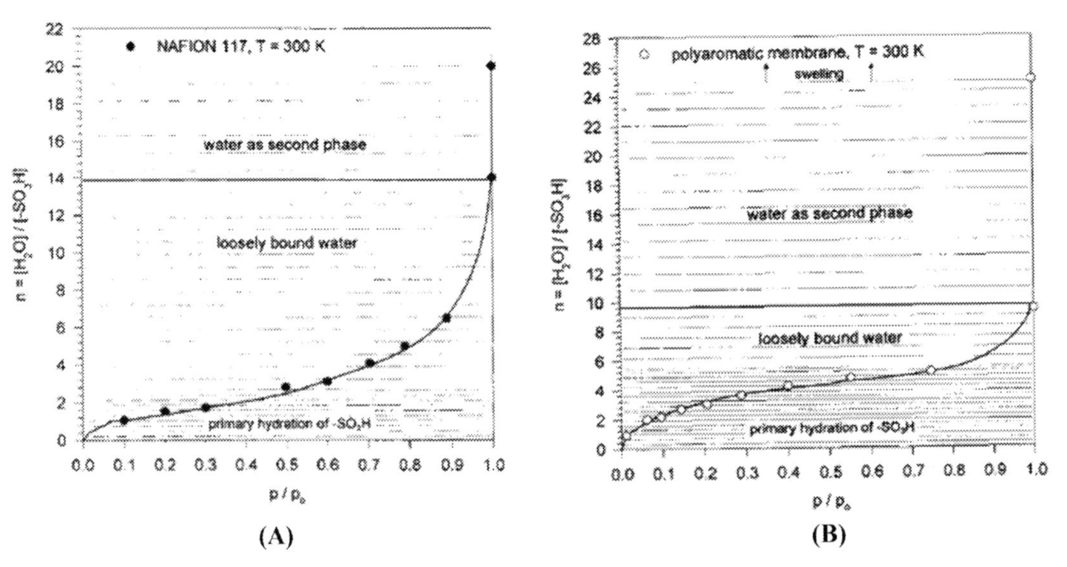

FIGURE 15.2 Water absorption isotherm for (A) Nafion and (B) sulfonated polyaromatic membranes. *Reprinted with permission from Kreuer, K.D. (1997). On the development of proton conducting materials for technological applications. Solid State Ionics, 97, 1–15.*

spectroscopy (Lu, Polizos, Macdonald, & Manias, 2008; Paddison, Bender, Kreuer, Nicoloso, & Zawodzinski, 2000), Fourier transform infrared spectroscopy in the mid-infrared (Hofmann et al., 2009; Veldre, Sala, Āboltiņa, & Vaivars, 2018) frequency range and molecular dynamics simulations (Bahlakeh, Hasani-Sadrabadi, & Jacob, 2016; Hofmann et al., 2009), terahertz spectroscopy probes (Devia, Raya, Shukla, Bhat, & Pesala, 2019), and neutron technique (De Caluwe, Baker, Bhargava, Fischer, & Dura, 2018; Page, Dura, Kim, Rowe, & Faraone, 2015).

Moreover, to predict the water distribution in the PEM, many water transport models were developed (Ge, Li, Yi, & Hsing, 2005; Khattra, Santare, Karlsson, Schmiedel, & Busby, 2014; Kim, Choi, Song, & Kim, 2016; Uddin, Saha, & Oshima, 2014). In particular, density functional theory, molecular mechanics, and dynamic molecular-based studies were carried out (Lopez-Chavez et al., 2015). Ionescu studied a one-dimensional mass transport PEFC model for evaluation of net water flux across the Nafion-type membrane. This model was implemented with the Comsol Multiphysics software based on the finite element method (Ionescu, 2020). Liso and Nielsen (2015) developed a new mathematical zero-dimensional diffusivity model for the water mass balance and hydration of a PEM, including all the crucial physical and electrochemical processes occurring in the membrane electrolyte and considering the water adsorption/desorption phenomena in the membrane.

15.2.2 Proton conduction

The proton conduction is fundamental for PEMs, and the proton transport is mainly associated with two distinct mechanisms: the vehicular (Dippel & Kreuer 1991) and Grotthuss (Agmon, 1995) one. Generally these mechanisms differ in the water contained into the ion clusters; in particular, the vehicular mechanism occurs when high-humidity and low-temperature conditions are combined. The water molecules contained into the electrolyte membrane form a cluster in which proton migration is possible. The conduction mechanism is water assisted and the water acts as a vehicle from one sulfonic group to another. Researchers have reported the formation of nano-channels (or clusters) with inner walls composed of sulfonic acid groups, which promote the high conductivity of Nafion in PEM applications. The proton conductivity decreases when temperature is increased and humidity is reduced, because the water clusters are no longer connected in these conditions. The schematic design of the vehicular mechanism in proton conduction has been shown in Fig. 15.3 (Peighambardoust, Rowshanzamir, & Amjadi, 2010).

Contrary to vehicular mechanism, low humidity and elevated temperatures are generally associated to the Grotthuss mechanism, where the proton transport occurs with the proton hopping from one functional group to the adjacent one. Under these conditions, protons are transported through long-range hydrogen bonding.

This transport involves the formation and breaking of hydrogen bonds with water molecules and neighboring sulfonic groups, which is simulated as a jump from one functional group to the other one.

Nonetheless, the transfer of a proton along the water chain does not produce the transfer of a unit charge, because the water chain remains in a polarized state. To accept another proton, the water chain must rotate and reorganize itself through the transport of an orientation defect. This two-step process can be also associated with the transfer of a negatively charged OH^- "proton hole" along the chain (Miyake & Rolandi 2016).

Fig. 15.4 depicts the Grothuss mechanism considering the water molecules interconnected via H-bonds (water or proton wire). The movement of H^+ from left to right is favored by an electrochemical gradient (Cukierman, 2006). The approach of H^+ (row 1 in Fig. 15.4) to the O of the first water molecule in the chain will conduct to the formation of a covalent OH-bond. The proton covalently linked to the O of first water molecule will be shared between first and second water molecules forming a protonated water dimer (Zundel's cation, $(H_5O_2)^+$, row 2 in Fig. 15.4). This hopping model will be propagated along the water wire (row 2 in Fig. 15.4). When the H^+ hops, the dipole moment of the water molecule donating the H^+ rotates. Once the H^+ releases the last water molecule in the water chain of Fig. 15.4 (row 2), the total dipole moment of the chain is reversed (row 3 in Fig. 15.4). If another H^+ is to be transferred in the same direction, the water

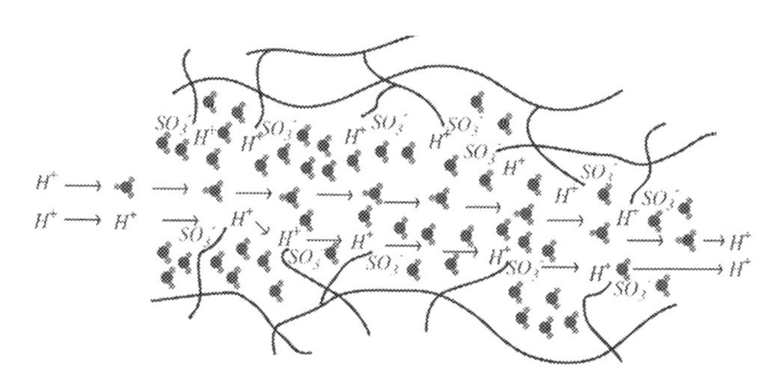

FIGURE 15.3 The schematic design of the vehicular mechanism as proton conduction in pristine membranes. *Reprinted with permission from Peighambardoust, S.J., Rowshanzamir, S., & Amjadi, M. (2010). Review of the proton exchange membranes for fuel cell applications. International Journal of Hydrogen Energy, 35, 9349–9384.*

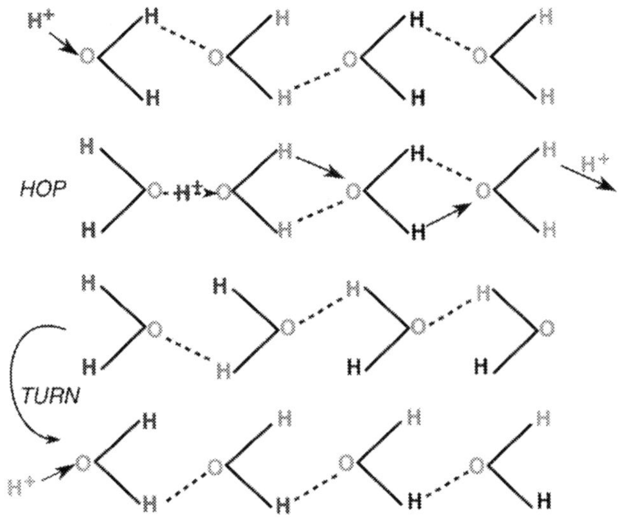

FIGURE 15.4 Schematic representation of a unidimensional chain of four water molecules interconnected via H-bonds (water or proton wire) for Grotthuss mechanism. *Reprinted with permission from Cukierman, S. (2006). Et tu, Grotthuss! and other unfinished stories. Biochimica et Biophysica Acta, 1757, 876–885.*

molecules need to rotate again to their original configurations (row 4, Fig. 15.4). The predominant idea was that the rate-limiting step of H^+ mobility in water was the rotation of water molecules (Fig. 15.4).

Noticeable discrepancies between the activation energies for water rotation and H^+ mobility at various temperatures were reported (Agmon, 1995). At temperatures higher than ~293K, a good agreement between those activation energies was found, while at lower temperature, the differences become increasingly larger. At relatively low temperatures, water molecule can be coordinated via H-bonds with up to four other water molecules. The rotation of water molecule at low temperatures requires the simultaneous breaking of four H-bonds in the first solvation shell of $(H_3O)^+$, and consequently, the activation energy exceeds that of H^+ mobility. On the contrary, at higher temperatures, water molecules lose H-bonds, decreasing the activation energy of water rotation and approaching the activation energy of H^+ mobility.

The proton conductivity of the membranes can be measured either through-plane or in-plane. Through-plane membrane conductivity is generally measured by electrochemical impedance spectroscopy (EIS) in the frequency range 0.01–100 kHz at signal amplitude of 100 mV. The membranes are sandwiched between two gas diffusion electrodes using a cell as described by Casciola, Donnadio, and Sassi (2013). Nyquist plots are used to analyze the impedance spectra. The membrane resistance is obtained by extrapolating the impedance data to the real axis on the high-frequency side. Equivalent circuit modeling of the EIS data is used to obtain the Ohmic membrane resistance, which is used to calculate the membrane ionic conductivity using self-developed fitting routines.

The in-plane proton conductivity is measured on the section of the membrane (longitudinal direction), without electrodes coupling, generally with a commercial cell (Bekktech). It is possible to determine the resistance with both EIS and DC current methods. The EIS measurements were carried out as reported above. A four-probe technique in DC current was used and from the slope of the I–V line, the resistance (R) value was calculated using Ohm's law. Then, proton conductivity of membrane, σ (S cm^{-1}), was calculated according to Eq. (15.6),

$$\sigma = L/RWT \tag{15.6}$$

where σ is defined as the reciprocal of R and L = 0.425 cm, the constant distance between the two Pt electrodes; R = resistance (Ω); W = sample width (cm); and T = sample thickness (cm). Carbone et al. (2018).

Ionic conduction in a solid material is due to a thermally activated process. Proton conductivity responds to both the Arrhenius law and the Vogel–Tamman–Fulcher equation. At temperatures below the glass transition temperature, T_g, the conductivity generally obeys the Arrhenius law. Guzman-Garcia, Pintauro, Verbugge, and Hill (1990).

$$\sigma = A/T\, e^{Ea/RT} \tag{15.7}$$

A is a constant, proportional to the number of charge carriers, E_a is the activation energy, T is the temperature, and R is the ideal gas constant.

The activation energy can be directly calculated using the Arrhenius equation and is used to understand which mechanism is predominant during proton conduction. Several papers report activation energy in the range 14–40 kJ/mol for Grotthuss mechanism and lower than 14 kJ/mol for the vehicular one (Paul, McCreery, & Karan, 2014; Smitha, Ridhar, & Khan, 2004).

To enhance the vehicle-type proton conduction in PEMs, hygroscopic oxides such as TiO_2, SiO_2, ZrO_2, etc. are commonly incorporated into PEMs owing to their enhanced water retention capacities (Laberty-Robert, Valle, Pereira, & Sanchez, 2011; Nagarale, Shin, & Singh, 2010). The incorporation of acid/base particles directly into base/acid polymer membrane provides membrane with acid/base pairs for proton hopping, with a more predominant Grotthuss-type proton conduction in PEMs (He et al., 2014; Zhang et al., 2016). $-SO_3H/-NH_2$ acid/base pairs have exhibited great prospects in improving proton conductivities of PEMs under low humidity (He et al., 2014; Wang et al., 2015).

The transport of hydroxide anions is the base mechanism of the AEMFC. The main transport mechanisms for OH^- through anion exchange consist of molecular diffusion or mass diffusion, structural diffusion, or Grotthuss mechanism, and surface-site hopping; migration and convection, but Grotthuss is considered to be the predominant contribution to hydroxide mobility through hydrated membranes (Castañeda Ramírez & Ribadeneira Paz 2018). In this mechanism for hydroxide, excess negative charge is transferred along a chain of water molecules through a series of processes of breaking and formation of O − H-bonds. To describe this phenomenon, it is not possible to use the conventional classical molecular dynamics approaches, neither the Ab initio molecular dynamics, but better results have been obtained by the development of a new multiscale reactive molecular dynamics model for hydroxide in aqueous solution (Chen, Tse, Lindberg, Knight, & Voth, 2016).

15.3 Polymer electrolyte membranes

As reported in the Introduction section, two types of PEFCs have been developed (Fig. 15.1): PEMFC and AEMFC. Both FCs use a solid polymeric membrane as electrolyte and although PEMFC are the most investigated, recently particular attention has been devoted toward AEMFC. This last type of FC is an alternative to the AFCs that use liquid solutions as electrolytes. AFCs show short lifetime because of the formation of carbonates (CO_3^{2-}) between hydroxides of the electrolyte and CO_2 and a weeping effect due to the permeation of liquid electrolyte to electrodes. Integrating the conductive species of the electrolyte into a rigid polymeric matrix the reactivity and production of carbonates is reduced and no weeping takes palce because the electrolyte is a solid. In the following segment, a description of the main protonic and anionic membranes used in the PEFC is reported.

15.3.1 Proton exchange membranes

PEMs exhibit excellent stability and processability, allowing the flexible design of electrochemical devices, especially FCs. The polymer membranes, such as PFSA-based membranes, work effectively in a limited temperature range (60°C–80°C) and require that the membrane is constantly hydrated with water, resulting in complex and expensive engineering solutions. A challenge in developing materials for PEMs is that these materials need to endure at prolonged exposure to the FC environment, such as oxidation, reduction, and hydrolysis. Moreover, it is desirable that the material will permit operation at a higher temperature (>90°C) and lower relative humidity ($RH < 50\%$) with a sufficiently high proton conductivity. The drastic conditions of T improve activity of catalyst and enhance tolerance toward the impurities, while reduced RH levels simplify the thermal and water management of the system. Under these conditions, PFSA polymers (in particular Nafion), which are the most used electrolyte for PEFC, undergo continuous swelling and de-swelling processes, which lower proton conductivity and mechanical integrity with consequent cell failure. Various solutions have been considered to overcome these problems, such as the use of membranes based on short side-chain PFSA (SSC-PFSA) or on thermostable polyaromatic polymers, membranes composed of acid–base mixtures, reinforced membranes, organic/inorganic membranes with hygroscopic and/or protonic conductive charges introduced into the polymeric matrix. Several approaches were taken and the developed membranes can be classified into three main classes: PFSA membranes; non-PFSA membranes and polymers functionalized with phosphoric acid (PA) or ionic liquids (ILs).

15.3.1.1 *Perfluorosulfonic acid membranes*

PFSA membranes consist of a polytetrafluoroethylene (PTFE) backbone with pendant side chains terminated by a sulfonic acid group. In their usual form, these polymer membranes require water for conductivity. As long as these PEMs are kept hydrated, they function well, but when they dry out, their resistance rises sharply. The best known example of this class of proton-conducting membranes is the Nafion membranes from DuPont de Nemours. The synthesis and preparation of PFSA membranes with various types of side chain and/or length, have been developed by different companies and have been continuously modified and improved. The so-called long side chains (LSCs), as well as DuPont, have been developed by companies such as Fumatech producing Fumion membranes, Asahi Chemical that produces the Aciplex membranes, and Asahi Glass Company producing the Flemion membrane. Membranes with SSC ionomers were introduced by Dow Chemical, but they stopped the production probably because of the high cost of the process with respect to the FCs market. Successively other companies, for example, Solvay (Aquivion), Fumatech (Fumapem FS), Asahi Chemical (Aciplex), and Asahi Glass (Flemion) have produced membranes based on SSC ionomers. The most used Nafion DuPont membranes are LSC (EW1100) extruded membranes (N117, N115, N1110), and solution cast membranes (NR211 and NR212). They vary for the thickness, from 89 to 254, and 25–50 μm for the Nafion Extruded and the Nafion Recast respectively. The IEC is 0.95–1.0 meqSO_3H/g. Their through-plane conductivity at 23°C and 100% RH are ranging 0.1–0.106 S/cm with a W_{up}, at 100% RH, varying from 20 to 50 wt.% depending on the temperature (https://nafionstore-us.americommerce.com/Shared/Bulletins/N115-N117-N1110-Product-Bulletin-Chemours.pdf, 2020; https://nafionstore-us.americommerce.com/Shared/P11_C10610_Nafion_NR-211__NR-212_P11.pdf, 2020). The series HP and HL are reinforced membranes through the addition of a microporous PTFE-rich support layer (http://www.nafionstore.com/Shared/Bulletins/HP-MEMBRANE.pdf, 2020; https://nafionstore-us.americommerce.com/Shared/P22_C10856_Nafion_XL_FuelCells_P22.pdf, 2020; Shi, Weber, & Kusoglu, 2016). The presence of reinforcement improves the mechanical resistance but reduces the proton conductivity.

Aciplex and Flemion are produced in both LSC and SSC forms. Aciplex S-1112 (1050 EW) and Flemion LSH-180 (1099 EW) present a proton conductivity of 0.13 S/cm at 25°C. At lower EW Aciplex S-1008 (950 EW) and Flemion LSH-120 (909 EW), the proton conductivity increases to 0.18 S/cm. For Flemion, λ value increases from 18 to 23. Moreover, the lowering of the EW increases water uptake and conductivity at low RH: at RH of 30%, the EW770 Asahi Glass membrane has a proton conductivity of 0.03 S/cm at 80°C, while an EW909 membrane has a proton conductivity of 0.017 S/cm in the same conditions. The mechanical strength clearly becomes lower when the equivalent weight of the membrane is lowered (Saito, Arimura, Hayamizu, & Okada, 2004).

The 3M membranes are 30 μm thick, have an EW of 980, and it is claimed by 3M to have better thermo-mechanical properties than Nafion. The proton conductivity is similar to that of Nafion, that is, 0.17 S/cm at 70°C and RH 100%. Its conductivity has the same dependency on RH as other PFSA membranes (de Bruijn, Makkus, Mallant, & Janssen, 2007). Recently, 3M has developed SSC membranes based on multiacid side-chain ionomers, reinforced with electrospun fibers. The ionomer is a perfluoroimmide (PFI) including one, two, or three imide groups and terminating with the sulfonic acid. A proton conductivity of 0.1 S/cm at 80°C and 40% RH was reported for PFICE-4 containing three imide groups (Yandrasits, 2015; https://www.hydrogen.energy.gov/pdfs/progress15/v_b_1_yandrasits_2015.pdf, 2020).

FuMaTech produces both LSC and SSC membranes. LSC membranes have EW ranging from 950 to 1000 with a proton conducibility of about 0.09 S/cm. SSC membranes are mainly reinforced membranes, but the high SO_3 groups concentration gives the membrane a proton conductivity of 0.15 S/cm (https://www.fumatech.com/EN/Onlineshop/Products%2bof%2bLSC-PFSA%2bpolymers/index.html, 2020).

The most interesting group of SSC-PFSA membranes are Aquivion, melt-extruded membranes based on SSC copolymer TFE-SFVE developed by Solvay. The most recent developed membranes have an EW of 870 and 980 with high proton conductivity 0.16 S/cm in standard conditions. These membranes are chemically stabilized to increase their lifetime (https://www.solvay.com/en/brands/aquivion-pfsa/technical-data-sheets, 2020).

The better performance of low EW SSC membranes in hydrogen FCs is chiefly the result of the higher IEC and more effective solvent permeation from the cathode to the anode side in the opposite direction of the electroosmotic drag.

Others fluorinate backbones such as poly(tetrafluoroethylene-*co*-hexafluoropropylene) (FEP), polyvinylidene fluoride (PVDF), poly(ethylene-alt-tetrafluoroethylene) (ETFE), which are manufactured in large quantities, can be modified into proton-conducting membranes. Radiation grafting for synthesis of proton-conducting membranes has been extensively studied by Scherer et al. of Paul Scherrer Institute in Switzerland (Rouilly, Kötz, Haas, Scherer, & Chapiró, 1993). Both ETFE and FEP have been used as base films, the FEP-based polymers appear to be the most advanced owing to the higher resistance to radical attack. Proton conduction is obtained by sulfonating the grafted membrane with chlorosulfonic acid, preferably at room temperature to maintain the mechanical strength of the polymer. The conductivity of the resulting membranes is 0.03 S/cm at 40°C and 0.04 S/cm at 60°C (Gubler et al., 2004).

Poly(α,β,β-trifluorostyrene) was studied at Ballard. The polymer membrane was considered to be a possible alternative for Nafion-type membranes, with sufficient chemical stability because of its fluorinated backbone and enabling FC performance comparable to or even better than using Nafion membranes. The membranes are also known as BAM-3G. FC testing with poly(α,β,β-trifluorostyrene) has proved its suitability at low-temperature FC conditions, that is, operation at 70°C–85°C and water-saturated conditions (Steck & Stone 1997).

One of the promising ways to develop new membranes consists in modifying PFSA by incorporating nanoparticles in order to enhance water retention at high operating temperature (>90°C) and low RH. Different inorganic compounds such as TiO_2, SiO_2, ZrO_2, and sulfated ZrO_2 and CeO_2 in different forms (powders, nanotubes, etc.) and/or novel materials with enhanced water retention capacity and/or proton conductivity (zeolites, MWCNTs, graphene oxide-GO, etc.) (Baglio et al., 2004; Chalkova, Fedkin, Wesolowski, & Lvov, 2005; Herring, 2006; Licoccia & Traversa 2006; Saccà et al., 2005; Saccà, Carbone, Gatto, Pedicini, & Passalacqua, 2018; Saccà et al., 2019; Sahu et al., 2007) were added to PFSA membranes to improve their properties under unconventional conditions ($T_{cell} > 80°C$; $RH < 100\%$). This concept was introduced by Watanabe (1995), allowing the cell to operate at 80°C in drastic humidification conditions. Furthermore, the direct insertion of materials with higher proton conductivity, such as ZrP, CeHSO4, ZrSPP, HPA (Alberti et al., 2007; Costamagna, Yang, Bocarsly, & Srinivasan, 2002; Kim et al., 2004; Ramani, Kunz, & Fenton, 2004) into PFSA membranes was also studied.

15.3.1.2 *Nonperfluorosulfonic acid membranes*

Aromatic hydrocarbons opportunely sulfonated can be used to produce promising alternatives to PFSA membranes for operation at elevated temperatures. These materials are generally composed of polyaromatic or polyetherocyclic repeat units that have a good thermal stability, low cost, simple processability, broad chemical tunability, even after a suitable functionalization, necessary to make them into a proton conductor and water absorbent.

The most studied polymers are polyarylenethers (PAEs), polyphenylenes (PPs), and polyimides PIs.

PAE materials such as polyetheretherketone (PEEK), polyether sulfone (PES), and their derivatives are attractive for use in PEMs because of their well-known oxidative and hydrolytic stability under harsh conditions and because many different chemical structures, including partially fluorinated materials, are possible. PAE polymers can be directly sulphonated by sulfuric and chlorosulfonic acids, sulfur trioxide, and complex formation with trimethylsilylchlorosulfonic acid, etc.

Recently, the sulfonated polyetherketone (sPEK) family has been the most widely studied; they are thermoplastic polymers composed of a nonperfluorinated aromatic chain, in which the 1,4-substituted phenilic groups are divided by ethers (E) and ketones (K) groups to give ether-rich polymers such as PEEK and PEEKK (polyetheretherketoneketone) or ketone-rich semicrystalline thermoplastic polymers such as PEK. Although the hydrocarbon membranes have certain good properties, they exhibit several drawbacks such as weak mechanical properties, low proton conductivity, and low fuel-cell performance. In this class of materials, the sulfonation degree depends on the number of aromatic rings linked to oxygen atoms that are available for sulfonation; thus polymers containing more E than K groups have a higher number of active sites for sulfonation. The most widely used and intensively studied poly (arylene ether)-based membrane is sPEEK because its original polymer, PEEK, is commercially available and has important properties similar to those of many classes of aromatic polymers. Proton-conducting membranes prepared from chloro-sulfonated (sulfonation degree SD 60%) PEEK with cardo group polymers (sPEEKWC) have shown a proton conductivity value equal to 6.7 10^{-2} S/cm at 120°C (Iulianelli et al., 2013). sPEEK with a 40%–60% of SD has an excessive swelling at $T > 80°C$, causing a reduction in mechanical properties and of its lifetime; on the contrary, the proton conductivity significantly increases. To balance these effects, the introduction of inorganic/organic components, acting as a mechanical reinforcement and as an additive hydrophilic agent, have been largely investigated. sPEEK membranes with different SD (35%–52%) demonstrated that the W_{up} and the swelling at 100°C increase about 100 times passing from 35% to 52% of sulfonation and the proton conductivity from 6.7×10^{-2} to 1.3×10^{-1} S/cm. By mixing different weight percentage of 3-aminopropyl functionalized silica ($SiO-NH_2$) the swelling was reduced through interactions between sulfonic and amino groups (Carbone, Pedicini, Saccà, Gatto, & Passalacqua, 2008). This effect is highlighted in membranes with 52% SD, where the reduction in water uptake and swelling does not interfere with the proton conduction. Recently, new SPEEK nanocomposite membranes were investigated by Simari et al., which contained highly ionic silica-layered nanoadditives. The presence of ionic-layered additives enhances the mechanical strength, water retention capacity, and transport properties. The membrane with 5 wt.% of nanoadditive showed a 10 times higher proton conductivity rate (0.0128 S/cm) under very harsh operative conditions (i.e., 90°C and 30% RH), with respect to pristine sPEEK (Simari, Enotiadis, & Nicotera, 2020).

Recently, Wang, Xu, Zang, and Wang (2019) used sulfonated poly(arylene ether ketone sulfone) containing amino groups (Am-sPAEKS) as the organic matrix and a titanium dioxide rich in the amino acid groups ($L-TiO_2$) as inorganic filler. The membrane with the highest $L-TiO_2$ doping content possessed the highest proton conductivities: 0.0323 S/cm at 30°C S cm^{-1} and 0.0879 S/cm 90°C S cm^{-1}.

Polyether sulfone polymers generally consist of aromatic rings alternated with ether and sulfone ($-SO_2-$) linkages. The most investigated polymers are PES composed of the basic units and polysulfone (PSU) containing 2-propylene in addition to ether and sulfone groups. Direct sulfonation, using chlorosulfonic or sulfuric acid as both solvent and sulfonating agent, can degrade the polymer backbone-producing membranes with poor properties and low thermal stability. To improve the mechanical stability and reduce the tendency of sPES and sPSU to swell in hot water has led to the development of methods based on cross-linking. On the contrary, direct sulfonation using the trimethylsilylchlorosulfonic acid complex produces more stable polymers, but for improving the stability and the membrane-forming properties, the salt form (Na^+) is preferred (Pedicini et al., 2008). Sulfonated polysulfone/zirconium hydrogen phosphate (sPSf/ZrP) composite membranes with different sulfonation degrees (20%–42%) and a constant concentration of ZrP (2.5%) have been found to be highly promising in PEFC using direct methanol as fuel (DM-PEFC) owing to their good characteristics of proton conductivity, water uptake, thermal resistance, oxidative stability, and methanol suppression (Ozden, Ercelik, Devrim, Colpan, & Hamdullahpur, 2017).

Sulfonated polyimides (SPI) polymers, especially sulfonated six-membered ring (naphthalenic) polyimides, have been proposed as promising candidates for PEMs owing to their excellent chemical and thermal stability, high mechanical strength, good film-forming ability, and low fuel gas (or liquid) crossover. Most of their physicochemical properties are different from Nafion. All of them yield a performance comparable to that of Nafion for direct hydrogen PEFC (DH-PEFC) or DM-PEFC, but they have a low water stability because of high sensitivity of the imide rings to hydrolysis under moist conditions and moderate temperatures ($>70°C$) (Zhang & Shen 2012). The conductivity of an SPI generally increases with the increase in the degree of hydration; however, its performance failed after 70 hours of operation. The most relevant problem is their thermo-hydrolytic stability. In fact, the hydrolysis of the SPI sequence leads to chain scissions with a consequent loss in mechanical properties. Limited data is available and the relationship between the ex situ tests and in situ FC characterization has not been reported (Marestin, Gebel, Diat, & Mercier, 2008)

Several polymers have been investigated in the PPs group. The direct sulfonation of the polyphenylquinoxalines (sPPQ) was carried out for the first time by Ballard Advanced Materials to give sulfonated PPQ at various EWs with a wide range of sulfonation levels. The prepared membranes have shown good mechanical properties in both the hydrated and dehydrated states, and during FC tests at 70°C, the performance was comparable to Nafion 117 (Dobrovol'skii et al., 2007). However, these membranes showed limited durability; in fact, only 350 hours of time tests were reported, because of progressive membrane brittleness. The second generation of BAM membranes involved several polymers including the sulfonated poly(substituted-phenylene oxide) (sPPO). For this membrane class, the lifetime increased compared to the previous sPPQ class but limited to 500 hours (Roziere & Jones 2003). Polyphenylensulfide also has been investigated as material for proton-conducting membranes. This polymer can be sulfonated in concentrated sulfuric acid. Despite the high proton conductivity, the system is water soluble at a degree of sulfonation less than 30%, so it is not indicated in the utilization of FC devices. Miyake et al. (2017) reported a novel polyphenylene-based PEM (sPP-QP) that exhibits one of the highest proton conductivities (0.22 S/cm) among the reported fluorine-free aromatic ionomer membranes even under low-humidity conditions (40% RH).

Polyphosphazenes (PPZs) are hybrid polymers that contain a backbone of alternating P and N atoms with two organic, inorganic, or organometallic side groups attached to each phosphorus atom. Some of the most thermally and chemically stable polyphosphazenes have aryloxy side groups that can be functionalized to introduce acidic units, including carboxylic, sulfonic, and phosphonic acid groups. The conductivities of the cross-linked and noncross-linked sPPZ membranes were found to be essentially identical, even if the water content was lower than Nafion 117 membranes (Wycisk & Pintauro 2008). Cross-linked membranes composed of highly sulfonated PEEK as the acidic component and PPZ backbone as the basic component were found to limit the water uptake and swelling. The introduction of sulfonated single-walled carbon nanotubes produced a proton conductivity of 0.132 S/cm at 80°C (Gao et al., 2017).

Most of the developed polymers have amorphous morphologies. Recently, it has been demonstrated that it is possible to reach high proton conductivity with a well-ordered, highly crystalline morphology. By placing sulfonic acid groups in determined positions in linear polyethylene crystalline structure containing water layers of subnanometer thickness was obtained. The resulting structure has proton conductivity very similar to Nafion 117 at 40°C and RH over 60%. The hydration number, λ, is comparable to Nafion 117 at low RH, while Nafion has higher λ at high RH (Trigg et al., 2018).

15.3.1.3 Polymers functionalized with phosphoric acid or ionic liquids

When sulfonated membranes are used for PEFC, the humidification is crucial for proton transport; therefore all the considered membranes are efficient at T not exceeding 130°C–140°C and RH ranging 20%–100%. To obtain more suitable membranes for high-temperature PEFCs ($T > 150$°C) and/or anhydrous or nearly anhydrous conditions, the replacement of water with nonvolatile solvents such as PA or ILs should be considered.

Acid–base complexes have been considered to be an alternative for membranes that can maintain high conductivity at elevated temperatures without undergoing dehydration effects. Generally, the acid–base complexes for FC membranes imply the incorporation of an acid component into a thermostable polymers containing basic sites (amine or amide, ether groups) such as polybenzimidazole (PBI), poly(ethylene oxide) (PEO), poly(vinyl alcohol) (PVA), poly(vinylpyrrolidone) (PVP), and others. This method has the advantage of producing membranes able to operate at a higher temperature (160°C for PBI) in nearly anhydrous conditions, but the main problem is the cleavage of the dopant acid during FC operations. The PBI/H_3PO_4 complex is the most interesting; the conductivity of doped PBI is independent of humidity, even if such complexes are sensitive to the doping level and temperature. At 450% doping and a temperature 165°C, the conductivity of PBI membrane was about 4.6×10^{-2} S/cm. It was also observed that at very high levels of doping (around 1600%), the conductivity could reach 0.13 S/cm (Smitha, Sridhar, & Khan, 2005). A way to improve the PA content and PA retention ability maintaining a good stability of PBI membranes is to form blend membranes. PBI–PAA [poly(acrylic acid)] blend membranes showed proton conductivities of 0.005 S/cm at 150°C in the anhydrous state and 0.5 S/cm at 150°C with increasing PA concentration (Taherkhani, Abdollahi, & Sharif, 2019). Blends of sulfonated polysulfones and PBI doped with H_3PO_4 showed improved mechanical properties and conductivities above 10^{-2} S/cm at 160°C and 80% RH, higher than for acid-doped PBI membranes under the same condition (Hasiotis, Deimede, & Kontoyannis, 2001).

During the last few years, several studies have been dedicated toward ILs as electrolytes in PEFCs to overcome the problems of volatile electrolytes. ILs are organic salts with melting points below or equal to room temperature. They have negligible volatility, nonflammability, high thermal and electrochemical stability, and ionic conductivity even under anhydrous conditions. Two general types of ionic liquids exist: aprotic and protic. Protic ionic liquids have a mobile proton located on the cation, and the reactivity of this active proton makes them suitable as electrolytes in FC applications The conductivities of ILs at RT range from 1.0 10^{-4} to 1.8 10^{-2} S/cm (Díaz, Ortiz, & Ortiz, 2014). As solidified electrolytes are preferable when employing ionic liquids as electrolytes, one approach is mixing them to a polymer. However, this technique often results in compromises between the desirable room temperature IL properties and the mechanical strength of the membranes. One of the most widely used polymers in polymer/ionic liquids blends is PVDF and its copolymers (Lee, Nohira, & Hagiwara, 2007). However, the most direct and innovative technique for the implementation of ILs in FCs is the use of polymerizable ILs. Polymerized ILs (PILs) include a wide variety of structures; they can be developed to form different systems, such as polycationic ILs, polyanionic ILs, polymer complexes, copolymers, and poly(zwitterion)s, depending on the final application of the polyelectrolyte. However, a progressive release of the PIL could affect the long-term operation of the PIL-based membranes. Similarly, Chu and coworkers (Chu, Lin, Yan, Qiu, & Lu, 2011) developed anhydrous proton-conducting membranes based on in situ cross-linking of polymerizable oils (styrene/acrylonitrile) containing polyamidoamine (PAMAM) dendrimer–based macromolecular protic ILs. These membranes showed a proton conductivity of 1.2 10^{-2} S/cm at 160°C and have better PIL retention.

15.3.2 Anionic exchange membranes

Several kinds of polymers are used for anionic membrane preparation and only few of them are specific for FC applications. Generally, the commercial products are aimed at a broader field of application such as electrolysis, electrodialysis, desalination, and batteries. They are composed of polyaromatic or polyaliphatic backbone with a quaternary ammonium functional group suitable for the ionic transport. Table 15.2 summarizes the commercially available membranes and some topic characteristics.

It is evident that the same manufacturer produces membranes based on the same polymer but with very different characteristics. For example FAA-3 membranes are developed for a variety of applications and are definitely used for AEMFC one. They show a wide range of thickness and IEC and consequently different conductivity.

Commercial FAA-3–50 membrane based on polyaromatic backbone is often used as a reference in several papers and several properties are reported in literature. For example, Marino, Melchior, Wohlfarth, and Kreuer (2014) reported that hydration plays a key role in the hydroxide conductivities and values of about 2 10^{-3} S/cm at water fraction 0.2 versus 5 10^{-2} S/cm at water fraction of 0.8 were measured.

TABLE 15.2 Commercially available anionic membranes.

Manufacturer	Membrane	(EW)	Thickness (μm)	IEC, meq/g	Ref.
FumaTech	Fumasep FAA-3–20	606–540	18–22	1.65–1.85	(https://www.fumatech.com/EN/Download/index.html, 2020)
	Fumasep FAA-3–50	540	45–55	1.85	
	Fumasep FAA-3-PK-75	833–714	70–80	1.2–1.4	
	Fumasep FAA-3-PK-130	909–714	110–130	1.1–1.4	
	Fumasep FAAM-15	–	13–17	–	
	Fumasep FAS-PET-75	–	70–80	–	
	Fumasep FAS-PET-130	1000–769	110–130	1.0–1.3	
	Fumasep FAB-PK-130	1428–1000	110–140	0.7–1.0	
	Fumasep FAS-30	625–500	25–35	1.6–2.0	
	Fumasep FAS-50	625–500	45–55	1.6–2.0	
	Fumasep FAD-PET-75	500–434	60–80	2.0–2.3	
	Fumasep FAD-55	500–400	50–60	2.0–2.5	
Carbon dioxide	Sustanion X37–50	909	50	1.1	(Kaczur, Yang, Liu, Sajjad, & Masel, 2018)
Ionomr	AemionAF1-HNN8–50-X	476–400	50	2.1–2.5	(https://ionomr.com/solutions/aemion, 2020)
	AemionAF1-HNN8–25-X	476–400	25	2.1–2.5	
	AemionAF1-HNN5–50-X	714–588	50	1.4–1.7	
	AemionAF1-HNN5–25-X	714–588	25	1.4–1.7	
	AemionAP1-HNN8–00-X	476–400	–	2.1–2.5	
Astom	Neosepta ASE	–	15	–	(http://www.astom-corp.jp/en/product/images/astom_hyo.pdf, 2020)
	Neosepta AHA	–	22	–	
	Neosepta ACS	–	13	–	
	Neosepta AFX	–	17	–	
	Neosepta ACM	–	11	–	
Orion	Durion	457	30	2.19	(https://www.agec.co.jp/eng/product/selemion, 2020)
Asahi Glass Co. Ltd.	Selemion AAV	–	120	–	(https://www.agec.co.jp/eng/product/selemion/membrane.html, 2020)
	Selemion ASVN	–	100	–	
	Selemion AHO	–	300	–	

Another paper Carbone, Campagna Zignani, Gatto, Trocino, and Aricò (2020) reports the influence of chemical treatment on the membrane properties. It was hypothesized that the ions confined in the hydrophilic domains have the same concentration of the solution, since the OH^- concentration in KOH solution was found to be close to the concentration of the solution in which it was immersed. This implies a good dissociation of ions promoting the diffusion into the water channels of the solid membrane. In fact, the diffusion coefficient value of 7.17 10^{-6} cm^2/s for the membrane sample after KOH uptake is only one order of magnitude lower than the value 5.3 10^{-5} cm^2/s reported for OH^- solutions under infinite dilution. The OH^- conductivity of 55 mS/cm was reached at 100°C, 100% RH, and activation energy of 27.4 kJ/mol was calculated meaning a Grotthuss mechanism.

Other kind of polymers produced by Ionomr are based on a benzimidazole backbone, and it was reported a maximum conductivity of 23 mS/cm for OH^- ions at 95% RH and 30°C and an activation energy of 25–26 kJ/mol, (Wright et al., 2016).

Despite the few varieties of commercial products, there are a lot of research groups moving on the development of stable anionic polymers to be used as a polyelectrolyte membrane. The polymer backbones are composed of different alkyl or aromatic polymers with a linear main chain or linked to a lateral chain. The most used functional groups are the quaternary ammonium (QA) linked to the main chain or the secondary one while few papers studied the functional groups based on phosphonium and ligand-metal complexes (Hagesteijn, Jiang, & Ladewig, 2018; Maurya, Shin, Kim, & Moon, 2015; Oshiba, Hiura, Suzuki, & Yamaguchi, 2017; Varcoe et al., 2014; You, Noonan, & Coates, 2020). Among QA, the most investigated are based on tetraalkylammonium, cyclic ammonium, tertiary diamines, imidazolium or (Benz) imidazolium, guanidinium, pyridinium, etc. The main drawback of QA-AEMs is poor chemical stability due to OH^- attack leading to ammonium group degradation and reduction of IEC (Hagesteijn et al., 2018). Moreover, the detrimental effect of carbonate and bicarbonate formation, when CO_2 from the ambient reacts with OH^-, leads to a reduction of IEC and conductivity (Wright et al., 2016; Ziv & Dekel 2018).

Examples of used polymers are polystyrene (PSt) cross-linked with divinylbenzene (DVB), PPO, polyarylene sulfone, PEEK, PBI, polyether imide, copolymers from vinyl monomers, and grafted perfluoropolymers.

Considering the high toxicity of reagents used for the functionalization reactions, the research is moving toward relatively green and environment-friendly synthesis based on the grafting process.

A recent study reported the equilibrium state and kinetics of vapor phase water uptake for AEMs considering different backbones (fluorinated and hydrocarbon-based backbones) and different functional groups (various cations as part of the backbone or as pendant groups) (Zheng et al., 2018). The variation in behavior found in this paper was explained considering the influence of RH on kinetics associated to a combination of diffusion, interfacial transport and swelling processes.

PVA is a polyhydroxypolymer and is considered one of the alternative used polymers with properties of easy preparation, biodegradability, and solubility in water. It is proposed as a "green" product. In addition, the presence of hydroxyl groups gives it good chemical stability and hydrophilicity. The PVA-based membranes need a cross-linking degree to avoid solubilization. Different cross-linkers were reported in literature, in particular potassium hydroxide, poly(acrylamide-*co*-diallyl dimethyl ammonium chloride), and poly(diallyl dimethyl ammonium chloride) that produces thermally stable membranes with ion conductivity ranging from $2.75\ 10^{-4}$ to 0.02 S/cm at room temperature (Pan, An, Zhao, & Tang, 2018).

Another class of polymers is the poly(arylene ether)s, including PPO, polyketones, polysulfones, and others.

A wide literature is based on this class and massive efforts have been made to improve the properties of the membranes produced.

A variety of functional groups were inserted in PPO-based membranes (Maurya et al., 2015), in particular membranes based on PPO-polyvinylbenzyl (PVB)-trimethyl ammonium with a IEC ranging 0.5–1.55 meq/g and a conductivity of 4–31 mS/cm at 25°C were reported. The change in the functional group in PPO-guanidinium produced a membrane with 0.37−2.69 meq/g and a corresponding conductivity of 11−71 mS/cm at the same temperature. Other compositions were investigated such as PPO-cross-linked-1,4-diazabicyclo[2.2.2]octane (DABCO) and PPO-benzimidazolium with membranes having IEC of 0.6−1.1 meq/g and 0.63−2.21 meq/g, respectively and a corresponding conductivity of 0.9−5.4 mS/cm and 10−37 mS/cm at 25°C.

Always considering the poly(arylene ether)s class, a paper reported the synthesis of aromatic multiblock copolymer membrane, based on poly(aryleneether)s with quaternized ammonio-substituted fluorene groups. High hydroxide-ion conductivity of 144 mS/cm at 80°C was obtained and retained for 5000 hours (Pan et al., 2018).

Research and studies on polyketones-based membranes have considered different functional groups. In particular, Maurya et al. (2015) reported PPEK-imidazolium with an IEC of 1.52−2.63 meq/g and a conductivity at 30°C of 28 mS/cm; PAEK-trimethyl ammonium with 1.32−1.46 meq/g and a conductivity of 12−23 mS/cm at 20°C; PEEK-trimethyl ammonium having 0.43−1.35 meq/g and a corresponding conductivity at 30°C of 0.5−12 mS/cm; PEEK-DABCO with 0.86−1.69 meq/g IEC and a conductivity of 18.4−47.8 mS/cm at 25°C; PEEK-imidazolium with a IEC ranging 1.56−2.24 meq/g and anion conductivity of 15−52 mS/cm at 20°C.

Another most studied subclass of poly(arylene ether)s is that of poly(ether sulfone)s. Also in this case a wide literature reports several functional groups bonded to different main chains producing membranes with IEC and conductivity ranging in a wide range.

Poly(bisphenol-A sulfone) (PSF)-based membranes are chemically and mechanically stable under highly basic conditions. Some of the functional groups comprise quaternary benzyl trimethylammonium (PSF-TMA^+OH^-), quaternary benzyl quinuclidum (PSF-$ABCO^+OH^-$), and quaternary benzyl 1-methylimidazolium (PSF-1 M^+OH^-). It was reported the same theoretical IEC of 1.8 mmol/g and an ionic conductivity at 50°C in liquid water of 17, 14, and 13 mS/cm, for PSF-TMA^+OH^-, PSF-$ABCO^+OH^-$, and PSF-1M^+OH^- AEM, respectively, (Kyung Cho et al., 2017)

Different properties of PES-imidazolium-based membranes are reported in the paper (Maurya et al., 2015) with an IEC of 1.45 meq/g and conductivity of 0.3 mS/cm at 20°C, while for PS-imidazolium an IEC of 1.39−2.46 meq/g and a conductivity of 16.1−20.7 mS/cm at the same temperature were reported (Maurya et al., 2015).

The transport properties of hydroxide conducting membranes were explored (Merle, Wessling, & Nijmeijer, 2011) controlling the degree of functionalization of polysulfone. The influence of the amount of ammonium groups and consequently the IEC, on parameters such as water uptake, ion conductivity, effective water self-diffusion coefficient, and pressure-driven water permeability was studied. It was concluded that contrary to the hydronium ions that are integrating part of the hydrogen bond of water, the hydroxyl anion tends to have stable solvation shells that reorganize the solvent molecules. In fact, it was reported that the transport coefficients of protons ($DH^+ = 9.3 \times 10^{-9}\ m^2/s$) and hydroxyl ions ($DOH^- = 5.3 \times 10^{-9}\ m^2/s$) measured in liquid water at 25°C and the transport processes based on diffusion or migration and Grotthusss, are comparable.

Another paper (Carbone et al., 2020) reports the synthesis of quaternized polysulfone via chloromethylation reaction. The influence of two different quaternary ammonium groups, trimethylamine (TMA) and triethylamine (TEA) on the membrane properties was studied and different properties of developed membranes were evidenced. The TMA-based sample with a higher IEC and hydrophilicity showed higher anion conductivity than TEA-based one with a maximum value of 44 mS/cm against 39.5 mS/cm at 70°C, respectively. The activation energies calculated for both membranes suggest a predominance of the Grotthuss conduction mechanism.

Another class of polymers different from polyaromatics is based on linear polymers comprising polyolefins and perluoropolymers with pendant chains attached via chemical reaction or grafting processes. A series of polyethylene backbones and imidazolium cations were prepared, and the corresponding membranes demonstrated the highest hydroxide conductivity of 49 mS/cm at 22°C and 134 mS/cm at 80°C with long-term alkaline stability (You, Padgett, MacMillan, Muller, & Coates, 2019).

In the class of fluorinated polymers, different combinations and results were reported (Maurya et al., 2015): PTFE/polyepichlorhydrin (PECH)-imidazolium with 1.31–1.64 meq/g of IEC, showed a conductivity of 14–18 mS/cm at 30°C; ETFE-PVB-trimethyl ammonium with 1.03 meq/g reached 27 mS/cm at 20°C; FEP-PVB-trimethyl ammonium having a IEC ranging 0.71–0.96 meq/g possessed a conductivity of 10–20 mS/cm; ETFE/PVB-DABCO-trimethyl ammonium with 1.67–2.11 meq/g had a conductivity at 30°C in the range 26–39 mS/cm.

The study of the last class of membranes based on the grafting process reported that the change of the grafting degree and the cross-linking degree is an important strategy to control the properties of the final membrane. In particular, it was studied a series of anion-exchange membranes obtained via chloromethylation and quaternization of the grafted copolymer films composed of styrene and UV-activated polymethylpentene. It was found that membranes with a higher content of the functionalized polymer demonstrate higher IEC, conductivity and swelling. For example, the hydration number increases twice by increasing of the grafting degree from 30 to 120. An increase in IEC from 1.05 to 1.90 mmol/g produces an increase in specific conductivity of the membranes from 7 to 14 mS/cm. Golubenko, Van der Bruggen, and Yaroslavtsev (2019)

15.4 Conclusion and future trends

FCs are an attractive power source because of their high efficiency and low emissions. Among the FCs types, the most interesting and recently studied is PEFC. In PEFCs, a solid IEM, which has the ability of ions transferring, is used as polymeric electrolyte. IEMs are typically composed of hydrophobic substrates, immobilized ion-functionalized groups, and movable counter-ions. Depending on the type of ionic groups, IEMs are broadly classified into PEMs for PEMFC and AEMs for AEMFC. The main transport phenomena that occur within an IEM are the transport of water and ions, and the two phenomena are strictly related. The maintenance of an optimal level of water is a critical issue in the membrane because sufficient water is needed to maintain the high conductivity. In fact the ions transport depends on the amount of water contained within the membrane, and the performance of the FC is directly controlled from those phenomena. There are four different types of water transport across the PEM: the back diffusion, the electroosmotic drag, the hydraulic permeation, and the thermo-osmosis flux. The first two are the principal phenomena and are strongly interrelated. The proton conduction is fundamental for PEMs.

Two proton transport mechanisms are widely accepted: the vehicular and Grotthuss mechanisms. The vehicular mechanism is usually applied in conditions where high humidity and low temperature are combined. Conditions involving low humidity and elevated temperatures are better modeled by the Grotthuss mechanism, where the proton transport is based on a so-called proton hopping process.

The transport of hydroxide ions is the base mechanism of the AEMFC. The main transport mechanisms for hydroxide ions through anion-exchange include molecular diffusion or a masse diffusion, structural diffusion or Grotthuss mechanism and surface-site hopping; migration and convection, but it is considered that Grotthuss mechanism has a predominant contribution to hydroxide mobility through hydrated membranes.

The PEMs exhibit excellent stability and processability, allowing the flexible design of electrochemical devices, especially FCs. Several approaches were taken and the developed membranes can be classified into three main classes: PFSA membranes; non-PFSA membranes and polymers functionalized with PA or ILs.

Although several kinds of polymers are used for anionic membranes preparation, only few of them are specific for FC applications. They are composed of polyaromatic or polyaliphatic backbone with a quaternary ammonium functional groups suitable for the ionic transport. Generally, the functionalization reactions to obtain more efficient AEM involve high toxicity reagent for this reason the research is moving toward relatively green and environment-friendly synthesis based on the grafting process.

References

Agmon, N. (2000). Mechanism of hydroxide mobility. *Chemical Physics Letters*, *319*, 247–252.

Agmon, N. (1995). The Grotthuss mechanism. *Chemical Physics Letters*, *244*, 456–462.

Alberti, G., Casciola, M., Capitani, D., Donnadio, A., Narducci, R., Pica, M., & Sganappa, M. (2007). Novel Nafion–zirconium phosphate nanocomposite membranes with enhanced stability of proton conductivity at medium temperature and high relative humidity. *Electrochimica Acta*, *52*, 8125–8132.

Baglio, V., DiBlasi, A., Aricò, A. S., Antonucci, V., Antonucci, P. L., Serraino, F., ... Traversa, E. (2004). Influence of TiO_2 nanometric filler on the behaviour of a composite membrane for applications in direct methanol fuel cells. *Journal of New Material and Electrochemical System*, *7*, 275.

Bahlakeh, G., Hasani-Sadrabadi, M. M., & Jacob, K. I. (2016). Exploring the hydrated microstructure and molecular mobility in blend polyelectrolyte membranes by quantum mechanics and molecular dynamics simulations. *RSC Advances*, *6*, 35517–35526.

Barbir, F. (2005). *PEM Fuel Cells: Theory and Practice* (pp. 75–84). Burlington, MA: Elsevier Academic Press.

de Bruijn, F. A., Makkus, R. C., Mallant, R. K. A. M., & Janssen, G. J. M. (2007). Materials for state-of-the-art PEM fuel cells, and their suitability for operation above 100°C. *Advances in Fuel Cells*, *1*, 235–336.

De Caluwe, S. C., Baker, A. M., Bhargava, P., Fischer, J. E., & Dura, J. A. (2018). Structure-property relationships at Nafion thin-film interfaces: Thickness effects on hydration and anisotropic ion transport. *Nano Energy*, *46*, 91–100.

Carbone, A., Campagna Zignani, S., Gatto, I., Trocino, S., & Aricò, A. S. (2020). Assessment of the FAA3–50 polymer electrolyte in combination with a $NiMn_2O_4$ anode catalyst for anion exchange membrane water electrolysis. *International Journal of Hydrogen Energy*, *45*(16), 9285–9292.

Carbone, A., Gaeta, M., Romeo, A., Portale, G., Pedicini, R., Gatto, I., & Castriciano, M. A. (2018). Porphyrin/sPEEK membranes with improved conductivity and durability for PEFC technology. *ACS Applied Energy Materials*, *1*, 1664–1673.

Carbone, A., Pedicini, R., Gatto, I., Saccà, A., Patti, A., Bella, G., & Cordaro, M. (2020). Development of polymeric membranes based on quaternized polysulfones for AMFC applications. *Polymers*, *12*(2), 283.

Carbone, A., Pedicini, R., Saccà, A., Gatto, I., & Passalacqua, E. (2008). Composite S-PEEK membranes for medium temperature polymer electrolyte fuel cells. *Journal of Power Sources, 178*, 661–666.

Casciola, M., Donnadio, A., & Sassi, P. (2013). A critical investigation of the effect of hygrothermal cycling on hydration and in-plane/through-plane proton conductivity of Nafion 117 at medium temperature (70–130°C). *Journal of Power Sources, 235*, 129–134.

Castañeda Ramírez, S., & Ribadeneira Paz, R. (2018). Hydroxide transport in anion-exchange membranes for alkaline fuel cells. In S. Karakuş (Ed.), *New Trends in Ion Exchange Studies*. IntechOpen. Available from 10.5772/intechopen.77148.

Chalkova, E., Fedkin, M. V., Wesolowski, D. J., & Lvov, S. (2005). Effect of TiO_2 surface properties on performance of Nafion-based composite membranes in high temperature and low relative humidity PEM fuel cells. *Journal of the Electrochemical Society, 152*, A1742.

Chen, C., Tse, Y.-L. S., Lindberg, G. E., Knight, C., & Voth, G. A. (2016). Hydroxide solvation and transport in anion exchange membranes. *Journal of the American Chemical Society, 138*, 991–1000.

Chu, F., Lin, B., Yan, F., Qiu, L., & Lu, J. (2011). Macromolecular protic ionic liquid-based proton-conducting membranes for an hydrous proton exchange membrane, application. *Journal of Power Sources, 196*, 7979.

Costamagna, P., Yang, C., Bocarsly, A. B., & Srinivasan, S. (2002). Nafion® 115/zirconium phosphate composite membranes for operation of PEMFCs above 100°C. *Electrochimica Acta, 47*, 1023.

Cukierman, S. (2006). Et tu, Grotthuss! and other unfinished stories. *Biochimica et Biophysica Acta, 1757*, 876–885.

Dadda, B., Abboudi, S., Zarrit, R., & Ghezal, A. (2014). Heat and mass transfer influence on potential variation in a PEMFC membrane. *International Journal of Hydrogen Energy, 39*, 15238–15245.

Dai, W., Wang, H., Yuan, X. Z., Martin, J. J., Yang, D., Qiao, J., & Ma, J. (2009). A review on water balance in the membrane electrode assembly of proton exchange membrane fuel cells. *International Journal of Hydrogen Energy, 34*, 9461–9478.

Devia, N., Raya, S., Shukla, A., Bhat, S. D., & Pesala, B. (2019). Non-invasive macroscopic and molecular quantification of water in Nafion® and SPEEK proton exchange membranes using terahertz spectroscopy. *Journal of Membrane Science, 588*, 117183–117193.

Dippel, T. H., & Kreuer, K. D. (1991). Proton transport mechanism in concentrated aqueous solutions and solid hydrates of acids. *Solid State Ionics, 46*, 3–9.

Dobrovol'skii, Y. A., Volkov, E. V., Pisareva, A. V., Fedotov, Y. A., Likhachev, D. Y., & Rusanov, A. L. (2007). Proton-exchange membranes for hydrogen-air fuel cells. *Russian Journal of General Chemistry, 77*(4), 766–777.

Díaz, M., Ortiz, A., & Ortiz, I. (2014). Progress in the use of ionic liquids as electrolyte membranes in fuel cells. *Journal of Membrane Science, 469*, 379–396.

E. Galitskaya, A.F. Privalov, M. Weigler, M. Vogel, A. Kashin, M. Ryzhkina. NMR diffusion studies of proton-exchange membranes in wide temperature range, Journal of Membrane Science, *596* (2020) 117691–117697

Gao, S., Xu, H., Luo, T., Guo, Y., Li, Z., Ouadah, A., … Zhu, C. (2017). Novel proton conducting membranes based on cross-linked sulfonated polyphosphazenes and poly(ether ether ketone). *Journal of Membrane Science, 536*, 1–10.

Ge, S., Li, X., Yi, B., & Hsing, I.-M. (2005). Absorption, desorption, and transport of water in polymer electrolyte membranes for fuel cells. *Journal of the Electrochemical Society, 152*(6), A1149–A1157.

Golubenko, D. V., Van der Bruggen, B., & Yaroslavtsev, A. B. (2019). Novel anion exchange membrane with low ionic resistance based on chloromethylated/quaternized-grafted polystyrene for energy efficient electromembrane processes. *Journal of Applied Polymer Science, 137*, 48656.

Gubler, L., Kuhn, H., Schmidt, T. J., Scherer, G. G., Brack, H. P., & Simbeck, K. (2004). Performance and durability of membrane electrode assemblies based on radiation-grafted FEP-*g*-polystyrene membranes. *Fuel Cells, 4*(3), 196–207.

Guzman-Garcia, A. G., Pintauro, P. N., Verbugge, M. W., & Hill, R. F. (1990). Development of a space-charge transport model for ion exchange membranes. *AIChE Journal, 36*, 1061–1074.

Hagesteijn, K. F. L., Jiang, S., & Ladewig, B. P. (2018). A review of the synthesis and characterization of anion exchange membranes. *Journal of Materials Science, 53*, 11131–11150.

Hasiotis, C., Deimede, V., & Kontoyannis, C. (2001). New polymer electrolytes based on blends of sulfonated polysulfones with polybenzimidazole. *Electrochimica Acta, 46*, 2401–2406.

Herring, A. M. (2006). Inorganic–polymer composite membranes for proton exchange membrane fuel cells. *Polymer Reviews, 46*, 245.

He, Y. K., Wang, J. T., Zhang, H. Q., Zhang, T., Zhang, B., Cao, S. K., & Liu, J. D. (2014). Polydopamine-modified graphene oxide nanocomposite membrane for proton exchange membrane fuel cell under anhydrous conditions. *Journal of Materials Chemistry A, 2*, 9548–9558.

Hickner, M. A., Ghassemi, H., Kim, Y. S., Einsla, B. R., & McGrath, J. E. (2004). Alternative polymer systems for proton exchange membranes (PEMs). *Chemical Reviewes, 104*, 4587–4611.

Hickner, M. A., Herring, A. M., & Coughlin, E. B. (2013). Anion exchange membranes: Current status and moving forward. *Journal of Polymer Science, PartB: Polymer Physics, 51*, 1727–1735.

Hofmann, D. W. M., Kuleshova, L., D'Aguanno, B., Di Noto, V., Negro, E., Conti, F., & Vittadello, M. (2009). Investigation of water structure in Nafion membranes by infrared spectroscopy and molecular dynamics simulation. *Journal of Physical Chemistry B, 113*, 632–639.

<https://www.fumatech.com/EN/Download/index.html> Accessed on November.2020.

Hu, J., Li, J., Xu, L., Huang, F., & Ouyang, M. (2016). Analytical calculation and evaluation of water transport through a proton exchange membrane fuel cell based on a one-dimensional model. *Energy, 111*, 869–883.

<http://www.astom-corp.jp/en/product/images/astom_hyo.pdf> Accessed on November 2020.

<https://www.nafionstore.com/Shared/Bulletins/HP-MEMBRANE.pdf> Accessed on October 2020.

<https://ionomr.com/solutions/aemion> Accessed on November 2020.

<https://nafionstore-us.americommerce.com/Shared/Bulletins/N115-N117-N1110-Product-Bulletin-Chemours.pdf> Accessed on October 2020.

<https://nafionstore-us.americommerce.com/Shared/P11_C10610_Nafion_NR-211__NR-212_P11.pdf> Accessed on October 2020.

<https://nafionstore-us.americommerce.com/Shared/P22_C10856_Nafion_XL_FuelCells_P22.pdf> Accessed on October 2020.

<https://www.agec.co.jp/eng/product/selemion/membrane.html> Accessed on November 2020.

<https://www.agec.co.jp/eng/product/selemion> Accessed on November 2020.

<https://www.fumatech.com/EN/Onlineshop/Products%2bof%2bLSC-PFSA%2bpolymers/index.html> Accessed on November 2020.

<https://www.hydrogen.energy.gov/pdfs/progress15/v_b_1_yandrasits_2015.pdf> Accessed on November 2020.

<https://www.solvay.com/en/brands/aquivion-pfsa/technical-data-sheets> Accessed on October 2020.

Ionescu, V. (2020). Water and hydrogen transport modelling through the membrane-electrode assembly of a PEM fuel cell. *Physica Scripta, 95*, 034006–034016.

Iulianelli, A., Gatto, I., Passalacqua, E., Trotta, F., Biasizzo, M., & Basile, A. (2013). Proton conducting membranes based on sulfonated PEEK-WC polymer for PEMFCs. *Journal of Hydrogen Energy, 38*, 16642–16648.

Kaczur, J. J., Yang, H., Liu, Z., Sajjad, S. D., & Masel, R. I. (2018). Carbon dioxide and water electrolysis using new alkaline stable anion membranes. *Frontiers in Chemistry, 6*, 263.

Khattra, N. S., Santare, M. H., Karlsson, A. M., Schmiedel, T., & Busby, F. C. (2014). Effect of water transport on swelling and stresses in PFSA membranes. *Fuel Cells, 15*(1), 178–188.

Kim, D. K., Choi, E. J., Song, H. H., & Kim, M. S. (2016). Experimental and numerical study on the water transport behaviour through Nafions 117 for polymer electrolyte membrane fuel cell. *Journal of Membrane Science, 497*, 194–208.

Kim, S., & Mench, M. M. (2009). Investigation of temperature-driven water transport in polymer electrolyte fuel cell: Thermo-osmosis in membranes. *Journal of Membrane Science, 328*, 113–120.

Kim, Y. T., Song, M. K., Kim, K. H., Park, S. B., Min, S. K., & Rhee, H. W. (2004). Nafion/ZrSPP composite membrane for high temperature operation of PEMFCs. *Electrochimica Acta, 50*, 645.

Kreuer, K.-D. (2013). Ion conducting membranes for fuel cells and other electro-chemical devices. *Chemical Materials, 26*, 361–380.

Kreuer, K. D. (1997). On the development of proton conducting materials for technological applications. *Solid State Ionics, 97*, 1–15.

Kreuer, K. D. (2001). On the development of proton conducting polymer membranes for hydrogen and methanol fuel cells. *Journal of Membrane Science, 185*, 29–39.

Kyung Cho, M., Lim, A., Lee, S. Y., Kim, H.-J., Yoo, S. J., Sung, Y.-E., ... Jang, J. H. (2017). A review on membranes and catalysts for anion exchange membrane water electrolysis single cells. *Journal of the Electrochemical Science Technology, 8*(3), 183–196.

Laberty-Robert, C., Valle, K., Pereira, F., & Sanchez, C. (2011). Design and properties of functional hybridorganic-inorganic membranes for fuel cells. *Chemical Society Reviews, 40*, 961–1005.

Lee, P. H., Han, S. S., & Hwang, S. S. (2008). Three-dimensional transport modeling for proton exchange membrane (PEM) fuel cell with micro parallel flow field. *Sensors, 8*, 1475–1487.

Lee, J. S., Nohira, T., & Hagiwara, R. (2007). Novel composite electrolyte membranes consisting of fluorohydrogenate ionic liquid and polymers for the unhumidified intermediate temperature fuel cell. *Journal of Power Sources, 171*, 535–539.

Licoccia, S., & Traversa, E. (2006). Increasing the operation temperature of polymer electrolyte membranes for fuel cells: From nanocomposites to hybrids. *Journal of Power Sources, 159*, 12–20.

Liso, V., & Nielsen, M. P. (2015). Modelling and validation of water hydration of PEM fuel cell membrane in dynamic operations. *ECS Transactions, 68*(3), 169–176.

Li, Q., He, R., Jensen, J. O., & Bjerrum, N. J. (2003). Approaches and recent development of polymer electrolyte membranes for fuel cells operating above 100°C. *Chemical Materials, 15*, 4896–4915.

Lopez-Chavez, E., Pena-Castaneda, Y., Gonzalez-Garcia, G., Perales-Enciso, P., Garcia-Quiroz, A., & Iran Diaz-Gongora, J. (2015). Theoretical methodology for calculating water uptake and ionic exchange capacity parameters of ionic exchange membranes with applications in fuel cells. *International Journal of Hydrogen Energy, 40*, 17316–17322.

Lu, Z., Polizos, G., Macdonald, D. D., & Manias, E. (2008). State of water in perfluorosulfonic ionomer (Nafion 117) proton exchange membranes. *Journal of the Electrochemical Society, 155*, B163.

Marestin, C., Gebel, G., Diat, O., & Mercier, R. (2008). Sulfonated polyimides. *Advanced Polymer Science, 216*, 185–258.

Marino, M. G., Melchior, J. P., Wohlfarth, A., & Kreuer, K. D. (2014). Hydroxide, halide and water transport in a model anion exchange membrane. *Journal of Membrane Science, 464*, 61–71.

Maurya, S., Shin, S.-H., Kim, Y., & Moon, S.-H. (2015). A review on recent developments of anion exchange membranes for fuel cells and redox flow batteries. *RSC Advanced, 5*, 37206.

Meng, H. (2006). A three-dimensional PEM fuel cell model with consistent treatment of water transport in MEA. *Journal of Power Sources, 162*, 426–435.

Merle, G., Wessling, M., & Nijmeijer, K. (2011). Anion exchange membranes for alkaline fuel cells: A review. *Journal of Membrane Science, 377*, 1–35.

Miyake, T., & Rolandi, M. (2016). Grotthuss mechanisms: From proton transport in proton wires to bioprotonic devices. *Journal of Physics: Condensed Matter, 28*, 023001.

Miyake, J., Taki, R., Mochizuki, T., Shimizu, R., Akiyama, R., Uchida, M., & Miyatake, K. (2017). Design of flexible polyphenylene proton-conducting membrane for next-generation fuel cells. *Science Advances, 3*, 1–8.

Nagarale, R. K., Shin, W., & Singh, P. K. (2010). Progress in ionic organic-inorganic composite membranes for fuel cell applications. *Polymer Chemistry, 1*, 388–408.

O'Dea, J. R., Economou, N. J., & Buratto, S. K. (2013). Surface morphology of Nafion at hydrated and dehydrated conditions. *Macromolecules, 46*, 2267–2274.

Oshiba, Y., Hiura, J., Suzuki, Y., & Yamaguchi, T. (2017). Improvement in the solid-state alkaline fuel cell performance through efficient water management strategies. *Journal of Power Sources, 345*, 221–226. Available from https://doi.org/10.1016/j.jpowsour.2017.01.111.

Ozden, A., Ercelik, M., Devrim, Y., Colpan, C. O., & Hamdullahpur, F. (2017). Evaluation of sulfonated polysulfone/zirconium hydrogen phosphate composite membranes for direct methanol fuel cells. *Electrochimica Acta, 256*, 196–210.

Paddison, S. J., Bender, G., Kreuer, K.-D., Nicoloso, N., & Zawodzinski, T. A. (2000). The microwave region of the dielectric spectrum of hydrated Nafion (R) and other sulfonated membranes. *Journal of New Materials for Electrochemical Systems, 293*. Available from https://www.researchgate.net/publication/238084563.

Page, K. A., Dura, J. A., Kim, S., Rowe, B. W., & Faraone, A. (2015). Neutron techniques as a probe of structure, dynamics, and transport in polyelectrolyte membranes. In G. Kearley, & V. Peterson (Eds.), *Neutron Applications in Materials for Energy. Neutron Scattering Applications and Techniques* (pp. 273–301). Cham: Springer International Publishing. Available from https://doi.org/10.1007/978-3-319-06656-1_10.

Pan, Z. F., An, L., Zhao, T. S., & Tang, Z. K. (2018). Advances and challenges in alkaline anion exchange membrane fuel cells. *Progress in Energy and Combustion Science, 66*, 141–175.

Paul, D. K., McCreery, R., & Karan, K. (2014). Proton transport property in supported Nafion nanothin films by electrochemical impedance spectroscopy. *Journal of the Electrochemical Society, 161*, F1395–F1402.

Pedicini, R., Carbone, A., Saccà, A., Gatto, I., Di Marco, G., & Passalacqua, E. (2008). Sulfonated polysulfone membranes for medium temperature in polymer electrolyte fuel cells (PEFC). *Polymer Testing, 27*, 248–259.

Peighambardoust, S. J., Rowshanzamir, S., & Amjadi, M. (2010). Review of the proton exchange membranes for fuel cell applications. *International Journal of Hydrogen Energy, 35*, 9349–9384.

Ramani, V., Kunz, H. R., & Fenton, J. M. (2004). Investigation of Nafion®/HPA composite membranes for high temperature/low relative humidity PEMFC operation. *Journal of Membrane Science, 232*, 31.

Ran, J., Wu, L., He, Y., Yang, Z., Wang, Y., Jiang, C., ... Xu, T. (2017). Ion exchange membranes: New developments and applications. *Journal of Membrane Science, 522*, 267–291.

Ran, J., Wu, L., Ru, Y., Hu, M., Din, L., & Xu, T. (2015). Anion exchange membranes (AEMs) based on poly(2,6-dimethyl-1,4-phenyleneoxide)(PPO) and its derivatives. *Polymer Chemisrty, 6*, 5809–5826.

Rouilly, M. V., Kötz, E. R., Haas, O., Scherer, G. G., & Chapiró, A. (1993). Proton exchange membranes prepared by simultaneous radiation grafting of styrene onto Teflon-FEP films. Synthesis and characterization. *Journal of Membrane Science, 81*(1–2), 89–95.

Roy, A., Hickner, M. A., Lee, H.-S., Glass, T., Paul, M., Badami, A., ... McGrath, J. E. (2017). States of water in proton exchange membranes: Part A - influence of chemical structure and composition. *Polymer, 111*, 297–306.

Roziere, J., & Jones, D. J. (2003). Non-fluorinated polymer materials for proton exchange membrane fuel cells. *Annual Review of Materials Research, 33*, 503–555.

Saccà, A., Carbone, A., Gatto, I., Pedicini, R., & Passalacqua, E. (2018). Synthesized yttria stabilised zirconia as filler in proton exchange membranes (PEMs) with enhanced stability. *Polymer Testing, 65*, 322–330.

Saccà, A., Carbone, A., Passalacqua, E., D'Epifanio, A., Licoccia, S., Traversa, E., ... Ornelas, R. (2005). Nafion–TiO_2 hybrid membranes for medium temperature polymer electrolyte fuel cells (PEFCs). *Journal of Power Sources, 152*, 16–21.

Saccà, A., Gatto, I., Carbone, A., Pedicini, R., Maisano, S., Stassi, A., & Passalacqua, E. (2019). Influence of doping level in yttria-stabilised-zirconia (YSZ) based-fillers as degradation inhibitors for proton exchange membranes fuel cells (PEMFCs) in drastic conditions. *International Journal of Hydrogen Energy, 44*, 31445–31457.

Sahu, A. K., Selvarani, G., Pitchumani, S., Sridhar, P., Shukla, A. K., & Sol-Gel, A. (2007). Modified alternative Nafion-silica composite membrane for polymer electrolyte fuel cells. *Journal of the Electrochemical Society, 154*, B123.

Saito, M., Arimura, N., Hayamizu, K., & Okada, T. (2004). Mechanisms of ion and water transport in perfluorosulfonated ionomer membranes for fuel cells. *Journal of Physical Chemistry B, 108*, 16064–16070.

Shi, S., Weber, A. Z., & Kusoglu, A. (2016). Structure/property relationship of Nafion XL composite membranes. *Journal of Membrane Science, 516*, 123–134.

Simari, C., Enotiadis, A., & Nicotera, I. (2020). Transport properties and mechanical features of sulfonated polyether ether ketone/organosilica layered materials nanocomposite membranes for fuel cell applications. *Membranes, 10*, 87.

Smitha, B., Ridhar, S. S., & Khan, A. A. (2004). Polyelectrolyte complexes of chitosan and poly(acrylic acid) as proton exchange membranes for fuel cells. *Macromolecules, 37*, 2233–2239.

Smitha, B., Sridhar, S., & Khan, A. A. (2005). Solid polymer electrolyte membranes for fuel cell applications—a review. *Journal of Membrane Science, 259*, 10–26.

Steck, A. E., & Stone, C. (1997). New materials for fuel cell and modern battery systems II. In O. Savadogo, & P. R. Roberge (Eds.), Proceedings of the second international symposium on new materials for fuel cell and modern battery systems (p. 266). Montreal, PQ, Canada: Ecole Polytechnique.

Taherkhani, Z., Abdollahi, M., & Sharif, A. (2019). Proton conducting porous membranes based on poly(benzimidazole) and poly(acrylic acid) blends for high temperature proton exchange membranes. *Solid State Ionics, 337*, 122–131.

Trigg, E. B., Gaines, T. W., Maréchal, M., Moed, D. E., Rannou, P., Wagener, K. B., ... Winey, K. I. (2018). Self-assembled highly ordered acid layers in precisely sulfonated polyethylene produce efficient proton transport. *Nature Materials, 17*, 725–731.

Uddin, K. M. S., Saha, L. K., & Oshima, N. (2014). Water transport through the membrane of PEM fuel cell. *American Journal of Computational and Applied Mathematics, 4*(6), 225–238.

Varcoe, J. R., Atanassov, P., Dekel, D. R., Herring, A. M., Hickner, M. A., Kohl, P. A., ... Zhuang, L. (2014). Anion-exchange membranes in electrochemical energy systems. *Energy & Environmental Science, 7*, 3135.

Veldre, K., Sala, E., Āboltiņa, E., & Vaivars, G. (2018). Hydration behaviour of sulfonated polyetheretherketone (SPEEK) membranes. *Key Engineering Materials, 762*, 220–225. Available from https://doi.org/10.4028/http://www.scientific.net/KEM.762.220.

Wang, J. T., Bai, H. J., Zhang, H. Q., Zhao, L. P., Chen, H. L., & Li, Y. F. (2015). Anhydrous proton exchange membrane of sulfonated poly(etheretherketone) enabled by polydopa-mine-modified silica nanoparticles. *Electrochimica Acta, 152*, 443–455.

Wang, Y., Xu, J., Zang, H., & Wang, Z. (2019). Synthesis and properties of sulfonated poly(arylene ether ketone sulfone) containing amino groups/functional titania inorganic particles hybrid membranes for fuel cells. *Journal of Hydrogen Energy, 44*, 6136–6147.

Watanabe, M. (1995). United States Patent 5.472.799.

Wright, A. G., Fan, J., Britton, B., Weissbach, T., Lee, H.-F., Kitching, E. A., ... Holdcroft, S. (2016). Hexamethyl-p-terphenyl poly(benzimidazolium): A universal hydroxide-conducting polymer for energy conversion devices. *Energy & Environmental Science, 9*, 2130.

Wycisk, R., & Pintauro, P. N. (2008). Polyphosphazene membranes for fuel cells. *Advances in Polymer Science, 216*, 157–183.

Xu, T., Wu, D., & Wu, L. (2008). Poly(2,6-dimethyl-1,4-phenyleneoxide)(PPO)—A versatile starting polymer for proton conductive membranes (PCMs). *Progress in Polymer Science, 33*, 894–915.

Yandrasits, M. (2015). V.B.1 New fuel cell membranes with improved durability and performance, FY 2015 Annual Progress Report, DOE Hydrogen and Fuel Cells Program.

You, W., Noonan, K. J. T., & Coates, G. W. (2020). Alkaline-stable anion exchange membranes: A review of synthetic approaches. *Progress in Polymer Science, 100*, 101177.

You, W., Padgett, E., MacMillan, S. N., Muller, D. A., & Coates, G. W. (2019). Highly conductive and chemically stable alkaline anion exchange membranes via ROMP of trans-cyclooctene derivatives. *Proceedings of the National Academy of Sciences, 116*(20), 9729–9734.

Zaffou, R., Kunz, H. R., & Fenton, J. M. (2006). Temperature-driven water transport in polymer electrolyte fuel cells. *ECS Transactions, 3*, 909–913.

Zamel, N., & Li, X. (2008). A parametric study of multi-phase and multi-species transport in the cathode of PEM fuel cells. *International Journal of Energy Research, 32*, 698–721.

Zawodzinski, T. A., Springer, T. E., Davey, J., Jestel, R., Lopez, C., Valerio, J., & Gottesfeld, S. (1993). A comparative study of water uptake by and transport through ionomeric fuel cell membranes. *Journal of Electrochemical Society, 140*, 1981–1985.

Zhang, H. Q., He, Y. K., Zhang, J. K., Ma, L. S., Li, Y. F., & Wang, J. T. (2016). Constructing dual-interfacial proton-conducting pathways in nanofibrous composite membrane for efficient proton transfer. *Journal of Membrane Science, 505*, 108–118.

Zhang, H., & Shen, P. K. (2012). Recent development of polymer electrolyte membranes for fuel cells. *Chemical Reviews, 112*, 2780–2832.

Zhao, Q., Majsztrik, P., & Benziger, J. (2011). Diffusion and interfacial transport of water in Nafion. *Journal of Physical Chemistry B, 115*, 2717–2727.

Zheng, Y., Ash, U., Pandey, R. P., Ozioko, A. G., Ponce-González, J., Handl, M., ... Dekel, D. R. (2018). Water uptake study of anion exchange membranes. *Macromolecules, 51*, 3264–3278.

Ziv, N., & Dekel, D. R. (2018). A practical method for measuring the true hydroxide conductivity of anion exchange membranes. *Electrochemistry Communications, 88*, 109–113.

Index

Note: Page numbers followed by "*f*" and "*t*" refer to figures and tables, respectively.

E

F

G

T

U

Printed in the United States
by Baker & Taylor Publisher Services